Cancer Diagnostics and

S. K. Basu • Chinmay Kumar Panda •
Subrata Goswami
Editors

Cancer Diagnostics and Therapeutics

Current Trends, Challenges, and Future Perspectives

Editors
S. K. Basu
Department of Computer Science,
Institute of Science
Banaras Hindu University
Varanasi, Uttar Pradesh, India

Chinmay Kumar Panda
Department of Oncogene Regulation
Chittaranjan National Cancer Institute
Kolkata, West Bengal, India

Subrata Goswami
Department of Pain and Palliative Medicine
ESI Institute of Pain Management
Kolkata, West Bengal, India

ISBN 978-981-16-4754-3   ISBN 978-981-16-4752-9 (eBook)
https://doi.org/10.1007/978-981-16-4752-9

© The Editor(s) (if applicable) and The Author(s), under exclusive license to Springer Nature Singapore Pte Ltd. 2022

This work is subject to copyright. All rights are solely and exclusively licensed by the Publisher, whether the whole or part of the material is concerned, specifically the rights of translation, reprinting, reuse of illustrations, recitation, broadcasting, reproduction on microfilms or in any other physical way, and transmission or information storage and retrieval, electronic adaptation, computer software, or by similar or dissimilar methodology now known or hereafter developed.

The use of general descriptive names, registered names, trademarks, service marks, etc. in this publication does not imply, even in the absence of a specific statement, that such names are exempt from the relevant protective laws and regulations and therefore free for general use.

The publisher, the authors, and the editors are safe to assume that the advice and information in this book are believed to be true and accurate at the date of publication. Neither the publisher nor the authors or the editors give a warranty, expressed or implied, with respect to the material contained herein or for any errors or omissions that may have been made. The publisher remains neutral with regard to jurisdictional claims in published maps and institutional affiliations.

This Springer imprint is published by the registered company Springer Nature Singapore Pte Ltd.
The registered company address is: 152 Beach Road, #21-01/04 Gateway East, Singapore 189721, Singapore

To the memory of Mrs. Nilanjana Basu

Preface

Oncology is the study, understanding, and treatment of cancer, which is complex, most dreadful, and almost incurable disease. It may affect any organ and spread to other sites of a living body over a period through the lymphatic and vascular channels. During initial spread of the disease, most of the individuals become asymptomatic. It is said that one in two men and one in three women are likely to be affected by one form of the disease over longer lifespan.

It is hard to find a person who has not have either a person affected or died from this disease amongst relatives, friends, and acquaintances. The unfortunate ones who first hear the occurrence of this disease to someone close or known to him/her becomes shattered and speechless, because there is no assurance of survival even after treatment to combat the disease. Along with the fear of losing the near/dear ones, the other major concerns evolve around heavy economic burden, pain, hopelessness, psychological imbalance, and so on. The line of treatment, the possible outcomes, and mental burden are too heavy to bear for a human being of flesh and blood. This causes a lot of confusion, stress, and anxieties for everybody connected with the affected person. Many of the issues involved and their scientific backgrounds are mostly not known to many.

No miracle or divine intervention is involved. Everything needs to be explained scientifically. What is possible and what is less likely to happen? Since the area is under the realm of many specialities, it is not possible to understand the issues and their scientific basis from one domain expert like surgeon, medical oncologist, or radiation oncologist. We need to hear from experts from multiple disciplines putting them on a common rational platform, which is less likely to find at the same place and time. Because of the patient load and other professional commitments, we hardly find someone to help us to make the best possible informed decision when the situation arises and arrange for the best care for the patient in need for therapeutic interventions. Several issues like cost, age, socio-economic factors, quality of treatment, and the quality of life after treatment need to be considered by the person, who is the interface between the patient and the medical care givers. It is by no way easy to make the best decision within the given parameters without some good understanding of the various facets of oncology. There is a need for such a volume contributed by experts from the related scientific disciplines.

This book gives technical outline of many scientific aspects of oncology. It includes chapters, most of which are written by specialists in the specific discipline and developed for readers with multitude of backgrounds, using methods which always think of developing the topic from simple to the advance level. The book does not expect the reader to be specialist in a specific domain of knowledge. Many of the issues involved, their scientific backgrounds in understanding the disease, coping with it, and appropriate treatments to be followed are not known to most of the people in the society. This book tries to fill some of those important gaps. There are professionals from a number of disciplines, like Bio-Medical, Medical Technology, and a few others, who are not trained in medical sciences but are required to understand selectively many of the processes, notions, and practices of oncologists. This volume will cater to their needs and help them acquire the necessary knowledge to further their endeavour in their fields in a confident way. This book tries to explain many of the scientific underpinnings of the disease, diagnostics, and treatment options related to cancer. Also, the book introduces to the authors themes like cancer genomics, immunotherapy, targeted therapy, stem cell therapy, oncology informatics, artificial intelligence, drug discovery, and others. The principles of working of important imaging and therapeutic machines like CT, PET scan, MRI, Linear Accelerator, and Protonmachines are discussed in different chapters. Radiomics for comprehensive and advance image analysis is also included.

The book comprises five parts: Part I provides a basic background of cancer in five chapters that discuss about the cellular structures and behaviour of normal cells and the tumours, providing a detailed insights about angiogenesis, cancer cell motility, and metastasis; Part II focuses on the cancer diagnostics and theory with five chapters under this section accumulating the diverse perspectives of engineers, physicists, biochemists, radiation oncologists, and pathologists for cancer diagnostics, along with giving a profound theoretical basis about MRI, CT, PET SCAN, and other radiological instruments which are of immense importance for cancer detection and diagnosis; Part III enlightens the cancer therapeutics, with five chapters under this section discussing about surgical and medical oncology for cancer treatment, effects of chemotherapy, telomerase and its therapeutic implications in cancer, and the cancer pain management; Part IV highlights some emerging trends in cancer research, and spans over six chapters which provide overviews about cancer stem cell research, translational research, immunotherapy, cancer cell lines, cancer genomics, radiomics, cancer informatics and drug discovery using artificial intelligence and information technology, as well as drawing a relationship between cancer with diabetes, a metabolic disease from which around 463 million adults are suffering currently worldwide; Part V presents some epidemiological and statistical figures to the readers, with three chapters providing a summary of diagnosis, intensity, treatment efficacy, and patient survival studies, discussing about the cancer epidemiology from Asian perspectives, and representing the North-East Indian scenario on cancer genomics and diagnostics.

Preface

This book is based on contributions by authors with different backgrounds trying to dissect different relevant issues so that knowledge seekers can get answers for a number of issues connected with oncology. This volume is of general pedagogic value and should not by any stretch of imagination be treated as a guide or text for students of oncology. This book is for the general readers having basic science, and wants to expand their knowledge base.

Varanasi, Uttar Pradesh, India S. K. Basu
Kolkata, West Bengal, India Chinmay Kumar Panda
Kolkata, West Bengal, India Subrata Goswami

Acknowledgements

We wish to express our sincere gratitude to the faculty members, research scholars, students, staffs and other personnel in Banaras Hindu University, Chittaranjan National Cancer Institute and the ESI Institute of Pain Management, Sealdah, India. We are immensely grateful to these institutes for extending necessary facilities and resources for preparation of the book.

Many individuals have encouraged and provided help during the preparation of the book. We wish to appreciate and acknowledge all of them. In particular, we would like to thank Mr. Sounak Sadhukhan and Ms. Debarpita Santra, PhD scholars of Prof. Swapan Kumar Basu, for their commendable and sincere co-operation during the entire journey.

We pay sincere regards to many authors and research workers whose papers/books have been freely consulted and referred to during the preparation of the book. We feel ourselves lucky to have sincere cooperation from all the contributors of the book.

We are very much thankful to Dr. Bhavik Sawhney, the Editor of Biomedicine, Springer Nature, Dr. Suraj Kumar, Production Editor, and the entire editorial team.

We are fortunate to have the complete support of our family members in the endeavour to publish the book. Special thanks go to Mr. Subhaditya Basu for his great support for completion of the book.

Contents

Part I Basic Background

1. **Fighting with Cancer: A Common Man's Dilemma** 3
   S. K. Basu, Santanu Basu, Subrata Goswami, and Saurabh Joshi

2. **Cell Biology and Cell Behavior in Cancer** 13
   Debarpita Santra

3. **Tumor Biology: An Introduction** 43
   Partha Nandi and Soumyabrata Roy

4. **Role of Angiogenesis in Tumors** 57
   Nidhi Gupta, Raman Kumar, and Alpana Sharma

5. **Biology, Chemistry, and Physics of Cancer Cell Motility and Metastasis** 81
   Sounak Sadhukhan and Souvik Dey

Part II Diagnostics and Theory

6. **MRI, CT, and PETSCAN: Engineer's Perspective** 113
   Subhankar Ghosh and Saurabh Pal

7. **Diffusion, MRI, and Cancer Diagnosis: Physicist's Outlook** 145
   Susanta Kumar Sen Gupta

8. **Oncology: Radiation Oncologist's View** 185
   Satyajit Pradhan, Ashutosh Mukherjee, Vinay Saini, Govardhan HB,
   Ankita Rungta Kapoor, Abhishek Shinghal, Lincoln Pujari,
   and Sambit Swarup Nanda

9. **Oncology: Biochemists' Perspective** 211
   Debolina Pal and Chinmay Kumar Panda

10. **Oncology: Pathologist's View** 239
    Anup Kumar Roy

Part III Cancer Therapeutics

11 Surgical Oncology: An Overview 261
Aseem Mishra and Vivekanand Sharma

12 Medical Oncology in Cancer Treatment 271
Mandira Saha

13 Chemotherapy Effects on Immune System 287
Debasish Hota and Amruta Tripathy

14 Telomerase and Its Therapeutic Implications in Cancer 303
Raman Kumar, Nidhi Gupta, and Alpana Sharma

15 Pain Management in Oncology 333
Subrata Goswami, Debolina Ghosh, Gargi Nandi,
Sayanee Mukherjee, and Biplab Sarkar

Part IV Emerging Trends in Cancer Research

**16 New Approaches in Cancer Research: Stem Cell Research,
Translational Research, Immunotherapy, and Others** 377
Soumyadeep Mukherjee, Ashesh Baidya, and Subhasis Barik

17 Cancer Cell Lines: Its Implication for Therapeutic Use 407
Sen Pathak

18 Genomics of Cancer 429
Avnish Kumar Bhatia

19 Diabetes and Cancer 443
Abhijit Chanda

20 Oncology Informatics, AI, and Drug Discovery 451
Debarpita Santra

21 Radiomics: Cropping More from the Images 461
Sounak Sadhukhan

Part V Epidemiology and Statistics of Cancer

**22 Statistics in Cancer: Diagnosis, Disease Progression, Treatment
Efficacy, and Patient Survival Studies** 473
Satyendra Nath Chakrabartty and Gopesh Chandra Talukdar

23 Epidemiology of Cancer: Asian Perspective Revised 489
Prasanta Ray Karmakar

24 Cancer Genomics and Diagnostics: Northeast Indian Scenario 509
Sharbadeb Kundu, Raima Das, Shaheen Laskar, Yashmin Choudhury,
and Sankar Kumar Ghosh

Editors and Contributors

About the Editors

S. K. Basu superannuated as a professor from the Department of Computer Science, Institute of Science, Banaras Hindu University, India. He served as emeritus professor in the School of Interdisciplinary Studies, S. N. Bose Innovation Centre, and Department of Computer Science and Engineering, Kalyani University, Kalyani, India. He obtained M.Tech and Ph.D degrees from Indian Statistical Institute, Kolkata, and Jadavpur University, Kolkata, respectively. He had attended various institutions in India and abroad. He obtained M.Sc degree in physics, specializing in biophysics. He had lectured extensively as a keynote and invited speaker in India and abroad. He had been a visiting scientist in different Indian institutes and abroad and was also the chairman of the Department of Computer Science, Banaras Hindu University. He had a teaching experience of more than 38 years and research experience of more than 32 years. During his career, he had supervised six Ph.D theses and two postdoctoral research works. His research interests were the diverse areas of parallel and distributed processing, simulation modelling, genetic algorithms, computer systems, wireless sensor networks, biodiversity, medical expert systems, simulation and modelling of tumour growth, etc. He had published more than 50 research articles in reputed internal journals and conferences. He had authored and edited a number of books. He also published two books and edited/co-edited a number of books.

Chinmay Kumar Panda is a NASI Senior Scientist Platinum Jubilee Fellow, Dept. of Oncogene Regulation, Chittaranjan National Cancer Institute, Kolkata, India. He superannuated as a Sr. Assistant Director Grade Scientist and Officer-in-Charge (Research) from the Chittaranjan National Cancer Institute, Kolkata, India, in 2019. He has done his Ph.D thesis work in the Chittaranjan National Cancer Institute, Kolkata, and received his Ph.D degree in biochemistry from the Calcutta University. His research interests include molecular genetics of cancer, regulation of gene expression, development of molecular markers for early detection and prognosis of cancer, and chemoprevention of cancer by natural products. He has done his postdoctoral training in Karolinska Institute, Stockholm, Sweden. He received different international fellowships like Swedish Institute fellowship and ICRETT fellowship of UICC. He is a fellow of The National Academy of Sciences, India (NASI) and West Bengal Academy of Science and Technology (WBAST). He is member of different scientific organizations like Society of Biological Chemists (India) (SBC(I)), Indian Association for Cancer Research (IACR), Indian Science Congress Association (ISCA), etc. He was president, New Biology Section of 106th ISCA Meeting in 2019. He is in the editorial board of different scientific journals like Scientific Reports and International Journal of Human Genetics. He has supervised number of Ph.D, DNB, and MD students for their thesis. He has over 200 publications in different peer-reviewed journals, reviews, and book chapters along with gene bank and GEO submissions.

Subrata Goswami FIPP from the World Institute of Pain, USA, and Ph.D (Science) from Jadavpur University, India, is the course director at the ESI Institute of Pain Management, Kolkata, India, since 2013. He is also the medical officer in charge at the Department of Pain Management, ESI Hospital Sealdah, India, since 2009. He is the honorary associate editor of the Indian Journal of Pain and the former national vice president of the Indian Society for the Study of Pain (Indian Chapter of IASP) during 2017–2018. He also served as the secretary of the West Bengal Chapter of IASP since 2015. He is the life member of the International Association for the Study of Pain, World Pain Institute (USA), Indian Society of Anaesthesiologists, Indian Society for the Study of Pain, and Indian Society of the Critical Care Medicine. He has notable academic and research activities in the field of pain medicine. He served as the principal investigator in two research projects funded by the National Jute Board, Govt of India (2014–2017), and the Department of Science & Technology, Government of West Bengal, India (2018–present), and as the co-investigators in other research projects. He authored more than 20 publications in journals of national/international repute, attended many national and international conferences and workshops as a faculty as well as one of the organizers, and also authored a number of chapters in pain management books.

Contributors

Ashesh Baidya Department of In Vitro Carcinogenesis and Cellular Chemotherapy, Chittaranjan National Cancer Institute, Kolkata, West Bengal, India

Subhasis Barik Department of In Vitro Carcinogenesis and Cellular Chemotherapy, Chittaranjan National Cancer Institute, Kolkata, West Bengal, India

Santanu Basu ESI Institute of Pain Management, ESI Hospital Sealdah Premises, Kolkata, India

S. K. Basu Department of Computer Science, Institute of Science, Banaras Hindu University, Varanasi, Uttar Pradesh, India

Avnish Kumar Bhatia ICAR-National Bureau of Animal Genetic Resources, Karnal, Haryana, India

Satyendra Nath Chakrabartty Indian Ports Association, Indian Maritime University, Noida, Uttar Pradesh, India

Abhijit Chanda Medica Superspecialty Hospital, Kolkata, West Bengal, India

Yashmin Choudhury Department of Biotechnology, Assam University, Silchar, Assam, India

Raima Das Department of Biotechnology, Assam University, Silchar, Assam, India

Souvik Dey Department of Pharmaceutical Technology, Jadavpur University, Kolkata, India

Debolina Ghosh ESI Institute of Pain Management, ESI Hospital Sealdah Premises, Kolkata, West Bengal, India

Sankar Kumar Ghosh University of Kalyani, Nadia, India
Department of Biotechnology, Assam University, Silchar, Assam, India

Subhankar Ghosh Department of Physics, St. Xavier's College, Kolkata, India

Subrata Goswami ESI Institute of Pain Management, ESI Hospital Sealdah Premises, Kolkata, West Bengal, India

Govardhan HB Department of Radiation Oncology, Kidwai Memorial Institute of Oncology, Bangalore, India

Nidhi Gupta Department of Biochemistry, All India Institute of Medical Sciences, New Delhi, India

Debasish Hota Department of Pharmacology, All India Institute of Medical Sciences, Bhubaneswar, Odisha, India

Saurabh Joshi ESI Institute of Pain Management, ESI Hospital Sealdah Premises, Kolkata, India

Ankita Rungta Kapoor Department of Radiation Oncology, Mahamana Pandit Madan Mohan Malaviya Cancer Centre, Varanasi, India

Prasanta Ray Karmakar Department of Community Medicine, Raiganj Government Medical College, Raiganj, West Bengal, India

Raman Kumar Department of Biochemistry, All India Institute of Medical Sciences, New Delhi, India
Baba Saheb Ambedkar Medical College, New Delhi, India

Sharbadeb Kundu Genome Science, School of Interdisciplinary Studies, University of Kalyani, Nadia, West Bengal, India

Shaheen Laskar Department of Biotechnology, Assam University, Silchar, Assam, India

Aseem Mishra Department of Head and Neck Surgical Oncology, Homi Bhabha Cancer Hospital, Varanasi, India

Ashutosh Mukherjee Department of Radiation Oncology, Mahamana Pandit Madan Mohan Malaviya Cancer Centre, Varanasi, India

Sayanee Mukherjee ESI Institute of Pain Management, ESI Hospital Sealdah Premises, Kolkata, West Bengal, India

Soumyadeep Mukherjee Department of In Vitro Carcinogenesis and Cellular Chemotherapy, Chittaranjan National Cancer Institute, Kolkata, West Bengal, India

Sambit Swarup Nanda Department of Radiation Oncology, Mahamana Pandit Madan Mohan Malaviya Cancer Centre, Banaras Hindu University, Varanasi, India

Gargi Nandi ESI Institute of Pain Management, ESI Hospital Sealdah Premises, Kolkata, West Bengal, India

Partha Nandi Department of Physiology, Government General Degree College, Jhargram, West Bengal, India

Debolina Pal Department of Oncogene Regulation, Chittaranjan National Cancer Institute, Kolkata, West Bengal, India

Saurabh Pal Department of Applied Physics, University of Calcutta, Kolkata, India

Chinmay Kumar Panda Department of Oncogene Regulation, Chittaranjan National Cancer Institute, Kolkata, West Bengal, India

Sen Pathak Department of Genetics, University of Texas and M.D. Anderson Cancer Centre, Houston, TX, USA
Department of Genetics, Southwest Foundation for Biomedical Research, San Antonio, TX, USA

Satyajit Pradhan Department of Radiation Oncology, Mahamana Pandit Madan Mohan Malaviya Cancer Centre, Banaras Hindu University, Varanasi, India
Mahamana Pandit Madan Mohan Malaviya Cancer Centre, Varanasi, India

Lincoln Pujari Department of Radiation Oncology, Mahamana Pandit Madan Mohan Malaviya Cancer Centre, Banaras Hindu University, Varanasi, India

Anup Kumar Roy Department of Pathology, Nil Ratan Sircar Medical College, Kolkata, India

Soumyabrata Roy Department of Microbiology and Immunology, Medical University of South Carolina, Charleston, SC, USA

Sounak Sadhukhan Department of Computer Science, Institute of Science, Banaras Hindu University, Varanasi, Uttar Pradesh, India

Mandira Saha Department of Radiation Oncology, HCG EKO Cancer Centre, Kolkata, India

Vinay Saini Department of Radiation Oncology, Mahamana Pandit Madan Mohan Malaviya Cancer Centre, Banaras Hindu University, Varanasi, India

Debarpita Santra Department of Computer Science and Engineering, Faculty of Engineering, Technology, and Management, University of Kalyani, Kalyani, West Bengal, India

Biplab Sarkar ESI Institute of Pain Management, ESI Hospital Sealdah Premises, Kolkata, West Bengal, India

Susanta Kumar Sen Gupta Department of Chemistry, Institute of Science, Banaras Hindu University, Varanasi, India

Alpana Sharma Department of Biochemistry, All India Institute of Medical Sciences, New Delhi, India

Vivekanand Sharma Department of Surgical Oncology, Homi Bhaba Cancer Hospital, Varanasi, India

Abhishek Shinghal Department of Radiation Oncology, Mahamana Pandit Madan Mohan Malaviya Cancer Centre, Banaras Hindu University, Varanasi, India

Gopesh Chandra Talukdar Indian Statistical Institute, Kolkata, India
ESI Institute of Pain Management, ESI Hospital Sealdah Premises, Kolkata, West Bengal, India

Amruta Tripathy Department of Pharmacology, All India Institute of Medical Sciences, Bhubaneswar, Odisha, India

Part I
Basic Background

Fighting with Cancer: A Common Man's Dilemma

S. K. Basu, Santanu Basu, Subrata Goswami, and Saurabh Joshi

Abstract

The impact of cancer diagnosis on the patient and the families is immense. The term cancer is quite frightening. It brings along a series of complicated thoughts about the disease, its course, treatment, associated difficulties, financial burden, social status, and existential issues. It traps the sufferers in a vicious loop of thoughts related to life, end of life, and thereafter. Another aspect that is mostly overlooked is the spiritual distress that comes along with the diagnosis of cancer. All these thoughts get amplified and increase psychological morbidity, which further gets compounded due to lack of knowledge and awareness about the disease and the ways to deal with it. This chapter broadly outlines the epidemiology and mechanism of the disease, types of cancer, various terminologies that are used, investigations that are undertaken, treatment options, and coping strategies during and beyond the process of disease and disability.

Keywords

Cancer types · Treatment options · Coping with cancer

S. K. Basu
Department of Computer Science, Institute of Science, Banaras Hindu University, Varanasi, Uttar Pradesh, India

S. Basu · S. Goswami (✉) · S. Joshi
ESI Institute of Pain Management, ESI Hospital Sealdah Premises, Kolkata, India

© The Author(s), under exclusive license to Springer Nature Singapore Pte Ltd. 2022
S. K. Basu et al. (eds.), *Cancer Diagnostics and Therapeutics*,
https://doi.org/10.1007/978-981-16-4752-9_1

1.1 Cancer

Globally, cancer is the second largest killer after heart diseases. A few million people are diagnosed every year with cancer. The number varies with geographical regions, regional economy, gender, age, etc. Cancer is not a single entity. It includes a multitude of diseases. All the diseases clubbed under the term cancer have one characteristic in common: uncontrolled growth and accumulation of abnormal cells. Collection of these abnormal cells beyond a certain size is visible. Prompt medical attention is needed to tackle this abnormal growth, as the tumor affects normal activity of the organ where it occurred or may spread to a distant site. One abnormal cell (derived from a normal cell having undergone mutation in its genetic material) will divide into 2 abnormal cells, 2 will divide into 4 abnormal cells, 4–8 abnormal cells, and so on. Every normal cell is supposed to undergo programmed death (*apoptosis*), but an abnormal cell loses capability for programmed death. Some of the common cancers for males are prostate, lung, colon, rectum, urinary bladder, etc. For females, the common cancers are breast, lung and bronchus, colon and rectum, uterine, ovary, etc.

Every cell growth is not cancer. One group is called *benign tumor*. It grows in a locally confined area. The other group is called *malignant tumor*. It can invade surrounding tissues, enter into the vascular system, and spread to distant part of the body through a process known as *metastasis*. Benign tumor is rarely life-threatening, whereas malignant tumors are often life-threatening. Benign tumors usually grow slowly, whereas malignant tumors may grow rapidly. Benign tumors are well-differentiated. State of differentiation (the process of becoming different by growth or development) in malignant tumors is variable.

There are a few distinguishing terms connected with tissue growth: *hypertrophy, hyperplasia, dysplasia,* and *neoplasia*. In hypertrophy, cell size increases, but the organization remains normal. In hyperplasia, cell number increases, but the organization remains normal. In dysplasia, the growth is disorganized. In neoplasia, growth is disorganized, but there is net increase in the number of dividing cells.

1.2 Types and Terminologies

Tumors can arise from different tissues and organs. Depending on its origin and cell types, some conventions are followed to name them. As shown in Fig. 1.1, they are classified into three main categories: *carcinomas, sarcomas, lymphomas* and *leukemias*. Cancer arising in cells covering layers over external and internal body surfaces (called epithelial cells) is called carcinomas. This constitutes roughly about 90% of all human cancers. Sarcomas are cancers of the supporting tissues such as bone, cartilage, blood vessels, fat, fibrous tissue, and muscle. They are the rarest occurring human cancers, about 1% of the total. The remaining cancers originate from cells of lymphatic and blood system. *Lymphoma* refers to tumors of lymphocytes (WBC) that grow as solid masses of tissue. *Leukemias* are cancers of blood cells, which proliferate in the bloodstream.

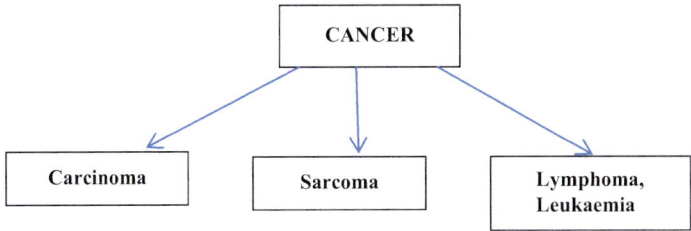

Fig. 1.1 Cancer types

Within each of these groups, individual cancers are named using prefixes that identify the involved cell type. For example, adenocarcinoma means carcinoma of the gland (adeno—meaning gland). Depending on the organ where it occurs, we may refer to the tumor as lung adenocarcinoma, colon adenocarcinoma, breast adenocarcinoma, colon adenocarcinoma, etc. If the tumor is benign, then these are referred to as *lung adenoma, colon adenoma, breast adenoma*, etc. There are a few exceptions to this nomenclature like melanomas are malignant tumors of pigmented cells, lymphomas are malignant tumors of lymphocytes, and myelomas are malignant tumors of the bone marrow cells.

Cells in a malignant tumor vary in appearance from those of benign tumor. When a sample is seen under a microscope, this difference provides the basis for cancer diagnosis. A doctor will usually cut a small piece of tissue from the suspected site of the tumor (e.g., fine needle aspiration cytology (FNAC)), which is examined by a competent pathologist to determine whether a tumor is present and whether it is benign or malignant, and the type of cell involved. This process is called *biopsy*. If the biopsy sample is collected by taking out a small bit of tissue from the tumor, it is called *incisional biopsy*, and if it is collected after taking out the whole tumor, it is called *excisional biopsy*. Pathologists assign some numerical grade to the tumor based on their microscopic appearance as there is variability among cancers, even if it involves the same cell type and organ.

When a tumor is ascertained to be malignant, oncologists decide the *staging* of malignancy through a scheme known as TNM system. T stands for tumor size, N for the number of lymph nodes positive for cancer cells, and M for metastasis, that is, has the cancer spread to other organs than the primary site of occurrence. The stages are generally 0, I, II, III, and IV. The more the value, the more advanced is the cancer. The treatment of cancer depends on the stage of the tumor. An advanced stage of the cancer tends toward more difficult case to provide remission.

1.3 Some Statistics

Though the world has seen major changes in the field of health care, cancer remains a leading cause of mortality, accounting for about 9.9 million deaths worldwide and 0.85 million deaths in 2020 in India (Sung et al. 2021; Chatterjee et al. 2016). It is anticipated that there will be more than 20 million new cancer cases worldwide by 2025, with 80% of these cases in low- and middle-income countries (LMICs) (Bray et al. 2015).

Grossly, throughout lifetime a man has probability of 0.5 for developing cancer and a woman has probability of 0.333 for developing cancer. Men are mostly affected by prostate cancer and women by breast cancer (Eyre et al. 2002). Different cancers affecting male population may be put in a descending sequence: prostate cancer < lung and bronchus < colon and rectum < urinary bladder. For female population, the descending sequence would be breast cancer < lung and bronchus < colon and rectum < uterine.

1.4 Role of Various Physicians and Nursing Care and the Available Treatment Options

The goal of cancer treatment is to cure or palliate the disease and improve survival and quality of life of the patient. Useful treatments destroy all cancer cells, prevent recurrence of the primary cancer, and balance likelihood of cure versus side effects of the treatment. An oncologist (medical, surgical, or radiation oncologist) makes a treatment plan based on the tissue type, age and physical fitness of the patient, status of certain organs like kidney, heart, lung functions, and aggressiveness of the cancer. The aggressiveness is decided by the size of the malignant tumor, how far it has spread, what is the chance of its recurrence, etc. People who are fit and healthy are able to cope with various therapies better. Chemotherapeutic and radiotherapeutic treatments alter the turnover of these cancer cells, but these agents might also interfere with the cell cycle of normal cells, and this may lead to certain adverse effects. The oncologist may like to know the personal, emotional, and financial status, family liabilities, etc., for making personalized treatment plan. Most cancer treatments involve surgery, chemotherapy, and radiotherapy or some combination of it. Surgery and radiotherapy are local treatments confined to the primary site of tumor occurrence. Teaming up with palliative medicine and pain management teams along with clinical psychologist, social counselor, and physiotherapist gives a good support system for the patient.

1.4.1 Surgery

This is the first choice, and it attempts to remove cancerous cells with a clear margin of the removed cancerous lump(s). From late 1800 till 1970, surgery was quite radical. The intent was to be to remove the tumor with many of the surrounding tissues to achieve better clearance of the disease. Approximately 90% of breast cancer patients underwent radical mastectomy (removal of breast) until 1970. This radical mastectomy was advocated by William Stewart Halsted of Johns Hopkins University under the assumption that cancer spread outward from the original growth and not through bloodstream. It was learnt later that this radical approach did not offer any extra advantage in terms of patient survival and so the conservative approaches in surgery were gradually adopted by many surgeons in various cancer centers.

Exploratory surgery to look for tumor growth inside the body has drastically reduced because of improved diagnostics techniques like computed tomography (CT scan), magnetic resonance imaging (MRI), ultrasonography (USG), and positron emission tomography (PET).

1.4.2 Chemotherapy

Chemotherapeutic agents are used to destroy cancer cells wherever they are in the body. For a particular patient, only a few are judiciously selected (a combination of drugs) by the attending medical oncologist. As the chemotherapeutic agents are introduced through the bloodstream, it quickly spreads in the body. There are several classes of chemodrugs such as *alkylating agents* like cyclophosphamide, *antimetabolites* like fluorouracil, *platinum drugs* like cisplatin, *mitotic inhibitors* like paclitaxel, and *antitumor antibiotics* like doxorubicin. These medicines are *cytotoxic* and targeted against rapidly growing cancer cells. It also damages cells that are not tumor-forming, if they are rapidly growing like in hair follicles, nails, bone marrow, lining of mouth, stomach, and intestines. These unintended destructions of normal cells are the *side effects of chemotherapy*, which may sometimes be serious in nature. Some of the common manifestations of this process are mucositis, hair fall, and low blood counts following chemotherapy. The challenge for the oncologist is to balance between the cancer-destroying capabilities of these drugs and their side effects.

When chemotherapy is used in the early stage of cancer, it may cause cessation of tumor growth, or it may limit the spread of cancer. It may also be used to shrink the size of the tumor before surgery. When the cancer is in an advanced stage, it may be used for purpose of palliation to reduce the symptoms of the patient and enhance quality of life. It may be used as the primary therapy **before** (*neoadjuvant*) other treatments like surgery or radiation therapy. It may also be used **after** (*adjuvant therapy*) surgery or radiation. Adjuvant chemotherapy targets stray cancer cells left in the body.

1.4.2.1 Radiotherapy
In this modality, X-rays or particulate radiation is used to kill the tumor cells. This modality can be used as the primary treatment or in combination with other modalities like surgery and chemotherapy. When it is used alone, it attempts to control the size of the tumor to reduce the pain or other palliative reasons. When it is used as an adjuvant treatment, it tries to destroy the remnant cancer cells, if any, after the primary treatment.

1.4.3 Cancer Pain Management and Palliative Care

A very important aspect of cancer care plan is the role of pain and palliative care services. Pain is a prominent symptom in majority of cancer patients. Other symptoms like nausea, vomiting, diarrhea, constipation, and sleepiness are mentioned elsewhere. Gradually, as the disease progresses and the symptom burden increases, the role of pain and palliative care team becomes more prominent as they focus on quality of life by providing optimal symptom control. The symptom could be any among the physical, psychological, social, and spiritual. The idea is to work around the concept of living well till we are alive and leaving well when the disease progresses and becomes refractory to treatment. Early and goal-based palliative care is the key. Good nursing care and maintenance of hygiene are important throughout the process of care.

1.5 Coping with Cancer Diagnosis and Treatments

Any sort of cancer diagnosis changes your thoughts, perceptions, and behavior. We suddenly realize that life is changed. We suddenly start realizing that there are many decisions to make, regarding the disease, its treatment, family issues, finances, etc. Moreover, the psychological and emotional burden of disease could be overwhelming. The natural response to cancer diagnosis is the negative thoughts and the ideas of death for many. The best way to deal with these thoughts is to gather information and work around our support systems. Information about the disease could be gathered from the authentic Internet sources, your own doctor, patient support groups, etc. We need to talk to the doctors and other members of the treating team, and talk to the family, friends, and children. Physical rehabilitation and psychological support play an important role throughout the cancer treatment to deal with day-to-day life issues and helps to adjust with the new way of life. Self-image, treatment-related physical and financial difficulties, relationships, and sexuality are important concerns. Whatever may be the prognosis, there are certain ways to deal with your thoughts, emotions, behaviors, and reactions to this disease and its treatment. Prioritizing quality of life is important.

1.6 Rays of Hope

It has become increasingly clear that cancer is a multifactorial complex illness, which involves various changes at the molecular and genetic levels. Thus, we need to assess and analyze the disease properly and accordingly target the molecule/structure/gene/cell sequence, etc., to get the desired response to treatment.

For example, based on PCR analysis of samples from colorectal premalignant polyp and carcinoma cell genomes, it is estimated that about 11,000 genomic alterations occur in a cancer cell. This large variation allows clonal heterogeneity seen in many cancer tissues. These tumor subpopulations interact with each other, thereby affecting their growth rate, chemosensitivity, and metastatic phenotype (Miller et al. 1981). New and important insights into the complexity of tumor progression lead to the development of novel treatment strategies. Cancer in different individuals behave differently; therefore, personalized or precision medicine allows the treating team to tailor the cancer treatment according to individual needs. For this, the gene changes or mutations specific to the malignancy under consideration are targeted. This targeted therapy allows targeting genes, molecules, proteins, blood vessels, hormones, signal transduction inhibitors, immune cells, etc., that help in growth and proliferation of target cells. Basic idea is to stop the action of the key molecule in the growth of cancer cells. Since the targeted treatments are matched to individual tumor types, the outcomes are better. Targeted therapies are broadly divided into two categories: one that enters the cells and works from inside the cell and the other that is big enough not to enter the cells but targets the receptors on the cell surface like monoclonal antibodies. Some examples of these targeted therapies are signal transduction/kinase inhibitors (imatinib, cetuximab, lapatinib); mTOR inhibitors (sirolimus, everolimus); hedgehog pathway inhibitor (vismodegib); immune system targets (rituximab, ipilimumab); angiogenesis targets (bevacizumab); hormonal targets (anastrozole, tamoxifen, bicalutamide); proteasome targets (bortezomib, carfilzomib); histone deacetylase targets (romidepsin); folate targets (pralatrexate); retinoic acid receptor targets (tretinoin, isotretinoin); EGFR inhibitors (cetuximab); and HER2 inhibitors (trastuzumab).

In addition to advancement in chemotherapeutics, radiotherapeutic treatments have also become more focused and sophisticated and are delivered through better and advanced machineries. These radiotherapy advancements include treatments like IMRT (intensity-modulated radiation therapy) where the radiation beam is targeted at the diseased area, causing least damage to the nearby noncancer tissue. IGRT (image-guided radiation therapy) uses imaging like MRI or CT to focus the radiation beam over the disease area. SRS (stereotactic radiosurgery) similarly delivers focused radiation to remove the disease, while sparing the normal tissue as much as possible, and is specifically useful in tumors involving brain and spine. Proton therapy is a type of radiotherapy that uses positively charged particles called protons to destroy cancer cells.

Surgical treatments for cancer have also taken big leaps in terms of robotic surgeries where the surgeons insert cameras and equipments into the body through very small holes; sit at a console; and, with the help of a viewfinder, they work with robotic arms through hand and foot controls. This helps in reaching body parts, which may otherwise be hard to reach. This leads to more precise and better tissue resections, cleaner margins with healthy tissues, lesser tissue damage, and early recovery.

One more important aspect of cancer care that cannot be ignored is the palliative medicine and pain management services, without which the loop of cancer care remains incomplete. Most of the cancer patients present with pain as a major symptom. Other symptoms include nausea, vomiting, breathlessness, and delirium. Management of pain and other symptoms remains important throughout the cancer treatment. Various nonpharmacological, pharmacological, and interventional strategies are available to manage these symptoms at various stages of the disease. Palliative and pain medicine are now gaining more and more attention and inclusion in cancer care. Cancer care sees a lot of transitions in process of care, and palliative medicine professionals help a lot during these transitions, especially the transition from hospital to hospice care.

Additionally, early and goal-based financial planning, which is a common need for all of us, becomes all the more important when it comes to disease like cancer where cost burden is significantly high. Asset allocation and rebalancing your expenses and funding, creating contingency funds, succession planning, etc., with the help of a financial/wealth planner who is exposed to and dealing with cancer care finances, are an indispensable element.

Integrative approach to cancer care, where different professionals from various fields and different hierarchies of healthcare integrate and work in cohesion, has contributed a lot to better care outcomes for the patients and their families.

1.7 Winning and Fighting Cancer

Many of the physicians have seen how the patients with the same age and illnesses undergoing similar treatment regimens have experienced different results. A lot of this depends on the will to live and fight that we can put up against the disease called cancer. This also stands true for other life-limiting illnesses. Patients and families with positive attitudes cope better with disease and the issues related to it and may respond better to treatments. Mindfulness and positivity are important elements in cancer combat. Positive attitudes and emotions help foster better adjustments to the stage and treatment of disease. It is important to understand that a patient is not alone and that he/she should not suffer in silence. Asking for and providing support to others, through physicians' own experiences, play an extremely important role. Diet, nutrition, physical exercises, rehabilitation, positive mental and physical health techniques, and strategies need to be followed. Existential and spiritual distress with anger at fate and God is a common occurrence and is manageable with relevant therapeutic medical and spiritual help. Friends, family, and finances are three important personal pillars apart from the possible treatments available for the disease.

1.8 Conclusion

Sir Francis Bacon published in his work, Meditationes Sacrae (1597), the saying: "knowledge itself is power," and this fits very rightly in the context of cancer care. Understanding the biology and stage of disease, knowing the various treatment options available at the relevant stages, and understanding systematic, logistical, and financial issues are the best possible way to self-help. With the advancements in the cancer treatments over the past two decades, the survival of people suffering from the disease has increased. Coping with the disease and its morbidities has also become better due to a lot of information, which is now available through print, Internet, social media, and support groups. With so much of information available these days, it becomes important to rationalize the information and discuss it with your treating team to understand what suits you the best. Cancer can happen to anyone so blaming the self and complaining will not help the cause; instead, what helps us is to understand the problem and find out ways to deal with the situation we are facing. The authors of the chapter and the book wish positivity, recovery, and good health to everyone who is fighting the disease.

References

Bray F, Znaor A, Cueva P et al (2015) Planning and developing population-based cancer registration in low- and middle income settings. IARC Technical Publication, USA, p 43

Chatterjee S, Chattopadhyay A, Senapati SN, Samanta DR, Elliott L, Loomis D, Mery L (2016) Cancer registration in India-current scenario and future perspectives. Asian Pac J Cancer Prev 17(8):3687–3696

Eyre HJ, Lange DP, Morris LB (2002) Informed decisions, 2nd edn. American Cancer Society, Lyon

Miller AB, Hoogstraten BFAU, Staquet MFAU, Winkler A (1981) Reporting results of cancer treatment. Cancer 47(1):207–214

Sung H, Ferlay J, Siegel RL, Laversanne M, Soerjomataram I, Jemal A, Bray F (2021) Global cancer statistics 2020: GLOBOCAN estimates of incidence and mortality worldwide for 36 cancers in 185 countries. CA Cancer J Clin 71(3):209–249

Cell Biology and Cell Behavior in Cancer

Debarpita Santra

> *Where a cell arises, there must be a previous cell, just as animals can only arise from animals and plants from plants.*
> *(Rudlof Virchow 1858).*

Abstract

Cell is the basic unit of a living organism. This chapter focuses on the structure and functionalities of a typical eukaryotic cell. Inside a eukaryotic cell, the organelles are arranged in an orderly fashion, maintaining a clear detachment from each other. Each of the organelles performs a particular cellular function and helps the cell to grow. A living organism grows gradually from its birth through cell divisions. A cell can produce replica giving birth to two new cells. During cell division, the hereditary materials are propagated into the offspring. The process is accomplished following some predefined rules. But unexpectedly, a cell may start disobeying the rules resulting in some genetic changes. A brief overview is given about how the cellular organelles are organized inside the cytoplasm, how a cell grows following a cyclic pattern, and how abnormal growth results in cancer.

Keywords

Cell biology · Cell cycle · Cell behavior · Cancer

D. Santra (✉)
Department of Computer Science & Engineering, Faculty of Engineering, Technology, and Management, University of Kalyani, Kalyani, West Bengal, India

© The Author(s), under exclusive license to Springer Nature Singapore Pte Ltd. 2022
S. K. Basu et al. (eds.), *Cancer Diagnostics and Therapeutics*,
https://doi.org/10.1007/978-981-16-4752-9_2

2.1 Introduction

Cell is an indispensable unit of life. More than three billion years ago, the living cells originated on earth through some molecular reactions in the atmosphere. Every cell is surrounded by a structure called the cell membrane or plasma membrane. This membrane serves as the clear boundary between the cell's inner and outer environments. A cell is well-equipped to execute all the functionalities of life. Superior organisms like animals and human beings are like cellular cities where many cellular groups exist and every group is focused to accomplish a specialized function. Each cellular group communicates with other groups and exchanges necessary signals to form a cellular city. In this chapter, we discuss briefly about cells, with an aim to analyze and understand the complexity of inner functioning of a complicated biological system. Later in this chapter, we will have a look at cancer biology in brief.

When life was created on earth, the metabolic reaction inside a cell was relatively simple. The oldest metabolic pathway was anaerobic due to nonexistence of oxygen in the atmosphere of the nascent earth. The metabolic pathways evolved with the gradual and incremental addition of new enzymatic reactions to the existing ones. The involved enzymes also underwent progressive adaptations with the evolution of organisms in different forms. With the molecular oxygen accruing in the atmosphere of the earth, some anaerobic cells became extinct; some cells somehow managed to live their anaerobic lifestyle; some became predators or parasites on aerobic cells; and remaining cells became adapted in the oxygenated atmosphere and are the present-day cells in the nature.

Cells vary from each other by their characteristics and shapes. There exist some cells that have no particular shapes and change their structures very quickly. Amoebae and rotifers are examples of these kinds of cells. Other cells have specific structural shapes. Also, some cells love to live in isolation (unicellular organism) and others prefer to live a colonized lifestyle, where one cell remains in the close vicinity of the other (multicellular organism). Despite these substantial differences, there are certain structural and functional features that are common in every cell. Based on these commonalities, the universe of cells is classified as *prokaryotic* and *eukaryotic cells*. A prokaryotic cell is formed of a single closed compartment surrounded by an outer membrane (cell/plasma membrane), and its internal organization is comparatively simple with no cellular organelle (nucleus) present at the center of the cell. Bacteria and blue-green algae are examples of prokaryotic cells. These cells can adapt themselves to the dynamically changing atmosphere. Prokaryotic cells are scattered everywhere; from 7 miles beneath the ocean to 40 miles above the earth, there exist millions of prokaryotic cells in the universe. On the other hand, a eukaryotic cell, which is also enclosed by cell membrane, has a definitive nucleus along with other organelles floating in the watery substance (cytoplasm) of the cell. All the organelles, unlike prokaryotic cells, are encircled by internal membranes. All the plants and animals that we see on the earth are made of eukaryotic cells. Fungi (both unicellular and multicellular) and protozoa (unicellular) are also formed of eukaryotic cell(s). Usually, the volume of a eukaryotic cell is 100 times larger than that of a prokaryotic cell, carrying larger quantity of intracellular materials. This chapter focuses on eukaryotic cell, which is the most basic entity of a human being.

Every eukaryotic cell has a protein-based internal skeleton called the cytoskeleton, which determines the shape of the cell, its ability for movement, and its internal organization for organelles. If we visualize a eukaryotic cell as a house, organelles are like distinct compartments in the house and each organelle accomplishes specific functionalities.

2.1.1 Eukaryotic Cell Structure

A eukaryotic cell consists of three major compartments: the cell membrane, the cytoplasm, and the nucleus. The nucleus along with some cellular organelles floats in the cytoplasm. While a cell takes food for its growth, it can also move and maintain itself, with performing some further activities like waste disposal and reprocessing of its parts (Song and Poo 2001). Description of each of the cell compartments is given below.

2.1.1.1 Cell Membrane

As a home is protected by walls, every cell is surrounded by a structure called the cell membrane or plasma membrane. The membrane separates a cell's inward environment from the outer one. A cell has a bilayer membrane made of fat molecules, which prevents water-loving stuffs from entering or evading the cell. Also, the membrane is stippled with proteins, which perform various tasks. Some of these proteins act as the gatekeepers who are responsible to decide which substances can or cannot cross the membrane. The membrane is absorbent to specific molecules depending on the requirements of the cell. In the presence of transport proteins, a set of specialized molecules enter the cell from outer environment through a selective pathway. There is one-to-one correspondence between a transport protein and a certain type of molecule. Figure 2.1 depicts this phenomenon by coloring a transport protein and the corresponding molecules with the same color. Some proteins work as markers, which identify whether the cell is part of the same or foreign organism. Some other proteins act as fasteners, which bind the cells together in a unit. Other membrane proteins are there to transmit and receive important signals from the cellular neighborhood or the outside environment. While small uncharged molecules like oxygen, carbon dioxide, nitric oxide, and water can pass through the membrane from extracellular fluid into the interior of the cell by simple diffusion, the charged ions are unable to penetrate the cell. Membrane proteins can travel freely move within the surface of the membrane. Constituent membrane proteins are often glycosylated as they contain some sugar residues on the side facing the external environment of the cell.

2.1.1.2 Cytoplasm

If a cell is a closed container, the cytoplasm is the watery matter inside the container. The sheath of the container is basically the cell membrane. Cytoplasm consists of

Fig. 2.1 Transport proteins in the cell membrane

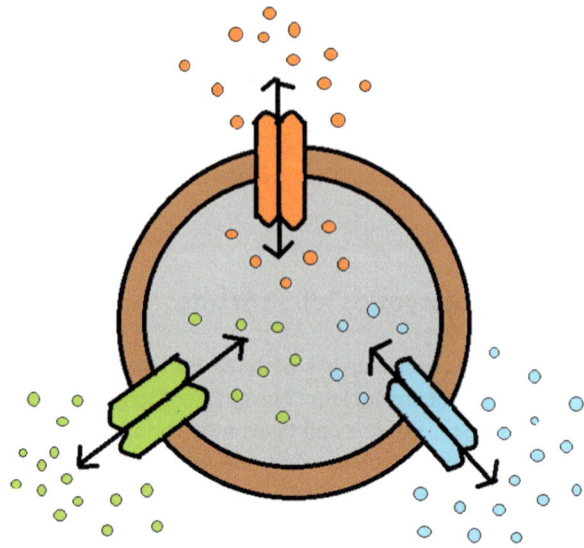

organic molecules and organelles, which can be visualized as the tools, appliances, and inner rooms of the cell. Inside the cytoplasm, different chemical reactions occur. The cytoplasmic elements take help of diffusion for traveling short distance within a cell. All the cellular functions for growth, expansion, and replication of a cell are performed inside the cytoplasm. Inside the cytoplasm, the concentrated, granular, and hard area near the boundary of cytoplasm is termed as *ectoplasm*, whereas the less concentrated area placed centrally is called as *endoplasm*. Maintenance of the high concentration of the constituents of the cytoplasm requires a lot of energy to be consumed by a cell for its survival. Now, we briefly discuss the constituents of cytoplasm.

Organic Molecules

Intracellular organic molecules are vital for proper functioning of a cell and may be of different types namely nucleic acids, carbohydrates, proteins, and lipids. The genetic code of a cell can be expressed using nucleic acids. There are two broad categories of nucleic acids: *deoxyribonucleic acid* (DNA) and *ribonucleic acid* (RNA). DNA holds all the information necessary for building and maintaining a cell. RNA plays several roles to express the information stored in DNA.

A second type of organic molecule is carbohydrate, which can be found in the form of starches and sugars in a cell. When simple carbohydrates fulfill the cell's instant energy requirement, complex carbohydrates act basically as the intracellular energy stores. Complex carbohydrates, which are available on a cell's surface, play a vital role in cell recognition.

Protein is the third type of intracellular organic molecule, which comprises many minor molecules named amino acids. Proteins serve both the catalytic and structural functions in the cell. The enzyme, another kind of protein, which is responsible for conversion of the cellular molecules into other forms for energy-related issues, develops supportive structures or throws out wastes. Along with nucleic acids, proteins can also preserve and express genetic material by helping replicate the genome and accomplish intense structural changes triggering cell division.

Another type of organic molecule is lipid or fat. The lipid molecules are the constituents of the plasma and intracellular membranes of a cell. These molecules basically can store energy and relay signals within the cells and from the bloodstream to a cell's interior.

Organelles

Organelles in a cell are basically a collection of molecules oriented in an orderly fashion. As each room in a house is separated from each other by a boundary wall, each of the organelles is detached from the remaining elements present inside a cell by intracellular membrane. Some of the organelles are bounded by double membranes, and the remaining are encircled by single membrane. The major bimembrane organelles are as follows: nucleus, mitochondrion, and chloroplast. Among these three, chloroplast is found in plant cells. All the double-membrane organelles contain the genetic material DNA. There are also many organelles in eukaryotic cell that are bounded by single membrane. These organelles are subdivided into different types based on their cellular functions. Some of the single-membrane organelles are as follows: peroxisomes, endoplasmic reticulum, Golgi apparatus, and lysosome.

Organelles with Double Membranes
Nucleus

The nucleus is the largest cellular organelle that is present at the center of most eukaryotic cells. It is quite natural to have a cell with only one spherical-shaped nucleus. A cell nucleus consists of a nuclear membrane, nucleoplasm, nucleolus, and chromosomes. The inner elements of nucleus are encapsulated by a bilayered membrane called nuclear membrane. The membrane is multiporous, with each of the pores being termed as nuclear pores. The nuclear pores allow some special molecules to travel between the nucleus and the cytoplasm. Nucleus controls the total activity of a cell like information storage, and retrieval and duplication of genetic information. The genetic information of a cell is stored in an element called chromatin, which is found in nucleus. The nucleus has a variety of subcompartments, which do not have membranes around them and are of varying functional capacity. The key functions of the cell nucleus include DNA replication and further controlling of gene expression during the cell cycle. We will discuss this in detail later. Activities like transcription and posttranscriptional processing of premessenger RNAs (mRNAs) occur inside nucleus. The translation event occurs inside cytoplasm, only after the mature mRNAs are transported into it. The nucleus, by allowing higher levels of gene regulation, supports functional segmentation inside the cell. We shall now discuss each of the four major parts of a nucleus.

Nuclear Membrane

The nuclear membrane is a double-layered boundary made of proteins and lipids. The inward substances of nucleus can communicate with the cytoplasm through the nuclear membrane. The intermembranous gap between the two membranes is termed as the perinuclear space. The width of the gap is generally about 20–40 nm.

Neoplasm

Similar to the cytoplasm of a cell, the nucleus contains neoplasm, karyoplasm, or nucleus sap, which is a transparent and dense fluid surrounded by nuclear membrane. The neoplasm makes room for the chromosomes and nucleoli. Substances such as nucleotides (required for DNA replication) and enzymes are present in the neoplasm. The liquefied area of the neoplasm is named as the *nucleosol* or *nuclear hyaloplasm*, whose main task is to act like a suspension matter for the organelles existing in the nucleus. The neoplasm determines the structural shape of the nucleus. It also permits movement of substances for cell metabolism and function.

Nucleolus

Occupying almost quarter portion of the nucleus, nucleolus is regarded as its largest component. This component consists of proteins and ribonucleic acids (RNA) and rewrites the ribosomal RNA (*r*RNA) for further making it combined with proteins. A continuous chain of nucleolar passages spanning from the neoplasm to the inner parts of the nucleolus is formed.

Chromosome

Chromosomes are the basic building blocks of life, which carry hereditary information in the form of genes. A chromosome is a compact structure formed by tight coupling of DNA (as chromatin) using histone proteins. The compactness property not only helps chromosomes to be fit inside the nucleus of the cell, but also helps in the organization of genetic material at that time when cell division occurs (Bruce et al. 2008). Every eukaryotic species contain a static chromosome number. While an asexually reproduced organism has the same chromosome numbers in every cell, a sexually reproduced organism has diploid ($2n$) chromosomes in body (somatic) cells and haploid ($1n$) chromosomes in sex cells (gametes) (Comai 2005). Figure 2.2 depicts the basic structure of a nucleus.

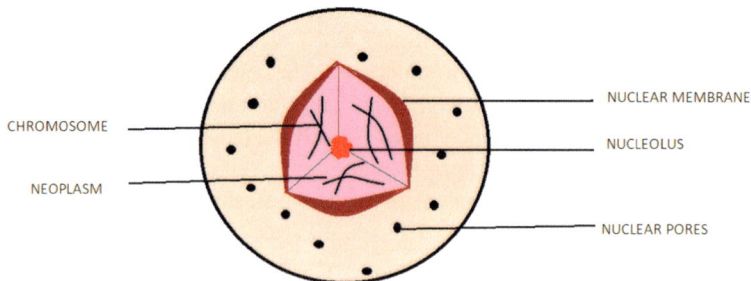

Fig. 2.2 Basic structure of nucleus

Mitochondrion

A mitochondrion is the prime energy resource of a cell. The equipments needed for a cell to perform different energy-supplying chemical reactions are supplied by mitochondrion. The main tasks of mitochondrion include supplying cellular energy, signaling, cellular differentiation, maintaining control of the cell cycle and cell growth, and cell death. Surrounded by a bilayer membrane made of phospholipid and proteins, a mitochondrion is composed of five distinct parts: the outer mitochondrial membrane, the inner mitochondrial membrane, the intermembrane space, the cristae, and the matrix (Mannella 2006). Figure 2.3 shows the structural components of a mitochondrion.

The exterior mitochondrial membrane encircling the whole organelle comprises some integral membrane proteins named *porins* helping in formation of passages that permit minute molecules with weight smaller than 5000 daltons to easily diffuse from inner side of the membrane to the outer side or vice versa. The outer membrane also holds a number of enzymes called *monoamine oxidase, kynurenine hydroxylase,* etc. These enzymes help in epinephrine oxidation, fatty acid elongation, tryptophan degradation, and so on.

The inward mitochondrial membrane holds a number of proteins made of a minimum of 151 different polypeptides. As porins are absent in the inner membrane, this membrane hardly comes in contact with all the molecules. There exist special membrane transporters whose task is to assist almost all ions and molecules in entering or exiting the covered zone of the mitochondrion. The gap residing between the external and internal membranes is termed as the *intermembrane space* or *perimitochondrial space*.

Numerous folds exist in the inner side of the internal mitochondrial membrane, which is also peppered with some small round-shaped bodies termed as F_1 *particles* or *oxysomes*. These kinds of folds are called as *cristae*. When the surface area of the

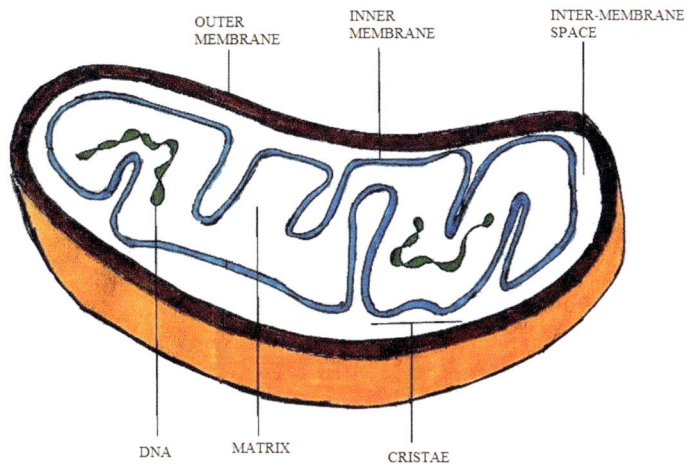

Fig. 2.3 Internal structure of a mitochondrion

internal mitochondrial membrane is expanded, the capability of producing ATP by cristae is improved.

Surrounded by the internal membrane, there is a particular space inside mitochondrion that contains almost 2/3rd of total mitochondrial proteins. This space is termed as matrix. While ATP is produced in the mitochondrion, the matrix is aids in synthase of ATP inside the internal membrane. The matrix accommodates a highly concentrated mixture of thousands of enzymes, specialized mitochondrial ribosomes, *t*RNA, and a number of replicated mitochondrial DNA genome.

Besides holding the previously specified substances, a mature cell's cytoplasm may store ergastic substances (nonprotoplasmic) like food items, crystals, and secretory or excretory materials.

Chloroplasts

Chloroplasts are visible only in photosynthetic protists and plant cells. Besides the presence of two enclosing membranes, there also exist internal membranes called *thylakoids* in these organelles. The thylakoids are made of proteins and other molecules, which enable the chloroplasts to capture light necessary for photosynthesis.

Organelles with Single Membrane

Peroxisome

Close to mitochondrion and chloroplasts (if present), a single-membrane organelle named *peroxisome* is found in the cell cytoplasm. Peroxisomes contain heterogeneous proteins and are responsible for variety of metabolic functions in an organism. These organelles are responsible for conversion of highly reactive molecule called *hydrogen peroxide* (H_2O_2) generated from water, after some chemical reactions have occurred inside the mitochondrion (Bolsover et al. 2011).

Endoplasmic Reticulum

The endoplasmic reticulum (Ron and Walter 2007) contains many tubules and sacs, which collectively make a network-like structure. The main functions of this organelle include production, processing, and transportation of proteins and lipids for its own membrane and other organelles like Golgi apparatus and lysosomes. The organelle is internally segmented into two parts: rough endoplasmic reticulum and smooth endoplasmic reticulum. The former segment has ribosomes (cell organelles consisting of RNA and proteins) attached to the cytoplasmic side of the membrane, while the latter segment has no such attached ribosomes.

Golgi Apparatus

Golgi apparatus (Beams and Kessel 1968) is positioned next to the cell center. This organelle has many flat sacs called *cisternae* and performs operations like creating, storing, and transporting certain cellular products, particularly those produced at endoplasmic reticulum. A cell contains high number of Golgi organelles if the cell has many secreting substances.

Lysosome

Lysosome (Duve 1963), which is an organelle of roughly spherical shape, contains acidic hydrolytic enzymes, which are made of proteins of endoplasmic reticulum and are surrounded by the vacuoles of Golgi apparatus. This organelle can digest cellular

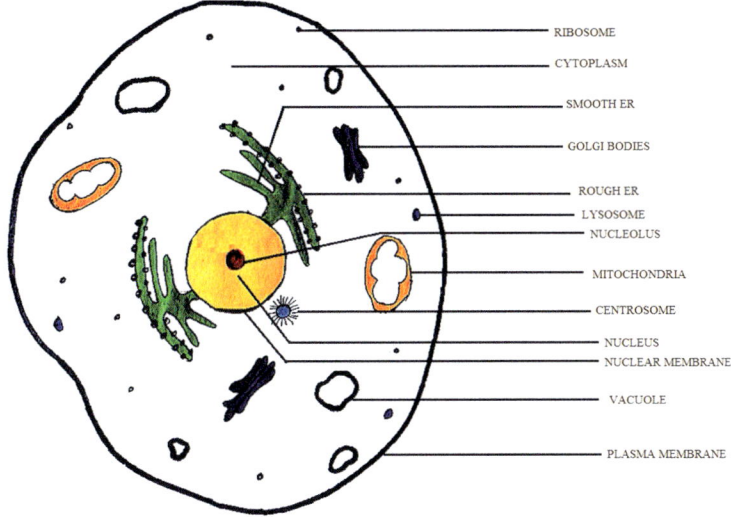

Fig. 2.4 Basic structure of a eukaryotic cell

debris and destroy foreign substances, microbes, cancer cells, and any other components that contain proteins, which are not healthy to body cells. Lysosome is primarily responsible for waste disposal of a cell. Figure 2.4 describes the basic eukaryotic cell structure.

2.1.2 Cell Cycle

It is a very obvious question how a living organism grows gradually from its birth. The answer is cell division. A cell gives birth to two daughter cells each time of cell division, and the offspring can grow themselves and take part in cell division. This iterative increment of cells forms a major cell population, which includes a single parental cell and its progeny comprising millions of cells. During cell division, it must be ensured that all the information embedded in the DNA of a cell is passed correctly to its offspring. This information is basically the instructions required to make or maintain a living organism. Every cell of a living organism contains these instructions encoded within the genes, which are basically the cellular elements that define the features of a species. So, for better conceptualization of cell cycle, first we should have a brief idea of DNA and chromosomes. As said earlier, chromosomes are basically thread-like structures present in the nucleus of a eukaryotic cell and become visible when the cell division starts.

In the year 1940, scientists recognized that DNA is responsible to carry the hereditary/genetic information from one generation to another of an organism. In the year 1953, two scientists James Watson and Francis Crick determined the structure of DNA. A gene of every cell on the earth is formed of DNA. In the following section, we discuss the basic structure of DNA and how hereditary information is captured in DNA.

2.1.2.1 Basic Structure of DNA and Its Function

A DNA molecule contains two long chains of polynucleotide, which are called *DNA chains* or *DNA strands*. This kind of formation with two linear strands running opposite to each other is termed as *double helix*. Every DNA strand is made of four different nucleotide subunits. One DNA strand is associated with the other using hydrogen bonds between the base proteins of the nucleotides. A nucleotide has sugar with five carbons, single or multiple phosphate groups, including one base made of nitrogen. An individual DNA nucleotide holds deoxyribose sugar, a phosphate group, and any base from adenine (*A*), guanine (*G*), cytosine (*C*), or thymine (*T*). Four DNA nucleotide subunits differ from each other only by the nucleotide base. In a DNA strand, the nucleotides share chemical bonds among each other following an alternating sugar–phosphate–sugar–phosphate pattern. Each DNA strand looks like a necklace threaded with four pendants *A*, *C*, *G,* and *T*. Four different nucleotides are also commonly called accordingly in the names of their bases. An interesting fact is that a DNA nucleotide has a knob and a hole. The phosphate acts as a knob. So, a complete DNA strand has some interlocking knobs with holes, keeping all of the nucleotide subunits properly aligned. If one end of the strand has a hole, the other end must have a knob. Figure 2.5 shows four different nucleotide blocks of DNA.

While the inner side of the DNA double helix pulls all the bases toward it, the outer side keeps the sugar–phosphate backbones. There exists a strict pairing strategy between two bases. While *A* can only pair with *T*, *G* would always attract *C*. With width of each base pair being alike, this kind of harmonizing coupling of the bases offers energetically most advantageous organization inside the double helix. The helix contains 10 bases per helical turn.

Encoding of genetic information in the form of nucleotide sequence alongside each strand is done by DNA. The gene stores information on how multicellular organisms grow. With each base corresponding to a letter, *A*, *C*, *T,* and *G* form a four-letter alphabet, which is basically a biological message embedded in the chemical structure of the DNA. Simply, we can say that a biological message imitates a nucleotide sequence. Every living organism has unique nucleotide sequences carrying unique biological message. The entire genetic information that a DNA gathers is termed as *genome*. If we write down in a white page texts corresponding to the nucleotide sequence of a small gene in a human being using the four-letter nucleotide alphabets, the whole information would occupy a quarter of the page. Moreover, the entire sequence of human genome would take more than 1000 voluminous books to be put down. This is very amazing to hypothesize that all the information is neatly wrapped into the nucleus of a eukaryotic cell. Here comes the inevitability of chromosomes.

2.1.2.2 Eukaryotic Chromosome

Compared to the diameter of a cell nucleus, that of DNA is almost 1000 times larger. So, accommodating all the genetic material into a small space is similar to

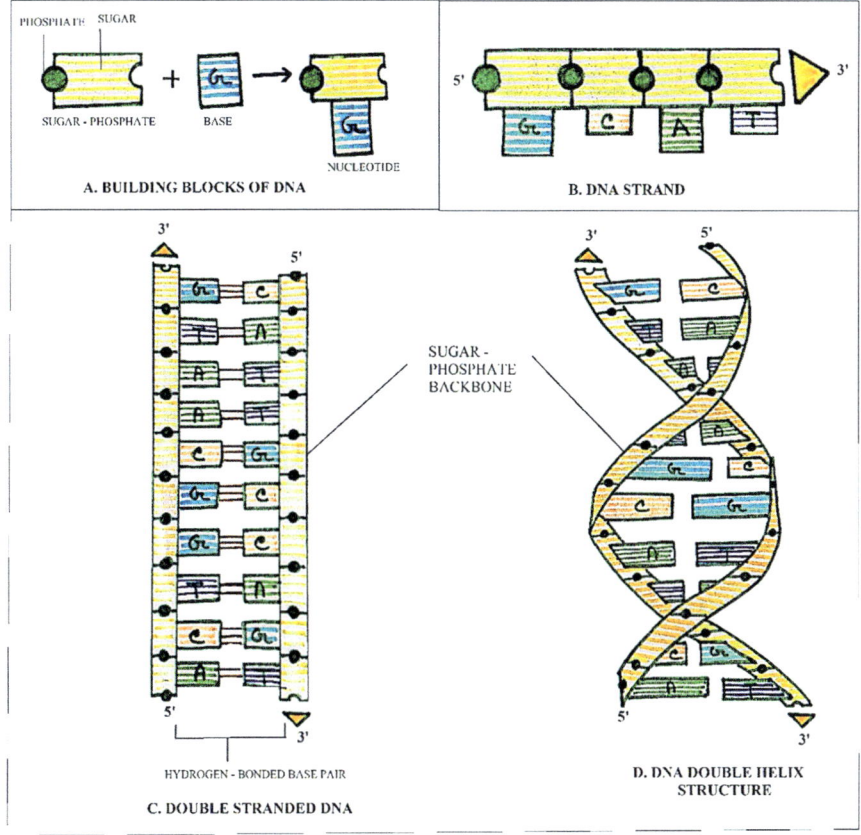

Fig. 2.5 Four different DNA nucleotide building blocks

encapsulating a 40-km-long enormously fine thread into a tennis ball. Chromosomes act as envelopes for the prolonged bistranded DNA molecules, and these envelopes fit readily inside the nucleus. During each cell division, the chromosomes are equally distributed into two progenies.

Making room for DNA molecules in chromosomes is not an easy task. Credit goes to some specialized proteins that fasten the DNA molecules to form a series of coils that are hierarchically organized, with no intermediate superfluous tangle. This kind of compaction helps enzymes and other proteins to access the DNA for its replication and repairing. There exist different chromosomes where DNA molecules are distributed. A human cell normally contains twenty-three pairs of chromosomes. Among them, twenty-two pairs look alike. These chromosomal pairs are called autosomes. Difference between males and females is found only due to the 23rd pair. This pair holds the sex chromosomes. A female has a pair of X chromosomes, while, on the contrary, a male has one X chromosome and one Y chromosome. The

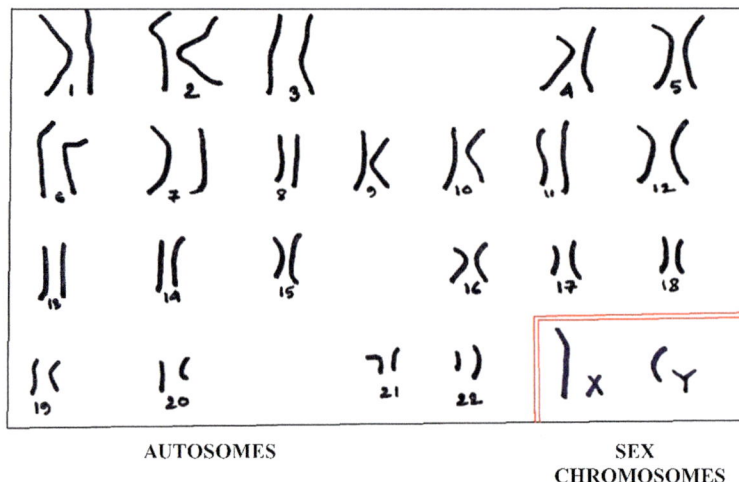

Fig. 2.6 Karyotype—the organization of human chromosomes

twenty-two autosomes are numbered by size. Figure 2.6 is a *karyotype*, which depicts how human chromosomes are lined up in pairs (Falini et al. 2005).

The complex between the DNA and proteins, which forms chromosomes within nucleus, is termed as chromatin (Kornberg 1977). The number of proteins that chromatin contains is almost twice compared with the protein count in DNA. Inside chromatin, there exist a number of small proteins—called *histones*—made of mostly basic amino acids (Cooper and Hausman 2004). Chromatin looks like droplet in a thread. The droplets are called *nucleosomes*. Every nucleosome contains eight histone proteins, with the DNA acting as a sheath around them. Then, a 30 nm spiral called *solenoid*, which comprises additional histone proteins, wraps the nucleosomes to give the shape of a chromatin (Kornberg 1977).

A gene resides within a chromosome. A gene is the functional element of heredity. More elaborately, a gene is a DNA segment carrying instructions to make specific proteins in most cases or to produce RNA molecules in some cases. The end yield (the proteins and RNA molecules) offers diversified catalytic and structural functions in the cell. In a human body, there are about 30,000 genes, with each gene having two replicas. Each of the replicas is inherited from each parent. It is seen that human beings have genes that are mostly common in all, but there are a very small number of genes ($<1\%$) diverging slightly from one person to the other. The same gene appearing in two distinct styles with little variation in the sequences of DNA bases offers unique physical features in every person (Fessele and Wright 2018).

Besides carrying genes, the chromosomes also are able to replicate themselves in presence of DNA molecules. During cell division, the copies of genes generated through chromosome duplication are distinguished and distributed reliably into the offspring. The process of chromosome duplication goes through a series of sequential phases—known as *cell cycle*.

2.1.2.3 Replication and Repairing of DNA

DNA replication (Kunkel and Bebenek 2000) process aids DNA in creating its own copy at the time of cell division. In this process, the genetic information embedded in the DNA of a cell is duplicated in its replica before the cell gives birth to two identical daughter cells. So, the DNA replication process can be considered as a complex semi-conservative process in which each parental DNA strand acts as a template for producing the new analogous daughter strand. This process, which is more complex compared with a simple enzymatic reaction, is mainly catalyzed by a special enzyme called *DNA polymerase* with the presence of other proteins. During the replication process, the *DNA polymerase* and the existing proteins become involved in selective removal of the mismatched bases to offer an almost accurate DNA replication.

The very first stage of DNA replication is to unwind the DNA double-helix structure to provide the single-stranded DNA templates with the help of special enzymes named *DNA helicases*. These enzymes disrupt the hydrogen bonding between the complementary bases of DNA namely *A–T* and *C–G* and a "**Y**"-shaped replication *fork* is created with the help of two separated single DNA strands. These strands work as templates for making new DNA strands. The two DNA strands are arranged in reverse directions, with one being oriented toward the replication *fork* (called the *leading strand*) and the other being positioned away from the *fork* (called the *lagging strand*). Because of the reverse orientation, the replication mechanisms are different for the two singly strands.

The leading strand attaches at its end an enzyme called *primase,* which is basically a small part of RNA. This enzyme acts as the catalyst for DNA synthesis. The leading strand also becomes accompanied by DNA polymerase and adds new complementary nucleotide bases to the lagging strand. This kind of replication process is called "continuous." On the contrary, the lagging strand follows a "discontinuous" replication process, in which many *primase* enzymes are attached at various points of the strand. Lots of DNA particles and chunks are added to the same strand. These chunks are named as *Okazaki* fragments, which become assembled at later times (Waga and Stillman 1998). Figure 2.7 depicts both the replication processes.

At the intermediate stage, the replication process finds that all the bases are properly aligned, that is, A is with T and C with G, the primase enzymes leave the DNA strands and make rooms for more complementary nucleotides to fill the gaps, in the presence of another enzyme called *exonuclease*. The new strand, which is a midway product of replication, is further checked to ensure that the new DNA sequence is correct. Finally, another enzyme named *DNA ligase* abstains the DNA sequence from any further modification. Two molecules with one new and one old chain of nucleotides are the end products of the DNA replication process. This is the justification behind naming the DNA replication process as semi-conservative. With

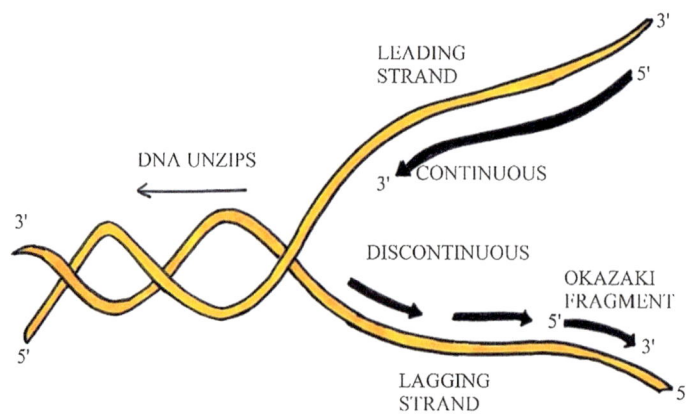

Fig. 2.7 DNA replication process

the termination of the replication process, a double helix is automatically formed by the new DNA.

From the discussions about DNA so far, it is clear to all that DNA is the repository of genetic information present in all living organisms. Sometimes, the genetic information is susceptible to copying errors, accidental damage, or permanent changes, causing the DNA unstable and disintegrated. These errors, if neglected and are not repaired on time, possibly lead to *mutations*. Mutation, in simple terms, can be defined as a change that occurred in the DNA sequence, due to mistakes made during DNA replication or due to many environmental factors like heavy exposure to UV rays and cigarette smoking (Campbell and Eichler 2013). The environmentally induced DNA damage may result in dreadful diseases like cancer. Also, DNA may be assaulted by some by-products of the metabolism process in every organism (Sancar et al. 2004). Identification of all the genetic errors and taking appropriate measures for correction of the errors are collectively called DNA repair (Hoeijmakers 2001).

Two general ways exist for repairing the DNA errors. One solution is to reverse the chemical reactions that lead to the DNA damage, and the other is to remove the damaged bases of DNA and replace them with the newly synthesized DNA. If these two repairing approaches fail, there exists another alternative approach, which motivates the cells to cope with the damage. The first strategy is meant for specific kinds of DNA damages such as *pyrimidine dimers*. *Pyrimidine dimers* are produced by UV light with wavelength below 340 nm. The *pyrimidine dimers* deform the DNA by partly unraveling and twisting it (Kripke et al. 1992). These kinds of damages are repaired by *photo-reactivation*. Photo-reactivation is a chemical process with the presence of visible light of wavelength of 300–600 nm and *photolyase* enzyme. The latter acts on dimers present in single- or double-stranded DNA. In the second approach, the damaged DNA bases or nucleotides are first identified and then thrown away. The resulting gaps are then occupied by the DNA polymerase, and the surface between the older DNA and newly synthesized DNA is sealed primarily by a special type of DNA enzyme called *ligase* (Sancar et al. 2004). With the help of *ligase*, the two DNA strands are joined to each other and a new chemical bond called p*hosphodiester* is produced. During this phenomenon, scientists have observed the presence of hydrogen groups in phosphoric acid and hydroxyl groups (Huang et al. 1992).

2.1.2.4 DNA Recombination

Now, it is clearly understood how the DNA sequences in cells are maintained over the generations with very small alteration. But the chromosomal DNA sequence differs with respect to time and occasionally can be re-oriented. The DNA rearrangement incorporates changes into particular gene pattern to allow evolution among the organisms living in a dynamically changing environment. These kinds of rearrangements are the result of a special mechanism called DNA recombination. The recombination can be homologous and nonhomologous. Between two isomorphic DNA molecules (i.e., with similar nucleotide sequences), homologous recombination occurs. This recombination allows cells and many organisms to repair DNA and make quick recovery from genetic accidents during the DNA replication.

On the other hand, nonhomologous recombination or, more specifically, the site-specific recombination occurs between DNA double helices that are dissimilar in nucleotide sequences. Among its many important cellular tasks, the most important one is to allow mobile genetic elements—short sequences of DNA—move everywhere within the genome in the cell and most of the sequences reside in the chromosomes of the host cell, with others, especially the viruses, being able to move out from the cell. Some genetic deviations are injected into the host genomes by the mobile genetic elements for maintaining the evolution.

2.1.2.5 DNA Transcription

Let us visualize the DNA as a book. There exists an easy mechanism on how the book is read. DNA transcription process provides this mechanism by converting the DNA into more portable set of instructions called RNA (Sobell 1985). For better understanding, let us take an example: Mr. A has left a message for Mr. B in Mr. B's voicemail. Mr. B has to write the matter in the message down on a paper. Rewriting of information on a paper is termed as transcription. The preliminary step in gene expression is DNA transcription, in which the DNA sequence of gene is duplicated into the analogous alphabets of RNA molecules. So, first we should have a brief idea about the structure of RNA molecule.

If DNA is compared to a main book, RNA is a reference book. RNA acts as a template to carry the same information as its DNA template, but the information is not kept stored for long time (Holley et al. 1965). With the help of a five-carbon ribose sugar, a nitrogen base, and a phosphate group, a ribonucleotide base of RNA is formed. There are four different ribonucleotide bases of RNA namely adenine (A), guanine (G), cytosine (C), and uracil (U). RNA molecules are coupled with each other through *phosphodiester* linkage (Konarska et al. 1981) between the phosphate of one nucleotide and the sugar of the other.

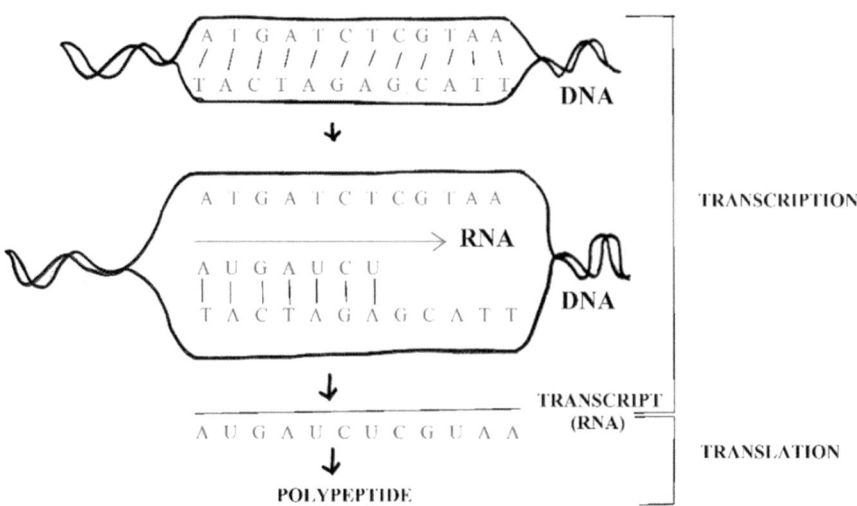

Fig. 2.8 Translation of RNA into proteins

There is some dissimilarity between DNA and RNA. Though DNA maintains a double-stranded structure, RNA follows a single-stranded configuration. Also, while measuring the stability of molecules, RNA is seen to show more instability and more propensities to degradation with respect to DNA due to the fact that RNA is made of ribose sugar rather than deoxyribose sugar that DNA holds. Three types of RNA namely *messenger* RNA (*m*RNA), *ribosomal* RNA (*r*RNA), and *transfer* RNA (*t*RNA) play vital roles in the DNA transcription process. The transcription process allows synthesis of RNA from DNA in the presence of an enzyme named *RNA polymerase*. The newly generated RNA sequences do not match with their DNA templates. After that, with the help of ribosomes, RNA is converted into proteins (polypeptides). This process is depicted in Fig. 2.8. While some of the RNA molecules remain as passive copies of DNA, most of them play active roles inside the cell. With some RNA molecules being responsible for switching on or off the genes, other molecules form the critical synthesis machinery in ribosomes.

Now, we again pay attention to our main topic of interest in this section—the cell cycle.

We have already learnt that the most fundamental functionality of cell cycle is to accurately duplicate the vast amount of DNA existing in chromosomes and then to correctly distribute the copies into the genetically identical offspring. Cells are created from cells, and the only way to generate more cells is by division of already existing cells. Most of us have watched in our childhood how a lowly worm caterpillar turns into a glorious creature butterfly. This is where the idea of cell cycle lies. The two biological processes of cell growth and cell division iterate continuously in every alive organism. During this process, as a cell first copies its contents and then divides into two, this event is known as cell cycle. The cell cycle is justified as a cycle, rather than just a linear pathway, as each of the two newly generated daughter cells repeats exactly the same process of cell cycle over and over from beginning.

A eukaryotic cell cycle is generally divided into two major phases: *interphase* and *mitotic* (*M*) phase. During interphase, the cell growth occurs and the DNA duplication process is performed. During the mitotic phase, the cell division occurs. So, starting from its birth time a cell must grow till it begins dividing itself. So, the phase of cell growth can be said as the preparation stage for the cell division. During interphase, the cell simply increases in size involving three steps: G_1 phase, S phase, and G_2 phase. During the G_1 phase (first *gap* phase; G = gap), the cell grows and becomes physically large, duplicates all the existing organelles inside the cell, and forms the molecular building blocks. These building blocks will be needed in next stages. In the S phase (S = synthesis), the cell prepares a complete copy of DNA within its nucleus. Besides copying DNA, the cell also duplicates *centrosome*—an organelle in cytoplasm that serves as main microtubule-organizing center (a structure from where all the microtubules emerge) in a eukaryotic cell. Centrosome acts as the regulator of cell cycle progression and helps DNA get separated in the M phase. During the G_2 phase, the cell keeps growing, creates organelles, and delivers proteins. The preparation for reorganizing the contents to be needed in M phase goes on in this phase. As soon as the G_2 phase ends, the M phase commences. There are some particular moments in the gap phases when the cell decides if it should proceed to the next stage or engage more time on preparation. Overall, during all of the interphase stages, the cell continues to perform gene transcription and protein synthesis, and also grows in size.

The cell enters the M phase when the chromosomes are condensed, after being replicated in the S phase. The condensation of chromosomes announces the ending of G_2 phase. The M phase sees the cell dividing the duplicated DNA and cytoplasm with an aim to form two new cells. This phase consists of two division-related events called *mitosis*—when the cell nucleus divides—and *cytokinesis*—when the cell splits into two daughter cells. The four different stages of *mitosis* include *prophase*, *metaphase*, *anaphase*, and *telophase*. As soon as the mitosis ends, the cytokinesis begins, with a very small overlap. The detailed description of the phases is provided in the next section. Figure 2.9 depicts the phases of cell cycle.

The outcome of completion of one cell cycle is the birth of two daughter cells. After the daughter cells are produced, some of the cells get divided rapidly with some other cells being divided slowly. There also exist some types of cells, which do not get divided at all. Those cells, which become divided quickly, face another round of cell division immediately after the previous cell division process ends. This characteristic is found in the cells of a tumor. Here, we will discuss the underlying process of cell division in more detail.

2.1.2.6 Cell Division

The cell division process occurs in the last phase (M phase) of the cell cycle, in which the nucleus along with the cytoplasm gets divided. There is a termination criterion that tells a cell when to stop this continual division process. Let us compare the physique of a human being, a mouse, and an elephant. While a mouse is small in size

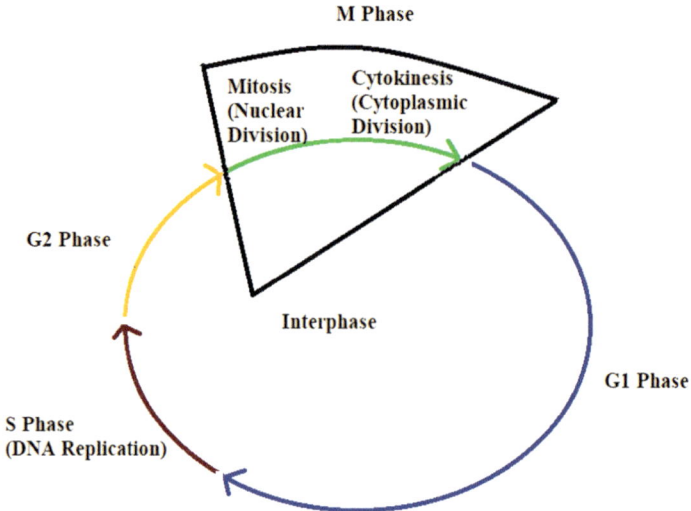

Fig. 2.9 Different phases of cell cycle

with respect to a human being, an elephant has a bigger shape. So, how this happens? Why does a human being not get shape of an elephant? In our cells, there are "cell cycle control genes" that not only determine our size, but also prevent the cells from being repeatedly divided, which might lead to cancer.

The *M* phase of a cell goes through two phases, with the first one being *mitosis* in which the nucleus gets divided and the other being *cytokinesis* when the cytoplasm becomes divided. The nuclear division is basically diving into the genetic material available in the nucleus. The nuclear division is performed in two steps—*mitosis* and *meiosis*. The two daughter cells that are generated due to nuclear division at the time of mitosis are genetically identical. More specifically, as a result of mitosis, each daughter cell obtains a perfect set of chromosomes, which are necessary for the correct functioning of the cell. When there exist cells with very few or too many chromosomes, there are chances that the cells either would not survive or will lead to cancer. On the other hand, meiosis lets the daughter cells obtain exactly half the genetic information of the parent cell. During cell division, the major works are performed in the mitosis phase.

Mitosis

When a cell undergoes mitosis, it goes through a set of steps. These steps are carefully organized so that the DNA of the cell does not get divided randomly and then distributed unevenly into the daughter cells. We have learnt in the previous section that mitosis consists of four basic phases: *prophase, metaphase, anaphase,* and *telophase*. These phases occur one by one. The cytokinesis step either occurs in anaphase or in telophase.

When interphase comes almost to the end (late G_2 phase) after the DNA copying is over, the chromosomes in the nucleus contain two connected copies of DNA

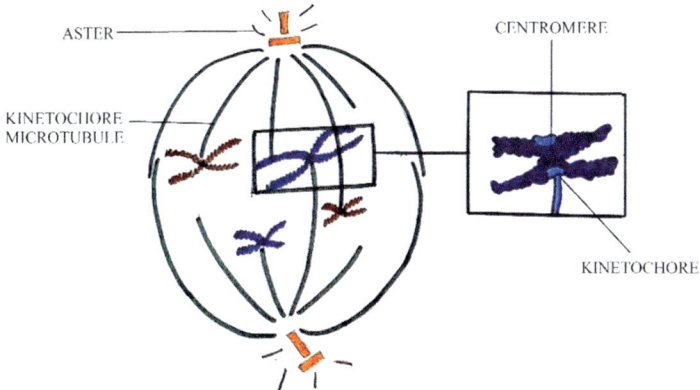

Fig. 2.10 Steps of prophase

called the *sister chromatids*. At this point of time, the chromosomes remain long, stringy, and decondensed. Each chromatid consists of a single, tightly coiled DNA molecule. A somatic cell, which is basically a body cell except egg and sperm, holds duplicate copies of every chromosome. For this reason, somatic cells are termed as diploid cells. Each such pair of chromosomes is termed as homologous pair, where one homolog originates from the mother and the other homolog from the father. Twenty-three homologous chromosome pairs are present inside a human body.

At the initial stage of prophase, the chromosomes begin condensing and a new structure called *mitotic spindle* is formed from the microtubules present in a cell. These kinds of spindles inside the centrosome are responsible for chromosome organization and let the chromosomes move around during mitosis. Then, the prophase allows disappearance of the nucleolus, a constituent of nucleus where ribosomes are made, giving an indication that the nucleus is now ready to break down. During the later stage of prophase, the chromosomes become fully condensed and very compact. With the bursting of the nuclear membrane, the chromosomes appear outside. The mitotic spindle grows in size and some of the existing microtubules begin capturing the chromosomes, forming a bonding between the microtubules and the chromosomes with the help of a special protein named *kinetochore*, which is present in the centrosome of each sister chromatid. The microtubules attached to the chromosomes are termed as k*inetochore microtubules*. Some microtubules that do not bind to kinetochores move toward the edge of the cell and form a structure named the *aster*. Figure 2.10 depicts the prophase.

During metaphase, the spindle aligns the captured chromosomes at the midline of the cell called the *metaphase plate*. At this moment, the two kinetochores of each chromosome remain attached to microtubules at the opposite poles of spindle. At the last stage of metaphase, the cell checks whether the chromosomes with kinetochores at the metaphase plate are appropriately attached to the microtubules. The point dedicated for performing these kinds of checks is called *spindle checkpoint*, which ensures even splitting of sister chromatids between the two daughter cells that are to

Fig. 2.11 Underlying steps of metaphase

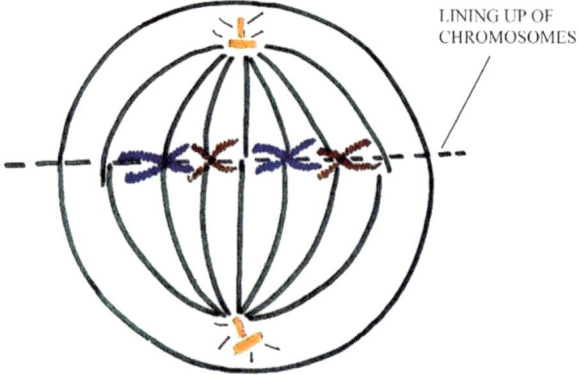

Fig. 2.12 Steps of anaphase

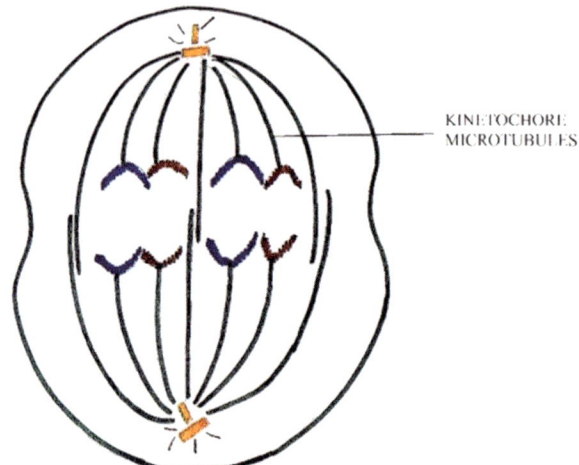

be separated in the next phase. If the cell identifies any misalignment or wrong attachments, the division process will halt until the problem is solved. Figure 2.11 shows the metaphase.

At the time of anaphase, the sister chromatids get separated from the "glue" protein (which is used to hold sister chromatids together) and become detached from each other. The alienated chromatids float toward the opposite ends of the cell. The microtubules disjointed from the chromosomes become longer and move aside by untying the poles and making the cell longer. Figure 2.12 describes the anaphase.

In telophase, the cell is almost ready to divide, and while the cytokinesis takes place, the cell begins to re-establish its normal structure. At this phase, the mitotic spindle decomposes into its constituents. With each set of chromosomes, two nuclei are formed. At that time, the nuclear membranes and nucleoli reappear. The chromosomes get back to their original form as they start to decondense. Figure 2.13 describes telophase.

Fig. 2.13 Steps of telophase

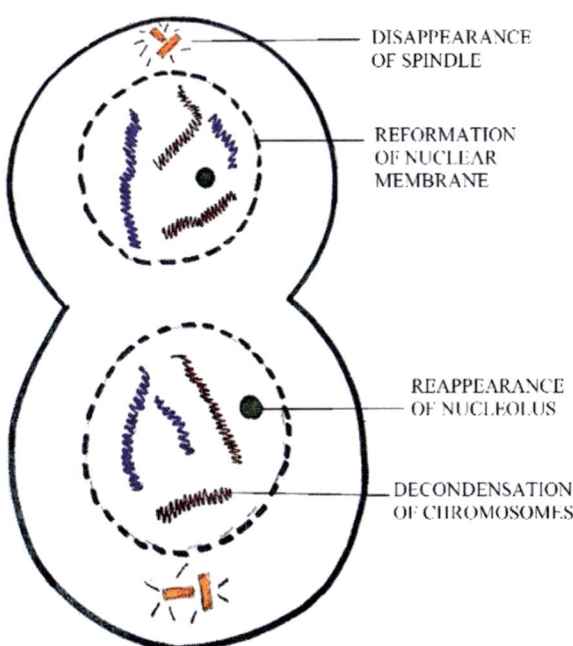

The process of cytoplasm division is known as cytokinesis, and this process runs simultaneously with anaphase or telophase of mitosis. More elaborately, the cytokinesis process starts at anaphase and runs through telophase. During this physical process, a ring of protein filaments is formed around the circular boundary of the cell just under the plasma membrane. This ring made of protein filaments is termed as the contractile ring. The contractile ring then starts shrinking and leads to creation of a *cleavage furrow* by pushing the plasma membrane toward the center (Satterwhite and Pollard 1992). The center pulls the sufficiently deep furrow of the plasma membrane of an animal cell. Now, the cell cytoplasm becomes divided into two halves (Alberts et al. 2015).

Meiosis

Meiosis is one type of cell division liable for cutting down the number of parental chromosomes by half and generating four gamete cells. Gamete cells are basically sex or reproductive cells, which are haploid in nature. Each gamete cell carries only one copy of each chromosome. Female gamete cells are called *ova* or *egg* cells, while male gamete cells are termed as *sperm* (Vacquier 1998). By producing the egg and sperm cells, Meiosis helps in sexual reproduction. When the sperm and egg cells are engaged in reproduction to produce a single cell, the number of chromosomes is reinstated in the offspring.

The parent cell that gets involved in meiosis is diploid in nature carrying two copies of each chromosome. After one round of DNA replication and two cycles of nuclear division, four haploid daughter cells are produced from the parent cell and each daughter cell contains half the number of chromosomes existing in the parent cell. The meiosis process follows two steps namely *meiosis* I and *meiosis* II. While meiosis I uniquely involves sex (germ) cells, meiosis II is the same as the mitosis process.

The initial phase of meiosis I is prophase I. In this initial phase, chromosomes are formed as the DNA and chromatin condense. The sister chromatids, which are pairs of replicated chromosomes, remain attached at centromere—a center point inside the cell. Microtubules are present at each pole of the cell. During metaphase I, identical chromosomes get aligned on either side of the central plate. Then, in anaphase I, the meiotic spindle fibers get contracted and attract the homologous chromosomes associated with two chromatids toward the poles of the cell. Telophase I is responsible for enclosing the chromosomes in nuclei. Telophase I is followed by cytokinesis similar to the mitotic cytokinesis, where the cytoplasm of the original cell is divided into two daughter cells.

Meiosis II is intended to the mitotic division of each of the haploid cells that are the outcomes of meiosis I. Meiosis II starts with prophase II where we can find condensed chromosomes with a new set of spindle fibers. The chromosomes move toward the central plate of the cell. Metaphase II finds that the centromeres of the paired chromatids are aligned central plates in both cells. During anaphase II, the chromosomes get separated at the centromeres. The spindle fibers attract the separated chromosomes toward each pole of the cell. At last, during telophase II, the chromosomes become enclosed in nuclear membranes. There also exists cytokinesis in meiosis II for the cytoplasmic division of two cells. The completion of meiosis II process also indicates the termination of entire meiosis process. The whole meiosis process ends with producing four haploid daughter cells. These four daughter cells would converge either into sperm cells or egg cells (Kleckner 1996).

2.1.2.7 Basics of Heredity

Heredity is basically passing the traits from parents to offspring through asexual or sexual reproduction. In asexual reproduction, a single organism is responsible to give birth to an offspring, and therefore, the genes of the single parent are inherited into the offspring. No fusion of gametes is found here, with no change in number of chromosomes. On the contrary, sexual reproduction involves two distinct gametes (one egg cell and one sperm cell) or reproductive cell. Meiosis process gives birth to gametes. These gametes possess half the number of chromosomes of parent cells. Heredity is a means by which an offspring can capture the parental genetic information.

Every organism has its unique characteristics like its structure, its nature of growth, its biochemical and physiological properties, and its behavior. The entire set of characteristics is called the unique *phenotype* of the organism. While some of the phenotypes of an organism are not inherited, the other traits that can be inherited are controlled by genes. On the other hand, *genotype* is defined as the collection of all the genes present in an organism's genome. The transmissible characteristics are carried forward from one generation to the next through the DNA. The parental genetic material is inherited in the child in the form of homologous chromosomes where, as we know, genes are encoded using unique combination of DNA sequences. A chromosome has a dedicated location called *locus* for a DNA sequence. The DNA sequence at a specific locus can vary from one individual to the other. These variations are scientifically termed as *alleles*. Sometimes, due to permanent alteration in the DNA sequence of the genome of an organism, new alleles can also be generated. This kind of permanent alteration in DNA sequence, which is commonly known as the outcome of erroneous DNA replication or some damage to DNA, is termed as *mutation*. Mutation also changes the phenotype of the organism.

2.1.2.8 Intracellular Signaling

Like a human being, a cell also leads a social life. It also needs to respond to its environment by interacting with other cells. The interaction occurs with the help of different signals. From being at the embryo stage until its death, a cell maintains coordination with other cells through transmission of different signals. At the embryo stage during the animal development, cells exchange signals among themselves to discuss their roles, their locations, their survival strategies, etc. There are specific signaling pathways in multicellular organisms.

In our daily lives, lots of information is exchanged for communication. At the time of communication, the signals corresponding to information change their forms from one to the other. For example, while two persons at two ends are talking over telephone, the sound waves of a person's voice are transformed into electrical signals that travel over the telephone wire. This kind of conversion is called *signal transduction*. Compared to this kind of communication, the cellular communication is much simpler, with a signaling cell producing a particular time of signal molecule to be detected by the signal cell. The target cells have receptor proteins that are responsible for recognizing and responding to the signal molecule. As soon as the target cells start to receive the incoming extracellular signal, the signal transduction begins and the incoming signal becomes converted into intracellular signal.

For sending signals, cells use many extracellular molecules like proteins, peptides, amino acids, nucleotides, and steroids. More often, the cells broadcast the signal to the whole body by secreting the signal molecules into bloodstream. These signal molecules are called *hormones*. Besides the broadcasting of signal when signal molecules enter the bloodstream, the signal molecules become spread in the neighboring region of cells through local diffusion. This type of less common signaling helps in wound-healing functions. Another kind of cell signaling is termed as *neuronal signaling* in which the neurons rapidly deliver messages to the target cells through private lines. The fourth kind of cell signaling excludes using secreted molecules during communication. Instead, the cells remain in direct contact with each other through the signaling molecules that are embraced in their plasma membranes. Binding between a signal molecule of the sender cell and the receptor molecule implanted in the plasma membrane of the target cell denotes the delivery of the message.

There are thousands of signals moving around everywhere in the cellular environment. But a cell should respond only to the intended signals to be used for the cell's functions, disregarding other signals. When a cell senses a signal, it first checks whether it possesses specific receptors for that signal. Without an appropriate receptor, the cell would not respond to an incoming signal. The cell only responds if an appropriate receptor is present. When a signal binds to a receptor protein, there may be a series of impacts on the target cells. For example, the shape of the cell can be altered; there may be change in cell's movement and metabolism; or change may be observed in gene expression. Also, sometimes a cell possesses a bunch of signal receptors. On receiving different kinds of signals simultaneously, the target cell becomes exposed to many cellular effects.

2.2 Introduction to Cancer Biology

A cell in a multicellular organism functions normally in accordance with some rules for cell growth and reproduction. But a cell can unexpectedly disobey the rules, dividing hysterically and invading other tissues. This unusual growth pattern is the result of genetic changes in a multicellular organism. The continuous uncontrolled growth of cells resulting in the development of tumor is known as *cancer*. A tumor that is confined to its original location is termed as *benign* tumor, whereas the tumor that is capable of invading surrounding tissues or the whole body is termed as *malignant* tumor. All cancer cells show abnormality in the process that regulates the normal cell division. Malignant tumor is caused by the cell division at very high frequency. Cancer is developed from the changes made by inherited mutations. These changes cause normal cells to behave abnormally. Before a normal cell turns into an invasive cancer, it has to go through a number of premalignant phases. The following chapters would discuss in detail the mechanisms behind formation of cancer. In this section, we just give an overview of the disease.

2.2.1 Factors Playing Behind Cancer

Formation of cancer depends on both external and internal factors. While external factors include mainly environmental issues like heavy exposure to sunlight and pollution, use of tobacco, exposure to chemical radiation, and infection from contagious viruses and bacteria, internal factors include inherited mutations, hormones, immune conditions, and mutations arising from metabolism.

Cancer is highly dependent on the lifestyle. For example, persons who smoke heavily have a greater chance of suffering from lung cancer. The persons who use tobacco snuff are more prone to nasal cancer. Consumption of alcohol also increases the chance of mouth and throat cancer. Imperfect diet also may lead to cancer; for example, intake of food items that are low in vegetables but rich in salt and nitrates may result in stomach or esophageal cancer, whereas high-fat, low-fiber diets may cause bowel, pancreatic, breast, and prostate cancer. Few communicable agents like *hepatitis B virus* (HBV), *human immunodeficiency virus* (HIV), *human papillomavirus* (HPV), and *helicobacter pylori* bacteria boost the possibilities of cancer. High exposure to ultraviolet radiation in sunrays and popularity of tanning salons worldwide make the people susceptible to skin cancer. One can prevent developing such types of cancer by living a healthy lifestyle and by taking vaccines for combating the infectious viruses.

Cancer may be developed in humans of any age-groups, but the older population, especially persons with age greater than 65 years, endures major risks of cancer. The incidence of occurrence of breast cancer, colorectal cancer, prostate cancer, and lung cancer grows with age, as in the old age, immune system becomes weakened; the rate of random genetic mutations rises; and besides hormonal changes, multiple genetic changes occur. In case of females, normal breast cells are stimulated by estrogen. Exposure to estrogen for a long time could be dangerous increasing the chances of breast or uterine cancer. For males, excessive and long-term exposure to testosterone leads to prostate cancer.

2.2.2 Hallmarks of Cancer

Six hallmarks of cancer are there which state what are the basic biological capabilities required for a tumor to have multistaged development. A tumor is a collection of complex tissues. These tissues are formed by many different cells having different kinds of interactions among each other. These six hallmarks collectively help in better understanding the biology of cancer (Hanahan and Weinberg 2011). The details about the six hallmarks are given below.

1. *Sustaining proliferative signaling*: In every cell division, cells in normal tissues wait for a growth-promoting signal or simply "Go" signal to enter into and proceed through the cell growth and cell division cycles. This is needed for maintaining the balance among the cell number, cell survival, energy metabolism, and the normal tissue. The receptors present at the cell surface release these kinds of signals following the branched intracellular signaling pathways. But the cancerous cells can carry on proliferative signaling. They may produce growth factor ligands (a molecule that binds to another molecule) themselves, follow the "Go" signal in terms of chemical message produced by their own, and continue the division process (Hanahan and Weinberg 2011).
2. *Evading growth suppressors*: In a normal human cell, some tumor-suppressive protein-coding genes are present for obstructing the cellular growth and proliferation. These tumor suppressor genes are activated by external and internal triggers in normal cells and lead to stoppage of the cell cycle at a certain time. More simply, like the "Go" signal, which indicates the start of cell division in a normal cell, there is another signal called "Stop," which tells when to halt the cell division process. Cancerous cells find their way to prevent the activation of tumor suppressor genes. This kind of inactivation can be because of mutations. So, the cancerous cells override the "Stop" signals to continue cell division (Gutschner and Diederichs 2012).

3. *Evading apoptosis*: Apoptosis is the antonym of cell growth. So, evading apoptosis means resisting cell death. Every normal cell has a "Self-Destruct" button to kill itself. A normal cell may switch on this button to react to stress with the help of apoptosis genes. If the functionality of these genes is lost by any means, the cells become cancerous with the capability of resisting their death, inhibiting the cell death signaling (Fernald and Kurokawa 2013).
4. *Enabling replicative immortality*: The number of cell division cycles performed by a normal is fixed. The *telomeres*—ends of chromosomes containing multiple repeats of nucleotide sequences "TTA GGG"—are responsible for imposing limits on replication. After each cell division in a normal cell, these telomeric repeats become shortened, finally resulting in cell death. The cancer cells are capable to evade the loss of telomeres with the help of a special enzyme called *telomerase* in 90% of cases. Telomerase adds telomeric repeats to the end of chromosomes, assisting cancer cells to keep on dividing indefinitely.
5. *Inducing angiogenesis*: During the growth and proliferation of cancer cells, the mass and size of the tumor increase. Like any normal tissue, tumors need oxygen and nutrients to sustain. Also, tumors should be capable to throw out their metabolic wastes and carbon dioxide. But the growth and proliferation of cancer cells are restricted by the natural diffusion limit of oxygen and nutrients. Cancer cells can overcome this problem through angiogenesis. Angiogenesis helps to form new blood vessels from the existing blood vessels, through which nutrients and oxygen are supplied to the tumors. These new blood vessels also allow tumors to evacuate the wastes. Angiogenesis is the elementary step for a benign tumor to become malignant. There is an "angiogenic switch" at "on" state, which enables continuous emerging of new blood vessels from the existing vessels to maintain tumor growth (Gutschner and Diederichs 2012).
6. *Activating invasion and metastasis*: Unlike any normal cell, cancer cells can spread the whole body through invasion and metastasis. Invasion enables cancer cells to penetrate the neighboring and adjacent tissues. Local invasion leads to the development of secondary tumors or metastasis. Metastasis describes the ability of cancer cells to intrude the lymphatic and blood vessels through which cancer cells are circulated elsewhere in whole body, and they invade normal tissues. The whole process of metastasis proceeds through an orderly and predictable manner similar to a cascade. The ability to move around from a primary site of disease to the other parts of the body is incorporated in the cancer cells by the mutations of the genes that control the production of proteins that normally bind cells to their adjacent tissues. Cancer cells perform some abnormal synthesis of enzymes, which break the bonds between the cells and their neighboring tissues, permitting the cancer cells to leave the primary site of the tumor (Gupta and Massagué 2006).

2.2.3 Classification of Cancer

There are hundreds of different kinds of cancer, which are categorized into six major categories based on the types of tissues from which they originate. There are six major categories of cancer namely *carcinoma, sarcoma, myeloma, leukemia, lymphoma,* and *mixed type*. In this section, we give brief description of each type of cancer.

Carcinoma: The body's epithelial tissue is the originator of this type of cancer. Epithelial tissue is observed everywhere in our body such as skin, covering and lining of all internal organs, and passageways like gastrointestinal tract. Carcinoma affects mainly breasts, lungs, colon, prostate, and bladder. Statistics show that carcinoma accounts for almost 80% to 90% of all cancer cases.

Sarcoma: Originated from the supportive and connective tissues such as bones, cartilage, tendons, muscle, and fat, this type of cancer is generally found to affect young adults. Sarcoma tumors are usually named after their originating tissues. For example, cancer in bone is termed as *osteosarcoma*; cancer in cartilage is known as *chondrosarcoma*; cancer in smooth muscle is called *leiomyosarcoma*; cancer in skeletal muscle is named as *rhabdomyosarcoma*; and cancer in blood vessels is known as *angiosarcoma*.

Myeloma: Myeloma, a special kind of blood cancer (Blimark et al. 2015), develops in the plasma cells of bone marrow. Plasma cell is a special type of white blood cell-containing bone marrow. Plasma cells are responsible for keeping the body's immune system strong by releasing antibodies called *immunoglobulins,* which fight against infection. In myeloma, the plasma cells act abnormally by dividing uncontrollably and produce an antibody called *paraprotein,* which is of no use.

Leukemia: The word "leukemia" means "white blood" in Greek. Leukemia, which is commonly known as blood cancer, originates in bone marrow. With the overproduction of immature white blood cells, which are not able to properly fight against cancer, leukemia makes the patients easily affected by viruses. White blood cells are affected very badly causing poor blood clotting and fatigue due to anemia.

Lymphoma: Lymphoma attacks the glands, organs (tonsil, spleen), and vessels and cleanses bodily fluids, creating infection-fighting white blood cells or lymphocytes. Other body parts like stomach, breast, or brain may be affected by lymphoma.

There are other types of cancer, which do not belong to any of the above categories. Some cancers of mixed types include adenosquamous carcinoma, carcinosarcoma, mixed mesodermal tumor, and teratocarcinoma.

2.2.4 Cancer Detection and Diagnosis

A tumor can be detected by common imaging techniques such as MRI (magnetic resonance imaging), X-rays (mammograms), CT scan (computed tomography), and ultrasound. The pathologist finds whether there are any kind of abnormalities in the cells in terms of their shape, size, and structure, especially the nucleus. A pathologist can understand whether the cells are normal by carefully observing the borders of the tumor. Only after carefully investigating the tumor cells, the pathologist comes into conclusion of whether the tumor is benign or malignant. The pathologist also states clearly whether the tumor is at early or late stage. Lymph node examination is also needed for diagnosis to know whether the other body areas have been affected by cancer.

Identification of tumor in a human being is easily performed by finding the existence of proteins in blood. An individual with tumor has frequent presence of proteins in blood, compared to a healthy person. But this kind of tumor marker is not ideal for cancer diagnosis, requiring more tests. While increase in prostate-specific antigen (PSA) indicates growth of tumor in prostate either in benign stage or in malignant stage, the presence of CA125 in ovarian cells may be an indication of ovarian cancer. Further confirmatory investigations are also required.

Early-stage cancer responds effectively to treatments. But detection of early-stage cancer is difficult as there are no typical symptoms. New molecular techniques are being developed to detect early-stage cancer. Scientists are utilizing technique like mass spectrometry to develop specific blood tests to identify a pattern of new proteins in the blood of individuals with a specific type of cancer. Scientists are trying to develop DNA microarrays to recognize the genes acting behind specific kinds of cancer cells (Chen et al. 2008). Clear idea about gene alterations would have a great impact on cancer detection and diagnosis.

2.3 Conclusion

This chapter provided an overview of cell biology and its behavior in cancer to the readers. This preliminary overview would make the readers familiar with the complex cellular phenomena for appreciating cancer biology, which is discussed in many places in the following chapters.

References

Alberts B, Bray D, Hopkin K, Johnson AD, Lewis J, Raff M, Roberts K, Walter P (2015) Essential cell biology. Garland Science

Beams HW, Kessel RG (1968) The Golgi apparatus: structure and function. Int Rev Cytol 23:209–276

Blimark C, Holmberg E, Mellqvist UH, Landgren O, Björkholm M, Hultcrantz M, Kjellander C, Turesson I, Kristinsson SY (2015) Multiple myeloma and infections: a population-based study on 9253 multiple myeloma patients. Haematologica 100(1):107

Bolsover SR, Shephard EA, White HA, Hyams JS (2011) Cell biology: a short course. John Wiley & Sons, Hobeken, NJ

Bruce A, Johnson A, Julian L, Martin R, Roberts K, Walter P (2008) Molecular biology of the cell. Garland Science, New York

Campbell CD, Eichler EE (2013) Properties and rates of germline mutations in humans. Trends Genet 29(10):575–584

Chen X, Ba Y, Ma L, Cai X, Yin Y, Wang K, Guo J, Zhang Y, Chen J, Guo X, Li Q (2008) Characterization of microRNAs in serum: a novel class of biomarkers for diagnosis of cancer and other diseases. Cell Res 18(10):997–1006

Comai L (2005) The advantages and disadvantages of being polyploid. Nat Rev Genet 6(11): 836–846

Cooper GM, Hausman RE (2004) The cell: a molecular approach

Duve DEC (1963) The lysosome concept. In: Ciba found. Symp. Lysosomes, vol 1, pp 1–35

Falini B, Mecucci C, Tiacci E, Alcalay M, Rosati R, Pasqualucci L, La Starza R, Diverio D, Colombo E, Santucci A, Bigerna B (2005) Cytoplasmic nucleophosmin in acute myelogenous leukemia with a normal karyotype. N Engl J Med 352(3):254–266

Fernald K, Kurokawa M (2013) Evading apoptosis in cancer. Trends Cell Biol 23(12):620–633

Fessele KL, Wright F (2018) Primer in genetics and genomics, article 6: basics of epigenetic control. Biol Res Nurs 20(1):103–110

Gupta GP, Massagué J (2006) Cancer metastasis: building a framework. Cell 127(4):679–695

Gutschner T, Diederichs S (2012) The hallmarks of cancer: a long non-coding RNA point of view. RNA Biol 9(6):703–719

Hanahan D, Weinberg RA (2011) Hallmarks of cancer: the next generation. Cell 144(5):646–674

Hoeijmakers JHJ (2001) DNA repair mechanisms. Maturitas 38(1):17–22

Holley RW, Apgar J, Everett GA, Madison JT, Marquisee M, Merrill SH, Penswick JR, Zamir A (1965) Structure of a ribonucleic acid. Science:1462–1465

Huang JC, Svoboda DL, Reardon JT, Sancar A (1992) Human nucleotide excision nuclease removes thymine dimers from DNA by incising the 22nd phosphodiester bond 5'and the 6th phosphodiester bond 3'to the photodimer. Proc Natl Acad Sci U S A 89(8):3664–3668

Kleckner N (1996) Meiosis: how could it work? Proc Natl Acad Sci U S A 93(16):8167–8174

Konarska M, Filipowicz W, Domdey H, Gross HJ (1981) Formation of a 2'-phosphomonoester, 3', 5'-phosphodiester linkage by a novel RNA ligase in wheat germ. Nature 293(5828):112–116

Kornberg RD (1977) Structure of chromatin. Annu Rev Biochem 46(1):931–954

Kripke ML, Cox PA, Alas LG, Yarosh DB (1992) Pyrimidine dimers in DNA initiate systemic immunosuppression in UV-irradiated mice. Proc Natl Acad Sci U S A 89(16):7516–7520

Kunkel TA, Bebenek K (2000) DNA replication fidelity. Annu Rev Biochem 69(1):497–529

Mannella CA (2006) The relevance of mitochondrial membrane topology to mitochondrial function. Biochim Biophys Acta Mol Basis Dis 1762(2):140–147

Ron D, Walter P (2007) Signal integration in the endoplasmic reticulum unfolded protein response. Nat Rev Mol Cell Biol 8(7):519–529

Sancar A, Lindsey-Boltz LA, Ünsal-Kaçmaz K, Linn S (2004) Molecular mechanisms of mammalian DNA repair and the DNA damage checkpoints. Annu Rev Biochem 73(1):39–85

Satterwhite LL, Pollard TD (1992) Cytokinesis. Curr Opin Cell Biol 4(1):43–52

Sobell HM (1985) Actinomycin and DNA transcription. Proc Natl Acad Sci U S A 82(16): 5328–5331

Song HJ, Poo MM (2001) The cell biology of neuronal navigation. Nat Cell Biol 3(3):E81–E88

Vacquier VD (1998) Evolution of gamete recognition proteins. Science 281(5385):1995–1998

Waga S, Stillman B (1998) The DNA replication fork in eukaryotic cells. Annu Rev Biochem 67(1): 721–751

Tumor Biology: An Introduction

Partha Nandi and Soumyabrata Roy

> *The gene that enables birds to learn songs can become cancer-causing. There is no normal physiological process that can't be bastardized by the disease.*
>
> Siddhartha Mukherjee (Oncologist and Author)

Abstract

About 200 different types of cells and their coordination make up and run the human body. And each of these cells is governed by the genetic information encoded in the DNA present in the cells' nuclei. Although the nucleotide sequence of DNA is well checked and maintained throughout one's life, mutations still occur that in certain instances cause diseases, cancer being one of them. The failure of the intricate genetic system that balances cell birth and cell death causes cancer. Cancer cells are clonal as a single common ancestral cell gradually accumulates mutation to form a tumor that over time develops malignancy. Growth factors are important components of tumor microenvironment that provide heterogeneity and autonomy to cancer cells, the properties that normal cells lack. Not only spontaneous mutations and genetic predisposition but also lifestyle, to a great extent, contributes to carcinogenesis. Tobacco smoke, UV

P. Nandi (✉)
Department of Physiology, Government General Degree College, Jhargram, West Bengal, India

S. Roy
Department of Microbiology and Immunology, Medical University of South Carolina, Charleston, SC, USA

rays, X-rays, agents that attack DNA, and change in its chemical structure are potential carcinogens. Viruses too can cause carcinogenesis. Whatever be the causes, they all affect the fundamental aspects of cellular function, including DNA repair, cell cycle regulation, apoptosis, and signal transduction. Thus, correct and effective management of this fatal disease is only possible if the underlying biological complicacies are well understood.

Keywords

Carcinogen · Mutation · Neoplasm · Oncogene

3.1 Introduction: Cancer Perspective from Fundamental Biology

In the sum of the parts, there are only the parts...., as Wallace Stevens stated, this can be well explained in the context of living organisms. Comprehending the diversity among organisms in practical terms is to understand the differences in structure, function, and behavior of cells that are the basic or fundamental units of a living organism. Every organism either consists of a single or multiple cells. The human body is composed of about 200 different types of cells, each with a specific function that works as a whole in a coordinated manner, to define the overall function of the organism. However, all cells share common fundamental properties that have been conserved throughout evolution. The discoveries of biochemists and molecular biologists suggest that all cells contain genetic information, generally in the form of deoxyribonucleic acid (DNA) and utilize the same basic principles of energy metabolism.

We now know that genes, which chemically are composed of DNA, ultimately define the biological structure and maintain the integration of cellular function. A cell, during reproduction, replicates its DNA and then passes a copy of the genetic information encoded in the form of DNA to its progeny, by dividing it into two daughter cells. The gradual understanding of this transmission of genetic information as a result of closer insight into the properties of genes, their composition, variation among them, and the role they play in heredity led to the emergence of altogether a separate branch of biology called *genetics*. However, the seed to the emergence of this discipline was sown much before genes contained in DNA were found to be the actual genetic material.

The true understanding of genetics in scientific terms began in a monastery garden in Central Europe, back in the 1860s, where an Augustinian monk named Gregor Johann Mendel, through a set of simple experiments on pea plants, determined the underlying principles of heredity and made some generalizations that later came to be known as *Mendel's laws of inheritance* namely the law of dominance and uniformity, law of segregation, and the law of independent assortment. The essence of all of them is based on the fact that each trait in an organism is controlled by *a pair of factors* (later termed *genes*), one *dominant* and the other *recessive*, and that during gamete formation members of a gene pair separate from each other. Although his works were published in 1866, its true significance was understood about three decades later when Hugo de Vries and other scientists rediscovered his research (Klug et al. 2016).

Owing to the discovery of chromosomes by Flemming in 1879, almost contemporaneously with the rediscovery of Mendel's work, two cytologists Sutton and Boveri observed that chromosomes and Mendelian factors have several properties in common and behaved as per Mendel's principles during gamete formation (*meiosis*). This parallelism called the *chromosome theory of inheritance* laid down the foundation for the physical basis of inheritance and it was further confirmed through subsequent works of Morgan and others that genes contained in chromosomes were the actual message bearers, transmitted through gametes, carrying hereditary information faithfully from generation to generation.

However, in the subsequent years, modification and extension of Mendelian principles and the study of interallelic interaction followed by elucidation of the structure of DNA in 1953 by Watson and Crick took us a bit closer to—*what could be a gene* in chemical terms. Further advancements in molecular biology and deciphering of the genetic code contained in the ATGC nucleotides' sequence in DNA led to better understanding of how the combined effect of genes and their products interacting with the environment control the *phenotype* of an organism as a whole and the biology of diseases in particular. As per the central dogma in molecular biology, the genetic code in a DNA sequence is utilized to synthesize mRNA (transcription), which in turn specifies the amino acid sequence in polypeptides and thus proteins (translation), whose specific molecular organization and function determine the phenotype of the organism. However, genetic code is not the absolute determiner of gene expression. Although all the cells in an organism contain the same DNA, they differ in the activation of selective genes at a specific time and in specific tissues in response to specific environmental stimuli. This differential gene expression could be manipulated at genomic, transcriptional, posttranscriptional, translational, posttranslational, and hormonal levels mediated through a variety of ways like DNA rearrangements, the effect of transcription factors, mRNA processing and export from the nucleus, gene silencing by noncoding RNAs, chromatin remodeling, and the effect of growth factors (Cooper 2000; Klug et al. 2016).

Although these processes are finely tuned and nevertheless DNA is a highly stable molecule, copied with high accuracy, changes in DNA structure do occur, which may alter the genetic information of the DNA, leading to mutations. On one hand, when mutation provides the raw material for evolution, when it harms the phenotype of an organism, it causes a genetic disorder or a disease.

Given the vast number of activities that need to be coordinated in every cell, it is not surprising that malfunctions occasionally arise. Cancer, the second leading cause of death after cardiovascular disease (Dagenais et al. 2020), is a prominent example of a disease that arises from such abnormalities in cell function. Cancer is not a single disease rather a hundred different diseases. All cells grow, may differentiate, and die after a predetermined time. When the intricate genetic control system that balances cell birth and death fails, cell immortality occurs that leads to cancer. According to the clonal evolution theory of cancer, most cancers come about from a progressive series of genetic changes. The mutations can be point mutations that directly affect the DNA sequence or as in most cases, as recent research has revealed, they could be chromosomal aberrations affecting the types and locations of chromatin modification, particularly DNA and histone acetylation patterns. Some of these mutations are results of epigenetic modifications as well that can be inherited from one cell to its progeny cells and may be present in either somatic or germ-line cells. However, these mutations occur predominantly in somatic cells. Thus, cancer although being a genetic disease is not always heritable. Only about 1% of cancers are associated with the germ-line mutations that increase a person's susceptibility to certain types of cancer. Whatever be the source or type of mutations, they affect the fundamental aspects of cellular function, including DNA repair, cell cycle regulation, apoptosis, and signal transduction.

3.2 The Nature of Cancer and Its Types

A multicellular organism starts its life from a *zygote*, which through a regulated and controlled manner of cell divisions develops into an embryo, finally becoming a fully grown organism. The mechanisms of cell growth and division are well checked and controlled throughout one's life unless rendered ineffective by some *tumorigenic effect*. When such a mutative condition leads to abnormal and uncontrolled cell divisions, the cells become congregated locally in a particular place to form a swelling called tumor. If the tumor remains less harmful, being restricted to the tissue of their origin, they are called benign. On the other hand, malignant tumors are those from which cells detach and migrate through blood or lymph systems to the other parts of the body, giving rise to secondary tumors. The second type is the cancerous one. Thus, apart from uncontrolled cell growth and division, the second most important fundamental property of cancer cells is *metastasis* during which the tumor cells undergo abnormal cytoskeletal changes, dissociate from the primary tumor secreting proteases that breach the components of *extracellular matrix* and *basal lamina,* and invade other tissues (loss of contact inhibition) (Wagener et al. 2017; Weinberg 2014).

Tumors are a mixture of cells, some of which act as *cancer stem cells*, each of which has the capability of both self-renewal and differentiation into a mature cell type. The tumor microenvironment contributes to the *heterogeneity* of cells within the tumor. However, all cancer cells are clonal; that is, they originate from a common ancestral cell that accumulates specific mutations. This means that a single mutation is not sufficient to transform a normal cell into a tumorigenic, malignant one. Cancer is initiated when a single mutated cell begins to proliferate abnormally, progressively gathering additional mutations, and the selection and proliferation of specific rapidly growing cells from the whole population ultimately cause malignancy.

Apart from spontaneous mutations and to some extent genetic predisposition to certain types of cancers, lifestyle has been found to directly or indirectly influence carcinogenesis, occupation, and diet being the two other main factors. Epidemiologists suggest that there is a 20-fold increased risk (Dela Cruz et al. 2011) of developing lung cancer in patients smoking tobacco that contains at least 60 chemicals that can alter DNA sequences. Similarly, alcohol consumption increases the risk of liver cancer (Marengo et al. 2016). Inhaling asbestos fibers may cause *mesothelioma*. Consumption of red meat and animal fat is associated with prostate, colon, and breast cancers (Kvale et al. 2017; Bray and Kiemeney 2017; World Cancer Research Fund/American Institute for Cancer Research WCRF/AICR 2018). The most potential chemical mutagens can be both naturally occurring (e.g., aflatoxin, component of a mold) (Lin et al. 2014) and synthetic compounds (e.g., nitrosamines and pesticides). These chemicals can directly act as mutagens or are converted to mutagenic compounds by cellular enzymes.

Most cancers fall into three major categories based on the embryonic tissue of their origin: carcinomas—malignancies of epithelial cell lining of various organs, like mouth, esophagus, intestines, and uterus, and also from the skin; sarcomas—solid tumors of connective tissues, arise from *mesodermal* cells of connective tissues like fibrous tissue and bone; and leukemias or lymphomas arising from blood-forming cells and cells of the immune system, respectively. However, some types of tumors do not fit into these major classifications. About 200 varieties have been described, whose properties and treatments are different. Leukemias of early childhood differ from adult leukemias in their properties and treatments.

Cancers are named according to the organ from which they arose. Retinoblastoma is mainly cancer of the eye, osteosarcoma of bone, and melanoma of skin pigment cells. The most common solid organ malignancies arise in the lung, breast, and gastrointestinal tract. Females have the highest lifetime risk of breast, lung, and bowel cancers; prostate, lung, and bowel cancers are prevalent in men. Cancers of lung, liver, stomach, and bowel are the most common causes of cancer death worldwide, accounting for more than four in ten of all cancer deaths. The rate of death varies greatly for different types of cancer. Lung cancer and pancreatic cancer are the worst, usually fatal within a year. But not all cancers are fatal, only one-fifth of breast cancer cases results in death (Bray et al. 2018).

3.3 Environmental Insults—Mutagens, Carcinogens, and Cancer-Causing Mutations

Sir Percivall Pott, a British surgeon, in 1775 became the first person to link malignancy with environmental carcinogen when he found an association between high incidences of an uncommon form of scrotal cancer called "chimney sweeps' carcinoma" and exposure to soot. This provided the first evidence of direct contact carcinogen to skin, i.e., soot, and indicated an occupational link to cancer (Pott 1775; 1974). Thereafter, isolated carcinogenic chemicals from soot along with several other compounds have been found to cause cancer in laboratory animals. Apart from these chemicals, many other types of agents including ionizing radiations and RNA- and DNA-containing viruses have been identified as potential carcinogens.

Ultraviolet radiation, the leading cause of skin cancer, may act as a direct mutagen. However, some carcinogenic chemicals, such as those present in soot or cigarette smoke, can act as direct mutagens or may be converted to potential mutagens by cellular enzymes. *The commonality in all these agents lies in the fact that all of them alter the genome.*

There are different ways by which a normal cell could acquire cancer-causing mutations. Environmental abuses like tobacco smoke, UV rays, X-rays, agents that attack DNA, and change in its chemical structure could result in such mutations. Mutations could arise from spontaneous errors during cell division. There is some chance of error during the DNA copying process (replication), such as an adenine (A) base is replaced by guanine (G) or cytosine (C). Some hereditary cancer syndromes, such as retinoblastoma and breast cancer, could be inherited from parents through mutant cancer genes. Still, inherited cancers probably account for 5–10% of all cases. In some cases, genes could also be carried into the cells through viruses. In all these cases, the result converged on the same pathological process—the inappropriate activation or inactivation of genetic pathways that controlled growth, causing the malignant, dysregulated cellular division that was characteristic of cancer.

Five leading behavioral and dietary risks, high body mass index, low fruit and vegetable intake, lack of physical activity, tobacco, and alcohol use, have been found to cause about one-third of deaths from cancer (Dela Cruz et al. 2011; Marengo et al. 2016; Kvale et al. 2017; Bray and Kiemeney 2017; World Cancer Research Fund/American Institute for Cancer Research WCRF/AICR 2018; Pott 1775; International Agency for Research on Cancer IARC 2014). Tobacco remains the leading factor in the etiology of 33% of cancers, including those of the HNSCC (head and neck squamous cell carcinoma), lung, nasopharynx, bladder, and kidney, and these could be prevented through reduction in smoking of the populace. Further 30% of cancers, including those of the breast, colon, esophagus, stomach, and liver, have been associated with diet, alcohol, obesity (high BMI), and lack of physical activity (sedentary lifestyle). Several modifications in lifestyle like reduced consumption of red meat, animal fat, and alcohol and increased intake of fiber, fresh fruit, and vegetable would prevent obesity and thus cancer. Vaccination and control of infections in cervix, stomach, liver, nasopharynx, and bladder may reduce the risk of about 15% of cancers.

3.4 Infectious Agents, Inflammation, and Cancer

The relationship between viruses and human cancer remained hypothetical until 1980 when Robert Gallo at the National Institutes of Health first found a *retrovirus* (human T-lymphotropic virus type 1) is responsible for a rare form of leukemia (Poiesz et al. 1980; Gallo et al. 1982). Work by Gallo's team and independent work in Japan by K Takatsuki showed that adult T-cell leukemia (ATL) was caused by HTLV-1 and the disease is highly prevalent in Japan and sporadic in most of the remaining parts of the world (Gallo et al. 1982; Yoshida et al. 1982). Some DNA viruses like *polyomavirus*, *simian virus* 40 (SV40), *adenovirus*, and *herpes-like viruses* can cause human cancers. However, about 15% of cancers worldwide are linked to other types of viruses. Mostly, these viruses, rather than acting as the sole cause of cancer, greatly increase the risk of its development. The human *papillomavirus* (HPV) provides a good illustration of relationship between viral infection and cancer. Transmission of this virus can occur through sexual activity, and its frequency is increasing in the population. Although the virus has been found to be linked with the development of about 90% of cervical cancers (Walboomers et al. 1999), the vast majority of women who have been infected with the virus will never develop this malignancy. HPV also induces the development of mouth and tongue cancers in both males and females. An effective vaccine against HPV is now available (Herrero and Murillo 2018). Other viruses like *hepatitis B virus* is associated with liver cancer (London et al. 2018), *Epstein–Barr virus*, associated with Burkitt's lymphoma (Brady et al. 2007), and a type of *herpes virus* (HHV-8), associated with *Kaposi's sarcoma (*Ablashi et al. 2002).

Helicobacter pylori, a bacterium living in the stomach, which is also responsible for ulcers, is associated with certain gastric carcinomas (Plummer et al. 2015). Recent evidence suggests that chronic inflammation triggered by the presence of the pathogen is linked to many of these cancers. For example, inflammatory bowel disease (IBD), a result of chronic inflammation, increases the risk of colon cancer (Rubin et al. 2012). Further findings of the general process of inflammation help elaborate its role in cancer development, which was much unexplored previously.

3.5 Cellular and Genetic Basis of Cancer

3.5.1 Comparison of Cancer Cell and Normal Cell

Cancer cell acquires a few noticeable properties, which are different from a normal somatic cell that favors them to become a neoplasm growth (Cooper 2000; Wagener et al. 2017; Weinberg 2014). Whether in vivo or in vitro, cancer cell loses their growth control and become malignant. These malignant cells can grow continuously on their own irrespective of the presence or absence of growth stimulatory or inhibitory signals that otherwise influence the growth of normal cells. When normal cells are grown in culture, they are usually supplemented by serum that contains essential growth factors, such as epidermal growth factor. But as cancer cells are independent of the regulatory mechanisms that govern normal cell proliferation and survival, they do not require growth factors for their proliferation. Normal cells, after dividing mitotically for a certain time, cease to continue their growth and division by undergoing senescence. On the contrary, the presence of telomerase in cancer cells (being one of the reasons for higher growth potential than normal cells) allows them to divide indefinitely, rendering them immortal. The absence of telomerase from most types of normal cells is thought to protect the body against tumor growth. Apoptosis is another important savior mechanism characterized by self-destruction of normal cells when the chromosome content in them is disturbed. In contrast, cancer cells fail to elicit apoptotic response even when their chromosome content becomes highly deranged.

3.5.2 Genetic View of Cancer

Single-gene mutations do not cause most common human diseases. Rather multiple genes spread diffusely throughout the human genome and determine the risk for a genomic illness. These diseases can be understood, diagnosed, or predicted only by understanding the underlying interrelationships between several independent genes.

3.5.3 Tumorigenesis

Cancer is an ultimate result of a multistep process that accumulates some sequential tumorigenic gene mutations. To initiate carcinogenesis successfully, cells require certain characteristics, collectively referred to as the "hallmarks" of cancer (Hanahan and Weinberg 2000). Cancer is an archetypal genomic disease, and its genetic nature had been known since 1872 when Hilario de Gouvea, a Brazilian ophthalmologist, had described a family in which a rare form of eye cancer, called retinoblastoma, coursed tragically through multiple generations (Monteiro and Waizbort 2007). Despite bad habits, bad recipes, neurons, obsessions, environments, behaviors, and other attributes shared by families, the familial pattern of the illness suggested the role of an *inherited factor* as proposed by deGouvea as the cause of retinoblastoma. Some light on the *inherited factors* in peas was already cast by an unknown botanist monk named Mendel seven years prior in a publication. However, de Gouvea had never encountered Mendel's paper or the word gene.

By the late 1970s, a full century after de Gouvea, scientists began to converge on the uncomfortable realization that cancers arose from normal cells that had acquired mutations in growth-controlling genes. These genes act as powerful growth regulators in normal cells. Hence, a wound in the skin, having healed itself, typically stops healing and does not morph into a tumor. Here, genes tell the cells in a wound when to start growing and when to stop. In cancer cells, these pathways were somehow disrupted. Start genes were shut off, and stop genes were ceased; genes that altered metabolism and identity of a cell were corrupted, resulting in a cell that did not know to stop growing.

Alterations of such endogenous genetic pathways caused cancer—a *distorted version of our normal selves*, as Harold Varmus, the cancer biologist (Varmus et al. 2016), but it was ferociously disquieting as for decades, scientists had hoped that some pathogen, such as a virus or bacterium, would be implicated as the universal cause of cancer and might potentially be eliminated via a vaccine or antimicrobial therapy. The gradual unveiling of correlation between cancer genes and normal genes threw open a central challenge of cancer biology. How the mutant genes are restored to their off or on states while they allow normal growth to proceed unperturbed remains the biggest conundrum of cancer therapy.

3.5.4 Oncogenes and Tumor Suppressor Genes

Basically, the two classes of genes—tumor suppressor genes and oncogenes—have been implicated in carcinogenesis (Cooper 2000; Klug et al. 2016). Tumor suppressor genes encode proteins that restrain cell growth and prevent cells from becoming malignant, thereby acting as cells' brakes. Oncogenes, on the other hand, encode proteins that may cause genetic instability, prevent apoptosis, promote metastasis and as a whole shatter the growth control, and promote the conversion of a cell to a malignant state. Most oncogenes act as accelerators of cell proliferation. *The existence of oncogenes and tumor suppressor genes suggests an elaborate system to positive and negative controls that maintain cell growth within normal limits.* Taken separately, the oncogenes and tumor suppressor genes provide alternate pathways of oncogenesis. *Mutation of oncogenes stimulates mitogenic pathways, resulting in cell proliferation. Loss of control of cell proliferation* can be brought about by mutations in tumor suppressor genes.

3.5.5 Genome-Wide Approach

It was remaining as a big question to the cancer biologist that how much such genes were involved in causing a typical human cancer, one gene per cancer, or, dozen and even hundred. In the late 1990s, at Johns Hopkins University, a cancer geneticist named Bert Vogelstein decided to create a comprehensive catalog of nearly all the genes implicated in human cancers (Vogelstein et al. 2013). *Vogelstein had already discovered that cancer arises from a step-by-step process involving the accumulation of dozens of mutations in a* cell. By acquiring sequential mutations, gene by gene, a cell leads to the dismantling of its growth regulatory pathways and proceeds toward cancer.

To cancer geneticists, these data suggested that the one-gene-at-a-time approach would be insufficient to understand, diagnose, or treat cancer. A fundamental feature of cancer is its enormous genetic diversity. Two specimens of breast cancer were removed from two breasts of the same woman at the same time and might have vastly different spectra of mutations and thereby behave differently, progress at different rates, and respond to different chemotherapies. To understand cancer, biologists would need to assess the entire genome of a cancer cell.

3.6 Temporal Variation of Cancer Incidence Rate in Different Populations

Several studies on migration have established that the frequencies of various cancers vary greatly between countries and also in ethnicity (Bray et al. 2018; International Agency for Research on Cancer IARC 2014; Herrero and Murillo 2018; London et al. 2018). Thus, not just genetic but environmental differences do affect the prevalence of various types of cancer; second-generation Japanese in California have a tenfold higher death rate from prostate cancer than do Japanese in Japan. Various epidemiological studies enlighten the effects for various carcinogenic agents. For example, meat and some high-calorie, fat-rich diet can be cancer-causing. Women from countries that consume high quantities of meat (12 pounds per day) possess tenfold higher risk of developing cancer. Japanese have a high level of stomach cancer correlated with consumption of fern fronds. Incidence rates of stomach cancer among first-generation Japanese migrants to Hawaii were interestingly decreased than the rates among Japanese living in Japan. Smoking increases the risk of lung cancer (Dela Cruz et al. 2011). In Norway, where only one-fourth as many cigarettes are smoked per person, rate of lung cancer is five times less than that in Britain. It increased much later in women than in men in whom it increased about 15-fold since 1930, when smoking became prevalent. Skin cancer correlates with excessive exposure to sunlight, especially for races with light skin pigmentation. Similarly, exposure to radiation may frequently give rise to leukemias. The risk of developing cancer can be decreased by avoiding smoking, a calorie and fat-rich diet, and excessive sun and radiation exposure.

3.7 Prevention and Early Detection

A closer insight into the causes of cancer certainly opens ways to prevention. Nevertheless, it cannot be completely eradicated unless the underlying biological complicacies of the disease are fully understood and handled effectively. Significantly effective and less expensive treatment, increased probability of survival, and less morbidity could be ensured to cancer patients if the disease is diagnosed early. Evidence-based prevention strategies can prevent the risk of about 30–50% of cancers. For example, about one million cancer cases per year could be possible to prevent by vaccination against HPV and hepatitis B viruses (Globocan 2018). In this way, early detection and adequate management of patients who develop cancer can cure many types of cancers. However, late-stage or inaccessible diagnosis or treatment is common. New research opens up early detection of various cancers, like mammography and BRCA screening for breast cancer; Pap smear test, which can detect precancerous cell changes of the cervix, has been primarily responsible for a 70% decline in uterine cancer deaths in the USA over the last five decades. In underdeveloped nations and among members of lower socioeconomic groups in developed countries, cervical cancer remains a major cause of female death. The availability of pathology services in public sector was reported from only 26% of low-income countries, which was far less than above 90% of high-income countries, which reported available treatment services. This says of the significant and increasing economic impact of cancer. The total annual economic cost of cancer in 2010 was estimated at approximately US$ 1.16 trillion (Plummer et al. 2016). Only 1 in 5 low- and middle-income countries have the necessary data to drive cancer policy.

3.8 Conclusion

Every cancer type requires a specific treatment regimen in terms of say surgery, radiotherapy, or chemotherapy. Thus, a correct diagnosis is crucial for an adequate and effective treatment. Prior to treatment and management of cancer patients, the goals of such undertakings must be determined and health services should be integrated and be made people-centric. Apart from the primary goal, i.e., to cure cancer or increase the patient's life span, improving the patient's quality of life should also be taken into consideration. For this, supportive or palliative care and psychosocial support are necessary. By early detection and best possible treatment, some of the most common cancer types, such as breast cancer, cervical cancer, oral cancer, and colorectal cancer, could be treated. Appropriate treatment can cure even metastatic conditions like testicular seminoma, leukemias, and lymphomas in children. Palliative care improves the quality of life of patients and their families rather than just curing the symptoms. Palliative care is effective not only for cancer patients but also for patients with other chronic fatal diseases, where there is little chance of cure. It can also relieve more than 90% of advanced-stage cancer patients from physical, psychosocial, and spiritual problems. Community-based and home-based healthcare strategies and palliative care for patients from low-income groups are some effective ways to relieve pain. Terminal phase cancer pain suffered by over 80% of patients could be treated by oral morphine (Howie and Peppercorn 2013).

Cancer has been a major focus of research for decades due to its impact on human health and the hope to develop a cure. On December 23, 1971, US President Richard M. Nixon signed the National Cancer Act to initiate *the war on cancer*. During the last five decades, over 100 billion dollars was sanctioned for the National Cancer Institute for an unprecedented expansion of basic biological research. In the years following the National Cancer Act, improved diagnosis and treatment have increased the five-year survival rate from 50% to 60%. Death rates for childhood cancers and cancers of the stomach, uterus, and colon have dropped dramatically. However, death rates from lymphoma, lung, prostate, liver, and brain cancers have risen markedly. Though these studies have broadened our way toward exploring the cellular and molecular basis of cancer, much success has not been achieved in the prevention or cure of most cancers. Although cancer is still not conquered, many believe that even difficult cancers will come under control with a new generation of diagnostics and treatments based on a detailed understanding of the inner working of tumor cells.

References

Ablashi DV, Chatlynne LG, Whitman JE, Cesarman E (2002) Spectrum of Kaposi's sarcoma-associated herpes virus, or human herpes virus 8, diseases. Clin Microbiol Rev 15(3):439–464

Brady G, MacArthur GJ, Farrell PJ (2007) Epstein–Barr virus and Burkitt lymphoma. J Clin Pathol 60(12):1397–1402

Bray F, Kiemeney L (2017) Epidemiology of prostate cancer in Europe: patterns, trends and determinants. In: Bolla M, van Poppel H (eds) Management of prostate cancer: a multidisciplinary approach. Springer-Verlag, Berlin, pp 1–11

Bray F, Ferlay J, Soerjomataram I, Siegel RL et al (2018) Global cancer statistics 2018: GLOBOCAN estimates of incidence and mortality worldwide for 36 cancers in 185 countries. CA Cancer J Clin 68(6):394–424

Classics in oncology. Sir Percivall Pott (1714 – 1788) (1974) CA Cancer J Clin 24(2):108–116

Cooper GM (2000) The cell: a molecular approach, 2nd edn. Sinauer Associates, Sunderland, MA

Dagenais GR, Leong DP, Rangarajan S, Lanas F et al (2020) Variations in common diseases, hospital admissions, and deaths in middle-aged adults in 21 countries from five continents (PURE): a prospective cohort study. Lancet 395(10226):785–794

Dela Cruz CS, Tanoue LT, Matthay RA (2011) Lung cancer: epidemiology, etiology, and prevention. Clin Chest Med 32(4):605–644

Gallo RC, Blattner WA, Reitz MS, Ito Y (1982) HTLV: the virus of adult T-cell leukemia in Japan and elsewhere. Lancet 1(8273):683

GLOBOCAN (2018) International Agency for Research on Cancer, GLOBOCAN 2018 accessed via Global Cancer Observatory

Hanahan D, Weinberg RA (2000) The hallmarks of cancer. Cell 100(1):57–70

Herrero R, Murillo R (2018) Cervical cancer. In: Thun MJ, Linet MS, Cerhan JR, Haiman CA, Schottenfeld D (eds) Cancer epidemiology and prevention, 4th edn. Oxford University Press, New York, pp 925–946

Howie L, Peppercorn J (2013) Early palliative care in cancer treatment: rationale, evidence and clinical implications. Therap Adv Med Oncol 5(6):318–323

International Agency for Research on Cancer (IARC) (2014) In: Stewart BW, Wild CP (eds) World cancer report 2014. International Agency for Research on Cancer, Lyon

Klug WS, Cummings MR, Spencer CA, Palladino MA (2016) Concepts of genetics, 10th edn. Pearson Education, Limited, India

Kvale R, Myklebust TA, Engholm G, Heinavaara S et al (2017) Prostate and breast cancer in four Nordic countries: a comparison of incidence and mortality trends across countries and age groups 1975–2013. Int J Cancer 141(11):2228–2242

Lin YC, Li L, Makarava AV, Burgers PM et al (2014) Molecular basis of aflatoxin-induced mutagenesis – role of the aflatoxin B1-formamidopyrimidine adduct. Carcinogenesis 35(7):1461–1468

London WT, Petrick JL, McGlynn KA (2018) Liver cancer. In: Thun MJ, Linet MS, Cerhan JR, Haiman CA, Schottenfeld D (eds) Cancer epidemiology and prevention, 4th edn. Oxford University Press, New York, pp 635–660

Marengo A, Rosso C, Bugianesi E (2016) Liver cancer: connections with obesity, fatty liver, and cirrhosis. Annu Rev Med 67:103–117

Monteiro AN, Waizbort R (2007) The accidental cancer geneticist: hilário de gouvêa and hereditary retinoblastoma. Cancer Biol Ther 6(5):811–813

Plummer M, Franceschi S, Vignat J, Forman D et al (2015) Global burden of gastric cancer attributable to helicobacter pylori. Int J Cancer 136:487–490

Plummer M, de Martel C, Vignat J, Ferlay J et al (2016) Global burden of cancers attributable to infections in 2012: a synthetic analysis. Lancet Glob Health 4(9):e609–e616

Poiesz BJ, Ruscetti FW, Gazdar AF, Bunn PA et al (1980) Detection and isolation of type C retrovirus particles from fresh and cultured lymphocytes of a patient with cutaneous T-cell lymphoma. Proc Natl Acad Sci U S A 77(12):7415–7419

Pott P (1775) Cancer scroti. In: Pott P (ed) Chirurgical observations. L. Hawes, W. Clarke, R Collins, London, pp 179–180

Rubin DC, Shaker A, Levin MS (2012) Chronic intestinal inflammation: inflammatory bowel disease and colitis-associated colon cancer. Front Immunol 3(107):1–10

Varmus H, Unni AM, Lockwood WW (2016) How cancer genomics drives cancer biology: does synthetic lethality explain mutually exclusive oncogenic mutations? Cold Spring Harb Symp Quant Biol 81:247–255

Vogelstein B, Papadopoulos N, Velculescu VE, Zhou S et al (2013) Cancer genome landscapes. Science (New York, N.Y.) 339(6127):1546–1558

Wagener C, Stocking C, Muller O (2017) Cancer signaling: from molecular biology to targeted therapy. Wiley-VCH, Weinheim

Walboomers JM, Jacobs MV, Manos MM, Bosch FX et al (1999) Human papillomavirus is a necessary cause of invasive cervical cancer worldwide. J Pathol 189(1):12–19

Weinberg RA (2014) The biology of cancer, 2nd edn. Garland Science, Taylor & Francis Group, New York

World Cancer Research Fund/American Institute for Cancer Research (WCRF/AICR) (2018) *Continuous Update Project Report: Diet, Nutrition, Physical Activity and Colorectal Cancer 2016*. World Cancer Research Fund International, London. aicr.org/continuous-up-date-project/reports/colorectal-cancer-2017-report.pdf

Yoshida M, Miyoshi I, Hinuma Y (1982) Isolation and characterization of retrovirus from cell lines of human adult T-cell leukemia and its implication in the disease. Proc Natl Acad Sci U S A 79:2031–2035

Role of Angiogenesis in Tumors

Nidhi Gupta, Raman Kumar, and Alpana Sharma

Abstract

Cancer is one of the deadly diseases marked by the uncontrolled proliferation and accumulation of abnormal cancer cells. The incidence of cancer and the mortality due to cancer is enormously rising worldwide. A variety of hallmarks have been associated with cancer, and out of which, angiogenesis is the highly relevant characteristic, which involves the development of new blood vessels from the existing vessel. This process of angiogenesis is kept under a tight regulation by the levels of positive and negative regulators of angiogenesis. The angiogenesis having a strong association with cancer is nowadays the key target of various researchers to identify an effective chemotherapy. The regulators of angiogenesis were targeted alone or in conjunction with other regulators or certain cytotoxic drugs, and the research has come up with the outcome that the combinatorial targeting of multiple angiogenic pathways and cancer-related signaling pathways is highly effective for the treatment of cancer. Keeping this point in mind, various such drugs are in clinical trials and some have been clinically approved by FDA for cancer therapeutics. This chapter therefore focuses on all the aspects related to angiogenesis, its mechanism in physiology or pathology, its regulators, and the drugs used nowadays targeting angiogenesis.

Keywords

Cancer · Angiogenesis · Anti-angiogenic agents

N. Gupta · R. Kumar · A. Sharma (✉)
Department of Biochemistry, All India Institute of Medical Sciences, New Delhi, India

4.1 Introduction

Cancer is the term, which was not common a few decades ago, but now it is quite familiar in the general population. The reason behind this drastic awareness is the prominent increase in the incidence of cancer worldwide including India. As per the statistical data, deaths due to this deadly disease are increasing rapidly as mortality rate due to cancer has been increased by approximately 6% between 2012 and 2014. As per Population Cancer Registry of ICMR, the incidence and mortality of cancer are highest in the northeastern region of the country (National Cancer Registry Programme 2013). Cancer could be defined as the uncontrolled proliferation of malignant tumor cells in the biological system, which could be the result of genetic alterations (such as mutations and chromosomal translocations), environmental factors, smoking, gender, and other risk factors. Cancer ranges from solid tumors to hematological malignancies depending upon their source, that is, either originating from any tissue or originating from bone marrow/blood.

Majority of cancers exhibit certain hallmarks as shown in Fig. 4.1, of which, one of the most relevant comprises of angiogenesis. Angiogenesis (or neovascularization) could be defined as the formation of new capillary vessels from pre-existing vascular structure. It is a fundamental process that affects physiologic reactions (e.g., wound healing, regeneration, and vascularization of ischemic tissues), and pathologic processes, such as tumor development and metastasis, diabetic retinopathy, and chronic inflammation. Every tumor requires an efficient

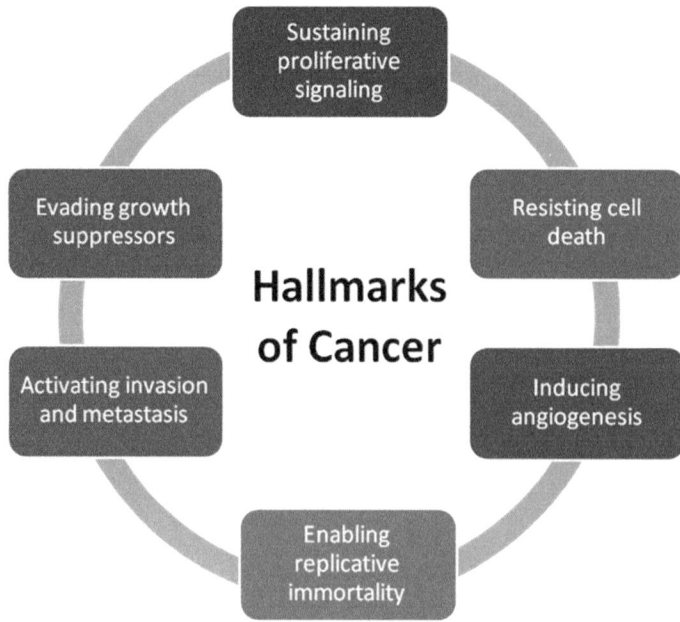

Fig. 4.1 Hallmarks of cancer

blood supply for their growth and development to meet the demand of oxygen and nutrients, and therefore, angiogenesis is closely associated with tumors (Hillen and Griffioen 2007). Therefore, research is being carried out focusing on the importance of angiogenesis in physiologic and pathologic processes including cancer, the mechanisms underlying these processes, and to target these factors by employing agents that possess pro- or anti-angiogenic activity. Recently, the importance of angiogenic markers has been reported as plausible markers for hematological malignancy, multiple myeloma (MM) (Gupta et al. 2020). Hence, this chapter has discussed the relevance of angiogenesis in the biological system, the factors responsible for their sustenance in conjunction with tumors, and finally the anti-angiogenic therapies identified for clinical settings to target various malignancies.

4.2 Mechanisms of Angiogenesis

Around 4000 BC, Egyptian physician believed that there were "vessels for every part of the body, which are hollow, having a mouth which opens to absorb medications and eliminate waste." Fortunately, the understanding of blood vessels has clearly improved since then. It has been known that during the course of embryonic development, blood vessels are assembled by vasculogenesis, which refers to the generation of blood vessels or vascular network from endothelial cell precursors (angioblasts), or from dual hemopoietic/endothelial cell precursors (hemangioblasts). The formation of blood vessels in adults is known as angiogenesis or neovascularization, which involves the development of blood vessels from the pre-existing vessels or by recruitment of endothelial progenitor cells (EPCs) from the bone marrow (Fig. 4.2). Neovascularization can also be distinguished into vasculogenic mimicry (formation of blood vessels by highly aggressive tumor cells) and vessel co-option (tumors obtain blood supply without sprouting of blood vessels by capturing host system) (Burri et al. 2004; Kumar et al. 2021).

Besides the contribution of different forms of neovascularization in growth of blood vessels in tumor, angiogenesis is considered an essential phenomenon as tumors may undergo cell death via apoptosis or necrosis in the absence of vascular support.

Neovascularization, including tumor angiogenesis, is basically a four-step process, which involves the following:

Vasodilation in response to nitric oxide and VEGF-induced increased permeability of the pre-existing vessel.

Matrix metalloproteinases (MMPs) mediate proteolytic degradation of the basement membrane of parent vessel followed by disruption of cell-to-cell contact between endothelial cells by plasminogen activator.

The endothelial cells then migrate toward the site of angiogenesis followed by proliferation and maturation into capillary tubes.

Recruitment of peri-endothelial cells (pericytes and vascular smooth muscle cells) to form the mature vessel.

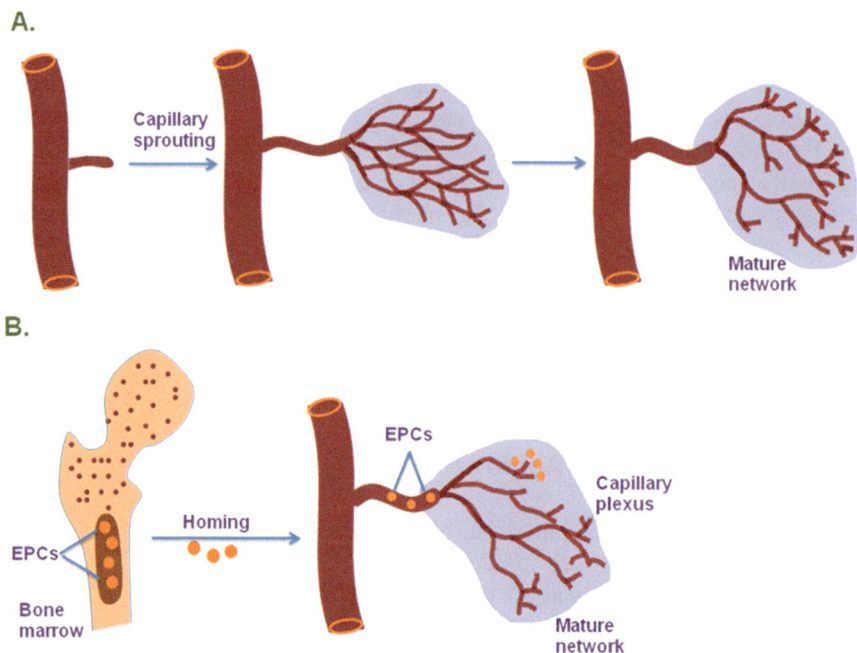

Fig. 4.2 Process of angiogenesis. (**a**) Angiogenesis from pre-existing vessels. (**b**) Angiogenesis by mobilization of EPCs from the bone marrow

4.2.1 Angiogenesis from Endothelial Precursor Cells (EPCs)

The angiogenesis can be initiated by the recruitment of EPCs into the tissues from the bone marrow (Fig. 4.2). EPCs may lead to re-endothelialization of vascular implants, which can further let neovascularization of ischemic organs, cutaneous wounds, and also tumors. The patients with ischemic conditions encompass higher number of EPCs in circulation, suggesting the contribution of EPCs in vascular function and also determining the probability of cardiovascular diseases. However, angiogenesis assists in promotion of tumor growth by providing adequate blood supply; hence, it could also be used as an indicator for assessing the tumor's metastatic potential, which is mainly associated with a poor prognosis (Lin et al. 2006).

4.2.2 Angiogenic Switch

The process of angiogenesis in physiologic or pathologic processes is determined by the balance between positive and negative regulators (or pro- and anti-angiogenic factors). Pro-angiogenic factors comprise those which regulate remodeling of extracellular matrix (ECM), changes in pericytes, proliferation, and migration of endothelial cells. These positive regulators include vascular endothelial growth factor

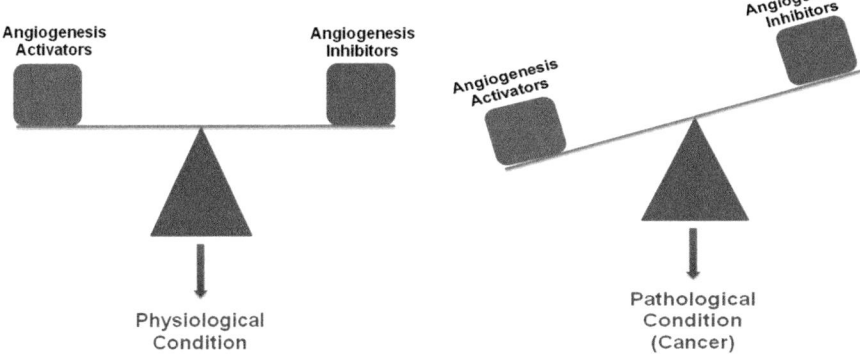

Fig. 4.3 Angiogenic switch

(VEGF), fibroblast growth factor (FGF), epidermal growth factor (EGF), platelet-derived growth factor (PDGF), and angiopoietin. On the other hand, anti-angiogenic factors, which inhibit the process of angiogenesis, comprise thrombospondin, vasostatin, endostatin, and angiostatin (Ferrara 2000). During tumor growth and development when proliferation is favored over apoptosis, balance between pro- and anti-angiogenic factors got disrupted with increase in pro-angiogenic factors called angiogenic switch (Carmeliet and Jain 2011; van der Meel et al. 2011). Hence, angiogenic switch has been considered as the rate-limiting step in the tumor propagation and metastatic pathway (Fig. 4.3). Moreover, tumor microenvironment (TME) and its associated factors determined the angiogenic switch (Gacche and Meshram 2013). Thus, the knowledge of TME in association with angiogenesis is crucial for the development of an effective and successful anti-angiogenic therapy for cancer. This chapter, therefore, focuses on the role and mechanism of angiogenesis in solid tumors and hematological malignancies along with the description of newly identified anti-angiogenic agents for cancer therapeutics.

4.3 Angiogenesis in Physiologic Processes

Angiogenesis is a complex process, which involves cumulative action of various vascular components that include degradation of vascular basement membrane, migration, and proliferation of endothelial cells and maturation into capillary tubes. This complex process employs pro-angiogenic molecules, which were otherwise present in quiescent state, but when angiogenesis is required for certain physiological processes such as reproduction, embryogenesis, organ differentiation, and tissue repair, these get upregulated (Chung et al. 2010; Folkman 2003). This upregulation is mainly occurred in the presence of external stimuli including hypoxia and mechanical stress in physiological conditions (Ferrara 2004). Hypoxia, which could be defined as the condition in which there is deficiency in the amount of oxygen entering the tissues, is one of the important factors in regulation of angiogenesis. It leads to the upregulation of hypoxia-inducible factor α (HIF-1α) (Chung et al. 2010; Ferrara 2004), soluble guanylate cyclases, and mammalian target of rapamycin (mTOR), which contributes to the development of vasculature and additional organs during embryogenesis. In addition to the role of angiogenesis during embryogenesis, it has a significant involvement in adults wherein it is responsible for the maintenance of vasculature in wound healing, ischemia, ovarian function, and endometrium proliferation during the reproductive cycle and in placenta formation (Chung et al. 2010; Hoff and Machado 2012).

4.4 Angiogenesis in the Pathogenesis of Cancer

The physiological angiogenesis or various processes associated with it could be initiated by certain signals including hypoxia and ischemia. However, due to the imbalance between pro-angiogenic and anti-angiogenic factors, and between angiogenic proteins of the tumor and the host, the angiogenic network is uncontrolled and remains upregulated during tumor formation and propagation. The hypoxic microenvironment, which facilitates the angiogenic network resulting in the sprouting of blood vessels from the surrounding tissues into the tumor, makes the conditions favorable for growth of tumors beyond 1–2 mm diameter.

On the contrary, pathological angiogenesis is described by the abnormal proliferation of endothelial cells leading to unusual morphology of tumor vasculature. The components that encompass tumor blood vessels such as pericytes, endothelial cells, smooth muscle cells, and basement membrane are also abnormal as the tight monolayers of endothelial cells, which are the characteristic of normal blood vessels, do not form. Moreover, the tumor vasculature becomes leaky contributing to the high interstitial pressure in most tumors. Furthermore, pericytes are loosely attached to endothelial cells in the tumor vasculature, which weaken the vessel wall and therefore increasing the risk of hemorrhage. In addition, the vascular basement membrane has irregular thickness and several surplus layers, and is also only loosely associated with endothelial cells (Baluk et al. 2005).

4.5 Angiogenesis: A Network of Signaling Pathways

The dynamic and multifaceted process of angiogenesis is the consequence of interplay between various pro-angiogenic and anti-angiogenic signals, which includes growth factors, angiopoietins, junctional molecules, oxygen sensors, and endothelial sensors (Carmeliet 2005). Initially, the VEGF family has been shown to play key involvement in tumor angiogenesis, but with ongoing research, other factors have also gained consideration.

4.5.1 Endogenous Pro-angiogenic Factors

With the advent of research, numerous proteins have been identified as pro-angiogenic factors such as vascular endothelial growth factor (VEGF), fibroblast growth factor (FGF), angiopoietins, and hepatocyte growth factor (HGF). Out of the above angiogenic activators, VEGF is the most established, studied, and potent factor in association with angiogenesis and is the target of majority of chemotherapeutics.

4.5.2 VEGF

The VEGF family and their receptors (VEGFRs) are receiving progressively more consideration in the area of neoplastic vascularization. VEGF is an important pro-angiogenic factor playing vital function in both physiological and pathological processes. Among the VEGF family, VEGF-A, VEGF-B, VEGF-C, and VEGF-E acting on their respective receptors allow proliferation of blood vessels, while VEGF-C and VEGF-D have association with lymphangiogenesis. VEGF-A has also been named as vascular permeability factor (VPF) as it is a highly specific mitogen for vascular endothelial cells leading to the stimulation of entire process required for neovascularization. Some angiogenic phenotypes can further be elicited by hypoxic condition resulting from the increasing distance between the growing tumor cells and the capillaries or from the inefficiency of new vessels (Nishida et al. 2006). It has been shown that hypoxia behaves as an activating factor for the expression of VEGF. The hypoxic condition has been observed to induce HIF-1α along with the simultaneous overexpression of VEGF in bladder cancer cells in vitro (Liu et al. 2017).

Tumor cells nourish the new blood vessels by producing VEGF and then secreting it into the surrounding tissue. When the tumor cells come in contact with endothelial cells, secreted VEGF binds to the receptor present on the surface of endothelial cells (Fig. 4.4). The binding of VEGF to its receptor activates various downstream proteins such as Akt, PI-3K, and ERK, which further transmits the signal into the nucleus of endothelial cells. The nuclear signal activates certain target genes required for the growth of new endothelial cells. Endothelial cells activated by VEGF produce matrix metalloproteinases (MMPs), which are responsible for the breakdown of extracellular matrix (ECM), which surrounds the cells, fills the spaces between them, and is usually made of proteins, proteoglycans, and polysaccharides. The degradation of ECM allows the migration of endothelial cells into the surrounding tissues. The migration of endothelial cells is followed by division of cells, which then further organize into hollow tubes, which gradually develop into a mature network of blood vessels (Nelson et al. 2000). Newly formed blood vessels need to stabilize or mature. Angiopoietin-1, angiopoietin-2, and their receptor Tie-2 can stabilize and govern vascular growth (Tournaire et al. 2004).

The expression of VEGF family was found to be upregulated in numerous cancerous tissues and the adjacent stroma in order to play an essential role in

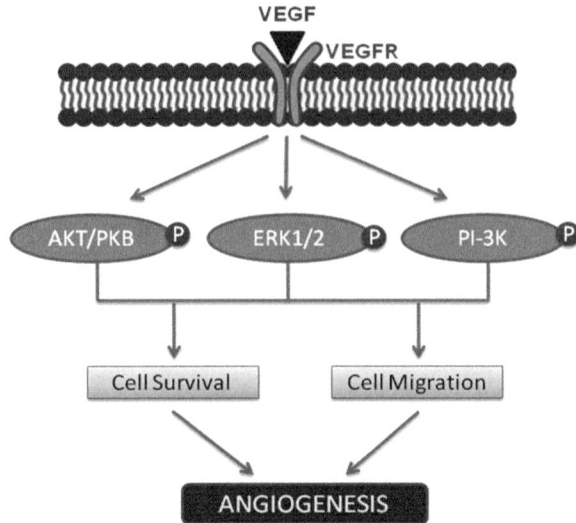

Fig. 4.4 Downstream signaling of VEGF in the promotion of angiogenesis

neovascularization (Folkman 1995; Luz et al. 2018; Yip-Schneider et al. 2020). Our laboratory has also worked on angiogenesis and the proteins associated with it. We have observed the overexpression of VEGF and other pro-angiogenic factors (angiopoietin-1, angiopoietin-2, and HGF) in multiple myeloma patients as compared to controls as depicted in Fig. 4.5 (Khan et al. 2013). In addition, we have observed significant upregulation of VEGF and Ang-2 levels in cervical cancer patients prior to treatment, but the levels decline significantly upon chemotherapy and teleradiation as shown in Fig. 4.6 (Sharma et al. 2017).

4.6 Angiogenesis Ahead of VEGF/VEGFR Pathway

Although VEGF is one of the important pro-angiogenic factors and signaling mediated by VEGF could promote progression of cancer cells, this signaling alone is not sufficient enough to promote tumorigenesis; hence, the involvement of a number of other signaling pathways in combination with VEGF/VEGFR signaling is now appreciated. The understanding of other pro-angiogenic signaling pathways has helped in deciphering the molecular mechanisms that could promote tumor angiogenesis and therefore could pave a way for the identification and development of novel anti-angiogenic therapies targeting these VEGF-independent pathways. Some of these pathways have been reported such as PDGF/PDGFR, FGF/FGFR, angiopoietin–Tie, and hepatocyte growth factor (HGF)/MET signaling pathways (Fig. 4.7), which has been discussed below in this chapter. Thus, regulation of angiogenesis in physiology or pathology could occur via VEGF-independent signaling pathways and hence might serve as an alternative mean for induction of tumor growth (Ferrara 2010; Zhao and Adjei 2015).

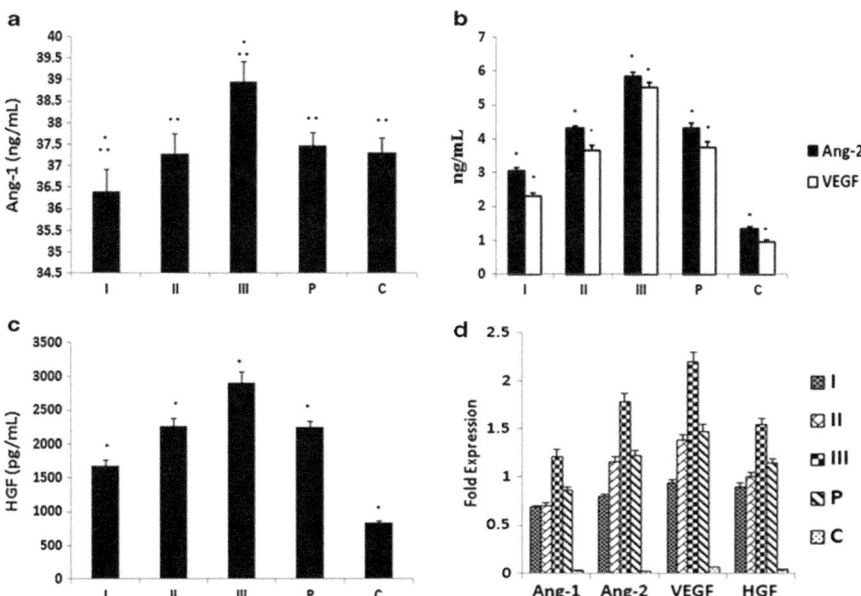

Fig. 4.5 Circulatory expression of (**a**) angiopoietin-1, (**b**) angiopoietin-2, VEGF, (**c**) HGF, and (**d**) relative mRNA expression of all the factors in multiple myeloma patients [adapted from Ref. (Khan et al. 2013)]

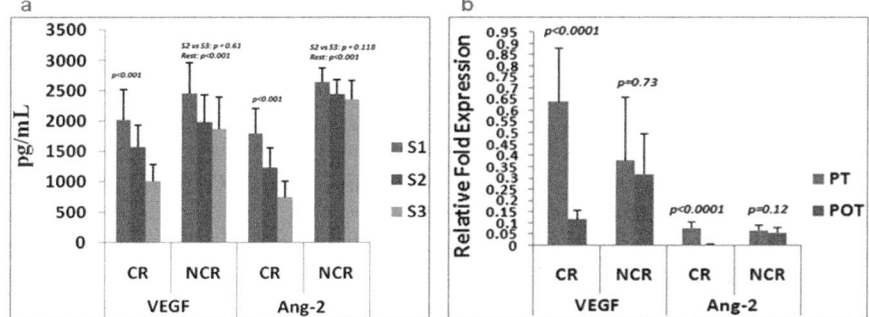

Fig. 4.6 (**a**) Circulatory levels, (**b**) relative mRNA expression of VEGF and Ang-2 in complete responders (CR) or nonresponders (NCR) in cervical cancer patients [S1: pretreatment levels; S2: postchemotherapy levels; S3: postradiotherapy, PT: pretreatment; POT: postchemotherapy + postradiotherapy] [adapted from Ref. (Sharma et al. 2017)]

4.6.1 Platelet-Derived Growth Factor (PDGF)/PDGFR

The PDGF family consists of PDGF-A to PDGF-D polypeptide homodimers (a protein composed of two polypeptide chains that are identical in the order, number, and kind of their amino acid residues) and the PDGF-AB heterodimer. These factors undergo downstream signaling by binding to PDGFR tyrosine kinase

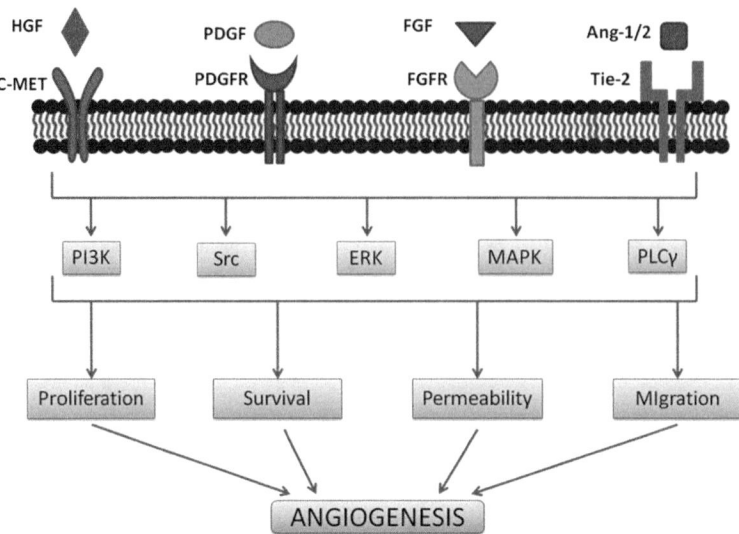

Fig. 4.7 Signaling of other pro-angiogenic factors (HGF, PDGF, FGF, angiopoietins)

receptors and activating molecules similar to those stimulated by VEGF (Heldin 2013; Wu et al. 2008). The stimulation of PDGF signaling has been shown to be involved in growth and progression of various cell types. It has been reported that overactivation of PDGF signaling alone or in combination with other angiogenic factors (such as FGF or VEGF) has been associated with angiogenesis in tumors (Jansson et al. 2018; Kerr et al. 2017). Furthermore, direct activation of PDGF signaling has been observed in multiple tumor types, and co-expression of PDGF and its receptor suggests a role for autocrine and paracrine activation (Tejada et al. 2006). Lin et al. recently reported the association of PDGF signaling with poor prognosis in oral squamous cell carcinoma (Lin et al. 2020). The effects caused by abnormal PDGF signaling in tumor angiogenesis comprise recruitment of pericytes to blood vessels, enhanced secretion of other pro-angiogenic factors, migration and proliferation of endothelial cells, sprouting, and tube formation in tumors, and promotion of lymphangiogenesis and subsequent lymphatic metastasis (Xue et al. 2011). The significance of PDGF signaling in relation to angiogenesis in tumor could further be supported by studies signifying that the antitumor efficacy of VEGF receptor blocking agents could be improved by the employment of PDGFR inhibitors (Lu et al. 2010). Further research is going on to have a thorough understanding of contribution of PDGF in tumor angiogenesis with the hope of developing more effective anti-angiogenic treatments that would reduce growth, maturation, and metastases of various tumor types.

4.6.2 Fibroblast Growth Factor (FGF)/FGFR

FGFs are heparin-binding growth factors, which mediate downstream signaling by binding to receptor tyrosine kinases (RTKs) (Beenken and Mohammadi 2009). FGFs and their receptors (FGFRs) are expressed ubiquitously and perform several functions, which include regulation of growth and differentiation of normal cells in addition to angiogenesis. Out of the various FGFR ligands known so far, FGF1 and FGF2 have been shown to possess pro-angiogenic potential by inducing migration and proliferation of endothelial cells (Sun et al. 2017). It has been observed that expression of FGF and FGFR increase in various cancers might lead to the stimulation followed by release of pro-angiogenic factors and therefore could enhance angiogenesis. An interaction between FGF and VEGF signaling has also been demonstrated to have a key involvement in tumor angiogenesis (Cao et al. 2012). FGF can act synergistically with VEGF to augment tumor angiogenesis; therefore, simultaneous targeting of both FGF and VEGF signaling pathways might be more effective in inhibition of angiogenesis and also tumor growth in comparison with targeting a single pathway alone. There are studies showing that advanced stage of pancreatic cancer models developed resistance for VEGFR-2 blockade therapy in vivo as tumor relapse suddenly following an initial period of remission with anti-VEGFR-2 treatment, which mediated growth suppression. Hence, FGFs are emerging as a newer and efficient phenomenon to combat resistance to VEGF inhibition (Casanovas et al. 2005).

4.6.3 Angiopoietins (Ang)/Tie2

Angiopoietins play a critical role in the maintenance of vessel quiescence and comprise a family of four ligands (Ang-1 to Ang-4). The two members of this family, Ang-1 and Ang-2, are best characterized in context of angiogenesis. These angiopoietins bind to the Tie receptor and undergo downstream signaling contributing to neovascularization in both physiological and pathological processes (Eklund and Saharinen 2013; Eklund et al. 2017). As per the literature, overexpression of Ang-2 in various cancers has been associated with progression and poor prognosis of the disease (Xu et al. 2017; Yang et al. 2017). We have also studied the expression of Ang-1 and Ang-2 in multiple myeloma patients and observed to have increased expression in the disease (Khan et al. 2013). We have also reported the diagnostic potential of Ang-2 in multiple myeloma (Fig. 4.8) (Joshi et al. 2011). As also discussed, we have found the overexpression of Ang-2 along with VEGF in cervical cancer patients, which declines after teleradiation and chemotherapy (Sharma et al. 2017).

The vascular stabilizing and destabilizing functions mediated by Ang1 and Ang2, respectively, prompt the researchers to attain better understanding of Ang–Tie system in inflammatory and neovascular diseases, associated with vascular leak and endothelial dysfunction. Targeting of Ang–Tie signaling system could provide

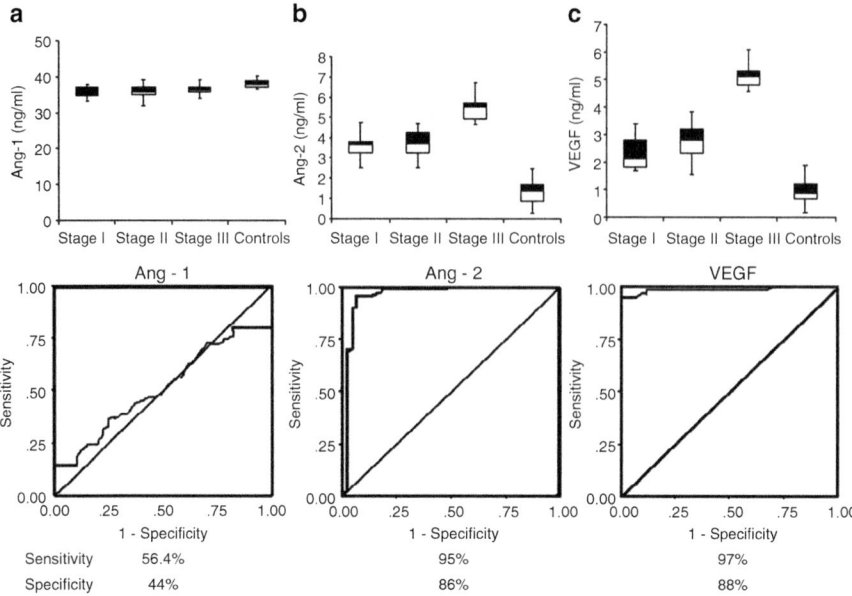

Fig. 4.8 Circulatory levels of (**a**) Ang-1, (**b**) Ang-2, and (**c**) VEGF along with their receiver operating characteristic curve in multiple myeloma patients [adapted from Ref. (Joshi et al. 2011)]

benefit when used either alone or in combination with VEGF inhibitors, which we will discuss in the consecutive sections.

4.6.4 Hepatocyte Growth Factor (HGF)/MET

Produced as a single-chain inactive precursor protein, HGF is a pleiotropic growth factor that binds to tyrosine protein kinase, c-MET. The signaling of HGF/c-MET is essential for the normal physiological processes, which includes normal cell proliferation, survival, and motility, and maintains homeostasis. In addition, this is also found to be involved in regulating angiogenesis and growth in variety of tumor types ranging from solid tumors to hematological malignancies (Graveel et al. 2013). The effects of HGF/MET on tumor angiogenesis could occur directly by activation of endothelial cells or indirectly by stimulation of the production of pro-angiogenic factors such as VEGF. Previously, our research group has observed elevated expression of HGF in patients with multiple myeloma (Khan et al. 2013). It has been reported that expression of MET increase in the bevacizumab-resistant glioblastoma in comparison with pretreated tumors from the same patients signifies that HGF/MET signaling compensates for the inhibition of VEGF and hence responsible for the resistance in these tumors (Jahangiri et al. 2013; Lu et al. 2012). These findings therefore suggest that this pro-angiogenic factor is equally important for angiogenesis, which could be further targeted for effective outcome.

4.7 Inhibitors of Angiogenesis

The initiation of blood vessel growth not only requires upregulation of the activity of pro-angiogenic factors but concomitantly requires the downregulation of inhibitors of angiogenesis. There are various negative regulators present naturally in the biological system, which include angiostatin, endostatin, and thrombospondin as described in Fig. 4.9 (Hoff and Machado 2012; Nishida et al. 2006).

4.7.1 Angiostatin

Angiostatin is composed of one or more fragments of plasminogen, which is produced by proteolytic cleavage. Studies showed that the fusion protein of angiostatin and endostatin possesses antitumor property against glioblastoma cells in vitro and in vivo. The virus-carrying endostatin–angiostatin fusion gene showed increased proliferation, decreased microvessel density, and enhanced survival and reduction in tumor volume (Zhang et al. 2014).

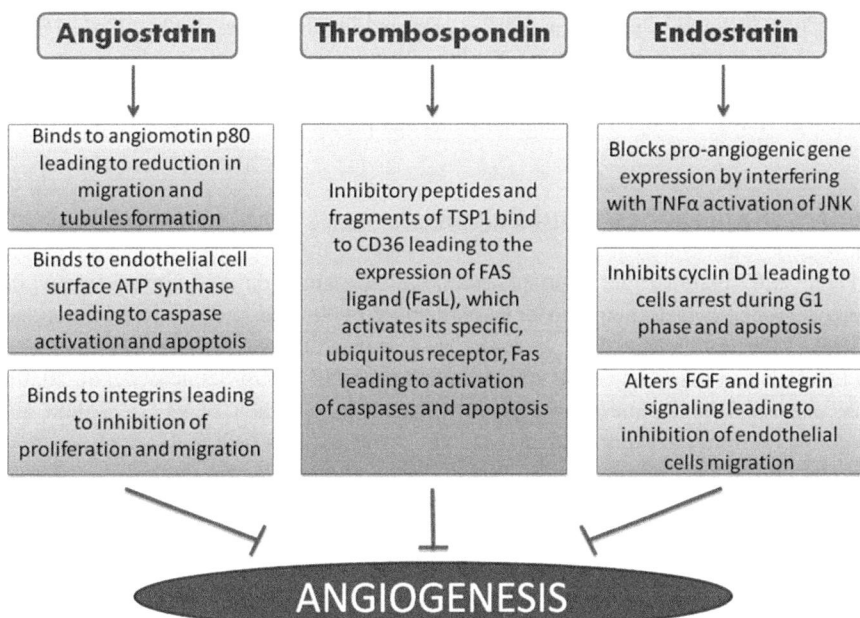

Fig. 4.9 Mechanism of endogenous anti-angiogenic molecules. *TSP* thrombospondin, *FAS* first apoptosis signal, *TNF* tumor necrosis factor, *JNK* Janus kinase

4.7.2 Endostatin

Endostatin is a naturally occurring anti-angiogenic factor, which is derived by the proteolytic cleavage of type XVIII collagen. It is capable of interacting with multiple cell surface molecules such as integrins with downstream inhibition of angiogenesis. The administration of recombinant endostatin along with chemotherapy has significantly enhanced progression-free survival and improves clinical benefit response in advanced sarcomas patients (Xing et al. 2017). Similar results were also reported in cancer xenograft models in which endostatin efficiently inhibited tumor angiogenesis in lung carcinoma models and reversed the immunosuppressive microenvironment in these systems (Liu et al. 2018).

4.7.3 Thrombospondins (TSPs)

TSPs are secreted glycoproteins found to be involved in inhibition of angiogenesis and therefore delay tumor progression. There are various members in this family of which TSP-1 is widely studied in association with tumor angiogenesis. There are few theories describing the mechanism of regulation of angiogenesis by TSP-1. Some studies stated that this regulation occurred by activation of transforming growth factor-β (TGF-β), while others proposed that TSP-1 binds to CD36 receptor via its inhibitory peptide leading to the activation of Fas ligand, which then activates Fas culminating to apoptosis of cells. TSP1 was reported to have positive correlation with p53 in various tumors as p53 overexpression increases the levels of TSP1, downregulates angiogenesis, and finally suppresses tumor growth (Giuriato et al. 2006; Kazerounian et al. 2008). These findings suggest an important association between TSPs, angiogenesis, and tumor development.

4.8 Angiogenesis and Cancer

There are innumerable studies available focusing on the involvement of pro-angiogenic factors in the development, growth, and progression of tumors. The VEGF family and their receptors were found to have elevated expression in numerous cancers (Khan et al. 2013; Sharma et al. 2017; Yip-Schneider et al. 2020). As discussed, angiogenesis not only signifies the involvement of VEGF as there are equally relevant pro-angiogenic factors whose overexpression has also been reported in various cancers (Graveel et al. 2013; Tejada et al. 2006; Xu et al. 2017). These angiogenic factors are known to affect the prognosis of certain carcinomas (Nishida et al. 2004; Shamsdin et al. 2019). These studies therefore emphasized on the fact that tumor aggressiveness could be reflected by the levels of pro-angiogenic factors in the respective tissue and thus might have prognostic value in the identification of the high-risk patients with poor prognosis. Neovascularization reduces a tumor's accessibility to chemotherapeutic drugs, thus leading to chemoresistance, which renders the patient unresponsive toward chemotherapy.

4.8.1 Anti-Angiogenic Treatment of Cancer

Surgery is widely being used for some localized benign solid tumors, but cytotoxic chemotherapy and radiotherapy are more appropriate treatment regimen for surgically unresectable tumors (Awada and de Castro 2005) and for hematological malignancies. Despite the availability of such sophisticated treatment strategies, nearly all of the advanced cancer cases relapse which generates an urge to identify some novel and effective treatment modality or drug for the better treatment and management of the tumors. Hence, targeting angiogenesis was the preliminary step in this course.

As discussed earlier, angiogenesis, one of the important hallmarks of cancer, is mainly regulated by the balance between pro-angiogenic and anti-angiogenic molecules and switch to angiogenesis marked the disturbance in their homeostasis. The downstream signaling of pro-angiogenic molecules favors the sprouting of blood vessels near the tumor tissue, which is further required for the growth and progression of tumor cells. Keeping the significance of angiogenesis for tumors in mind, it could be proposed that employment of anti-angiogenic drugs might reduce the growth of tumor cells, which could improve the quality of life of cancer patients leading to the reduction in mortality from tumors.

The process of angiogenesis is complex involving multiple steps, hence, the inhibitors of angiogenesis which are currently under clinical trials have been classified into several classes and are as follows:

Proteases inhibitors which hamper the synthesis of MMP and thus reduce the breakdown of matrix inhibitors which reduce the migration and proliferation of endothelial cells;
Angiogenic growth factors inhibitors;
Inhibitors with unique mechanisms;

Though the effect of anti-angiogenic drugs was found to be detrimental for cancer cells and showed satisfactory results for combating the tumor, the long-term effect of these has not been commented so far (Mayer 2004; Nishida et al. 2006). However, it could be proposed that these drugs when used in combination with chemotherapy or radiotherapy tend to increase survival. There are reports showing that considering both anti-angiogenic drugs and chemotherapy/radiotherapy (cytotoxic treatment) in combination could be much more beneficial for combating the cancer than when used alone. Treatment with cytotoxic drugs or therapy acts directly on cancer cells to suppress their growth, whereas anti-angiogenic drugs inhibit the source by which cancer cells get their nutrition and oxygen to ultimately suppress their growth and proliferation (Jain 2005). Therefore, combination of both these treatments would serve dual purpose with their final goal to destroy the cancer cells and hence would be highly beneficial for the treatment of cancer.

VEGF being one of the most relevant pro-angiogenic factors is the key target for the treatment of most of the solid tumors. But, anti-angiogenic therapy is not only confined to solid tumors as it has shown promising results in hematological malignancies too. Thalidomide was the first angiogenic inhibitor approved by

FDA in 2006 for the treatment of multiple myeloma. This drug acts by inhibiting VEGF secretion from the bone marrow endothelial cells. Followed by this, lenalidomide, an analog of thalidomide has been taken into consideration for its improved safety and efficacy. In addition, bortezomib, a proteasome inhibitor, which inhibits VEGF secretion from endothelial cells, has also been approved by FDA for multiple myeloma treatment.

4.9 Combinatorial Targeting of Multiple Pro-angiogenic Pathways: An Alternative to Overcome Resistance

Though anti-angiogenic therapies targeting VEGF/VEGFR signaling pathway were found to be effective to combat the cancer, as discussed earlier, most of the advanced cancer cases tend to relapse by developing resistance against that therapy. It has been mentioned in the previous section that combination of anti-angiogenic and cytotoxic drugs could possess a heightened response against cancer cells. However, there is another difficulty in association with this approach as the resistance against a particular anti-angiogenic therapy might be because of utilization of alternate compensatory pro-angiogenic signaling pathways to recruit vasculature (Welti et al. 2013). Hence, multiple angiogenic pathways could also be targeted together for an effective outcome. In support of this hypothesis, studies show that placental growth factor (PlGF) promotes pro-angiogenic signaling when VEGF-A is blocked, hence contributing to resistance. A phase II clinical trial of FOLFIRI (cytotoxic drug-containing folic acid, fluorouracil, and topoisomerase inhibitor) and bevacizumab (VEGF-A inhibitor) in metastatic colorectal cancer (mCRC) patients lead to increase in the plasma levels of VEGF-C, VEGF-D, and PlGF which suggests that these pro-angiogenic factors rise in order to compensate for the effect of VEGF-A blockade (Kopetz et al. 2010). This compensatory elevation of VEGF-C and VEGF-D in the absence of VEGF-A has also been explained in a study in which it has been reported that VEGF-C and VEGF-D bind to VEGFR-2 and VEGFR-3 receptors when VEGF-A is blocked by drugs, which might be sufficient to promote angiogenesis followed by progression of tumor (Clarke and Hurwitz 2013). A report showed that expression of platelet-derived growth factor A (PDGFA) increased in mCRC cells resistant to treatment with bevacizumab and their levels were observed to be negatively associated with patients' survival (Makondi et al. 2018). It is not only the enhanced signaling of PDGF-A which comes into consideration while blockage of VEGF-A by bevacizumab as there are other pro-angiogenic molecules that overexpressed as a result of VEGF-A inhibition. Certain clinical studies stated the enhanced levels of FGF and PDGF-A in circulation of cancer patients receiving VEGF-A inhibition therapy as treatment. These increased levels also correlated with progression of disease and poor patients' survival affirming the significance of FGF and PDGF signaling in promoting resistance in cancer cells against anti-VEGF chemotherapy. Upregulation of FGF and PDGF factors in bevacizumab chemotherapy has been reported in various tumors including pancreatic cancer. In a recent study, mechanistic interactions between VEGF, PDGF, and FGF and remodeling

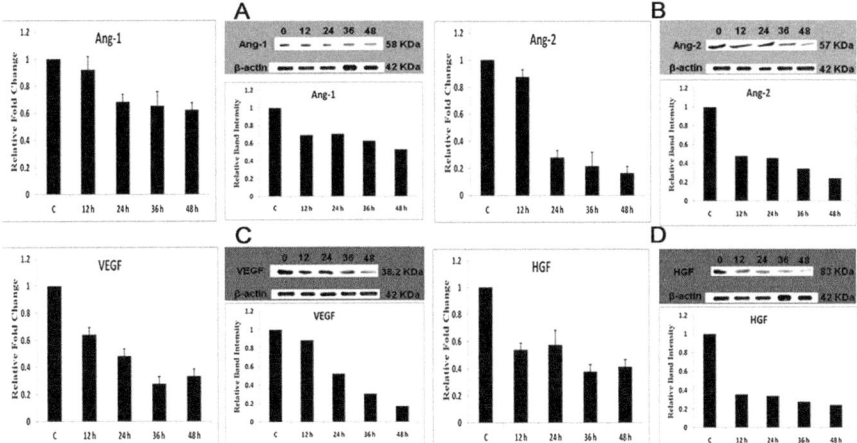

Fig. 4.10 Bar graphs showing the effect of CBPE treatment on mRNA (left panel) and protein (right panel) levels of angiogenic factors in multiple myeloma cells in time-dependent manner. (**a**) Ang-1; (**b**) Ang-2; (**c**) VEGF; and (**d**) HGF [adapted from Ref. (Khan et al. 2016)]

factors have been reported in breast cancers and fibrosarcoma. The dual targeting of VEGF and PDGF showed superior response against FGF+ tumors for which any specific therapy is not known (Hosaka et al. 2020). These findings suggest that VEGF is not the sole pro-angiogenic molecule responsible for angiogenesis and others are equally important as VEGF, FGF, and PDGF act in a synergistic action for regulating neovascularization. A multi-targeted approach to treatment is believed. Hence, targeting of multiple pro-angiogenic factors in combination could be highly effective to limit the development of resistance, maximize antitumor efficacy, and for controlling tumor growth and progression.

In our laboratory, we have also observed the multi-angiogenic factors targeting by a herbal cinnamon bark powder extract (CBPE) in multiple myeloma as shown in Fig. 4.10. Treatment with cinnamon extract caused myeloma cell death, reduced proliferation, and significantly decreased molecular expression of pro-angiogenic factors VEGF, Ang-1, Ang-2, and HGF (Khan et al. 2016).

4.10 Current and Emerging Multi-Targeting Anti-Angiogenic Agents

As discussed earlier that multiple angiogenic pathways act in synergy to regulate angiogenesis, hence, drugs simultaneously targeting multiple signaling pathways have been identified in order to increase antitumor efficacy and to compensate for the rising relapse rate. Cancer treatments available nowadays target angiogenesis by inhibiting VEGF, FGF, PDGF, and/or other angiogenesis signaling pathways in combination with cytotoxic drugs and have been employed worldwide for the treatment of a range of cancers (Fig. 4.11). These agents include sorafenib, sunitinib,

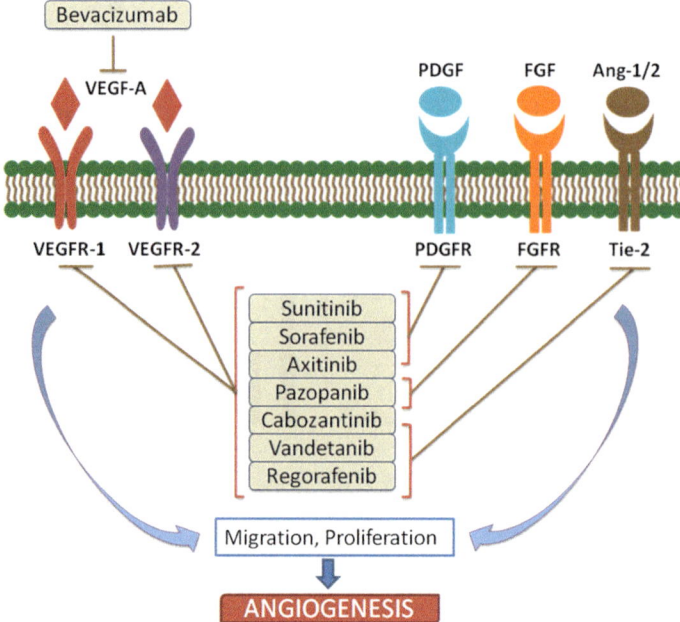

Fig. 4.11 Targets of anti-angiogenic drugs employed in clinical trials [adapted from Ref. (Clarke and Hurwitz 2013)]

axitinib, and pazopanib. First three drugs (sorafenib, sunitinib, and axitinib) are tyrosine kinase inhibitors (TKIs), which simultaneously target multiple pathways including VEGF, PDGF, and other cancer-related signaling pathways. On the other hand, pazopanib inhibits FGF in addition to the inhibition of VEGF, PDGF, and other signaling pathways (Hamberg et al. 2010; Majumder et al. 2013). These drugs have been clinically approved by FDA for the treatment of a variety of cancers ranging from solid tumors (includes renal carcinoma and thyroid cancer) to hematological malignancies (such as leukemia) (Röllig et al. 2015).

Apart from the above-mentioned drugs, there are some more inhibitors having potential of targeting additional relevant cancer-related pathways. These recently FDA-approved agents include cabozantinib and vandetanib showing antitumor property with enhanced effectiveness. Cabozantinib inhibits not only VEGFR family but also angiopoietin signaling, c-Met, and c-KIT, while vandetanib targets Src and EGFR signaling along with VEGFR family and angiopoietin signaling. Both these drugs have been approved by FDA for the treatment of advanced medullary thyroid carcinoma, while cabozantinib has also been approved for the treatment of advanced renal cell carcinoma in 2017. Moreover, another tyrosine kinase inhibitor (regorafenib) has been found to inhibit VEGFR-1 to VEGFR-3, PDGFR-a and PDGFR-b, FGFR-1 and FGFR-2, TIE2, c-KIT, and RET. This inhibitor had been approved by FDA in 2012 for the treatment of metastatic colorectal carcinoma (mCRC) patients who were not showing remission after receiving chemotherapy of cytotoxic drug, and anti-VEGF and anti-EGFR drugs. Moreover, in 2017, this drug has also been approved for the treatment of hepatocellular carcinoma (HCC). Regorafenib has shown effective results in mCRC and HCC models by preventing cancer cell growth and metastasis. This suggests the use of regorafenib as an antitumor agent, which simultaneously targets angiogenesis and cancer signaling pathways in colorectal and hepatocellular carcinoma (Cyran et al. 2013; Zhang et al. 2021).

4.11 Our Research Endeavors

Angiogenesis is one of the important hallmarks associated with cancer and has been an emerging area of research worldwide. Our research group has been engaged in cancer biology from the past two decades, and we have made significant contributions in this area. We have studied the involvement of angiogenesis along with other relevant phenomena in various cancers such as multiple myeloma, cervical cancer, and bladder cancer. We have observed significantly high expression of pro-angiogenic factors, that is, VEGF, Ang-2, and HGF in circulation and at molecular level in multiple myeloma and cervical cancer. We have reported the diagnostic potential of Ang-2 in multiple myeloma. We have also investigated the changes in angiogenic factors after chemotherapy and radiotherapy and found significant decline in the levels of these factors upon treatment in cervical cancer. We have also assessed the anti-angiogenic and antitumor property of a herbal cinnamon bark powder extract in myeloma cell lines in vitro. We have observed downregulation of VEGF, Ang-1, Ang-2, and HGF at molecular level upon treatment with cinnamon extract in dose-dependent manner. This finding suggests the multi-targeting approach of cinnamon extract in multiple myeloma. These results cumulatively emphasize on the role of angiogenesis in various tumors ranging from solid to hematological malignancies. Their targeting could prove to be the better therapeutics for cancer treatment in future.

4.12 Conclusions

This chapter has emphasized on most of the significant points in perspective of the role of neovascularization in cancers. To summarize, angiogenesis, one of the important hallmarks of cancer, could be defined as the development of new blood vessels from the existing ones, which is required for physiological homeostasis and pathological processes. This phenomenon is mainly the result of balance between pro-angiogenic and anti-angiogenic factors. Their imbalance leads to pathological conditions including cancer wherein pro-angiogenic factors get upregulated while anti-angiogenic factors undergo downregulation. This imbalance has been reported in plethora of cancers and showed association with the progression of cancer leading to poor overall survival. The remedy for this problem is adoption of anti-angiogenic drugs, which target upregulated pro-angiogenic factors but targeting single factor showed the development of drug resistance due to utilization of alternate pathways. Hence, it opens a new avenue for the utilization of multi-targeting therapy for better treatment of cancer. Thus, after taking into consideration the importance of multi-targeting approach for better treatment of cancer, several such drugs have been tested in cancer models and are in clinical trials, while some have been approved by FDA for cancer chemotherapy.

Moreover, angiogenesis is merely one of the hallmarks of cancer as there are other factors available known for their pro-tumorigenic property. Therefore, targeting only angiogenesis would not be sufficient enough to combat the disease. Hence, targeting multiple properties of cancer using combination of drug regimen would aid in designing better therapeutic modality for the treatment of cancers. With the advancement in cancer research, it is hoped that an appropriate chemotherapy might improve the quality of life of patients and aid them to fight the disease in future with minimal side effects.

References

Awada A, de Castro G (2005) An integrated approach for tailored treatment in breast cancer. Ann Oncol 16(Suppl 2):203–208

Baluk P, Hashizume H, McDonald DM (2005) Cellular abnormalities of blood vessels as targets in cancer. Curr Opin Genet Dev 15:102–111

Beenken A, Mohammadi M (2009) The FGF family: biology, pathophysiology and therapy. Nat Rev Drug Discov 8:235–253

Burri PH, Hlushchuk R, Djonov V (2004) Intussusceptive angiogenesis: its emergence, its characteristics, and its significance. Dev Dyn 231:474–488

Cao R, Ji H, Feng N, Zhang Y, Yang X, Andersson P, Sun Y, Tritsaris K, Hansen AJ, Dissing S et al (2012) Collaborative interplay between FGF-2 and VEGF-C promotes lymphangiogenesis and metastasis. Proc Natl Acad Sci U S A 109:15894–15899

Carmeliet P (2005) Angiogenesis in life, disease and medicine. Nature 438:932–936

Carmeliet P, Jain RK (2011) Molecular mechanisms and clinical applications of angiogenesis. Nature 473:298–307

Casanovas O, Hicklin DJ, Bergers G, Hanahan D (2005) Drug resistance by evasion of antiangiogenic targeting of VEGF signaling in late-stage pancreatic islet tumors. Cancer Cell 8:299–309

Chung AS, Lee J, Ferrara N (2010) Targeting the tumor vasculature: insights from physiological angiogenesis. Nat Rev Cancer 10:505–514

Clarke JM, Hurwitz HI (2013) Understanding and targeting resistance to anti-angiogenic therapies. J Gastrointest Oncol 4:253–263

Cyran CC, Kazmierczak PM, Hirner H, Moser M, Ingrisch M, Havla L, Michels A, Eschbach R, Schwarz B, Reiser MF et al (2013) Regorafenib effects on human colon carcinoma xenografts monitored by dynamic contrast-enhanced computed tomography with immunohistochemical validation. PLoS One 8:e76009

Eklund L, Saharinen P (2013) Angiopoietin signaling in the vasculature. Exp Cell Res 319:1271–1280

Eklund L, Kangas J, Saharinen P (2017) Angiopoietin-tie signalling in the cardiovascular and lymphatic systems. Clin Sci (Lond) 131:87–103

Ferrara N (2000) Vascular endothelial growth factor and the regulation of angiogenesis. Recent Prog Horm Res 55:15–35; discussion pp. 35-36

Ferrara N (2004) Vascular endothelial growth factor as a target for anticancer therapy. Oncologist 9 (Suppl 1):2–10
Ferrara N (2010) Pathways mediating VEGF-independent tumor angiogenesis. Cytokine Growth Factor Rev 21:21–26
Folkman J (1995) Angiogenesis in cancer, vascular, rheumatoid and other disease. Nat Med 1:27–31
Folkman J (2003) Fundamental concepts of the angiogenic process. Curr Mol Med 3:643–651
Gacche RN, Meshram RJ (2013) Targeting tumor micro-environment for design and development of novel anti-angiogenic agents arresting tumor growth. Prog Biophys Mol Biol 113:333–354
Giuriato S, Ryeom S, Fan AC, Bachireddy P, Lynch RC, Rioth MJ, van Riggelen J, Kopelman AM, Passegué E, Tang F et al (2006) Sustained regression of tumors upon MYC inactivation requires p53 or thrombospondin-1 to reverse the angiogenic switch. Proc Natl Acad Sci U S A 103: 16266–16271
Graveel CR, Tolbert D, Vande Woude GF (2013) MET: a critical player in tumorigenesis and therapeutic target. Cold Spring Harb Perspect Biol 5
Gupta N, Sharma A, Sharma A (2020) Emerging biomarkers in multiple myeloma: a review. Clin Chim Acta 503:45–53
Hamberg P, Verweij J, Sleijfer S (2010) (Pre)clinical pharmacology and activity of pazopanib, a novel multikinase angiogenesis inhibitor. Oncologist 15:539–547
Heldin CH (2013) Targeting the PDGF signaling pathway in tumor treatment. Cell Commun Signal 11:97
Hillen F, Griffioen AW (2007) Tumor vascularization: sprouting angiogenesis and beyond. Cancer Metastasis Rev 26:489–502
Hoff PM, Machado KK (2012) Role of angiogenesis in the pathogenesis of cancer. Cancer Treat Rev 38:825–833
Hosaka K, Yang Y, Seki T, Du Q, Jing X, He X, Wu J, Zhang Y, Morikawa H, Nakamura M et al (2020) Therapeutic paradigm of dual targeting VEGF and PDGF for effectively treating FGF-2 off-target tumors. Nat Commun 11:3704
Jahangiri A, De Lay M, Miller LM, Carbonell WS, Hu YL, Lu K, Tom MW, Paquette J, Tokuyasu TA, Tsao S et al (2013) Gene expression profile identifies tyrosine kinase c-Met as a targetable mediator of antiangiogenic therapy resistance. Clin Cancer Res 19:1773–1783
Jain RK (2005) Normalization of tumor vasculature: an emerging concept in antiangiogenic therapy. Science 307:58–62
Jansson S, Aaltonen K, Bendahl P-O, Falck A-K, Karlsson M, Pietras K, Rydén L (2018) The PDGF pathway in breast cancer is linked to tumor aggressiveness, triple-negative subtype and early recurrence. Breast Cancer Res Treat 169:231–241
Joshi S, Khan R, Sharma M, Kumar L, Sharma A (2011) Angiopoietin-2: a potential novel diagnostic marker in multiple myeloma. Clin Biochem 44:590–595
Kazerounian S, Yee KO, Lawler J (2008) Thrombospondins in cancer. Cell Mol Life Sci 65:700–712
Kerr DA, Busarla SVP, Gimbel DC, Sohani AR, Nazarian RM (2017) mTOR, VEGF, PDGFR, and c-kit signaling pathway activation in Kaposi sarcoma. Hum Pathol 65:157–165
Khan R, Sharma M, Kumar L, Husain SA, Sharma A (2013) Interrelationship and expression profiling of cyclooxygenase and angiogenic factors in Indian patients with multiple myeloma. Ann Hematol 92:101–109
Khan R, Sharma M, Kumar L, Husain SA, Sharma A (2016) Cinnamon extract exhibits potent anti-proliferative activity by modulating angiogenesis and cyclooxygenase in myeloma cells. J Herbal Med 6:149–156
Kopetz S, Hoff PM, Morris JS, Wolff RA, Eng C, Glover KY, Adinin R, Overman MJ, Valero V, Wen S et al (2010) Phase II trial of infusional fluorouracil, irinotecan, and bevacizumab for metastatic colorectal cancer: efficacy and circulating angiogenic biomarkers associated with therapeutic resistance. J Clin Oncol 28:453–459

Kumar V, Abbas AK, Aster JC, Turner JR, Robbins SL, Cotran RS (2021) Robbins & Cotran pathologic basis of disease

Lin EY, Li J-F, Gnatovskiy L, Deng Y, Zhu L, Grzesik DA, Qian H, Xue X, Pollard JW (2006) Macrophages regulate the angiogenic switch in a mouse model of breast cancer. Cancer Res 66: 11238–11246

Lin LH, Lin JS, Yang CC, Cheng HW, Chang KW, Liu CJ (2020) Overexpression of platelet-derived growth factor and its receptor are correlated with oral tumorigenesis and poor prognosis in oral squamous cell carcinoma. Int J Mol Sci 21(7):2360

Liu C, Shui CL, Wang Q, Luo H, Gu CG (2017) Mechanism of hif-1α mediated hypoxia-induced permeability changes in bladder endothelial cells. Braz J Med Biol Res 51:–e6768

Liu X, Nie W, Xie Q, Chen G, Li X, Jia Y, Yin B, Qu X, Li Y, Liang J (2018) Endostatin reverses immunosuppression of the tumor microenvironment in lung carcinoma. Oncol Lett 15:1874–1880

Lu C, Shahzad MMK, Moreno-Smith M, Lin YG, Jennings NB, Allen JK, Landen CN, Mangala LS, Armaiz-Pena GN, Schmandt R et al (2010) Targeting pericytes with a PDGF-B aptamer in human ovarian carcinoma models. Cancer Biol Ther 9:176–182

Lu KV, Chang JP, Parachoniak CA, Pandika MM, Aghi MK, Meyronet D, Isachenko N, Fouse SD, Phillips JJ, Cheresh DA et al (2012) VEGF inhibits tumor cell invasion and mesenchymal transition through a MET/VEGFR2 complex. Cancer Cell 22:21–35

Luz CCF, Noguti J, Araújo L, Simão Gomes T, Mara G, Silva MDS, Artigiani Neto R (2018) Expression of VEGF and Cox-2 in patients with esophageal squamous cell carcinoma. Asian Pac J Cancer Prev 19:171–177

Majumder S, Piguet AC, Dufour JF, Chatterjee S (2013) Study of the cellular mechanism of Sunitinib mediated inactivation of activated hepatic stellate cells and its implications in angiogenesis. Eur J Pharmacol 705:86–95

Makondi PT, Lee CH, Huang CY, Chu CM, Chang YJ, Wei PL (2018) Prediction of novel target genes and pathways involved in bevacizumab-resistant colorectal cancer. PLoS One 13: e0189582

Mayer RJ (2004) Two steps forward in the treatment of colorectal cancer. N Engl J Med 350:2406–2408

National Cancer Registry Programme (2013) Three-year report of population based cancer registries: 2009–2011. NCDIR-ICMR, Bangalore

Nelson AR, Fingleton B, Rothenberg ML, Matrisian LM (2000) Matrix metalloproteinases: biologic activity and clinical implications. J Clin Oncol 18:1135–1149

Nishida N, Yano H, Komai K, Nishida T, Kamura T, Kojiro M (2004) Vascular endothelial growth factor C and vascular endothelial growth factor receptor 2 are related closely to the prognosis of patients with ovarian carcinoma. Cancer 101:1364–1374

Nishida N, Yano H, Nishida T, Kamura T, Kojiro M (2006) Angiogenesis in cancer. Vasc Health Risk Manag 2:213–219

Röllig C, Serve H, Hüttmann A, Noppeney R, Müller-Tidow C, Krug U, Baldus CD, Brandts CH, Kunzmann V, Einsele H et al (2015) Addition of sorafenib versus placebo to standard therapy in patients aged 60 years or younger with newly diagnosed acute myeloid leukaemia (SORAML): a multicentre, phase 2, randomised controlled trial. Lancet Oncol 16:1691–1699

Shamsdin SA, Mehrafshan A, Rakei SM, Mehrabani D (2019) Evaluation of VEGF, FGF and PDGF and serum levels of inflammatory cytokines in patients with glioma and meningioma in Southern Iran. Asian Pac J Cancer Prev 20:2883–2890

Sharma M, Khan R, Aggarwal M, Sharma A (2017) Modulatory effects of chemoradiation on angiogenic factors and laminin in cervical cancer: link with treatment response. Asian Pac J Cancer Prev 18:2937–2944

Sun Y, Fan X, Zhang Q, Shi X, Xu G, Zou C (2017) Cancer-associated fibroblasts secrete FGF-1 to promote ovarian proliferation, migration, and invasion through the activation of FGF-1/FGFR4 signaling. Tumor Biol 39:1010428317712592

Tejada ML, Yu L, Dong J, Jung K, Meng G, Peale FV, Frantz GD, Hall L, Liang X, Gerber H-P et al (2006) Tumor-driven paracrine platelet-derived growth factor receptor alpha signaling is a key determinant of stromal cell recruitment in a model of human lung carcinoma. Clin Cancer Res 12:2676–2688

Tournaire R, Simon M-P, le Noble F, Eichmann A, England P, Pouysségur J (2004) A short synthetic peptide inhibits signal transduction, migration and angiogenesis mediated by Tie2 receptor. EMBO Rep 5:262–267

van der Meel R, Symons MH, Kudernatsch R, Kok RJ, Schiffelers RM, Storm G, Gallagher WM, Byrne AT (2011) The VEGF/Rho GTPase signalling pathway: a promising target for anti-angiogenic/anti-invasion therapy. Drug Discov Today 16:219–228

Welti J, Loges S, Dimmeler S, Carmeliet P (2013) Recent molecular discoveries in angiogenesis and antiangiogenic therapies in cancer. J Clin Invest 123:3190–3200

Wu E, Palmer N, Tian Z, Moseman AP, Galdzicki M, Wang X, Berger B, Zhang H, Kohane IS (2008) Comprehensive dissection of PDGF-PDGFR signaling pathways in PDGFR genetically defined cells. PLoS One 3:e3794

Xing P, Zhang J, Yan Z, Zhao G, Li X, Wang G, Yang Y, Zhao J, Xing R, Teng S et al (2017) Recombined humanized endostatin (Endostar) combined with chemotherapy for advanced bone and soft tissue sarcomas in stage IV. Oncotarget 8:36716–36727

Xu Y, Zhang Y, Wang Z, Chen N, Zhou J, Liu L (2017) The role of serum angiopoietin-2 levels in progression and prognosis of lung cancer: a meta-analysis. Medicine (Baltimore) 96:e8063

Xue Y, Lim S, Yang Y, Wang Z, Jensen LDE, Hedlund E-M, Andersson P, Sasahara M, Larsson O, Galter D et al (2011) PDGF-BB modulates hematopoiesis and tumor angiogenesis by inducing erythropoietin production in stromal cells. Nat Med 18:100–110

Yang P, Chen N, Yang D, Crane J, Yang S, Wang H, Dong R, Yi X, Xie L, Jing G et al (2017) The ratio of serum Angiopoietin-1 to Angiopoietin-2 in patients with cervical cancer is a valuable diagnostic and prognostic biomarker. PeerJ 5:e3387

Yip-Schneider MT, Wu H, Schmidt CM (2020) Novel expression of vascular endothelial growth factor isoforms in the pancreas and pancreatic cystic lesions. Biochimie 181:234–239

Zhang G, Jin G, Nie X, Mi R, Zhu G, Jia W, Liu F (2014) Enhanced antitumor efficacy of an oncolytic herpes simplex virus expressing an endostatin-angiostatin fusion gene in human glioblastoma stem cell xenografts. PLoS One 9:e95872

Zhang N, Zhang S, Wu W, Lu W, Jiang M, Zheng N, Huang J, Wang L, Liu H, Zheng M et al (2021) Regorafenib inhibits migration, invasion, and vasculogenic mimicry of hepatocellular carcinoma via targeting ID1-mediated EMT. Mol Carcinog

Zhao Y, Adjei AA (2015) Targeting angiogenesis in cancer therapy: moving beyond vascular endothelial growth factor. Oncologist 20:660–673

Biology, Chemistry, and Physics of Cancer Cell Motility and Metastasis

Sounak Sadhukhan and Souvik Dey

Abstract

A malignant tumor has ability to invade the nearby tissues and colonize distant sites (organs) through the lymphatic/vascular system. From intravasation of cancer cells into the vascular system to extravasation from the vascular system to a new organ involve basic processes and concepts of biology, chemistry, and physics. Cancer cells migrate either singly or collectively using actin deformation proteins (RhoGTPase family), which help to create cell protrusions to move to different parts of a cancer-affected body. For moving to secondary locations, cancer cells breach the surrounding extracellular matrix and the basement membrane of vascular systems with the help of tissue-degrading proteins (proteolytic enzymes and matrix metalloproteinases).

Once tumour cells enter into the vascular system, they are controlled by the blood flow pattern, blood vessels' diameter, shear flow, and intercellular adhesion. They are also affected by hemodynamic forces, immunological stress, collisions with the blood cells, and the endothelial cells of the vessel wall. The shear flow in the blood vessels influences the rotational and translational motion of circulating tumor cells. These two motions decide the orientation of the cell with respect to the receptor–ligand interactions with the vessel's wall. During circulation, cancer cells bind to platelets, leukocytes, and fibrin with the help of adhesion proteins. Circulating cancer cells, which survive from fluid shear force and immune surveillance, arrest the microvascular endothelium of a secondary location using physical occlusion and/or cellular adhesion. The probability of cell arrest depends on the collision rate between the membrane-bound receptors and

---

S. Sadhukhan (✉)
Department of Computer Science, Institute of Science, Banaras Hindu University, Varanasi, India

S. Dey
Department of Pharmaceutical Technology, Jadavpur University, Kolkata, India

© The Author(s), under exclusive license to Springer Nature Singapore Pte Ltd. 2022
S. K. Basu et al. (eds.), *Cancer Diagnostics and Therapeutics*,
https://doi.org/10.1007/978-981-16-4752-9_5

endothelial ligands, and residence time of the cell. The chances of cell arrest are much higher if the shear force is at intermediate level. The location of secondary site for a tumor cell line is preferential and not random. This chapter provides an outline of the processes involved in cancer cell motility and metastasis based on the basic concepts of biology, chemistry, and physics.

Keywords

Cancer cell motility · Single cell motility · Collective cell motility · Metastasis · Intravasation · Extravasation

5.1 Introduction

An adult human body may contain nearly 100 trillion ($\approx 10^{14}$) cells. The growth of multicellular human body occurs due to cell divisions, which are aptly controlled and regulated by some factors. Noncancerous cells obey the cellular signals that instruct whether a cell should divide, differentiate into another cell, or die. Due to some inheritable (successive rounds of mutation) or epigenetic changes at the time of cell division, some cells may be transformed and divide at abnormally faster rate. The abnormal growth of cells in an uncontrollable and uncoordinated manner forms a lump/tumor, which may lead to cancer. If this proliferation is allowed to continue and spread, it may become life-threatening to the host.

Tumor development and its progress are a combination of several correlated biochemical processes. Genetic irregularities that affect mainly *oncogenes* and *tumor suppressor genes* lead to tumor development. While responsibility of oncogenes is expressing growth factors, which stimulate mitosis and lead to cell survival and proliferation, tumor suppressor genes restrain cell cycle and lead to *apoptosis* (programmed cell death) if any cellular abnormality arises. When oncogene is affected by gain in functions during mutations and becomes either constitutively expressed or overexpressed, the cell begins to proliferate regardless of the amount or even the absence of growth factors, resulting in uncontrolled cell growth and proliferation (Croce 2008). On the other hand, if a cell becomes damaged or mutated, the tumor suppressor gene arrests the progression of the cell cycle in order to carry out DNA repair or to induce apoptosis. When tumor suppressor genes lose its functionality, the cells are able to avoid apoptosis, enabling the propagation of mutated and damaged DNA to the daughter cells (Barnes et al. 1993). A successive number of mutations in oncogenes and tumor suppressor genes initiate the development of carcinogenesis.

The growth of a tumor may be classified into three distinct phases: avascular phase, angiogenesis, and vascular phase. In avascular phase, tumor does not have any blood supply, and it has to depend on nutrients and oxygen transported to and from extracellular matrix (ECM) through passive diffusion (Sutherland 1988) and consists of approximately 10^6 to 10^7 cells. The avascular nodules grow to 1–2 mm^3 in volume. Being deprived of vital nutrients, cells near the center of a tumor may die and create a necrotic core of dead cells (Sherratt and Chaplain 2001). Proliferative cells are present on the surface area of a tumor. Between the necrotic and the proliferative layers, another layer of cells exists (quiescent cells), which are recruited into the proliferative layer as required (Ang and Tan 2008).

To continue its growth, the tumor needs nutrient in proportion to its volume, but its ability to absorb nutrient is proportional to its surface area (Orme and Chaplain 1996). This phenomenon limits maximum size of tumor to which it can grow before it experiences nutrient deficiency. So, it diffuses a number of chemical substances (*tumor angiogenesis factors* (TAFs)) into the nearby blood vessels. These diffusible TAFs stimulate the endothelial cells of local vessels to form capillary sprouts that grow toward the tumor. The newly developed sprouts are often fused together to form loops (anastomoses) that are also fused with the other loops and build a complex network of vessels. The newly developed networks eventually supply necessary nutrients to the tumor for continuing growth. This phenomenon is known as *angiogenesis* or *neovascularization*.

After angiogenesis, tumor enters into the vascular phase, and it rapidly increases in mass. This massive growth in tumor is self-limiting due to breakdown of the vascular system and again a necrotic core is developed at the center of a tumor surrounded by a layer of proliferative and quiescent cells (Paweletz and Knierim 1989). Thus, in order to support constant progression, the vascular system untiringly modifies itself, and this process lasts until a tumor is removed or the host dies.

A malignant tumor is distinguished from a benign tumor by its ability to invade nearby tissues. The rapidly growing tumor cells (in vascular phase) create tremendous pressure to the adjacent cells and the ECM and penetrate into the nearby tissues (invasion). Some of the tumor cells may detach from its origin and spread through the blood/lymphatic vessels to the distant parts of the body (metastasis). The circulating tumor cells form distant colonies depending upon its ability to induce new blood vessels from the nearby tissues to sprout toward the tumor. This results in adequate blood supply and microcirculation in the tumor cells.

The rest of the chapter elaborates the processes of cancer cell motility and metastasis in detail and is organized as follows: Sect. 5.2 gives the biological backgrounds; Sect. 5.3 illustrates the biological and chemical processes of cancer cell motility and metastasis; and Sect. 5.4 describes the physical interactions and mechanical forces involved in cancer development and its progression. Section 5.5 concludes the chapter. Some of the biological terms are briefly described in the appendix.

5.2 Biological Background

Cells are the unit of life. Most of the activities in living organisms are performed inside the cells. A cell contains a nucleus, bounded by cytosol and wrapped in a bilipid membrane. Cell membrane protects the cell from the surrounding environment, and it is porous to passive diffusion of small size molecules like oxygen and glucose. It helps to maintain pH and other chemical balance within the cell by actively pumping other molecular species and also exchange complex signals with the microenvironment. These complex signals play a key role to maintain equilibrium with the surrounding environment; that is, the cell cycle speed is regulated by growth-inhibitory signals, which are further controlled through the interaction of cyclin and cyclin-dependent kinases (CDKs). Surface receptors regulate gene expression through complex signals. Also, the gene expression pattern dictates the production and balance of proteins. Hence, a complex interaction exists between a cell and its surroundings that control the cell cycle and cell growth (Clyde et al. 2006).

5.2.1 Equilibrium in Cell Population

The population of each cell type in a tissue is maintained by balancing cell proliferation and apoptosis. If differentiated cell dies, somatic (adult nongermline) stem cells divide asymmetrically into a stem cell and a progenitor cell. The progenitor cell then further splits or terminally differentiates into the preferred cell type, which migrates to the correct place and replaces the dead cell. The whole process is regulated and controlled by a complex system of *biochemical signals* that are monitored by all the cells in the structure. Whenever a signal is received, each cell activates the corresponding receptors on its surface. The activated receptors are responsible for activation and/or deactivation of genes within the nucleus. The transformed genes then start to express proteins within the cells that control the cell cycle and all other cell functions. Stromal cells (connective tissue cells of any organ) play a major part in creating and responding to growth factors and biochemical signals (Zipori 2006).

5.2.2 Oncogene and Tumor Suppressor Genes

Gene carries genetic information of growth, development, functioning, etc. Gene expression pattern expresses the behavior of a cell, which receives growth signals and growth inhibitory signals. These two signals are vital to maintain the health of a tissue. *There are mainly two types of genes that are responsible for promoting the cell cycle advancement: oncogenes and tumor suppressor genes.* While the former is responsible for generating and endorsing the cell cycle progression, the later guarantees proper DNA repair, react to the growth inhibitory signals, and may control apoptosis due to some special circumstances. The malfunctioning DNA is the key factor for developing carcinogenesis (Hanahan and Weinberg 2000). A single erroneous mutation is enough to affect the functionality of an oncogene (Malumbres and Barbacid 2003), or neutralize the functionality of tumor suppressor gene (Horowitz et al. 1989). Sometimes, at the time of cell division (during mitosis phase), any kind of errors can produce a faulty oncogene. Thus, signals change their routes to the oncogene, which regulate the activity. There are other possibilities too. Errors may occur at the time of cell division if the offspring get extra copies of the oncogenes or fewer copies of the tumor suppressor genes from the parent cell (Castro et al. 2006). The increment in the functions of the oncogenes and decrement in functions of tumor suppressor genes in the daughter cell can significantly damage the normal behavior and accelerate the chances of completing a multistep carcinogenesis path (Quon and Berns 2001). *Latest study* (Kashkin et al. 2013) *says that key sponsor of unrestricted cell growth is the over- and underexpressed genes.* As the gene expression product transfers from one generation of a cell to its next generation, the fundamental changes in gene expression potentially affect the malignant transformation of a cell, similar to a genetic mutation. When a cell with defective genes malfunctions and interacts with the body in an abnormal, proliferative manner, we term it cancer.

5.2.3 Cellular Adhesion

Human body is constructed with a community of cells. Each cell occupies a suitable place to perform its tasks (white blood cell is an exception). Every cell has its own address or area code system-written domain that can either interact with the cellular address molecules (CAMs) of the same kind (hemophilic binding) or interact with the other kind of CAMs or the extracellular matrix (heterophilic binding). In hemophilic binding, receptor molecules on the cell surface knot with similar target molecules of adjacent cells (in cell–cell adhesion), the ECM, or the basement membrane in the microenvironment (in cell–ECM or cell–basement membrane adhesion), while in heterophilic binding, surface adhesive molecules of one type knot with dissimilar molecules in the ECM or the basement membrane or the adjacent cells.

Adhesion is crucial to multicellular structure and cell motility. *There are mainly three types of cellular adhesive forces: cell–cell adhesion, cell–basement membrane adhesion, and cell–matrix adhesion.* Cell–cell adhesion controls the tissue structure, while cell–basement membrane adhesion and cell–matrix adhesion are necessary for adhesive friction during cell motility. Cell–cell adhesion helps a cell to cooperate and attach to its adjacent cells and act as intermediator for communication among molecules of the cell surface. It controls the epithelial cell growth and cell division. The proliferation rate is increased to fill up the gap in the epithelium due to apoptosis and the rate is reduced whenever the epithelium is fully populated (Hansen and Bissell 2000). These interactions and the ability to transfer signals among cells are essential for the survival of the tissues.

Cell–matrix and cell–basement membrane adhesive forces are transient or temporary such as adhesion between cells of the immune system or the interactions involved in tissue inflammation. *Cell–basement membrane and cell–matrix adhesion govern the traction during cell motility.* After losing the adhesive interaction with the basement membrane, epithelial cells are often converted to a specialized type of apoptosis, named anoikis (Frisch and Ruoslahti 1997). Anoikis helps to suppress overgrowth of separated cells into the lumen. ECM controls the growth, differentiation, and apoptosis of stromal cells (Selam et al. 2002). Cellular adhesive force governs the multicellular arrangement and cell motility. *The loss of adhesive forces between cells can result in uncontrollable cell growth and carcinogenesis.* Cell–cell adhesion is an important factor in normal tissues, but in the cancerous environment, cell–cell adhesion molecules seem to disappear or compromised. Losing cell–cell adhesion ability between cells seems to be the first step in cancer invasion.

5.2.4 Cell Signaling

Cell signaling regulates the basic functionalities of cells and synchronizes all cell activities. The basic developments like tissue repair, wound healing, and tissue homeostasis all depend on it. A cell can be able to receive a signal from the surrounding and produce an appropriate reply to the microenvironment; it is the basis of the cellular developments. Malfunctions in cell signaling system are the key factors for the diseases such as cancer, diabetes, and autoimmunity (Wang et al. 2013).

Cell signals can be categorized into two groups: mechanical type of signals and biochemical type of signals. In the case of mechanical type signal, forces are generated by the cell and it is applied on the surrounding cells. Also, these forces can be sensed and responded by the other cells. In case of biochemical signals, signals are the molecules of proteins, lipid, gases, and ions. *Based on the distance between the sender and the receiver cells, biochemical signals can also be classified as intracrine, autocrine, juxtacrine, paracrine, and endocrine.* In a multicellular organism, signaling between cells transmits either through extracellular space (paracrine signals over short distances, endocrine signals over a longer distance) or through direct contact (juxtacrine signals). *Cells receive information from its surroundings through some proteins known as receptors.* Notch is one of the cell surface proteins that act as a receptor. A set of signaling proteins interact with the *Notch receptors* and generate response to the cells. *The receptor activation molecules (ligand receptors) can be classified as hormones, neurotransmitters, cytokines, and growth factors.* Hormones are the main signaling molecules of the endocrine system; however, they often control each other's secretion through local signaling. Neurotransmitters (including neuropeptides and neuromodulators) are signaling molecules of the nervous system, and cytokines (paracrine or juxtacrine in nature) are signaling molecules of the immune system. Signaling molecules communicate with a target cell (through ligands to cell surface receptors, and/or by entering through the cell membrane (endocytosis)) for intercrine signals. Ligand receptors are responsible for cell signaling mechanisms and communications. Notch acts as a receptor for ligands. Some receptors are cell surface proteins, but others are found inside the cells.

Several signaling factors activate surface receptors, which regulate gene expressions. Internal chemical molecules (e.g., oxygen) also affect gene expressions. Failures in these signaling processes are involved in several cancers. A few examples of signaling are hypoxia-induced factor (HIF)-1α signaling, epidermal growth factor

(EGF) signaling, and E-cadherin/β-catenin signaling. These signals are directly or indirectly responsible for tumor growth and progression. HIF-1α is generated inside cells and is affected by the presence of oxygen (Semenza 2001). A cell enters into hypoxic state due to insufficient oxygen levels. *Whenever a cell enters into hypoxic state, HIF-1α activates the target genes, which are responsible for the increment of motility rate, secretion of angiogenic factors, anaerobic glycolysis, and reduction in cell–cell and cell–matrix adhesion, and also reduces the effectiveness of apoptotic signals* (Pouysségur et al. 2006). Epidermal growth factor (EGF) activates the EGF receptors. When two activated EGF receptors bind together, it can transmit signals that increase the secretion of HIF-1α, cell proliferation, and cell motility and reduce apoptosis (Allen et al. 2006). Among the signal receptors, some of them can perform in multiple roles, like E-cadherin acts as arbitrating in the intracellular domain as it binds to α-catenin (using β-catenin) to mechanically bind an adhered cell to its actin cytoskeleton, and also suppress cell cycle progression (Seidensticker and Behrens 2000). *These signaling systems play a crucial role to sustain tissue microarchitecture* (Wei et al. 2007).

5.3 Cancer Cell Motility and Metastasis: Biology and Chemistry

While the biological notion is required to understand the processes behind tumor formation, growth, progression, and transformation into cancer, chemistry helps to recognize the microscopic interactions of thousands of genes, molecules, and conditions that lead to aggressive cellular growth and the chemical processes involved at the time of cancer cell motility and metastasis. On the other hand, physics expresses the key physical and mechanical forces applied on the cancer cells and their role in cancer cell motility and metastasis. It also explains the physical properties of cell interaction, the role of inflammation in cancer, and connective tissue microenvironment.

5.3.1 Cancer Cell Motility

In a biological environment, different types of cell movements can be seen. Cell movement is the fundamental property of living cells that have been adapted throughout the development of multicellular organisms including immune response, embryogenesis, and adult tissue homeostasis. Cells move due to some external signals in the direction of higher concentration of the substrate, this is, a primitive feature of biological systems. It is called *chemotaxis*. Motile cells require mechanical interaction between the microenvironment and the cell membrane bulges. Individual cells may move through the stroma in an amoeboid motion by squeezing between the ECM fibers or by extending a finger-like pseudopod that forms focal adhesions with the ECM to create space for motion, and is accomplished by forming tiny invadopodia on the filopodium surface that secretes proteases to degrade the ECM (Weaver 2006).

In case of cancer cells, though the mechanism is not fully understood, it can be said that oncogenes are responsible for this phenomenon. The proteins, that is made by the oncogenes carry a wrong signal to misguide the nucleus that the cell is appropriately attached to its neighbors, though actually, it is not. Thus, the cell tries to stop its own growth and die due to apoptosis. Epithelial cells are the most common cancer sources in the human body. It is separated from the rest of the body by a basement membrane and a skinny layer of the specialized ECM. Basement membrane develops an obstacle that most of the normal cells cannot breach, but the cancer cells can break the wall, penetrating it by secreting metalloproteinases enzyme (Liotta 1992). *Basement membrane and other extracellular lamina are dissolved by the matrix metalloproteinases* (MMPs). After puncturing the basement membrane, tumor cells shortly meet another basement membrane, surrounding blood vessels/lymphatic vessels. MMPs also break the second basement membrane by penetrating the wall and forming a finger-like shape to get access to the bloodstream directly. After breaching the basement membrane of blood vessels/lymphatic vessels, cancer cells are free to spread throughout the whole body through those vessels, though the tumor cells must undergo some changes/transformation (till now not clearly understood) to adapt, survive, and proliferate in the new locality. If the invasion occurs within the primary organ, then tumor cells do not require any transformation. Eventually, a tumor cell may lodge in a capillary. If it penetrates the capillary wall again, it can create a secondary tumor. Perhaps less than 1 of 10,000 tumor cells is able to escape a primary tumor and survive to colonize at another location. So, the cell motility mechanism is primitively knotted with cell survival, which is the key to metastasis.

Tumor cell motility and invasion to the other organ or blood/lymphatic vessels are the pillars of metastasis. *Motile cancer cells remodel their cell–cell or cell–matrix adhesion and their actin cytoskeleton, which are involved in cell signaling.* There can be various ways of cancer cell migration depending on cell types and degree of differentiation. These different types of cell migration are controlled by various mechanisms. The main factor of these mechanisms is a dynamic reformation of the actin cytoskeleton, and this reformation yields the necessary force for cell migration. The actin reformation is governed by RhoGTPase family, which includes Rho, Rac, and Cdc42. The RhoGTPase carries the extracellular chemotactic signals to activate the downstream enzyme family: Wiskott–Aldrich syndrome proteins (WASPs) that are the major factors in cell migration. These WASPs initiate the development of protrusion in the cellular structures that are involved in the motility process and degradation of the ECM.

5.3.1.1 RhoGTPase Family

Rho, Rac, and Cdc42 are members of RhoGTPase family. They are the key factors for cell migration and invasion, as it has the ability to alter the polymerization of actin. They are activated by guanosine triphosphate (GTP)-binding proteins. Their stimulation is controlled by guanine nucleotide exchange factors (GEFs), GTPase-activating proteins (GAPs), and guanine nucleotide dissociation inhibitors (GDIs). GEFs stimulate RhoGTPase through the interchange of Rho-bound guanosine

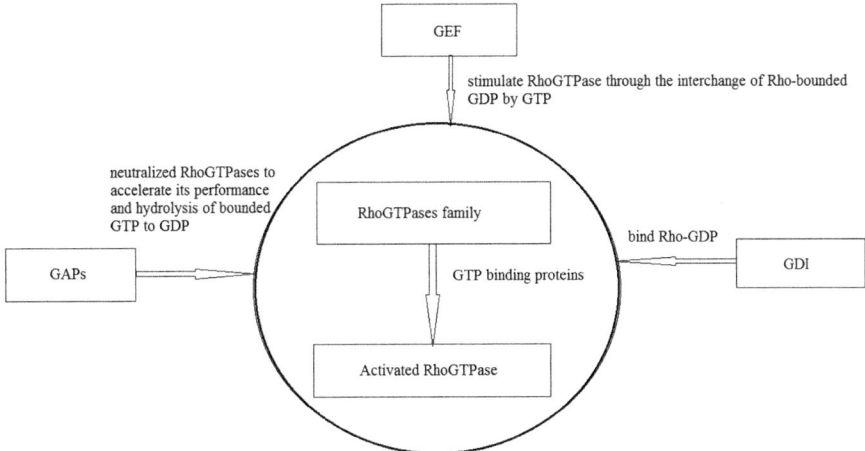

Fig. 5.1 Activation of RhoGTPase family

diphosphate (GDP) by GTP, and it is further neutralized by GAPs that accelerate the performance of RhoGTPase and hydrolysis of bounded GTP to GDP. Lastly, GDIs bind to Rho-GDP and avoid their contact with RhoGEFs (Fig. 5.1). RhoGTPases regulate the effector protein serine/threonine kinases, P21-activated kinases (PAK1–3), and myosin light-chain kinase (for Rac1); PAK1–6, WASP, N-WASP, and mDia2 (for Cdc42); and ROCK kinases I and II, Citron, and mDia 1 and mDia2 (for RhoA). Activated Rac1 and Cdc42 initiate the reformation of the actin cytoskeleton at the leading edge (Small et al. 2002).

Actin polymerization impulses at the foremost edge of the cell membrane and forms finger-like protuberances (filopodia) or sheet-like bulges (lamellipodia) (Fig. 5.2a, b). These protuberances provide locomotive forces, which help cells to move to a different location. On the other hand, RhoA controls the contractile actomyosin filaments using downstream enzyme ROCK/Rho-kinase. Actomyosin endorses locomotive forces in the cell body and the trailing edge (Fig. 5.2c). While Rac1 and Cdc42 motivate the development of cellular bulge at the leading edge, RhoA makes retraction at the trailing edge. These synchronized efforts make a cell to move toward a direction (Fig. 5.2d).

5.3.1.2 Single Cell Motility

There are many types of cell motility observed in cancer cells (Friedl and Wolf 2003), and these movements are simultaneous. *In epithelial tissues, the cell moves as a sheet-like formation.* This type of motility in epithelial cells is not only seen in angiogenesis but also seen in other biological phenomena including wound healing (Friedl 2004). At the time of epithelial cell differentiation, the utility of cadherin is inhibited that leads to abrogation cell–cell adhesion. The suppression of cell–cell adhesive force separates the cells from each other and makes them ready for individual or collective migration (Lozano et al. 2003). *This phenomenon is known*

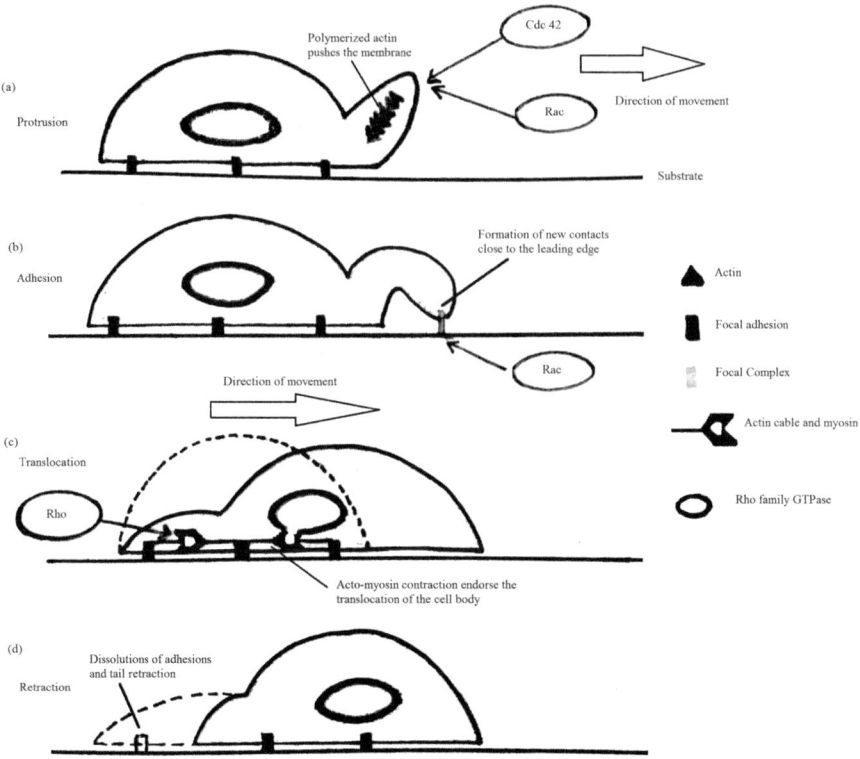

Fig. 5.2 Cell migration consists of these four processes: (**a**) protrusions, (**b**) adhesion, (**c**) translocation, and (**d**) retraction (Yamazaki et al. 2005)

as epithelial-to-mesenchymal translation (EMT). Cancer cells are able to migrate and invade in the absence of EMT too (Friedl 2004). The reason behind EMT, cell motility, and invasions is the damage of the cell–cell adhesion molecules. E-cadherin is the key element of epithelial adherens junctions. With the loss of E-cadherin functions in mesenchymal cell–cell adhesion, N-cadherin increases (cadherin switch) leading to cell motility and invasion (Lehembre et al. 2008). This functional loss can happen due to germline and somatic gene mutations, chromosomal abnormality, transcriptional repression, DNA hypermethylation of E-cadherin gene (cdh1), signal transduction of a large number of growth factors, and Notch signaling. Several growth factors like transforming growth factor β, hepatocyte growth factor, epidermal growth factors, fibroblast growth factors, and Notch signaling stimulate the E-cadherin loss. *Also, hypoxic condition in tumor cells influence them for migration and spreading.* These changes in the microenvironment stimulate one or several receptors of E-cadherin gene expression (*e.g.*, Snail1, Snail2, ZEB1, ZEB2, E47, and Twist). Binding of the transcriptional repressor to the E-cadherin gene promoter causes silencing of the gene (for details, one may refer to Herranz et al. 2008; Hou et al. 2008).

Cell motility can be either single (amoeboid or mesenchymal) or collective (in cell sheets, strands, tubes, or clusters). As an example, cells in colorectal cancer transform into motile form and can roam alone after losing E-cadherin, whereas squamous cell carcinomas migrate and invade collectively. *Different factors like extracellular protease activities, integrin-mediated cell–matrix adhesion, cadherin-mediated cell–cell adhesion, cell polarity, and cytoskeletal organization decide the different types of cell motility and invasion.* Single or collective cell migration can be seen in stromal cells. In EMT, during embryonic development in morphogenic procedures, loss of epithelial polarity or gain of mesenchymal morphology is responsible for single cell migration.

Tyrosine kinases at the endosome receptor arbitrate motogenic activation of Rac. To limit Rac to the plasma membrane and to induce the development of migratory actin flange, Rab-5-dependent *endocytosis* is needed. The control of endocytosis on migrating cells is needed for the fast recovering of β1-integrin at pseudopodal membrane flanges and the retaining of a pool of β1-integrin at the cell front by the GTPase Rab25. Mutant p53 endorses the transferring of β1-integrin and of EGFR, which increases the β1-integrin/EGFR signaling, cell invasion, and metastasis (Muller et al. 2009). p120-catenin is present at the cell membrane of epithelial cells, and it regulates the activity of RhoA and Rac1 with E-cadherin. RhoA degrades cell–cell adhesion. p120-catenin-reliant RhoGEF enzymes, for example, Vav2 or RhoGAP enzymes, are required for activation and deactivation of RhoA. p190-RhoGAP activates Rac1 for stabilizing E-cadherin junction through suppression of the actions of IQGAP1 (a Rac1 effector protein and a mediator of E-cadherin endocytosis). Cytoplasmic p120-catenin endorses the activity of Rac1 at the cell membrane, and it leads to the formation of lamellipodia (Fig. 5.3a, b). So, cytoplasmic p120-catenin and membrane-sequenced p120-catenin both collaborate with each other during EMT that leads to cell motility. Rac1 suppresses the actions of RhoA through the generation of responsive oxygen species (increases the expression of Snail1) that also inhibit lower molecular weight protein: tyrosine phosphatase (it stimulates p190RhoGAP). All the above-mentioned activities may be found in the single cell migration. (For the detailed study, one may refer to Kleer et al. 2006).

Single cell motility is classified into *mesenchymal cell migration* (Fig. 5.3a) and *amoeboid migration* (Fig. 5.3b) (Friedl and Wolf 2010), and anyone or both of it can be seen in cancer cells. Mesenchymal cell motility is usually exercised by spindle-shaped, fibroblast-like cells (such as fibroblasts, endothelial cells, smooth muscle cells, and cancer cells). It can be categorized by *Rac-induced cell protrusion*-driven cellular movements and *actin polymerization-based* cellular movements. Amoeboid cell migration can be seen among round-shaped cells such as hematopoietic stem cells and leukocytes. The driving force of amoeboid cell movement is RhoA/ROCK-mediated bleb-like protrusions with active myosin/actin contractions.

5.3.1.3 Motility of Collective Cells

In collective cell migration (Fig. 5.4), cells preserve their cell–cell junctions and move with the connection to their originating tissue or separately in the form of a sheet, strands, tubes, and clusters. Migrating cells develop a cone or finger-like cell

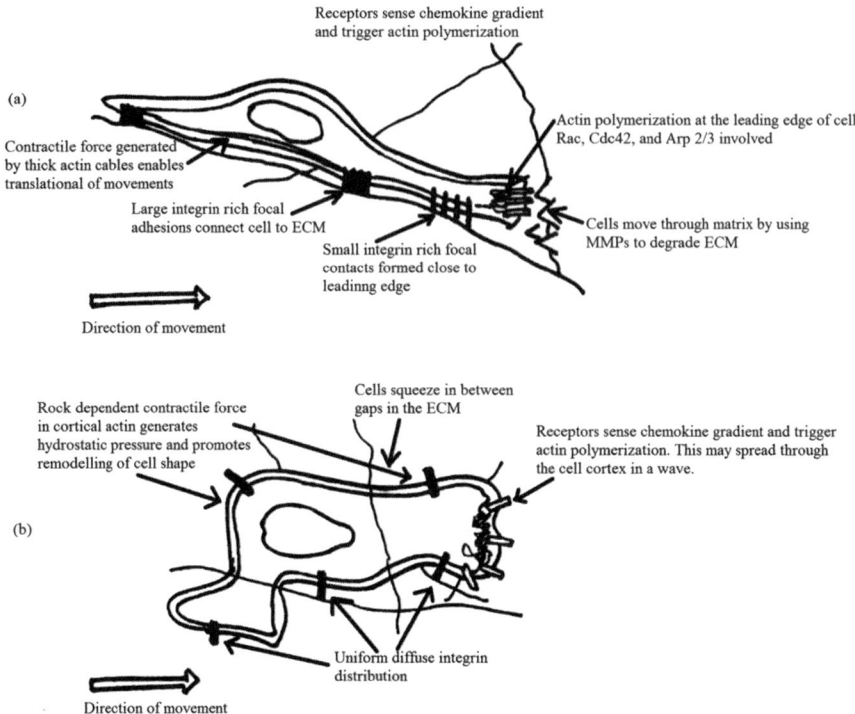

Fig. 5.3 Cell motility: (**a**) mesenchymal cell migration (**b**) amoeboid motility (Sahai 2005)

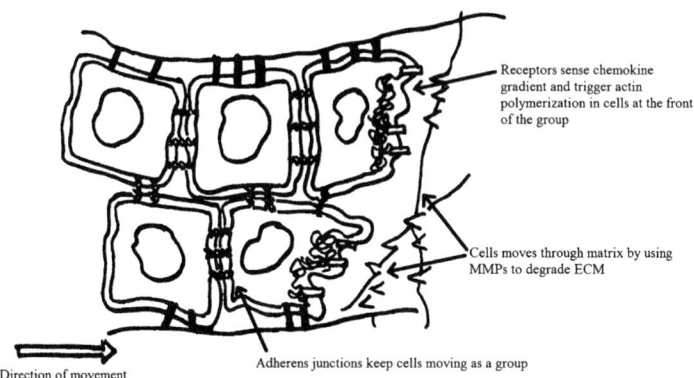

Fig. 5.4 Collective motility is characterized by collection of cells invading with many still retaining cell–cell contacts (Sahai 2005)

membrane bulge; it then intravasate and circulate like cell clusters that are very proficient in embolizing (it refers to the path and accommodation of an embolus within the bloodstream) lymphatic or blood vessels. Collectively migrating cells differ from the solitary migrating cells in many ways, like collectively migrating cells only use their pulling forces on its neighboring cells that are connected through adhesion junctions instead of reacting to their cellular trails. Though collective migrating cells have few similarities with solitary migrating cells on the cellular level, like both of the migratory processes develop protrusions that use β1-integrin and/or β3-integrin to develop focal adhesive force connected with actin cytoskeleton, and use actin contractile for cell motility and local shrinkage. Collective migration can be seen in many physical morphogenic processes (it causes an organism to develop its shape), where cells move in groups. Motile cells use cell membrane bulges that help to distinguish motile cells from their adjacent followers that move in a particular direction.

Collective migration needs synchronization between cell–cell adhesion and cell contractibility (Montell 2008). Collective migration is seen in various types of cancers (like breast cancer, prostate cancer, lung cancer, and melanoma). Motile cells collectively penetrate the ECM; the leading cells develop an invasion path using β1-integrin-mediated focal adhesions and MMPs to rupture the collagen fibers. It forms a tube-like structure in which the following cells can migrate. Invasive migration and proteolytic matrix remodeling control collective cell motility. It can be explained in detail through the inspection of cell surface glycoprotein podoplanin (38 kilodalton) that define various cell types like kidney podocytes, lymphatic endothelial cells, and platelets and various cancer types including squamous cell carcinoma of oral cavity, lung, and skin. In the above-mentioned cancer types, collectively motile cells form a finger or cone-like shape to penetrate the neighboring tissue.

5.3.2 Metastasis

If a normal cell has undergone repetitive genetic and epigenetic changes, then it will transform into cancerous with uncontrolled proliferation potential. These features of the cancerous cells are passed on to its daughter cells. This uncontrolled proliferation helps to produce primary heterogenetic tumor (different cells showing distinct morphological and phenotypic profile). The cells of a developing tumor ultimately undergo *metaplasia*, *dysplasia*, and then *anaplasia*. This results in malignant phenotype. Cancer cells break away from a primary tumor and degrade the surrounding ECM for entering the vascular and lymphatic systems. These circulating tumor cells develop a second site of tumorigenesis.

Using the tissue-degrading proteins, the cancerous cells can be able to penetrate the ECM and escape from its primary location. These circulating tumor cells may be able to breach the lymphatic or blood vessels walls and then reach to other sites of the body through the lymphatic channels or blood vessels (lymphatic spread/hematogenous spread). After reaching the secondary site, they might repenetrate the vessel wall and continue to grow leading to a cancer cell colony. *This new tumor is known as a metastatic/secondary tumor.* From carcinogenesis to metastasis, a series of interconnected biochemical processes are involved (Fidler 1990). The

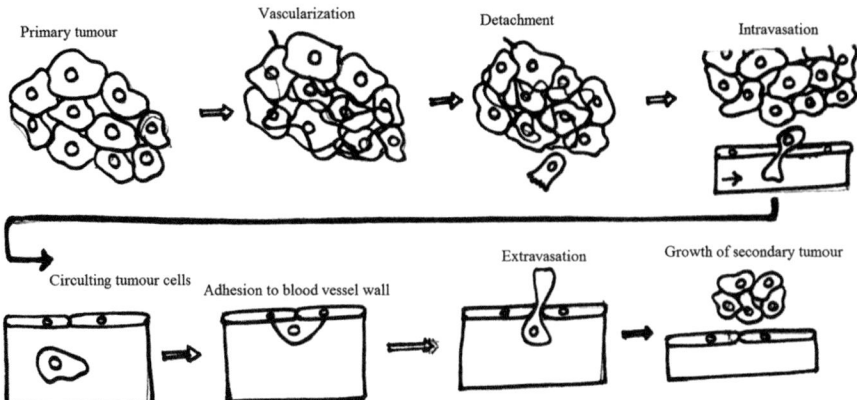

Fig. 5.5 Metastatic process leading to secondary tumor development (Wirtz et al. 2011)

processes are as follows: primary tumor formation and its growth, vascularization, invasion, and entry of tumor cells into blood/lymphatic vessels. Interaction of tumor cells with fibrin and platelets, sticking on capillary bed of secondary organs, and extravasation into the secondary organs are also involved. It is important to state that the cells of this secondary tumor are the same as the cells of a primary tumor.

Not all malignant tumors give rise to metastasis. Basal cell carcinoma of the skin is capable of local invasion, but it rarely produces secondary growths. Most other cancers give rise to metastasis at some point in their evolution. Though metastasis has a great clinical importance, *till now this process is very poorly understood* (Kaplan et al. 2006). The time period between the initial diagnosis of primary cancer and the first detection of metastasis varies widely from patient to patient. This phenomenon is very common in case of breast cancer. Different tumors have different metastatic capabilities. The presence of malignant cells in circulation does not necessarily lead to metastasis. Researches showed that after 24 h in circulation, less than 1 percent of cancerous cells are viable and less than 0.1 percent survived to form metastasis (Fidler 1970).

Spread of cancer cells takes place through three different routes: *transcoelomic, lymphatic, and hematogenous*. In transcoelomic spread, malignant cells reach into body cavities by crossing the surface of the peritoneal, pleural, pericardial, or subarachnoid areas. In lymphatic spread, tumor cells move into the lymph nodes and then spread into other parts of the body, and in the hematogenous spread, tumor cells spread through blood vessels. While the lymphatic spread is the usual route for carcinoma metastasis, it is an unusual route for sarcoma metastasis (suitable route is blood vessels). On the contrary, hematogenous spread is also suitable for a few types of carcinoma like renal cell carcinoma. Lymphatic vessel is the most common pathway for the initial spread of carcinomas. The vessel walls of the lymphatic system provide less resistance to the tumor cells during spread. Cancerous cells very easily break these vessel walls and reach the regional lymph nodes. Within the lymph nodes, the malignant cells can halt and proliferate further and develop a secondary tumor/metastatic tumor. The malignant cells may move away from the lymph nodes and enter into the bloodstream, ultimately reaching other tissues (Fig. 5.5).

The most common organs affected by metastasis are lungs, liver, brain, and the bones (Berenson et al. 2006). Different types of cancer tend to spread to specific organs and tissue in preferential way at a rate that is higher than expected. There is a famous *Paget postulate* (Paget 1989) regarding this phenomenon, known as *seed and soil hypothesis*. It says that cancer cells spread to different organs throughout the body but grow only in "fertilized soil." The "fertilized soil" defines the presence of favorable growth factors and absence of anti-growth factors for metastasis.

5.3.2.1 Proteolytic Enzymes

During cancer cell invasion and metastasis, a number of tissue walls require to be breached. These walls are made with basement membrane or interstitial connective tissues. In invasion and metastasis, basement membrane is invaded at least thrice: first, when the cancerous cells invade these walls to escape from its primary location; and second during entering into and third at time of exiting from the blood vessels. *The most important elements of basement membrane are collagen* (type IV), *laminin, proteoglycans* (heparin sulfate and chondroitin sulfate), *entactin,* and *osteonectin; the key elements of interstitial connective tissue are collagen fibers* (type I, II, III, or interstitial collagen), *glycoproteins* (e.g., fibronectin), *proteoglycans, and hyaluronic acid.* During degradation of walls, cancer cells secrete a number of enzymes that degrade proteases like urokinase form of plasminogen activator (uPA), cathepsin B (CB), cathepsin D (CD), and various metalloproteases that perform as a cascade (see appendix).

Several proteases are involved in tissue degradation processes. Though these processes partly look like normal cell invasion, proteolysis is self-limiting and organized, whereas in cancer the regulatory mechanisms are absent. There is a correlation between the different proteases and metastatic attributes of a tumor. According to Duffy (1992), malignant cells' capability to metastasize not only depends on the levels of the different proteases but also depends on the density of appropriate endogenous substrate. *An inversely proportional relation is present between the metastatic potential and the density of a specific substrate as the low level of substrate may boost the metastatic process, whereas the higher level may suppress metastasis, though in the case of some proteases, higher density of substrate helps to upregulate metastasis* (Nekarda et al. 1994).

Proteases are responsible for accelerating cancer cell spread; most of the events would have been estimated at the invading point of tumor where damage of normal tissue has occurred. That is why certain proteases are found in those sites with highest density. As an example, in Lewis lung tumors, uPA is found in high density in the area of tumor where invasive growth can be seen and no tissue damage has occurred. In some areas where invasive growth with tissue degradation has taken place, that contain uPA of high density with PAI-1 of low density. In general, proteases help to spread cancer cells through accelerating the destruction of ECM.

As ECM is developed with different elements, various types of proteases are needed to accomplish the metastasis and also to activate several precursors, for example, plasmin, cathepsin, and a trypsin kind protease activate pro-uPA, and plasmin activates some metalloproteases. Additionally, proteases have some extra utility to boost invasion and metastasis, like uPA and CD, both are the mitogens and they help to accelerate cell division, which indirectly endorses cell motility and metastasis. Also, proteases activate the positive growth factor and inactivate the metastasis suppressor to help metastasis.

5.3.2.2 Adhesion Proteins

At the time of cancer cell spread, tumor cells divide repeatedly and develop new attachments with their neighboring structures. Thus, in the early phase, invasive and metastatic cells degrade the adhesion with the surrounding cells and release themselves from the neighborhood to stick with the basement membranes. *During circulation within blood/lymphatic vessels, cancer cells bind to platelets, leukocytes, and fibrin, and later, they adhere to microvascular endothelium in the secondary organ. The chemical substrates that permit these different cell–cell or cell–matrix attachments are called adhesion proteins.* Adhesion proteins can be categorized into different groups: integrins, cadherins, selectins, CD44, and immunoglobulin superfamily (see appendix).

Some of the adhesion proteins help cancer spread by decreasing expressions, while others accelerate metastasis by increasing expressions. Specifically, integrins, E-cadherin, and CD44 directly interfere in metastasis. Scientist has shown that peptides with RGD sequence try to suppress metastasis in animal system, which proves that some integrin proteins interfere in cancer spread (Ruoslahti 1992). It has also been found that for different integrins, transfection (the process of inserting genetic material) of various cell types with cDNA (complementary DNA) raises the metastatic phenotype of inheritor cells. *Evidence has been found for negative correlation between the expression of E-cadherin and cancer cell spread*; invasion can be prevented through E-cadherin transfection with cDNA and antibodies with the power to inactivate E-cadherin causing the invasive phenotypes. These facts help to conclude that loss of E-cadherin raises the chances of metastasis, and CD44 helps in cells spread. It has been discovered that isoforms of CD44 are involved in metastasis process (Günthert et al. 1991).

Cell motility is very important as without cell motility neither invasion nor metastasis occur. Motility in cancer cells was discovered using time-lapse cinematography that was applied on V2 malignant cells. *In time-lapse cinematography, it was observed that those malignant cells continuously moved at an average speed of 6-7 μm/min.* This speed is the same as the normal leukocyte, but it is more than the speed of normal fibroblasts, histiocytes, or epithelial cells. Cell motility is governed by several features that are known as "motogens." Some of the motogens also act as growth factors or vice versa (Table 5.1). Motogens are also involved in angiogenesis; among them, few motivate cell motility and inhibit growth, whereas the others inhibit motility as well as growth. There are some factors that act as motility factor primarily, including scatter factors, autocrine motility factors (AMFs), and migration

Table 5.1 Examples of peptides that regulate growth and motility

Response	Peptide
Cell motility stimulation and growth stimulation	PDGF-AA and BB TGF-beta EGF/TGF-alpha Basic FGF Acidic FGF IGF-1 PD-ECGF G-CSF/GM-CSF Bombesin
Cell motility stimulation and growth inhibition	TGF-beta IL-6 EGF TNF-alpha
Cell motility inhibition and growth inhibition	TGF-beta TNF-alpha INF-gamma

stimulating factors (MSFs). Scatter factor, AMF, and MSF all are peptides. *Scatter factor* is produced by fibroblast; AMF is actually refined from human melanoma cells; and MSFs are produced by embryo and certain tumor-linked fibroblasts. AMFs and MSFs both can govern the movement in an autocrine manner, but scatter factors act in a paracrine way. Plasminogen activator (μPA) receptors also play an important role in cell motility (Gyetko et al. 1994).

5.4 Physical and Mechanical Processes in Cell Motility and Metastasis

In metastasis, a series of steps are involved, which include cancer cell motility through blood vessels, lymphatic vessels, and the tissues at the secondary sites (organ). The ability to handle each phase and advancement toward the development of secondary tumor and its growth are completely dependent upon the mechanical forces and physical interactions between cancerous cells and its microenvironment. During the invasion through the basal membrane into the blood vessels, cancer cells must undergo elastic distortions to penetrate endothelial cell–cell joint. The interplay between the tumor cell velocity and adhesion helps to bind the cancerous cells to the vessel walls that lead to the formation of a secondary tumor and its growth (Wirtz et al. 2011).

5.4.1 Physical Interaction in Cell Motility

Loss of E-cadherin and cytokeratins causes reduction in adhesive forces between cells and morphological changes from cuboidal to mesenchymal. During this phase, cells undergo dramatic changes like, detachment of cells from the primary tumor, and acquisition of motile phenotypes as their physical and mesenchymal properties are altered. After leaving the primary site, the migrating cells find the ECM, which is structurally complex and rich in collagen I fiber and fibronectin (Hotary et al. 2003). As the ECM consists of enhanced collagen deposition and lysyl oxidase (LOX) cross-linking of collagen, it is more rigid in nature than the normal tissue. These collagen cross-linking improves integrin signaling and bundling of fibers, which promote cell proliferation and invasion. Despite recent advances, scientists have very little information regarding the molecular and physical mechanisms of cell motility and metastasis.

Stress fibers are either local to the cell cortex or emanates from the nucleus in the direction of the plasma membrane to develop pseudopodial protrusions (Bloom et al. 2008). In the case of tumor cells, cell–matrix adhesion molecules are absent in cells and matrix junctions and stress fibers are also reduced and re-localized. *At the time of cell motility, cells generate contractile forces through cellular protrusions that influence the bundling of collagen fibers.* These collagen bundles increase the available surface area, which helps to form larger adhesion (Smith et al. 2007). Collagen fibers help to form small-sized dynamic integrin clusters with less than a few seconds of life. Due to the absence of stress fibers and cell–matrix adhesion molecules in the junction of tumor cells and matrix, tumor cells and epithelial or endothelial cells cannot form wide lamellipodium and associated filopodial protrusions at the cell boundary. It is visible in 3D traction microscopy; cells, present in the matrix, only pull the local fibers, but it never pushes the local matrix (Bloom et al. 2008; Legant et al. 2010). These pseudopodial bulges pull the leading and trailing cell edges. Considerable matrix traction only happens in the neighborhood of pseudopodial protrusions (each cell contains one to five protrusions at a time) (Bloom et al. 2008). Though pseudopodia released from collagen fibers are asymmetric and develop a defect in the matrix during cancer cell migration, the pseudopodial protrusions at the trailing edge of cancer cell are released initially and then the back of the cell in the forward direction is dragged (biased motion). The cellular traction force on collagen may activate the MMPs (Ellsmere et al. 1999). Pseudopodial protrusion movement in the matrices is regulated by cell–matrix adhesion components like, scaffolding protein p130CAS and mechanosensing protein zyxin. p130CAS mediates a high number and high growth rate of protrusions. Scientists (Fraley et al. 2010) have shown that a strong correlation exists between the numbers of protrusion per unit time and the growth rate of protrusions and tumor cell motility in the matrices.

Experiments have been performed on mouse mammary tumors indicating that a few tumor cells diffuse from the primary tumor sites and go through a highly directed migration away from the tumor guided by collagen fibers (Sahai 2005). *Intravital microscopy of GFP-labeled breast cancer cells in mice supports the fact that cancerous cells circulate as individual cells toward blood capillaries, whereas they migrate collectively in the direction of lymphatic vessels* (Giampieri et al. 2009). Such type of collective migration requires the dominance of actomyosin contractility at intercellular adhesions (Hidalgo-Carcedo et al. 2011).

5.4.2 Cellular Mechanics in Intravasation

Tumor cells undergo shape changes during entry into and exit from the vascular system. This is driven by cytoskeletal remodeling for penetrating endothelial cell–cell junctions. Cytoplasm is a complex component of any cell; it is a rubber-like liquid material (viscous), and its distortion rate is very high (Wirtz 2009). The elasticity property of cytoplasm measures the ability to rebound from an applied force, whereas viscosity reflects its ability to go through under external shear of a flow. The nuclear lamina underlying the nuclear envelope, chromatin structure, and linkers of the nucleus and cytoskeleton (LINC) together determine the elastic property of the nucleus. *But, the role of LINC and nuclear lamina in cell migration is still not properly understood.*

Cancerous cells are much softer than normal cells and have strong correlation with metastatic potential (Cross et al. 2007). Also, the softer cytoplasm of cancerous cells correlates with a less-structured cytoskeleton. Physical properties of the microenvironment are very crucial for cancer cell migration as they control the cell movement (Lee et al. 2006). Some optimal mechanical properties like rigidity or softness are the key during cell migration through matrix or incursion into endothelium. If the surface is too rigid or too soft, cells are able to distort the cross-linked collagen fibers to migrate proficiently. However, individual cells of a specific cell type are heterogeneous and have shown a variation in mechanical properties, which illustrates that *invasion and intravasation* (Fig. 5.6) *capable cells are likely to preserve this phenotype over several generations.* The attributes of cancer cells alter dynamically during the metastasis to survive in the new environment of the vasculature system and the stromal space. These alterations in mechanical attributes might also be controlled by biochemical gradients (Tseng et al. 2004), interstitial flows (Swartz and Fleury 2007), and endogenous electric fields (Mycielska and Djamgoz 2004).

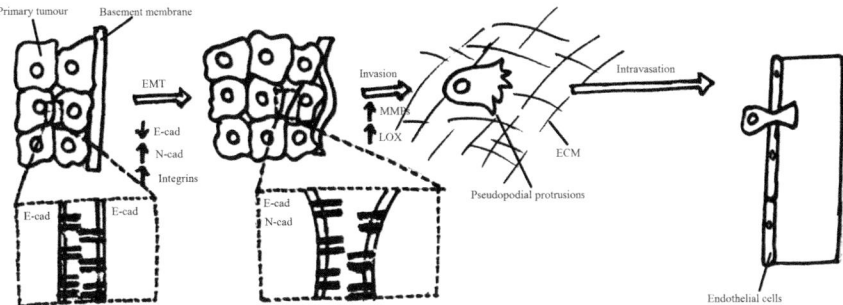

Fig. 5.6 Invasion and intravasation mechanism (Wirtz et al. 2011)

5.4.3 Forces Acting on the Circulating Tumor Cells

During migration through the vascular system, cancerous cells are affected by hemodynamic forces, immunological stress, and also collisions with the blood cells and the endothelial cells of the vessel wall. The survival of these circulating tumor cells and establishment of metastatic foci are largely controlled by these stresses. The circulating cancer cells, which succeed in dealing with the fluid shear and immune observation of the body, will be stuck to the vascular endothelium (Fig. 5.7) of a distant site (organ) and exit from the vascular systems successfully to go into the tissue.

When the tumor cells enter into the vascular system, it is regulated by a number of physical and mechanical attributes: *the blood flow pattern, blood vessels' diameter, and complicated interaction between shear flow and intercellular adhesion.* The rotational and translational motion of circulating tumor cells is influenced by the shear flow, and therefore, these two motions determine the orientation with the receptor–ligand interactions, which lead to adhesion. Also, shear flow influences distortion of circulating tumor cells on the way to the vessel walls. Though the amount of these effects and motivation on occlusion and adhesion remain to be determined; *there is very less information regarding the properties of shear flow on the capability and proliferation of circulating tumor cells.*

5.4.4 Extravasation of Circulating Tumor Cells

Whenever circulating cancerous cells enter into the blood vessels with smaller diameter than the cancerous cells, then an arrest can occur by physical occlusion. In general, circulating cancer cells of epithelial origin are more than 10 μm in size. So, physical occlusion occurs in the vessels with less than 10 μm in size.

At the time of extravasation, the tumor cell circulating through a larger blood vessel needs the adhesion to stick to the vessel wall. *The probability of arrest of a cell at a large vessel is proportionate to the product of multiplication of the collision rate between membrane-bound receptors and endothelial ligands, and residence*

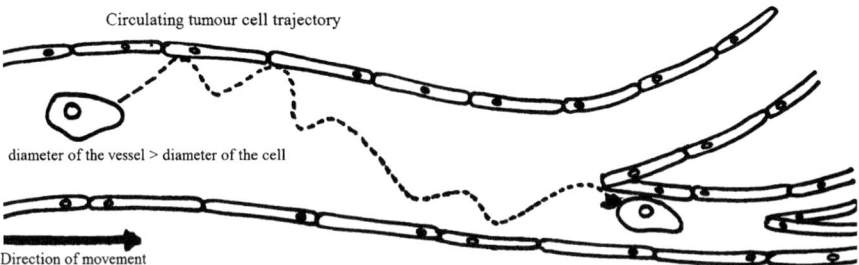

Fig. 5.7 Circulating tumor cell getting arrested under the influence of local flow pattern and collisions with wall (Wirtz et al. 2011)

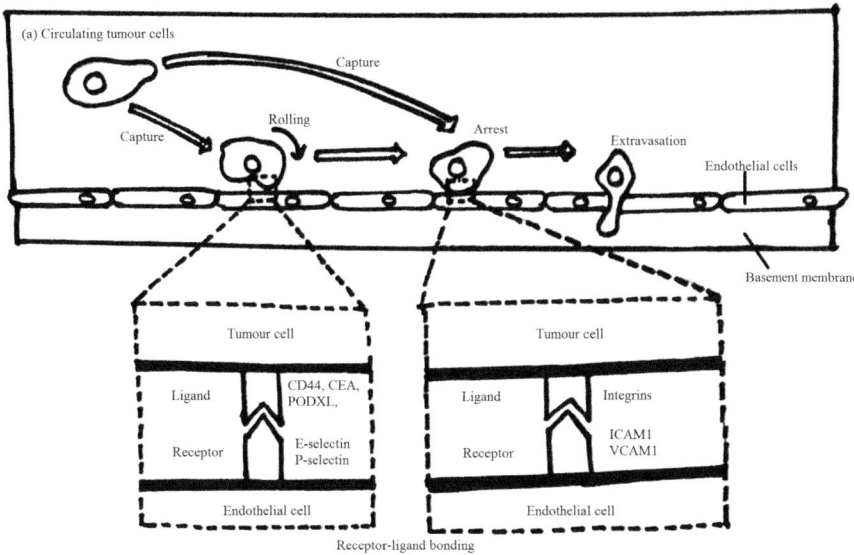

Fig. 5.8 Capture and arrest of circulating tumor cells through transient and/or persistent adhesion (Wirtz et al. 2011)

time of the cell (Zhu et al. 2008). *The residence time of a cell is dependent on the ligand–receptor adhesive forces between the vessel wall and the circulating cell and the shear fluid force applied to the cell. If the shear force increases, then the collision frequency with the endothelium increases, but the residence time is decreased. The total adhesive force depends on the strength of the ligand–receptor bond and the number of ligand–receptor bonds involved.*

When a cell is moving along the vessel wall, it has two velocities: translational velocity and angular velocity. The translational velocity is always greater than the angular velocity. As a result, a sliding movement can be seen in contrast to the motionless vessel wall. This sliding movement raises the chances of meeting a single receptor on a circulating tumor cell and ligands on the vessel wall. In the case of rotational motion, it brings consecutive receptors on the circulating tumor cell surface into interaction with ligands on the vessel wall (Fig. 5.8).

The chances of cell arrest are much higher if the shear force is at the intermediate level. A bond at a particular shear stress level and cellular adhesion pattern will be formed or not is mainly regulated by the kinetic and other micromechanical properties of receptor–ligand. As an example, to form receptor-mediated cell adhesion in the presence of shear force, a comparatively fast *on* rate and relatively slow *off* rate are seen. Fast *on* rate helps receptor–ligand binding within a short span of interaction time, whereas slow *off* rates allow sustainable bond lifetime. Receptor–ligand bonds (such as selectins and their ligands) show high micromechanical strength; fast *on* rates and comparatively fast *off* rates influence binding in the presence of shear force. Molecules with slower *on* rates (such as integrin) can be involved at a very low shear force only after selectin-mediated cell binding or in the absence of selectin-dependent interactions. Integrins are engaged in the circulation of tumor cells and also regulate angiogenesis and metastatic growth.

5.4.4.1 Receptor–Ligand Interactions of Circulating Tumor Cells

Circulating tumor cells sometimes escape immune surveillance and exit from the circulatory system (Fig. 5.9). Activated platelets release a number of chemical substrates to promote angiogenesis and metastatic growth (Pinedo et al. 1998). To promote vascular hyperpermeability and extravasation, platelets secrete a number of chemical factors like VEGF at the points of attachments to the endothelium (Nash et al. 2002). Circulating tumor cells often capture the endothelium of the secondary organ with the help of polymorphonuclear leukocytes (PMNs). Close association with PMNs and metastatic tumor cells during tumor cell arrest and extravasation has been observed (Crissman et al. 1985). By directly binding to the vascular endothelium with the help of selectin-mediated tethering, circulating tumor cells behave like neutrophils. This binding can happen using cell rolling followed by strong cell adhesion. In fact, P-, L-, and E-selectins help cancer metastasis through arresting the circulating tumor cells in the microvasculature.

5.4.4.2 Physics behind the Location of Metastatic Site

The patterns of metastasis can be explained with two different hypotheses: *seed and soil hypotheses* (discussed in Sect. 5.3.2), and *mechanical hypotheses* and both of them have complementary roles in influencing the location of metastasis. According to *seed and soil hypothesis*, the chances of metastasis occurring at a particular site are proportional to the multiplication of three components: (i) *probability of adhesion due to the collision at a particular site*, (ii) *probability of extravasation at that site*, and (iii) *probability of colonization*. On the other hand, according to the *mechanical hypothesis*, cell catching, its extravasation, and colonization occur in a cascade; so, the chances of metastasis occurring at a particular site is proportional to the multiplication of three components: (i) *probability of encountering a vessel with diameter less than the cell diameter*, (ii) *probability of extravasation at that site*, and (iii) *probability of colonization*. Chances of metastasis occurrence in a specific organ

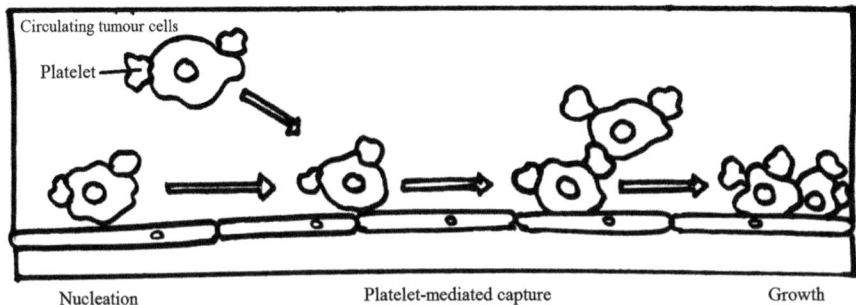

Fig. 5.9 Circulating tumor cell arrest through platelet-mediated capture (Wirtz et al. 2011)

have common elements related to extravasation and colonization, and both of them are dependent on local microenvironment.

According to *mechanical hypothesis*, the probability of occurrence of metastasis at a site is dependent on the blood flow pattern (Weiss 2000). Blood is pumped from the heart to different organs through the arterial system and is subsequently returned to the heart. If a tumor cell is found in a capillary with smaller diameter, then the chances of the cell catching through physical occlusion are higher for that site.

For metastasis to occur, a tumor cell must extravasate and colonize some neighboring tissue. At the time of circulation, tumor cells often collide with the vessel wall and have potential to stuck in that location due to adhesion. If the cell is stuck in that location for sufficiently long time, then it might extravasate. The chances that the residing time is sufficiently long are dependent on shear stress. Receptor–ligand adhesion varies depending on organs (Trepel et al. 2002).

Metastatic tumor cells need an optimal shear stress to stuck sufficiently long time in a particular position. If the stresses are high (12 dyn per cm^2) like in arterial circulation, then the immune system destroys the tumor cells due to cell cycle arrest (Chang et al. 2008). On the other hand, if the stress is at a lower level like in venous system, it can create an opposite effect on intercellular signaling and tumor cell function.

5.5 Conclusions

Cancer cells can move to other locations through several ways either singly or collectively. They breach the basement membrane of the host tissue using metalloproteinase enzymes and also the nearby blood vessels using proteolytic enzymes. After breaching these membranes, they are free to spread to other organs in the body. In this journey, tumor cells undergo some transformation to adapt to the new environment. During motility, cell modifies their cell–cell or cell–ECM junctions. The primary reason of the cell movement is dynamic reformation of actin cytoskeleton (using RhoGTPase family), which provides the required force during cell motility. During actin reformation, cell membrane forms finger-like filopodia or lamellipodia protrusions. These protrusions help the cell to move to a different location. Also, RhoA regulates actomyosin filaments, which promote locomotive forces in the cell body. While Rac1 and Cdc42 motivate the growth of cellular protrusions at the leading edge, RhoA makes retraction at the trailing edge. These cooperative efforts make a cell to move.

The migrating tumor cells reach other sites of a body through the blood/lymphatic vessels. When a tumor cell is in the vascular system, its movement is regulated by different factors like blood flow pattern, diameter of blood vessels, shear flow of fluids, and intercellular adhesion. The properties of cancer cells (in metastasized condition) dynamically change to survive in the new environment of the vasculature system and are affected by hemodynamic forces, immunological stress, and collisions with the blood cells and the endothelial cells of vessel wall. The rotational and translational motions of circulating tumor cells are influenced by the shear flow;

these two motions determine the orientation with respect to the receptor–ligand interactions and lead to adhesion. A sliding movement of cancer cells can be seen in contrast to the motionless vessel wall that raises the chances of meeting between a single receptor on a circulating tumor cell and the ligands on the vessel's wall.

Circulating tumor cell in a larger blood vessel needs the requisite adhesion to stick to the vessel wall. The chance of cell arrest at a large vessel depends on the collision rate between the membrane-bound receptors and the endothelial ligands, and the residence time of the cell. With increasing shear force, the collision frequency with the endothelium increases and the residence time decreases. The total adhesive force depends on the strength of the ligand–receptor bond and the number of ligand–receptor bonds involved. Tumor cells can exit from the vascular system, but it first binds to the vessel's wall through physical occlusion and/or cellular adhesion. The cells that succeed to survive with the fluid shear and immune observation of the body will be stuck to the microvascular endothelium of a secondary organ and exit from the vascular system successfully (extravasation) and enter into another tissue by invading the vessel's wall and the tissue membrane. After reaching the secondary site, it penetrates the secondary organ and continues its growth leading to a new colony of cancerous cells. This cycle continues till the host dies or the cancer cells are completely removed from the body.

Acknowledgement We are thankful to Debarpita Santra, PhD scholar, Department of Computer Science and Engineering, University of Kalyani, West Bengal, India; Soumyabrata Roy, postdoctoral research associate at School of Medicine, University of California, Irvine, USA; and Subhasis Barik, Senior Scientific Officer-II, Chittaranjan National Cancer Institute, Kolkata, West Bengal, India, for their cooperation and help.

Appendix

Biological Notes

Anaerobic glycolysis:	It transforms glucose into lactate in the presence of limited amount of oxygen
Anaplasia:	Cells with the least cellular differentiation lose the morphological properties of mature cells and destroy the orientation with the other cells and endothelial cells
Autocrine signals:	Cells secrete hormones that bind to its corresponding receptors on the same cell
Cadherins:	Cadherins are calcium-based transmembrane, cell–cell adhesive glycoprotein
Cathepsin B:	It is normally found in lysosomes and activated by cathepsin D-like enzymes or metalloprotease and can also activate some of the collagenases and uPA
Cathepsin D:	It is a lysosomal protease. Cathepsin D behaves like a mitogen for estrogen-deprived breast cancer cell lines and accelerates the damage of the ECM

(continued)

CD44:	It acts as receptors for hyaluronate and lymphocyte homing receptors and is coded by a single gene located on chromosome 11 in human cells
Cell differentiation:	In this process, a less specialized cell transforms into a more specialized type of cell. It changes a size, shape, metabolic activity, and responsiveness to signals of a cell
Complementary DNA:	It is DNA synthesized from a single-stranded RNA template in a reaction catalyzed by the enzyme reverse transcriptase and is often used to clone eukaryotic genes in prokaryotes
Dysplasia:	It is abnormal growth of a tissue
Endocytosis:	Transporting molecules into a cell by engulfing it is known as endocytosis; when molecules cannot pass through the membrane, cells use endocytosis
Extracellular matrix:	It is a group of extracellular molecules providing structural and biochemical support to the adjacent cells (tissue)
Extravasation:	It is the process during metastasis in which cells are come out from blood vessels and develop a colony into secondary tissue
Filopodia:	It is narrow cytoplasmic projections beyond the lamellipodia of migrating cells
Glycoprotein podoplanin:	It upregulates in the invasive front of several human cancers. It has been associated with EMT and increased cell migration and tissue invasion
Hemodynamic forces:	It is the dynamics of blood flow within circulatory systems
Homeostasis:	It is the activity of an organism to autoregulate and maintain its internal environment in a stable state
Immunoglobulin superfamily:	Proteins belonging to this family consist of an immunoglobulin domain that is made of 90–100 amino acid molecules decorated within a sandwich of two antiparallel strands
Immunological stress:	It is the status of human immune system when it is affected by bacteria, virus, and endocrine.
Integrins:	Integrins are a set of heterodimeric glycoproteins
Intracrine signals:	Signaling molecules binding to intracellular receptors
Intravasation:	It refers to cancer cells entering through the basal membrane into blood or lymphatic vessels
Lamellipodia:	It refers to large cytoplasmic projections found primarily at the leading edge of migrating cells
Matrix metalloproteinases:	It is a group of enzymes responsible for damaging most of the extracellular matrix proteins during organogenesis, growth, and normal tissue turnover
Metalloproteases:	The metalloproteases can be categorized as interstitial collagenases, type IV collagenases, and the stromelysins. Interstitial collagenases accelerate the damage of type I, II, and III collagens. It helps to accelerate the degradation of different elements of ECM
Metaplasia:	It is a change in the nature of a tissue that is not normal for that tissue
p120-catenin:	It has multiple roles like controlling cadherin stability, adhesion-induced signaling, and cancer progression
p130CAS:	It regulates varieties of signal pathways related to cell adhesion, migration, and invasion
p190-RhoGAP:	It controls cell spreading and migration

(continued)

p53:	It is a gene that codes for a protein that controls cell cycle and hence functions as tumor suppressor
Plasminogen activators:	It acts as catalyst during the conversion of inactive plasminogen to the active plasmin. It can be available in two forms: Tissue-type plasminogen activators and urokinase-type plasminogen activators into different genes
Polymorphonuclear leukocytes:	It is a kind of immune cells having small particles with enzymes released during infections, allergic reactions, and asthma
RGD sequence:	The tripeptide Arg-Gly-asp (RGD) is the amino acid sequence within the ECM protein fibronectin-mediating cell attachment
Selectins:	The proteins belong to selectin group contain lectin-type domain. These types of proteins regulate heterophilic communications between endothelium and blood cells
Vascular hyperpermeability:	It defines the capacity of a blood vessel wall to permit the flow of small molecules or even cells in and out of the vessel

References

Allen JW, Khetani SR, Johnson RS, Bhatia SN (2006) In vitro liver tissue model established from transgenic mice: role of HIF-1alpha on hypoxic gene expression. Tissue Eng 12(11):3135–3147

Ang KC, Tan LS (2008) A numerical approach to modelling avascular tumour evolution with white noise. ANZIAM J 50:C569–C582

Barnes DM, Dublin EA, Fisher CJ, Levison DA, Millis RR (1993) Immunohistochemical detection of p53 protein in mammary carcinoma: an important new independent indicator of prognosis? Hum Pathol 24(5):469–476

Berenson J, Rajdev L, Broder M (2006) Pathophysiology of bone metastases. Cancer Biol Ther 5(9):1078–1081

Bloom RJ, George JP, Celedon A, Sun SX, Wirtz D (2008) Mapping local matrix remodeling induced by a migrating tumor cell using three-dimensional multiple-particle tracking. Biophys J 95(8):4077–4088

Castro P, Soares P, Gusmao L, Seruca R, Sobrinho-Simoes M (2006) H-RAS 81 polymorphism is significantly associated with aneuploidy in follicular tumors of the thyroid. Oncogene 25(33): 4620–4627

Chang SF, Chang CA, Lee DY, Lee PL, Yeh YM, Yeh CR, Cheng CK, Chien S, Chiu JJ (2008) Tumor cell cycle arrest induced by shear stress: roles of integrins and Smad. Proc Natl Acad Sci U S A 105(10):3927–3932

Clyde RG, Bown JL, Hupp TR, Zhelev N, Crawford JW (2006) The role of modelling in identifying drug targets for diseases of the cell cycle. J R Soc Interface 3(10):617–627

Crissman JD, Hatfield J, Schaldenbrand M, Sloane BF, Honn KV (1985) Arrest and extravasation of B16 amelanotic melanoma in murine lungs. A light and electron microscopic study. Lab Invest 53(4):470–478

Croce CM (2008) Oncogenes and cancer. N Engl J Med 358(5):502–511

Cross SE, Jin YS, Rao J, Gimzewski JK (2007) Nanomechanical analysis of cells from cancer patients. Nat Nanotechnol 2(12):780–783

Duffy MJ (1992) The role of proteolytic enzymes in cancer invasion and metastasis. Clin Exp Metastasis 10(3):145–155

Ellsmere JC, Khanna RA, Lee JM (1999) Mechanical loading of bovine pericardium accelerates enzymatic degradation. Biomaterials 20(12):1143–1150

Fidler IJ (1970) Metastasis: quantitative analysis of distribution and fate of tumor emboli labeled with 125I-5-iodo-2′-deoxyuridine. J Natl Cancer Inst 45(4):773–782

Fidler IJ (1990) Critical factors in the biology of human cancer metastasis: twenty-eighth GHA Clowes memorial award lecture. Cancer Res 50(19):6130–6138

Fraley SI, Feng Y, Krishnamurthy R, Kim DH, Celedon A, Longmore GD, Wirtz D (2010) A distinctive role for focal adhesion proteins in three-dimensional cell motility. Nat Cell Biol 12(6):598–604

Friedl P (2004) Prespecification and plasticity: shifting mechanisms of cell migration. Curr Opin Cell Biol 16(1):14–23

Friedl P, Wolf K (2003) Tumour-cell invasion and migration: diversity and escape mechanisms. Nat Rev Cancer 3(5):362–374

Friedl P, Wolf K (2010) Plasticity of cell migration: a multiscale tuning model. J Cell Biol 188(1):11–19

Frisch SM, Ruoslahti E (1997) Integrins and anoikis. Curr Opin Cell Biol 9(5):701–706

Giampieri S, Manning C, Hooper S, Jones L, Hill CS, Sahai E (2009) Localized and reversible TGFβ signaling switches breast cancer cells from cohesive to single cell motility. Nat Cell Biol 11(11):1287–1296

Günthert U, Hofmann M, Rudy W, Reber S, Zöller M, Haußmann I, Matzku S, Wenzel A, Ponta H, Herrlich P (1991) A new variant of glycoprotein CD44 confers metastatic potential to rat carcinoma cells. Cell 65(1):13–24

Gyetko MR, Todd RF, Wilkinson CC, Sitrin RG (1994) The urokinase receptor is required for human monocyte chemotaxis in vitro. J Clin Invest 93(4):1380–1387

Hanahan D, Weinberg RA (2000) The hallmarks of cancer. Cell 100(1):57–70

Hansen RK, Bissell MJ (2000) Tissue architecture and breast cancer: the role of extracellular matrix and steroid hormones. Endocr Relat Cancer 7(2):95–113

Herranz N, Pasini D, Díaz VM, Francí C, Gutierrez A, Dave N, Escrivà M, Hernandez-Muñoz I, Di Croce L, Helin K, De Herreros AG (2008) Polycomb complex 2 is required for E-cadherin repression by the Snail1 transcription factor. Mol Cell Biol 28(15):4772–4781

Hidalgo-Carcedo C, Hooper S, Chaudhry SI, Williamson P, Harrington K, Leitinger B, Sahai E (2011) Collective cell migration requires suppression of actomyosin at cell–cell contacts mediated by DDR1 and the cell polarity regulators Par3 and Par6. Nat Cell Biol 13(1):49–59

Horowitz JM, Yandell DW, Park SH, Canning S, Whyte P, Buchkovich K, Harlow ED, Weinberg RA, Dryja TP (1989) Point mutational inactivation of the retinoblastoma antioncogene. Science 243(4893):937–940

Hotary KB, Allen ED, Brooks PC, Datta NS, Long MW, Weiss SJ (2003) Membrane type I matrix metalloproteinase usurps tumor growth control imposed by the three-dimensional extracellular matrix. Cell 114(1):33–45

Hou Z, Peng H, Ayyanathan K, Yan KP, Langer EM, Longmore GD, Rauscher FJ (2008) The LIM protein AJUBA recruits protein arginine methyltransferase 5 to mediate SNAIL-dependent transcriptional repression. Mol Cell Biol 28(10):3198–3207

Kaplan RN, Rafii S, Lyden D (2006) Preparing the "soil": the premetastatic niche. Cancer Res 66(23):11089–11093

Kashkin KN, Chernov IP, Stukacheva EA, Kopantzev EP, Monastyrskaya GS, Uspenskaya NY, Sverdlov ED (2013) Cancer specificity of promoters of the genes involved in cell proliferation control. Acta Naturae 5(3):79–83

Kleer CG, Teknos TN, Islam M, Marcus B, Lee JSJ, Pan Q, Merajver SD (2006) RhoC GTPase expression as a potential marker of lymph node metastasis in squamous cell carcinomas of the head and neck. Clin Cancer Res 12(15):4485–4490

Lee JS, Panorchan P, Hale CM, Khatau SB, Kole TP, Tseng Y, Wirtz D (2006) Ballistic intracellular nanorheology reveals ROCK-hard cytoplasmic stiffening response to fluid flow. J Cell Sci 119(9):1760–1768

Legant WR, Miller JS, Blakely BL, Cohen DM, Genin GM, Chen CS (2010) Measurement of mechanical tractions exerted by cells in three-dimensional matrices. Nat Methods 7(12):969

Lehembre F, Yilmaz M, Wicki A, Schomber T, Strittmatter K, Ziegler D, Kren A, Went P, Derksen PW, Berns A, Jonkers J (2008) NCAM-induced focal adhesion assembly: a functional switch upon loss of E-cadherin. EMBO J 27(19):2603–2615

Liotta LA (1992) Cancer cell invasion and metastasis. Sci Am 266(2):54–63

Lozano E, Betson M, Braga VM (2003) Tumor progression: small GTPases and loss of cell–cell adhesion. BioEssays 25(5):452–463

Malumbres M, Barbacid M (2003) RAS oncogenes: the first 30 years. Nat Rev Cancer 3(6): 459–465

Montell DJ (2008) Morphogenetic cell movements: diversity from modular mechanical properties. Science 322(5907):1502–1505

Muller PA, Caswell PT, Doyle B, Iwanicki MP, Tan EH, Karim S, Lukashchuk N, Gillespie DA, Ludwig RL, Gosselin P, Cromer A (2009) Mutant p53 drives invasion by promoting integrin recycling. Cell 139(7):1327–1341

Mycielska ME, Djamgoz MB (2004) Cellular mechanisms of direct-current electric field effects: galvanotaxis and metastatic disease. J Cell Sci 117(9):1631–1639

Nash GF, Turner LF, Scully MF, Kakkar AK (2002) Platelets and cancer. Lancet Oncol 3(7): 425–430

Nekarda H, Siewert J, Schmitt M, Ulm K (1994) Tumour-associated proteolytic factors uPA and PAI-1 and survival in totally resected gastric cancer. Lancet (London, England) 343(8889): 117–117

Orme ME, Chaplain MAJ (1996) A mathematical model of the first steps of tumour-related angiogenesis: capillary sprout formation and secondary branching. Math Med Biol 13(2):73–98

Paget S (1989) Distribution of secondary growths in cancer of the breast. Lancet I 571

Paweletz N, Knierim M (1989) Tumor-related angiogenesis. Crit Rev Oncol Hematol 9(3):197–242

Pinedo HM, Verheul HMW, D'amato RJ, Folkman J (1998) Involvement of platelets in tumour angiogenesis? Lancet 352(9142):1775–1777

Pouysségur J, Dayan F, Mazure NM (2006) Hypoxia signaling in cancer and approaches to enforce tumour regression. Nature 441(7092):437–443

Quon KC, Berns A (2001) Haplo-insufficiency? Let me count the ways. Genes Dev 15(22): 2917–2921

Ruoslahti E (1992) Control of cell motility and tumour invasion by extracellular matrix interactions. Br J Cancer 66(2):239–242

Sahai E (2005) Mechanisms of cancer cell invasion. Curr Opin Genet Dev 15(1):87–96

Seidensticker MJ, Behrens J (2000) Biochemical interactions in the wnt pathway. Bioch Biophys Acta Mol Cell Res 1495(2):168–182

Selam B, Kayisli UA, Garcia-Velasco JA, Arici A (2002) Extracellular matrix-dependent regulation of Fas ligand expression in human endometrial stromal cells. Biol Reprod 66(1):1–5

Semenza GL (2001) HIF-1, O2, and the 3 PHDs: how animal cells signal hypoxia to the nucleus. Cell 107(1):1–3

Sherratt JA, Chaplain MA (2001) A new mathematical model for avascular tumour growth. J Math Biol 43(4):291–312

Small JV, Stradal T, Vignal E, Rottner K (2002) The lamellipodium: where motility begins. Trends Cell Biol 12(3):112–120

Smith ML, Gourdon D, Little WC, Kubow KE, Eguiluz RA, Luna-Morris S, Vogel V (2007) Force-induced unfolding of fibronectin in the extracellular matrix of living cells. PLoS Biol 5(10):e268

Sutherland RM (1988) Cell and environment interactions in tumor microregions: the multicell spheroid model. Science 240(4849):177–184

Swartz MA, Fleury ME (2007) Interstitial flow and its effects in soft tissues. Annu Rev Biomed Eng 9:229–256

Trepel M, Arap W, Pasqualini R (2002) In vivo phage display and vascular heterogeneity: implications for targeted medicine. Curr Opin Chem Biol 6(3):399–404

Tseng Y, Lee JS, Kole TP, Jiang I, Wirtz D (2004) Micro-organization and visco-elasticity of the interphase nucleus revealed by particle nanotracking. J Cell Sci 117(10):2159–2167

Wang K, Grivennikov SI, Karin M (2013) Implications of anti-cytokine therapy in colorectal cancer and autoimmune diseases. Ann Rheum Dis 72(Suppl 2):ii100–ii103

Weaver AM (2006) Invadopodia: specialized cell structures for cancer invasion. Clin Exp Metastasis 23(2):97–105

Wei C, Larsen M, Hoffman MP, Yamada KM (2007) Self-organization and branching morphogenesis of primary salivary epithelial cells. Tissue Eng 13(4):721–735

Weiss L (2000) Patterns of metastasis. Cancer Metastasis Rev 19(3–4):281–301

Wirtz D (2009) Particle-tracking microrheology of living cells: principles and applications. Annu Rev Biophys 38:301–326

Wirtz D, Konstantopoulos K, Searson PC (2011) The physics of cancer: the role of physical interactions and mechanical forces in metastasis. Nat Rev Cancer 11(7):512–522

Yamazaki D, Kurisu S, Takenawa T (2005) Regulation of cancer cell motility through actin reorganization. Cancer Sci 96(7):379–386

Zhu C, Yago T, Lou J, Zarnitsyna VI, McEver RP (2008) Mechanisms for flow-enhanced cell adhesion. Ann Biomed Eng 36(4):604–621

Zipori D (2006) The mesenchyme in cancer therapy as a target tumor component, effector cell modality and cytokine expression vehicle. Cancer Metastasis Rev 25(3):459–467

Part II
Diagnostics and Theory

MRI, CT, and PETSCAN: Engineer's Perspective

Subhankar Ghosh and Saurabh Pal

Abstract

We first discuss historical developments in tomographic imaging used for medical diagnosis. To discuss magnetic resonance imaging, we introduce the fundamental ideas of magnetic moment, nuclear magnetic moment, and total magnetization. Then, we clarify the notions on classical precession and the principle of magnetic resonance signal detection and explain spin–lattice and spin–spin relaxation. Further, the relevant details of MRI machine and its components like magnetic gradient coils, signal generation, and pulse sequence–k-space imaging are discussed in brief form.

We then introduce computed tomography (CT) based on X-rays. Basic concept of X-ray, different generations of CT, EBCT, helical CT, etc., are introduced. Various clinical applications of CT scan image are discussed in brief. The last item discussed is positron emission tomography scan (PET scan), its basic principle, and utilities. The concluding remarks regarding these imaging techniques and comparison in medical diagnosis are made.

Keywords

MRI · CT · PET scan · X-rays

S. Ghosh (✉)
Department of Physics, St. Xavier's College, Kolkata, India
e-mail: subhankar@sxccal.edu

S. Pal
Department of Applied Physics, University of Calcutta, Kolkata, India

© The Author(s), under exclusive license to Springer Nature Singapore Pte Ltd. 2022
S. K. Basu et al. (eds.), *Cancer Diagnostics and Therapeutics*,
https://doi.org/10.1007/978-981-16-4752-9_6

6.1 Introduction

This is a composite chapter discussing the underlying principles of three very important machines: magnetic resonance imaging, computed tomography, and positron emission tomography. These are widely used in complex medical diagnostics, especially for the detection of malignancies in different organs of human body. No medical center treating cancer can afford not to use the images produced by these machines. The presentation in this chapter is as follows. We first give historical developments in tomographic imaging in medical diagnosis. We then give introduction to magnetic resonance imaging, the machine.

Computed tomography (CT) is a widely adopted imaging technology, making the technology a reliable tool for noninvasive detection of unwanted growth and its progression in living bodies. This forms the second part of this chapter. Discussion is restricted to basic principle of computed tomography, development of CT scanning technology, and its clinical application.

Positron emission tomography (PET) is a very effective imaging technique used to explore the functionality of the different tissues and organs in the human body. This type of scan requires that a small amount of radioactive material (tracer) be delivered into the patient's bloodstream through intravenous injection. Once the tracer has spread through the body, the patient is taken to the scanning machine to have the procedure done.

This chapter is organized as follows: Sect. 6.2 deals with tomographic imaging in medical diagnosis, Sects. 6.3 and 6.4 deal with MRI, Sects. 6.5 through 6.10 deal with CT, Sects. 6.11 and 6.12 deal with PETSCAN, and Sect. 6.13 concludes the chapter.

6.2 Tomographic Imaging in Medical Diagnosis: Historical Developments

Whenever light (an electromagnetic wave) falls on any object, it gets reflected or refracted with part absorption in general. Both the reflection and the refraction help in the formation of an image. The image is sometimes real (can be seen on a screen) or virtual at times. This image formation and detection by eyes are daily experience for all of us. But when a small object is to be seen, we like the magnified view. This magnification was achieved historically by using lenses, mirrors, and other optical instruments. The simple microscope was an important beginning in this regard and seen as of paramount importance in medical diagnosis. Subsequent improvement in the form of compound microscope helped in the detection of disease-causing agents and many physiological studies.

In the twentieth century, mankind was gifted with two important discoveries in physics: the invention of X-ray by German Physicist Wilhelm Conrad Röntgen and the discovery of the electron by British Physicist J. J. Thomson. X-ray could be used to directly image the human body parts like bones. On the other side, limitation of the magnification (≤ 1000 X) for the optical microscope was overcome by using

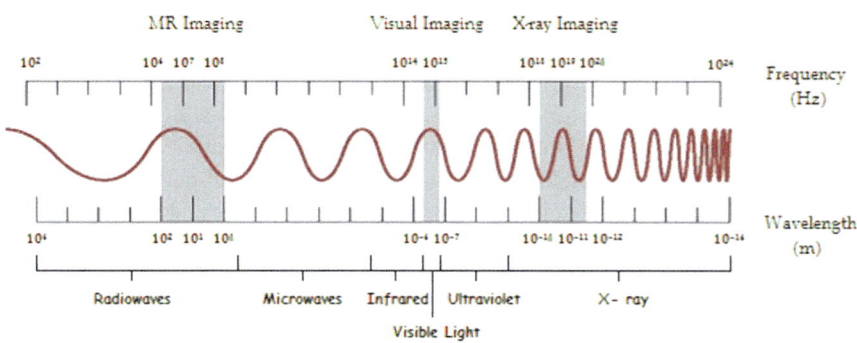

Fig. 6.1 Different ranges of the electromagnetic spectrum used for various imaging systems

electrons as beam in the revolutionary imaging technique known as electron microscope. Throughout the last few decades, the imaging techniques gradually matured in terms of precision, resolution, portability, and affordability in perfect harmony with the progress through inventions in the broad fields of biology, chemistry, and physics.

It is well known that the electromagnetic energy is a result of two mutually perpendicular fields, namely electric and magnetic fields that travel together in free space with the speed of light (~1,86,000 miles per second or 3,00,000 kilometers per second). As different amounts of electromagnetic energy are associated with the different ranges of the electromagnetic spectrum, most imaging techniques fall at different ranges of the electromagnetic spectrum depicted by Fig. 6.1.

As the history says, the X-ray, ultrasonography (USG), and nuclear medicine diagnostic tools supplement the sophisticated ones like magnetic resonance imaging (MRI), computed tomography (CT), positron emission tomography (PET SCAN). Some of the later techniques carry the common term as "Tomography," which is a combination of the Greek word *"tomos"* that means slice or section and English word *"graphy"* that means technique.

In this chapter, we discuss the basic principles behind the three important imaging techniques (MRI, CT, and PET scan) and how MRI, CT, and PETSCAN help in detecting cancer cells. Also, another point needs special mention here. We have restricted the basic principles to classical mechanics and physics although the fundamental ideas like the magnetic moment of proton and its behavior in an external magnetic field (MRI), the annihilation of electron–positron pair, and resulting photons (PET) are quantum phenomena and demand familiarity in the area of quantum mechanics. But there is no harm in projecting classical analogies for the above-mentioned imaging techniques.

6.3 Magnetic Resonance Imaging (MRI)

The principal components of an MRI process (Liang and Lauterbur 2000; Brown and Semelka 2003) are the following steps: It stimulates a signal from the object or the patient body parts using both the magnetic field and the radiofrequency pulse. It then reads the experimental data using magnetic gradients and puts it into k-space related to the frequency domain. Thereafter, the frequency domain data get transformed into spatial domain that helps in resulting in an observable image.

6.3.1 The Machine: A Gigantic Chamber

An MRI machine consists of primarily the following parts: the magnet, the radiofrequency coils, the gradient coils, and the imaging accessories as shown in Fig. 6.2.

The magnet is the heart of a MRI scanner controlling the image quality and the cost decision. Magnetic field is measured in terms of two popular units: tesla (T) as the SI unit and gauss (G) as the C.G.S unit (1 tesla $=$ 10 kilogauss). The earth's magnetic field is of the order of 0.6 T. There are normally three types of magnet that can be used here: permanent (up to 0.4 T), resistive (up to 0.6 T) using coils with current and superconducting (0.5–3 T—clinical limit) using superconducting coils with cryogens (to maintain very low temperature). Superconducting magnets are the common one with the five important aspects of field strength, homogeneity, shimming, sitting ease, and cryogen consumption.

The patient has to lie by being as still as possible on the table that enters through the chamber assembly. Now to appreciate the MRI process in totality, we digress a bit for firstly understanding the MRI signal followed by the concepts over the magnetization vector and pinpointing the signal using magnetization gradient and

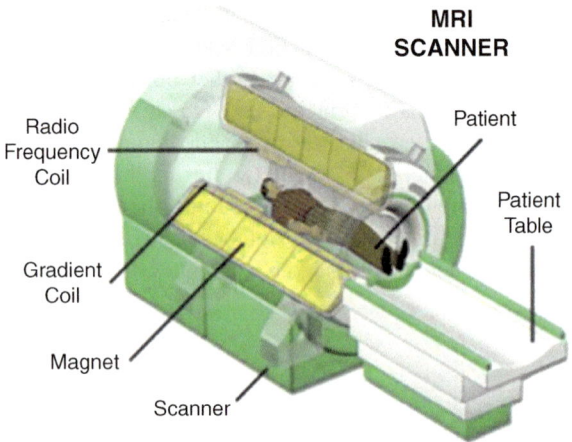

Fig. 6.2 MRI machine diagram

creating suitable signal, and finally the pulse sequence and collecting the emitted signal with the idea on *K*-space.

6.3.2 Magnetic Moment

It is now well known that the atom of any element is composed of fundamental particles like negatively charged light particle "electron," positively charged heavy particle "proton," and electrically neutral but heavy particle "neutron." A charged particle gives rise to an electrostatic or electric field. But if it moves with an acceleration or rotates with a frequency, the same particle generates a magnetic field. Thus, an electron possesses an angular momentum as it revolves around the central tiny nucleus (protons and neutrons) and a magnetic moment also.

A bar magnet (Fig. 6.3) is made up of two poles: north and south at the ends (strictly speaking, located inside at approximately 0.83 of its geometric length).

The magnetic moment (**M**) is the product of magnetic pole strength (*m*) and the length (*l*) or separation of the poles; that is, $M = m \times l$. The bar magnet experiences a *torque or turning moment* in an external magnetic field **B** and is expressed as $\tau = M \times B$.

The electron also spins about its own axis like a top (modern idea about spin based on quantum mechanics does not require this similarity: Spin is an intrinsic property of particles like charge and mass). Thus, the electron acquires a spin-based angular momentum and the corresponding magnetic moment also. They are related by a simple relation **μ** or $M = (e/2\,m)\,L = \gamma\,L$, where *e* is the electron charge, *m* is the mass of the electron, and **L** is the angular momentum. The factor of $\gamma = (e/2\,m)$ is known as *gyromagnetic ratio* (sometimes called as magnetogyric ratio) for the electron. The γ value for the electron is 1.76×10^8 MHz / T (megahertz per tesla).

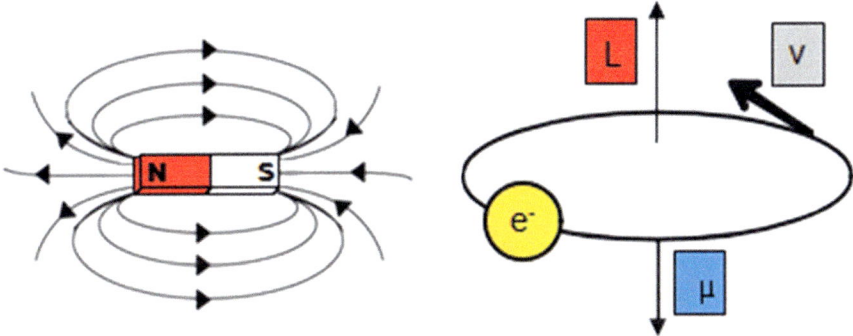

Fig. 6.3 Lines of force of a bar magnet and electron as a magnet

6.3.3 Nuclear Magnetic Moment

From the basics of atomic structure, we know that an atom of every element contains a tiny nucleus with positive charge at the core surrounded by the electrons, which revolve in different orbits. The nucleus consists of mainly two types of particles—positively charged protons and neutral neutrons. A proton is considered as a charged particle spinning around an internal axis of rotation with certain angular momentum and also a magnetic moment for which it can be treated as a very tiny bar magnet (see Fig. 6.4).

Unlike the electron, the magnetic field or magnetic moment due to the spin of the proton coincides with the sense and direction of the angular momentum on the spin axis. The proton magnetic moment is given by $\boldsymbol{\mu} = \gamma \boldsymbol{L}$. The gyromagnetic ratio (γ) has a value of 42.575 MHz / T. Thus, the sensitivity of detection in magnetic resonance is determined by the strength of the magnetic moment, which varies for different types of nucleus. Hydrogen is the first element in the modern periodic table, and its nucleus contains a single proton and, consequently, the highest magnetic moment. Further, it is found abundantly in organic tissues or matters. It is therefore the best choice as the nucleus used in the MRI process.

When there is no external magnetic field, as shown in Fig. 6.5, the magnitude of the magnetic moment of every proton (the hydrogen nucleus) is fixed in our body and their orientations are completely random. Therefore, the net magnetization or the vector sum of all the individual magnetic moments in human body is zero. However, when we apply a static magnetic field and electromagnetic radiation, the spin angular momentum of the hydrogen atoms interacts and produces a finite magnetization.

With the external constant magnetic field ($\boldsymbol{B_0}$) applied, the magnetic moment now can have only two possible discrete values in the direction of $\boldsymbol{B_0}$ following the principles of quantum mechanics. Therefore, it results in the magnetic moments being aligned at an angle of $\theta = 54.7°$ with respect to B_0, either in the same favorable direction or in the opposite direction (Fig. 6.5). The first configuration is known as the parallel and the latter as the antiparallel configuration for the magnetic moments.

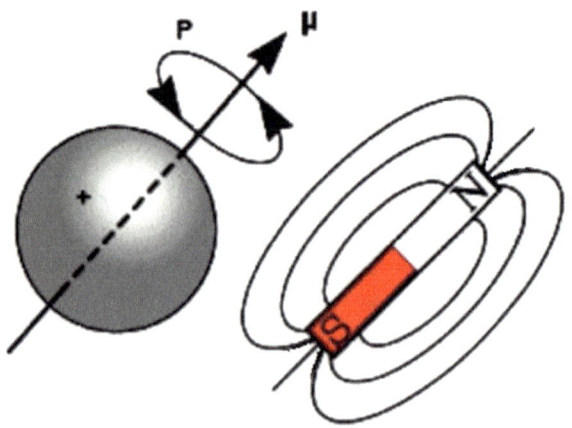

Fig. 6.4 Proton or nucleus behaves like a bar magnet

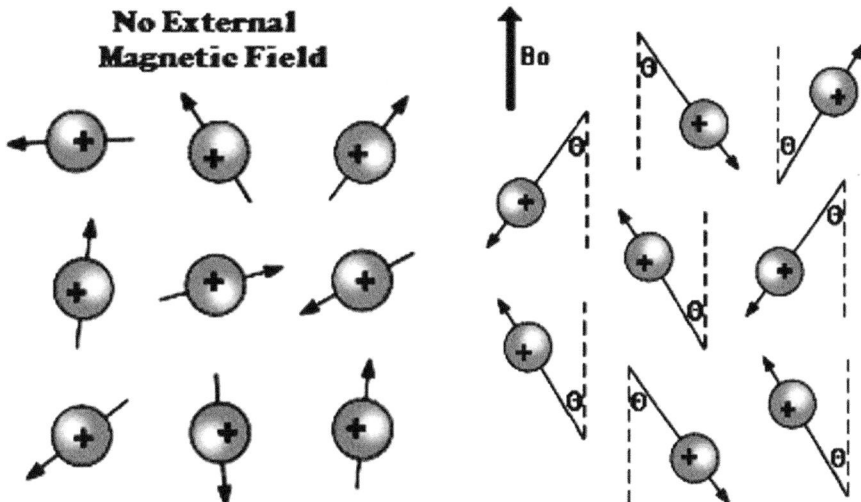

Fig. 6.5 Orientation of the proton magnetic moments

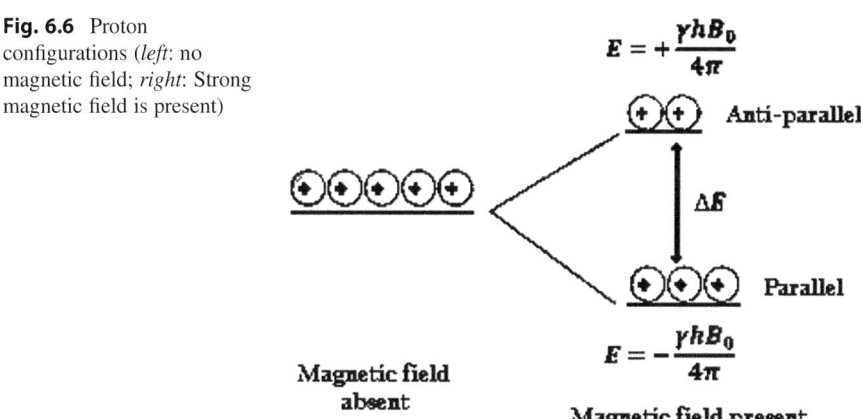

Fig. 6.6 Proton configurations (*left*: no magnetic field; *right*: Strong magnetic field is present)

It is found that the relative number of hydrogen protons in the parallel and antiparallel configurations depends upon the value of external constant magnetic field. Quantum mechanical calculations show that the protons in the parallel configuration have a lower energy state in comparison with those in the antiparallel state. Thus, in the absence of the external magnetic field, the protons have the same energy because of their random orientations. But with a strong magnetic field, the single energy level gets split into two energy levels depending on their parallel or antiparallel states. Here, the energy difference between the two energy states depends on the value of the external constant magnetic field. The energy difference (ΔE) between the above two energy states is shown in Fig. 6.6. It is important to note that only the difference between the number of parallel and antiparallel and not the total number of protons is the guiding factor in the principle of the MRI process.

At any finite temperature, the two orientations of parallel or antiparallel are almost equally populated. So, the thermal energy fluctuations minimize the energy difference (ΔE) and result in a net bulk magnetization vector M. Thereafter, with the electromagnetic radiation present, the protons can absorb or emit photons with an energy equal to ΔE and change its orientation.

6.3.4 Classical Precession

The motion of the proton magnetic moments can most easily be described using classical mechanics. The external magnetic field B_0 attempts to align the proton magnetic moment with itself, and this action creates a torque (C) given by the vector cross-product of the magnetic moment (P) and the magnetic field B_0 that means $C = P \times B_0$.

As shown in Fig. 6.7, the external constant magnetic field B_0 influences motion of the proton to result in precession around the axis of the magnetic field being inclined at a constant angle of $\theta = 54.7°$. Here, the torque that is defined as the rate of change in the angular momentum is responsible for the proton to precess around B_0 at a frequency ω_L known as Larmor frequency. This frequency termed is named after renowned Irish physicist Joseph Larmor. It is directly proportional to the strength of the magnetic field, i.e., $\omega_L = \gamma B_0$.

For the hydrogen atom, the value of γ is 42.575 MHz / T. For a 1.0 T MRI machine, the Larmor frequency for hydrogen is 42.575 MHz. With a smaller magnetic field of 0.5 T, the Larmor frequency for hydrogen is approximately equal to 21.29 MHz. But For a 1.5 T MRI scanner, it becomes equal to 42.575 MHz / T × 1.5 T = 63.86 MHz. In comparison, the 3.0 T MRI scanner has the Larmor

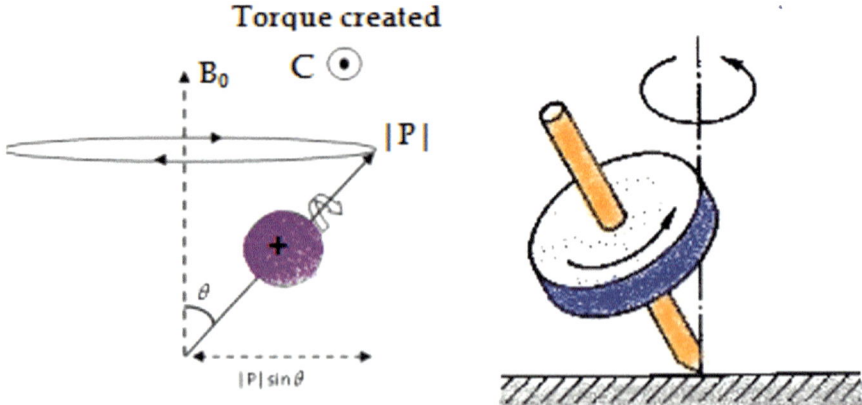

Fig. 6.7 (*Left*) A proton spinning about an internal axis is also precessing about the vertical axis (the direction of B_0); (*Right*) Classical precession of a spinning top

frequency of 127.72 MHz. So, it is clear that the magnetic field strength and the Larmor frequency are directly proportional to each other. One can observe that the magnetic resonance frequencies are close to the radiofrequency range (RF range) of the electromagnetic energy spectrum.

6.3.5 Total Magnetization

The net magnetization can be calculated by the superimposition of several proton magnetic moments in a simple vector form. The vector sum of the components in a Cartesian coordinate system has only a vertical or z-component as in Fig. 6.8, while the vector sum of the components on the x- and y-axes is zero. It is therefore clear that the net magnetization has only a z-component nonvanishing.

6.3.6 Radiofrequency Pulse: Related Magnetization

An NMR signal in a MRI scanner is normally detected through a condition known as resonance condition. To understand resonance in our day-to-day world, we can consider the case of mechanically vibrating systems. The resonance condition is basically a state of alternating absorption and dissipation of energy. The stretched wire on a guitar can be plucked at any point to set up transverse vibration, but the sound may not be loud enough to be heard. A hollow wooden box with a hole beneath the striking point of the guitar string provides a resonating system to boost the intensity of the sound. It can be analytically shown that if the frequency of the forcing vibration is the same as the natural frequency of the body, then resonance

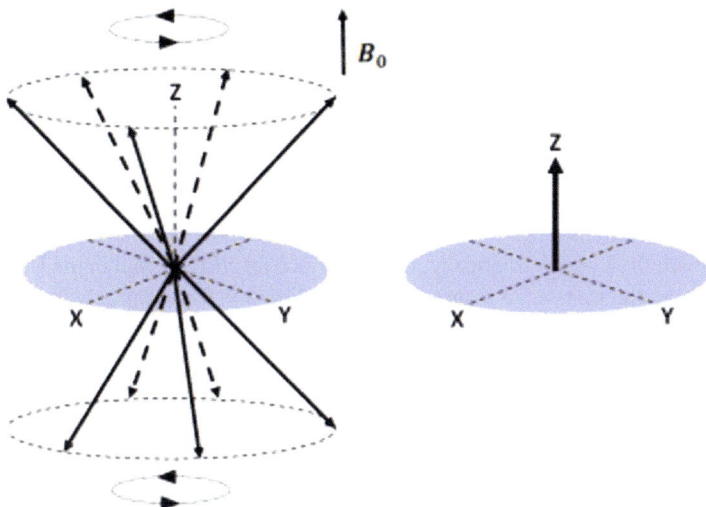

Fig. 6.8 (*Left*) Individual magnetization vectors and (*right*) their vector sum

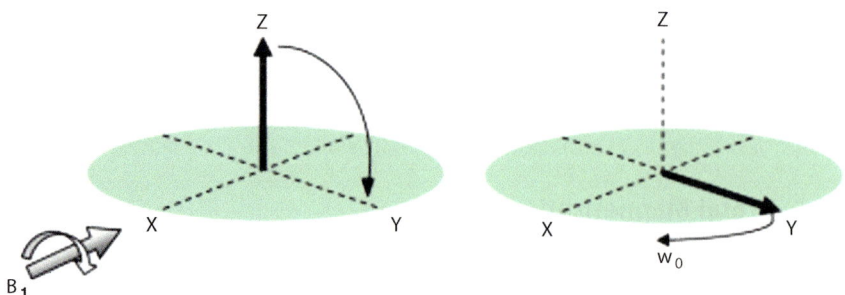

Fig. 6.9 (*Left*) RF pulse applied and (*right*) RF pulse is switched off

with large amplitude takes place. The same is the case of electrical resonance circuit. If the frequency of an alternating voltage/current source is the same as the natural frequency of a LCR (impedance, capacitor, resistor) circuit, current resonance can take place.

Here, in the MRI machine, the energy is supplied as a radiofrequency (RF) field with a certain frequency. When a RF pulse is applied, energy is absorbed and it gets dissipated through the relaxation processes. The energy levels for hydrogen atoms or protons in a magnetic field are similar to the multilevel energy values in a semiconductor. The generation of a MR signal requires an energy necessary to stimulate transitions between the energy levels. It can be seen that the frequency of the RF field is the same as the precession frequency or Larmor frequency. Let us now consider the application of a RF radiation with Larmor frequency to a nonmagnetic material in the presence of an already existing static magnetic field. The magnetic field component associated with the RF pulses is denoted by $B1$, which lies in a plane normal to B_0, as shown in Fig. 6.9. If we follow the same classical treatment of the proton precession, the new magnetic field $B1$ creates a torque on the proton. This causes the net magnetization vector M to rotate through a certain angle away from the B_0 axis toward the X-Y plane as shown in Fig. 6.9. This angle of rotation is known as the i_p angle. This angle is directly proportional to both the applied RF field intensity and the time of duration of $B1$. One can now easily guess the basic point: If the RF field component $B1$ lasts for a proper duration, the magnetization vector M can be made to rotate onto the transverse plane. That means, while in the transverse plane and rotating with the Larmor frequency, M will induce an NMR signal in the RF receiver coil placed in the transverse plane as shown in Fig. 5.9. This signal helps to observe the characteristics of M in the transverse plane and constitute the basis of MR signal detection. The nomenclature that follows is such the radiofrequency field that brings the magnetization vector M into the transverse plane is known as the 90° pulse. Further, an i_p angle of 90° results in maximizing the M_y component of magnetization. In contrast, a pulse of 180° produces no M_y magnetization but rotates the net magnetization M_0 from $+z$- to the $-z$-axis. The i_p angle of 90° is very important as it produces the strongest NMR signal with the magnetization vector M rotating in the transverse plane. But, the i_p angle of 180° is primarily important when applied in the spin-echo imaging technique. Here, the reversal of the direction of M is made when it is on the transverse plane.

6.4 MR Signal Detection

Generally, the MRI detector is made up of a pair of conductive loops like copper wires that are mutually perpendicular to each other and are placed close to the subject (patient). From the basics of electromagnetic induction principles (Faraday's law), one can understand the development of a voltage (V) induced in each of these loops. The voltage has a value that is directly proportional to the rate of change in the magnetic flux (magnetic field intensity multiplied by the area) with time. Figure 6.10 shows magnetization vector M (precessing in the XY-plane) producing a voltage or the MR signal. It is obvious that the vertical or z-component of magnetization M does not precess, and hence, it is not effective in the generation of any voltage or signal.

6.4.1 MR Signal Intensity

The intensity of the MR signal received is determined by the following factors.

- It is proportional to the number of hydrogen atoms or protons in the subject. Thus, it is representative of the number of protons in each voxel of the image.
- As the value of magnetization M_0 is proportional to the external magnetic field B_0, a 2 tesla MRI system has twice the M_0 of a 1 tesla system.
- Further, the induced voltage is directly proportional to the precession frequency or to B_0, and consequently, the MR signal is proportional to the square of the B_0 field. That is why the scientists are putting more efforts toward MRI systems with higher fields applied.

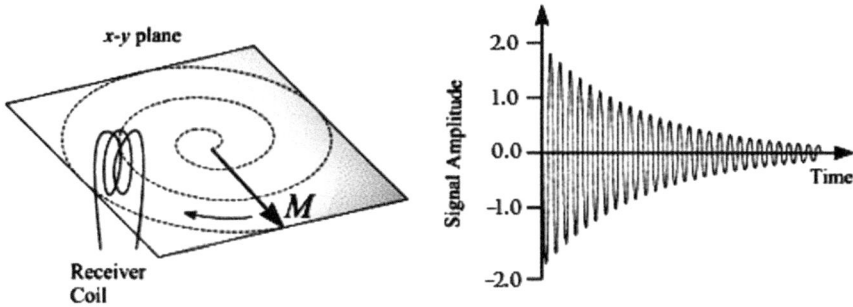

Fig. 6.10 MR signal produced through the effect of electromagnetic induction

6.4.2 Relaxation: Spin–Lattice (T1) and Spin–Spin (T2)

Now, let us try to understand the relaxation process as conceived in the MRI machines. In the presence of a strong constant magnetic field B_0, the equilibrium magnetization state corresponds to a z-component (M_z) equal to M_0 and both the transverse components of M_x and M_y being equal to zero (Fig. 6.11). As the RF pulse with a suitable frequency is applied, it adds energy to the whole system and, consequently, results in a nonequilibrium state. As soon as the pulse gets switched off, the system relaxes back to the thermal equilibrium state. Relaxation is a common term in many of the excitation phenomena in the physical processes. If we consider an active electrical circuit containing a combination of resistance and capacitance (RC) or resistance and inductance (RL) and apply an impulse voltage pulse across to any of them, the circuit develops time-varying voltages across each of the resistance, inductance, and capacitance. After a certain time span usually signified by the time constants, these time-varying voltages get back to their previous values before the application of the pulse. Similarly, with the application of a 90° RF pulse, the magnetization vector M rotates in the transverse plane with the Larmor frequency. It then decays to zero gradually as shown in Fig. 6.10.

Two different relaxation times generally govern the return of the z-component and the x- and y-components to their equilibrium values. Firstly, the $T1$-relaxation mechanism controls only z-component of the magnetization vector M and is also known as the spin–lattice relaxation. Secondly, the $T2$ relaxation affects only x- and y-components of magnetization and is referred to as the spin–spin relaxation. It thus governs the return of the M_x and M_y components of magnetization to their thermal equilibrium values of zero. Mathematically, the MR relaxation process is analyzed by applying the first-order differential equations (Bloch equations) and their solutions show how the M_z component changes with the i_p angle of an RF pulse at a time t after the application. It has been found that different tissues in human and other bodies have different values of $T1$ relaxation time. But the $T1$ relaxation time gets significantly changed for the dead or diseased tissues. Needless to say is that the

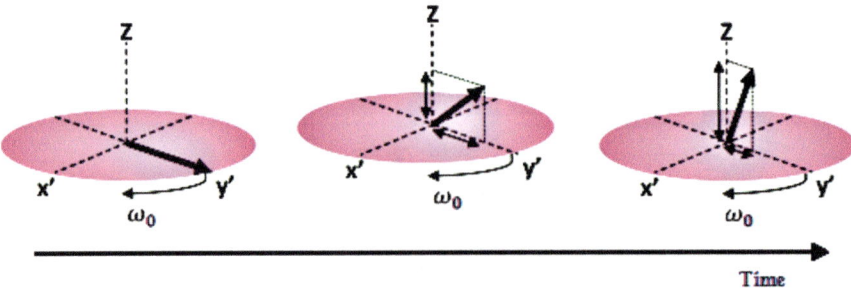

Fig. 6.11 (Left) Orientation of the magnetization vector after a 90° RF pulse is applied about the x-axis. (Center) $T1$ and $T2$ relaxation of the magnetization results in an increased M_z component and reduced M_y component, respectively, at a certain time after the pulse has been applied. (Right) The values of M_z and M_y components become M_0 and zero, respectively, at a later time

resulting differences are the basis for introducing contrast into the MR image. Again, for the second relaxation time T2, similar observations of different values for different tissues like healthy and dead or deceased in the body are used to differentiate between them in the final clinical images available.

6.4.3 Magnetic Gradient Coils

To translate the theoretical basis of magnetic resonance (MR) into a powerful imaging technique requires the use of magnetic gradient coils. The idea is based on the central understanding that if the applied magnetic field varies spatially within the body, it results in a spatial variation of the resonant frequencies, which, in turn, can help us in producing an image of high quality. Further, as the spatial variations have to be made dynamically, three separate "gradient coils" are incorporated into the experimental design of an MRI scanner. It is an engineering marvel that these gradient coils are designed as linear to ensure that the magnetic field varies linearly with respect to spatial location.

The gradient coils need to be fabricated such that there is no extra contribution to the magnetic field at the center of the Cartesian coordinate system ($z = 0$, $y = 0$, $x = 0$), also known as isocenter of the gradients. Thus, the magnetic field there is just equal to B_0 (Fig. 6.12). Then, the three separate magnetic field gradients are produced by passing a high value of direct current through separate coils with the help of high-power gradient amplifier and can be turned on and off very quickly with computer control. Conventionally, the y-axis points to the anterior/posterior direction and the x-axis corresponds to the left/right direction of the patient as shown in Fig. 6.13.

Three types of field gradients (Fig. 6.14) are generally used in an MRI scanner namely slice selection along the z-axis, phase encoding along the x-axis, and frequency encoding along y-axis. The amplitude and duration of these field gradients are critically important in reading the information in k-space. Points in k-space are generated by manipulating these gradients.

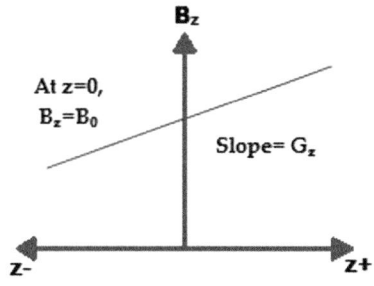

Fig. 6.12 The resulting magnetic field B_z as a function of the z-direction

Fig. 6.13 Different axes for the gradient coil

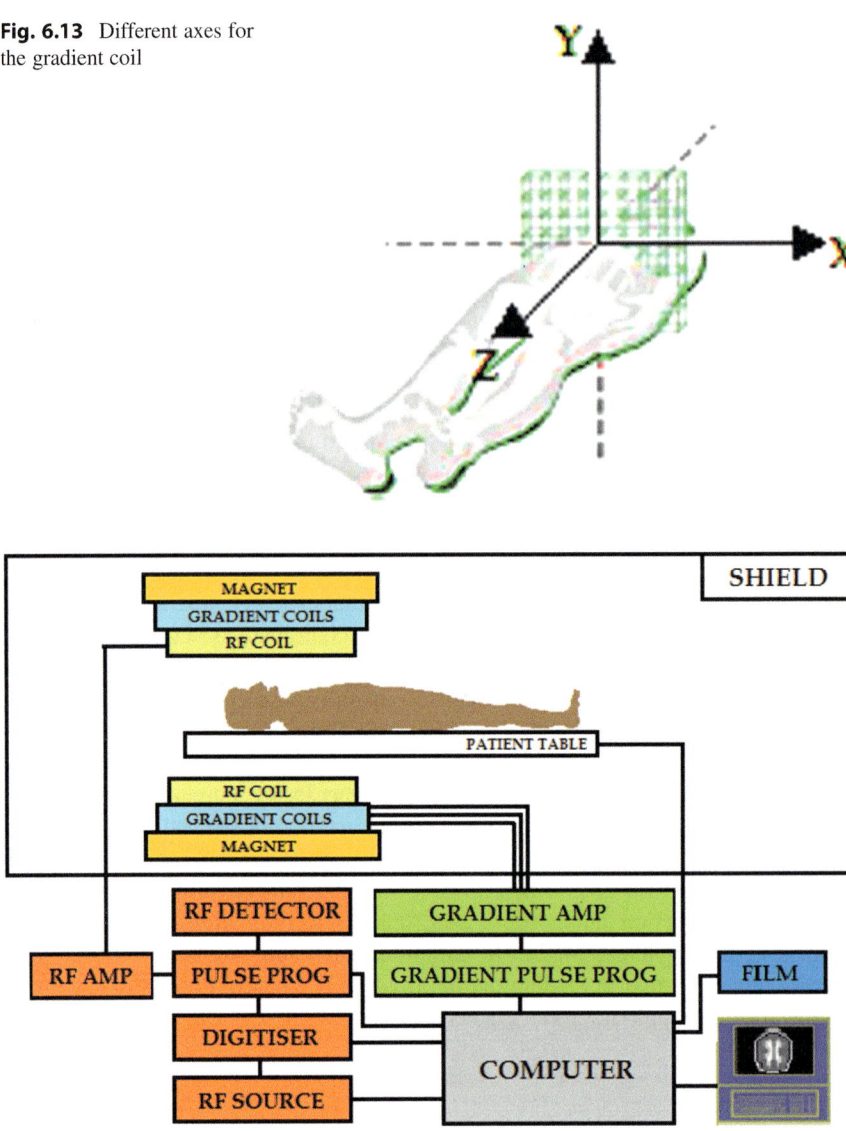

Fig. 6.14 Block diagram of a basic MRI hardware

6.4.4 *K*-Space and Image Formation

In general, the symbol "K" or "k" stands for momentum of a quantum particle or wave. It is conjugated to the real space denoted by X or x. On the other side, frequency (n) has inverse of time (t) dimension. A well-known technique in mathematics is the Fourier transformation that translates k to x or n to t. Thus, MRI data in space X can be saved in the k-space. There are several reviews available till now over the techniques to fill points in k-space. The k-space can then be easily transformed into an image of high-contrast through a Fourier transform technique applied.

6.4.5 Pulse Sequence

At last, it is pertinent to just mention that the pulse sequence generally shows the timing of RF pulses and gradients. It determines the type of image as PD (proton density), $T2$-weighted, and $T1$-weighted to help in gathering better clinical insights on the nature of cells in human and other bodies.

6.5 Computed Tomography

Computed tomography (CT) is a widely adopted imaging technology used to inspect internal morphological structure of human organs and in industrial applications with high precision. This makes the technology a reliable tool for noninvasive evaluation of structural defects in engineering field and detection of unwanted growth and its progression in living bodies (Hounsfield 1973; Webster 2010). We will restrict our discussion to basic principle of computed tomography, development of CT scanning technology, and its clinical application. CT scan is based on imaging with X-radiation, and it has made a revolutionary change in clinical diagnosis of tumors, bone structure and deformation study, and selection of radiotherapy doses in cancer treatment.

6.5.1 Basics of X-ray

German scientist Conrad Röntgen discovered a new type of radiation in the year 1895 while he was studying experimentally different phenomena inside gas discharge tubes. This radiation was able to penetrate opaque tissues and objects so that the inner structural image like bone or skeleton for human body can be obtained. This development dramatically changed the medical diagnosis procedure. Being previously unknown, the ray was named as X-ray.

X-ray generator has a X-ray tube comprising of cathode and anode (Fig. 6.15). Cathode generates the electron beam that is collected by the anode. During the travel, electrons collide with the target and as a result a small fraction of the energy is liberated as X-ray, while the major component is converted to heat. The generated X-radiation is collimated and allowed to fall on the target or subject through the collimator aperture.

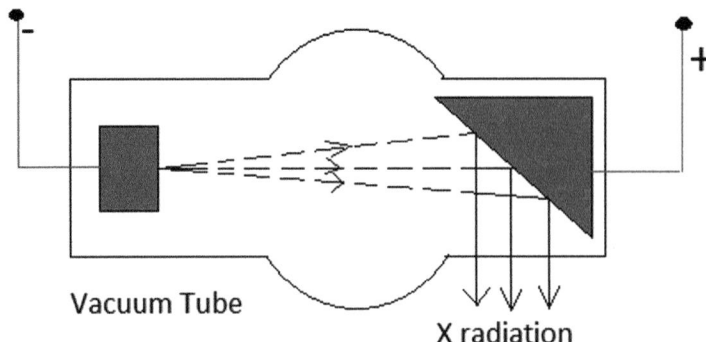

Fig. 6.15 X-ray production technique

Fig. 6.16 Multidirectional X-ray transmission measurement through a slice of the subject

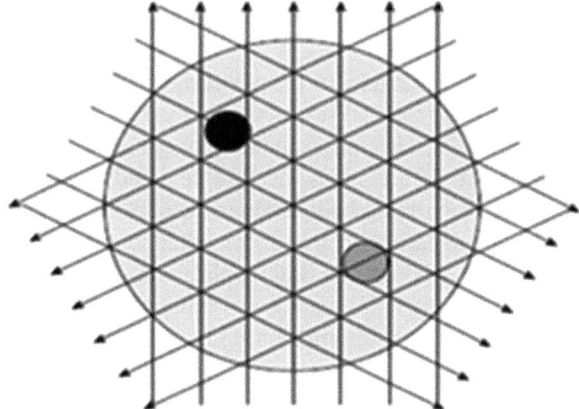

6.5.2 Diagnostic Application of Radiography

It is a well-known fact that different components of human body having different densities and thus possessing different absorption coefficients for X-ray. Therefore, an incident X-ray undergoes varying amounts of absorption in the organs while penetrating the human body. Thus, the outgoing radiation has spatial intensity variation as in the case of image. Multidirectional scanning through a slice is shown in Fig. 6.16. A suitable device that is capable to capture the intensity distribution can be employed to form the image of the section of the body. It is noticed that metallic foreign bodies, bones, and air cavities appear prominent in such imaging as the relative density of these sections is different from their neighboring tissues. But other organs of the body do not show much difference in intensity. Better image quality, clarity, and multidirectional view are possible to observe if special techniques like tomography are adopted.

6.5.3 Limitations of Radiography

There are some basic limitations of using conventional X-ray radiography (Hounsfield 1973; Barrett and Sweindell 1981). In general, two-dimensional X-ray plates have limited application in diagnosing 3D organs of the body as the dimensions of 3D objects are confusing in X-ray plate. Moreover, due to the limited dynamic range of the radiographic films, only larger intensity gradients are better visualized. So, the sections comprising soft tissues are difficult to estimate in radiography. Two closely spaced different organs with similar physical density can appear as a unit or the smaller organ may be lost from the image plate behind the larger one. Most of the body organs do not show much difference in intensity. Better image quality, clarity, and multidirectional view are possible to observe if special techniques like tomography are adopted.

6.6 Basic Principle of Computed Tomography

In CT scan, each slice of the subject is exposed in the radiation beam, which attenuates the radiation during its passage through the section. Governing factors of attenuation of the beam are incident intensity, composition, and density gradient of the medium through which the radiation is passed (Barrett and Sweindell 1981; Littleton 1976). This operation is performed over the entire section of interest through different angles, and the final image is reconstructed using specialized computer-based reconstruction algorithm (Cormack 1973).

In the earlier version of the CT scanners, the subject is scanned by a narrow X-ray beam and the detector is placed in the opposite side of the source beam to capture the transmitted radiation. If the subject is inhomogeneous and the beam is polychromatic, the intensity of the attenuated X-radiation is.

$$I = I_0 e^{-\sum_{i=1}^{n} \mu_i x_i}$$

where I_0 is intensity of input beam, μ_i is the attenuation coefficient, and n stands for the scanned region with varying attenuation coefficients.

Different orientations of the beam source and the detector in a scan plane provide different attenuation coefficients in the transmission plane. The grayscale image is formed by assigning different brightness to each predefined range of the attenuation coefficient.

6.6.1 Image Viewing System

Formation and interpretation of the image are based on a derived parameter called CT number computed from the attenuation coefficients of the region which the beam passes through during imaging. CT number, also known as *Hounsfield number* for any linear attenuation coefficient μ, is defined as.

$$\text{CT number} = 1000 \frac{(\mu - \mu_W)}{\mu_W}$$

where μ_W stands for the linear attenuation coefficient of water. The normalized value of CT number in most CT units is restricted between -1000 for air and $+1000$ for bone (Cierniak 2011; Rao and Guha 2001).

Attenuation coefficients for different body tissues slightly differ from each other. The differential attenuation coefficient is thus highly correlated with the physical density. Hence, the image reconstructed as a result of computed tomography can be considered as the mapping of densities with respect to that of water. Typical CT numbers of various body tissues are as follows: fat–60, brain white matter—30, brain gray matter—38, blood—42, muscle—44, etc.

6.7 Components of CT Scanner

A typical CT scanner has the following components: gantry, X-ray source, X-ray detector, and data acquisition system and finally the computer and operator console. The synchronized operation of the components develops the required image information of clinical relevance. Brief description as given below of the physical components of CT scanner may be pertinent to understand its functioning. The scan room arrangements are shown in Fig. 6.17.

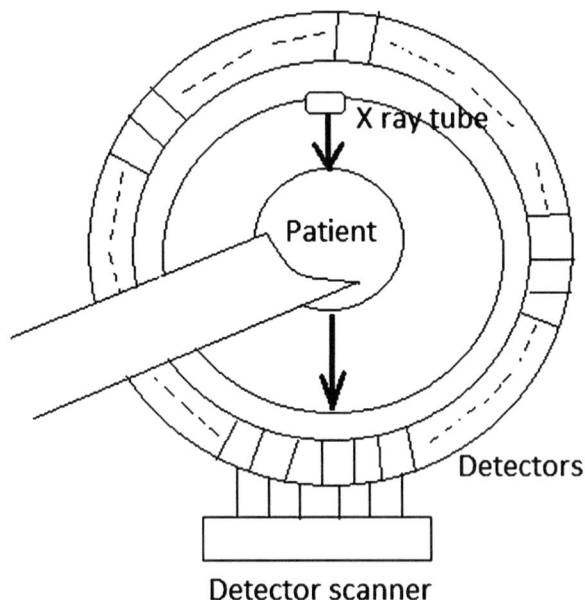

Fig. 6.17 Major components of scan room arrangement

Fig. 6.18 X-ray source and bowtie filter

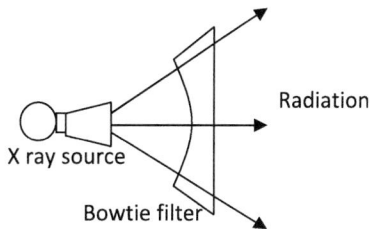

Gantry: It is the patient end component of the CT scanner, where the units like X-ray source and detector system are fitted along the periphery of the ring-type unit (Fig. 6.17). The patient table is mounted at the aperture of the gantry where the patient is allowed to lie down during scanning. The mechanical arrangement is provided to rotate the X-ray source and detector system such that the radiation source can scan the section of interest and the related output radiation be captured by the detectors.

X-ray source: The technique for generating X-radiation is introduced in Sect. 6.5. The collimated radiation is allowed to fall on the section of interest of the body with full alignment with the detectors. The section of the beam away from the detectors' scope is restricted by the collimator. The low-energy beam (soft X-ray, having low frequency) that cannot penetrate the body is filtered by the grid. Further filtration of the beam is performed to reduce the radiation exposure of the patient and to minimize beam hardening issues as well. This, in turn, improves the CT image quality. X-ray filter is designed to provide uniform radiation to cross section of the body. As human body has circular cross section and the thickness reduces gradually toward the periphery, filters are designed to reduce X-ray intensity toward the periphery with respect to the central area. Such filter is shown in Fig. 6.18, which is named as bowtie filter due to its specific shape.

6.7.1 X-Ray Detector and Data Acquisition System

The emerging radiation flux is absorbed in the detector array with rapid response time. The detector system rotates synchronously with the source and converts the absorbed beam into a proportional detector signal (Webster 2006). The resolution of the image mainly depends on data acquisition rate, detector geometry, and the size of the focal spot of the source. Mainly three types of detectors are used in CT scanners:

Scintillation detector: Hounsfield originally used a detector made up of sodium iodide (NaI) scintillation crystal combined with a photomultiplier tube (PMT). The scintillation crystal absorbs the transmitted X-ray beam, which is converted into light photons. PMT receives the visible light through a quartz or glass window covering the surface of photocathode. It, in turn, emits electron proportional to the light quanta. The electrons are multiplied in the series of metal channel dynode. Finally, the anode at the end of the dynode series generates current proportional to the absorbed radiation. The schematic operation is shown in Fig. 6.19.

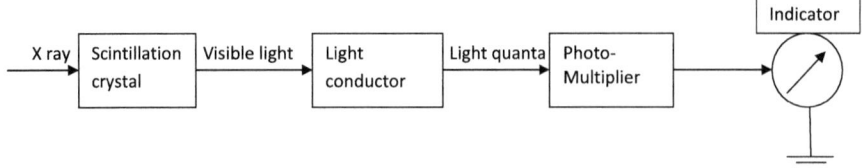

Fig. 6.19 Scintillation crystal and photomultiplier

Xenon gas ionization detector: In these detectors, the X-ray is allowed to pass through a chamber filled with inert gas like xenon. The energy of the incident radiation ionizes the gas. Resulting electrons are collected by a set of electrodes. The current flowing through the output circuit is proportional to the incident radiation. In order to increase the interaction between the gas and the incident radiation and thus to increase the output signal level, operating pressure in the xenon chamber is kept high.

Scintillare: It contains scintillation and photodiodes in order to collect the beam. The scintillation detector is small dimension such that it permits good resolution. The optical signal from the scintillation detector is allowed to fall on the photodiode through optical fiber. Photodiode generates the electrical signal, which is sent to the computer for data reconstruction using array processors.

Computer and operator console: The attenuation of the source radiation is detected by the X-ray detector and data acquisition system, and the resulting electrical signal is received by the computer, which digitally reconstructs the image using the projections from all slices. The overall control of operation like start, progress, and end of scan is also preserved by the operator's console. Patient information and other clinical information are also handled in this section.

6.8 Reconstruction of CT Image

The scanning data are required to be processed using specialized mathematical algorithms to reconstruct the image of the scanned section (Ledley et al. 1974; Hendee and Ritenour 2002). In the iterative mode of data reconstruction, one initial pattern of the image is guessed and the reconstructed data are mathematically compared with the guess for minimum deviation. In this mode, the manner of variation of the model each iterative step is also noted along with the voxel-based X-ray attenuation coefficients. The deviation between the model and the measured pattern is minimized by modifying the pixel intensity values along the path of the beam for each iterative step. The intensity pattern is a complicated function of X-ray direction, absorption coefficient, detector location, and projection angle (Goldman 2007). This intensity function is represented by *radon transform* of the attenuation coefficients over the entire scan plane. Explicit mathematical formulation is required for reconstruction of the image by inversion of the *radon transformation*. A point in image space is represented by a sinusoidal pattern in a plane created by the projection angle and the detector number.

6.9 Evolution of CT Scanning Technology

Hounsfield first incorporated a computer with X-ray imaging technology for scanning the brain in 1971; however, tomographic techniques were in use since the 1930s. Because of the ability to demonstrate anatomical information with final detail, computer-assisted tomography (CAT) attracts the interest of the clinicians. With the development of data acquisition hardware electronics and data processing technology, gradually the scan time and the information quality were enhanced with the newer generation of CT system. Depending upon this, the technology is mainly divided into four generations:

6.9.1 First-Generation CT (Translate–Rotate Mode)

In first-generation CT system, pencil-like narrow X-radiation beam was used with a single detector mechanically linked to each other. The source–detector system translates along the patient to cover his/her full width and then is rotated by $1°$. Each translation took about 4–5 s. Thus, the entire scan is completed with the data captured from $180°$ of rotation. The average scan time is 25–30 min using this technique. A later design consists of a detector pair to handle two slices at the same time. Hounsfield's first brain scanning system uses a water bath where the patient's head covered with an elastic membrane is place such that the X-radiation passes through the water to reduce the dynamic range of radiation. However, this is not suitable for whole-body scanning. Figure 6.20 shows the imaging technique of first-generation CT scanners.

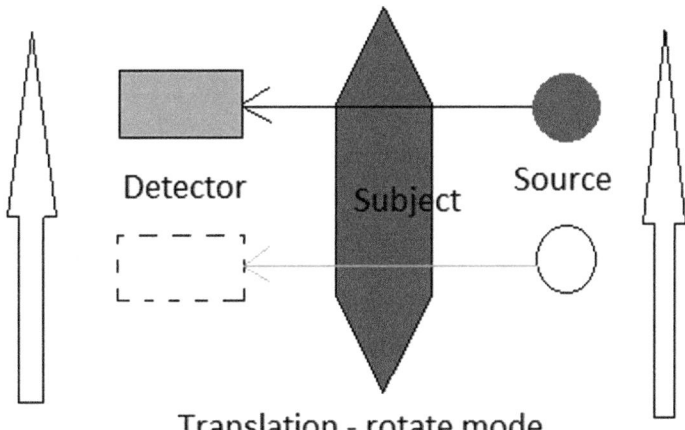

Fig. 6.20 Translate–rotate mode of first-generation CT scanning

6.9.2 Second-Generation CT (Multidetector Translate–Rotate Mode)

Second-generation CT was the outcome of the research toward minimization of the scan time. For that, additional detectors were used in the form of an array in the scan plane and a narrow fan beam of X-ray replaced the pencil beam. Because of the presence of detector array, the rotational angle is increased up to 20°–30°, which considerably reduced the scan time. As human is capable to hold the breath for at least 10 s, it was possible to capture chest CT image using this variant of CT scanner; however, the motion of heart causes noise in chest imaging. Figure 6.21 represents the scanning technique of second-generation CT scanners. This generation of CT scanners introduced table movement and laser-based slice indication and reconstruction technique based on Fourier transformation.

6.9.3 Third-Generation CT (Rotate–Rotate Mode)

Here, the rotate–rotate mode of scanning is adopted as now the gantry is equipped with few hundreds of detectors covering the total width of patient synchronized with a fan beam of X-ray (Fig. 6.22). With a very large data set in the output side, this technique requires efficient computing facilities. The scan time is reduced round 2 s in this orientation. This type of scanner is still in use in most of the clinical applications. However, this rotate–rotate mode generates the ring artifact in the

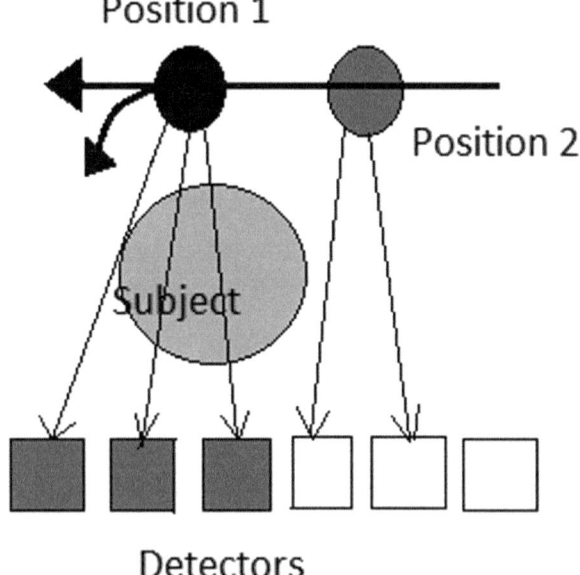

Fig. 6.21 Translate–rotate mode with multidetector for second-generation CT scanning

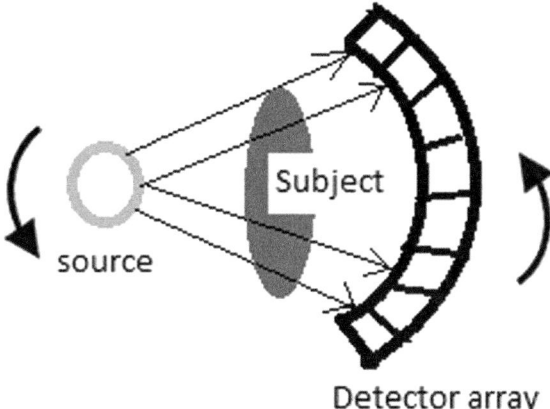

Fig. 6.22 Rotate–rotate mode of third-generation CT scanning

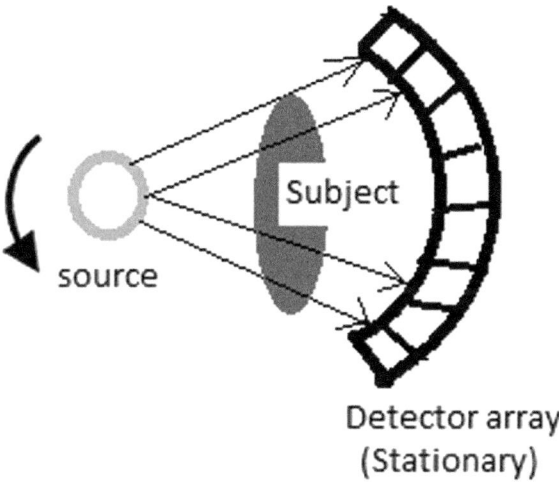

Fig. 6.23 Fixed–rotate mode of third-generation CT scanning

reconstructed image, which was minimized but not completely eliminated using xenon detectors.

6.9.4 Fourth-Generation CT (Fixed–Rotate)

The fourth-generation CT is developed specially to eliminate the ring artifact. In this case, more numbers of detectors up to few thousands are fixed on the gantry and a fan beam is used as X-ray source. Fixed detectors receive the attenuated radiation from the rotating source along the ring as shown in Fig. 6.23. Here, each detector is exposed with a complete fan beam as the beam passes across each detector.

Further developments including oscillating ring of detectors, steerable electron beams, 2D detector arrays, and helical data acquisition are sometimes given generation numbers, but not applicable in regular manner.

6.9.5 Electron Beam CT (Fixed–Fixed)

This provides a relatively new technology especially rapid cardiac imaging for screen coronary artery disease where the scan time is further reduced to synchronize the system with cardiac dynamics. Fixed set of detectors system with movable source is used, which encircles the patient (Goldman 2007). Source used here is an electron gun instead of conventional X-ray. The source is rotated electronically around the subject in EBCT.

6.9.6 Helical CT

In helical type, CT scanning patient faces a translation movement along the gantry and the X-ray source moves along a helical path around the patient. This technology was introduced in 1989 and adopted throughout for CT scanning (Lee et al. 2007). The helical motion of the tube is supported by a sliding contact-type arrangement such as slip ring. With this revolutionary up-gradation, the data acquisition time is reduced significantly. This results in a 3D image data of the slice per rotation of the tube, which requires efficient reconstruction and interpolation to estimate the actual image.

6.10 Clinical Application of CT Scan Image

As mentioned in the earlier sections, computed tomography not only captures images of different internal structures but also can differentiate the organs like lung, liver, and colon from other internal organs and fat. CT imaging has significant application in identifying large masses like tumors and metastatic growth at different volumes of human torso (Bar-Shalom et al. 2003). CT imaging and reconstruction procedures can also point out toward the shape, size, volume, and location of the growth. Besides, CT images can reveal brain hemorrhage and clot, blood vessel defects, and enlargement of different organs. Recent advances in image processing technology enable us to develop assistive diagnostic technologies for automatic detection of tumors or unnatural growths from the CT image. The general steps of CT image processing are given in Fig. 6.24.

To reduce the risk of radioactive exposure, radiation dose is reduced within a certain limit, which degrades the quality of image. The subject movement, respiration, etc., also introduce noise in the image. Image smoothing and enhancement retrieve the blurred image by filtering. Gaussian filter, bilateral filter, Gabor filter, Weiner filter, threshold-based enhancement, etc., are the most popular techniques for

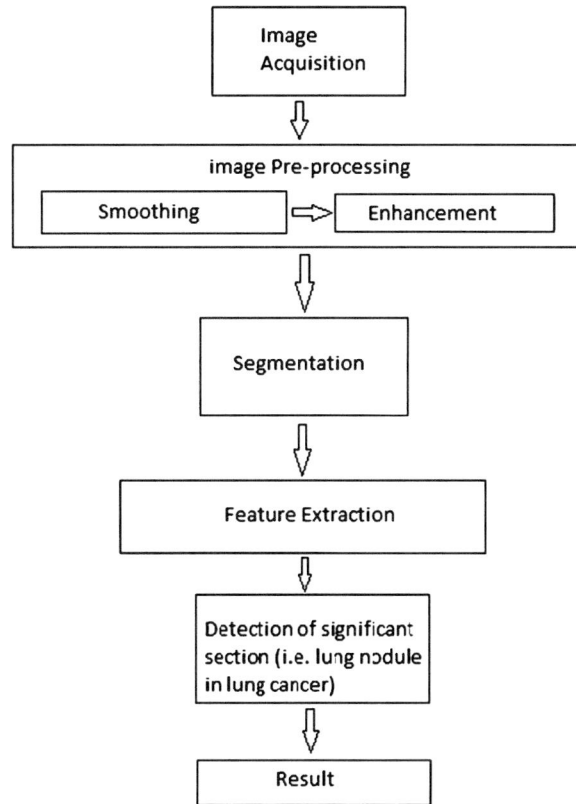

Fig. 6.24 Steps for CT image analysis

image denoising (Bar-Shalom et al. 2003). Segmentation is the technique to partition the image into a number of sections to isolate the tissues of organs of interest like lung and colon from the adjacent mass. It is generally done by grouping the pixels (or voxels) of similar intensity of predefined range and isolating them from the other pixels (or voxels) to form the object of interest. Automated classification of segmented image is done by classifiers based on advanced mathematical techniques such as fuzzy logic, artificial neural network, and support vector machine. Actually, pixel-based input image information is too large to be dealt with by the classifier due to computational complexity. A smarter approach is to extract relevant features from the image and design a classifier. For example, dimensional features of lung nodules can be used for staging the tumor. Also, the gradual variation in volume of the tumors in malignancy is helpful to select the radiotherapy doses for a patient.

Use of CT is also prevalent in orthopedic study for bony structures analysis and bone density study, imaging of complex joints like the spine, shoulder, or hip for fracture healing, deformation study, etc.

6.11 PET Scan

In the physicist's world, particles and their antiparticles form a fascinating view of how matter is made of. As we move from the molecule to the atom and then in the central tiny-sized nucleus surrounded by the revolving electrons, we see a totally different and interesting behavior of the subatomic particles. Soon after the firm foundation of quantum mechanics, Paul Dirac, the British Nobel winner physicist, predicted the existence of antiparticles theoretically in 1928, and their existence was confirmed experimentally. The particle known as *positron* is basically anti-electron or the antiparticle or the antimatter counterpart of the electron. The *positron* has an electric charge of $+1\ e$, a spin of 1/2 (the same as that of an electron), and has the same mass as an electron. Carl Anderson, the American Physicist, discovered positron. Thereafter, many other antiparticles like antiproton and antineutron were successfully discovered.

In the year 1896, Henry Becquerel, a French Physicist, discovered *radioactivity* that relates to the emission of alpha, beta, and gamma rays from certain substances or nuclei like uranium. Madame Curie discovered radium, another well-known radioactive element. Later, alpha rays were identified as helium nucleus, beta (β^-) as electron, and gamma as photons. Positron emission is basically a beta-plus decay (β^+ decay), another form of radioactive decay. Here, a proton inside a radionuclide nucleus is converted into a neutron while releasing a positron and an electron neutrino (ν_e). Positron emission is mediated by the *weak force*, one of the four fundamental forces in nature.

6.12 PET Scan: Basic Principle

A PET scan is an imaging technique that can help us to know how well the tissues and organs in our body are working, especially metabolism, oxygen use, and blood flow to the organs. PET stands for positron emission tomography (Strunk et al. 2018). Fig. 6.25 shows what a PET scan machine looks like.

This type of scan requires that a small amount of radioactive material be delivered into the patient's bloodstream through an intravenous injection. This radioactive material is known as a tracer. Once the tracer has time to work through the body, the patient is taken back to the scanning machine to have the procedure done. The PET scan is a gigantic machine containing a tunnel as in MRI. The patient has to lie on a hard table that can move his or her body into the tunnel. Once the body enters the tunnel, the PET reads signals sent from the tracer. The machine converts these signals into 3D pictures, which are further interpreted by a specialist (radiologist). The tracer when injected into the body is active only for the duration of 2–10 h. The radiation eventually loses its potency and leaves the body normally through urine. Patients are advised to drink plenty of water after the procedure to help speed up this process.

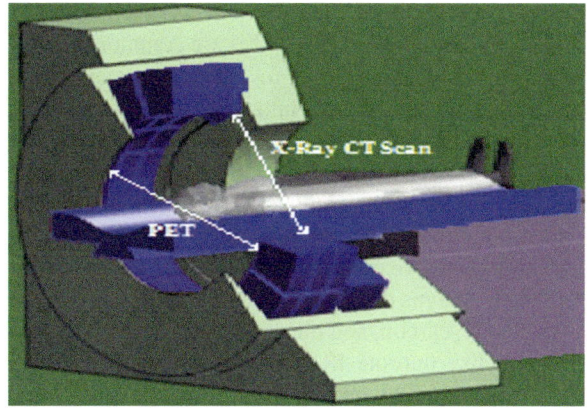

Fig. 6.25 PET scanner: a marvel realized by the synergy of different imaging methods

6.12.1 PET Scan Utilities

PET scans can directly show how our internal organs are working and indicate the presence of any abnormality. So, it is a very effective test to look for the diseases and functionality of the organs. It helps in detecting abnormalities or lumps or masses inside the body, checking whether cancer treatments have been effective or the disease has recurred, any disorder in brain such as tumors, seizures and Alzheimer's disease, heart disease, weakened heart muscles, blockages and decreased blood flow, etc.

6.12.2 Early Detection of Malignant Tumors/Cancers

PET scan today helps doctors across the globe to execute accurate diagnoses of patients before a disease spreads irreversibly. It can further help to monitor continuously the recovery from surgical operations. During cancer treatment, oncologists use PET scans to identify any potential tumor inside the body. Later, they can see the progress of their patients with treatments like radiotherapy or chemotherapy. The detailed images obtained with the help of PET can guide the doctors in assessing the condition of the patients properly. It can help also to stop inefficient treatments, replacing usual surgery with radiotherapy or the future course of actions to be taken.

Beta-plus radioelements, emitters of positive electrons or positrons, are used as the radioactive tracers in this machine. It is a common fact that positrons annihilate by emitting two back-to-back gamma rays or photons within a short time. Thereafter, we need to detect simultaneously these gamma or photon pair so as to locate the specific zone of emission close to the fixation of the radioactive tracer. With this detection of a large number of photon pairs, we can map the points in our cells containing radiotracers in order to screen the hot spots with greater resolution. In this context, we use a term known as *scintigraphy*. It is a technique using the scintillation counter or similar detectors with a radioactive tracer to produce n very high-quality image of a body organ or a record of its functioning. Normally, isotopes like iodine 131 (^{131}I) or technetium 99 (^{99}Tc) are commonly used in scintigraphy. Radioactive iodine 131 is generally used for medical purposes. If a small dose of iodine 131 is swallowed, it gets absorbed into the bloodstream of the gastrointestinal (GI) tract and concentrated from the blood by the thyroid gland, where it begins destroying the gland's cells. On the other hand, the isotope technetium 99 can be easily detected in the body by medical techniques. The reason is that it emits 140.5 keV gamma rays, which are similar to the conventional X-rays. Further, its half-life for gamma or photon radioactive emission is about 6 h, which means half of the initial number of radioactive atoms decays in about 6 h. Generally, the positron emitters are elements like oxygen, carbon, and nitrogen that are abundantly available in the body. Further, halogens like fluorine 18 or bromine 76 can easily be attached to other molecules. This means that chemicals involved in the human metabolism can be attached to different radioactive elements and made to flow inside the body. For example, the way human tissue absorbs glucose can be properly mapped after an injection of a compound FDG (fluorodeoxyglucose), which contains radioactive element fluorine 18 (^{18}F). The absorption of glucose can monitor a person's health as it can reveal important information on how different tissues work in our body organs like brain, heart, or any malignant tumor.

It is a commonly known fact that water molecules form a major constituent of the human body. The molecular structure of water is simple with two hydrogen atoms covalently bonded to a single oxygen atom. The water molecule can be marked with oxygen 15 and can be used to follow the path of blood flow inside the brain. Thus, it can provide a greater understanding of the way it works. These explorations have paved the way for new insights into neurology, psychology, linguistics, and the cognitive sciences. Unfortunately, beta-positive isotopes do not exist in nature and must be created in the laboratory. It is now an established fact that the half-lives of these artificial positron emitters are very short. Oxygen 15 has a period of 2 min, while the amount of fluorine 18 halves in just under 2 h. These isotopes must therefore be inserted into the body as soon as possible after they are created in the laboratory of the hospitals having a cyclotron facility or other mini-accelerators. Another important fact is that FDG is the most widely used one among all the chemicals marked with positron emitters. It has a comparatively long half-life, and so, it can be easily transported from its production site to diagnostic centers in the close-by area. Reportedly, FDG has been used in PET scan for the full-body scan performed in clinical oncology to determine the spread of cancer (Oehr et al. 2004).

6.13 Conclusion

MRI is safe as far as the latest study reports suggest except for the persons who carry non-MRI-compatible implants and metallic components near some vital organs inside the body. It is a common observation that the much-needed MRI is sometimes avoided by the people as the small imaging space leads to a claustrophobic feeling and therefore nervousness. Further, recent studies have established that patients with kidney problems may need extra monitoring during MRI scan. Anyway, MRI is undoubtedly a better imaging technique having no radiation exposure but high-quality resolution although it is a relatively time-taking process.

Computed tomography has a significant position in the history of clinical diagnostic imaging. In terms of engineering knowledge and its application, it achieves a superior height by collectively applying X-ray imaging, use of rotary radiation detector arrays, mathematical methods for reconstruction of the image, and computer capabilities as a whole to produce a novel way of noninvasive analysis of internal organs of human body. Actually, CT became exceptionally useful because of its ability to generate better contrast of image of adjacent but different mass of tissues of different organs, which are neither readily accessible nor observable by other imaging techniques. The main advantages of computed tomography are most detailed and informative among all imaging techniques (refer to Fig. 6.26), the process is rapid and painless, clear, and specific information about the morphological structure of the growth based on which a reliable clinical decision can be suggested, and large portion of the subject can be viewed at the same time. But like any other radiation type imaging, computed tomography has also some disadvantages: Risks due to higher dose of radiation are greater than other imaging types and which is greater for children, injection of a contrast medium (dye) can cause kidney problems, and iodine-enriched contrast dye may cause allergic reaction in patients. Cancer detection and preparation of treatment protocol have become easier with the development of CT images. Moreover, the effectiveness of treatment can also be studied by consulting the CT images successively captured within the duration of the treatment.

The latest technological advance in the field of medical imaging for cancer screening has been achieved by combining the two methods of "*computed tomography*" and "*positron emission tomography.*" This composite method being much informative helps in diagnosis otherwise difficult with other single imaging techniques like CT or MRI. Thus, an X-ray scanner is also attached to the PET cameras. The images produced by scanner are very precise from an anatomical point of view. But it cannot provide much information on cell metabolism like the changes within the substances they are made of. In contrast, PET images are sensitive to the metabolism and so functional in projecting more information. But these images suffer from the qualities of sharpness and precision from an anatomical point of view. So, the CT images can be combined with the functional images from PET scan in a hybrid device. The combination of these two images can precisely locate an injury or malfunction. This helps optimize a protocol during radiotherapy. So, it makes an operation more efficient, avoiding unnecessary surgical procedures, or reduces the use of invasive procedures such as biopsies. As the CT scanner and the PET camera are side by side and the patient does not need any movement move between the two tests, the merging of two images is of excellent quality. Due to the high speed of the scan usually less than 30 min, the results can be processed quickly compared to a day in the past days. Thus, it helps in drawing a more reliable and accurate diagnosis for the sick patients and in cases of suspected cancer. Finally, the distinguishing features of the three different imaging methods are summarized in the following table (Ref. Thedens 2013).

Modality	Approximate resolution	Source of energy	Image contrast of the source	Advantages	Limitations
CT	0.2–1.0 mm	X-ray	Tissue density	Fast, high resolution, and 3D reconstruction	Poor soft tissue contrast, radiation dose
MRI	0.3–1.0 mm	Radiofrequency	Multiple	Soft tissue contrast, multiple contrast methods	Scan times, metal, patient comfort
PET	4–7 mm	Positrons/ gamma rays	Tissue biochemistry	Tissue function information	Low resolution, radioactive elements

References

Barrett HH, Sweindell W (1981) Radiological imaging-the theory of image formation, detection, and processing, vol 2. Academic Press, New York

Bar-Shalom R et al (2003) Clinical performance of PET/CT in evaluation of cancer: additional value for diagnostic imaging and patient management. J Nucl Med 44(8):1200–1209

Brown MA, Semelka RC (2003) MRI: basic principles and application, 3rd edn. Wiley – Liss, Hoboken, NJ

Cierniak R (2011) X-ray computed tomography in biomedical engineering. Springer-Verlag London Limited, London

Cormack AM (1973) Reconstruction of densities from their projections, with applications in radiological physics. Phys Med Biol 18(2):195–207

Goldman LW (2007) Principles of CT and CT technology. 35:115–128

Hendee WR, Ritenour ER (2002) Medical imaging physics, 4th edn. Wiley-Liss, Inc., New York

Hounsfield GN (1973) Computerized transverse axial scanning (tomography), part 1: description of system. Br J Radiol 46:1016–1022

Liang Z-P, Lauterbur PC (2000) Principles of magnetic resonance imaging: a signal processing perspective. SPIE Optical Engineering Press, New York

Ledley RS, DiChiro G, Luessenhop AJ, Twigg HL (1974) Computerized transaxial x-ray tomography of the human body. Science 186:207–221

Lee KS, Yi CA, Jeong SY et al (2007) Solid or partly solid solitary pulmonary nodules: their characterization using contrast wash-in and morphologic features at helical CT. Chest 131:1516–1525

Littleton JT (1976) Tomography: physical principles and clinical applications. Williams and Wilkins

Oehr P, Biersak HJ, Coleman RE (2004) PET and PET-CT in oncology. Springer-Verlag, Germany

Rao CR, Guha SK (2001) Principles of medical electronics and biomedical instrumentation. Universities Press, Hyderabad

Strunk A., Gazdovich J., Redouté O., Reverte J. M., Shelley S. & Todorova V., 2018 Model PET scan activity. The physics teacher 56, 278https://doi.org/10.1119/1.5033868

Thedens DR (2013) PhD and International Center for Postgraduate Medical Education. www.icpme.us

Webster JG (2006) Encyclopedia of medical devices and instrumentation, vol 2, 2nd edn. John Wiley & Sons, Inc., Hoboken, NJ

Webster JG (2010) Medical instrumentation application and design, 4th edn. John Wiley & Sons, Inc., Hoboken, NJ

Diffusion, MRI, and Cancer Diagnosis: Physicist's Outlook

7

Susanta Kumar Sen Gupta

Abstract

A very elegant application of nuclear magnetic resonance (NMR) phenomenon is imaging the ^{1}HNMR signals emitted from water molecules of human cells and tissues induced by the decay of bulk transverse magnetization on turning off the RF pulse. MRI is a noninvasive in vivo radiological modality to diagnose soft tissues. The success of MRI of soft tissue hinges on how good the image of a tissue is weighted by MRI parameters: ^{1}H spin density (PD), longitudinal relaxation time (T_1), and transverse relaxation time (T_2). Further, any one of PD, T_1, and T_2 parameters is a potential source of image contrast for distinguishing two tissues or the healthy and diseased states of a tissue. MRI of a tissue weighted predominantly by any one of PD, T_1, or T_2 parameters is achieved by *spin-echo* (SE) imaging pulse sequence by appropriately tuning relative values of echo time (TE) and repetition time (TR) of the pulse sequence. The T_2-weighted image of a tissue is preferred over its PD- or T_1-weighted one for its ability to provide images of distinctly superior clarity.

The clarity in image contrast is furthermore enhanced by exploiting differential self-diffusion of ^{1}H spins in molecules of H_2O across different cells and tissues. This has led to an incredible advancement in MRI: diffusion MRI (d-MRI), which can image how water molecules are diffusing in cells and tissues of our organs. This noninvasive and in vivo radiological tool has proven its immense potentiality in neuronal and oncologic imaging. Diffusion MRI technique is implemented by utilizing a method based on a bipolar pulsed-field gradient spin-echo (PFG-SE) pulse sequence.

Since diffusion rather self-diffusion of H_2O molecules in cells and tissues unlike in a glass of water is significantly restricted due to obstacles and barriers

---

S. K. Sen Gupta (✉)
Department of Chemistry, Institute of Science, Banaras Hindu University, Varanasi, India

© The Author(s), under exclusive license to Springer Nature Singapore Pte Ltd. 2022
S. K. Basu et al. (eds.), *Cancer Diagnostics and Therapeutics*,
https://doi.org/10.1007/978-981-16-4752-9_7

there, the coefficient of diffusion for tissue H_2O is generally, as anticipated, is significantly lower than its true diffusion constant (D) and is given a new nomenclature: *"Apparent diffusion coefficient"* (ADC). ADC of H_2O for a given tissue is evaluated from at least two echo signals at different gradient parameters of the PFG-SE pulse sequence. Depending on the level of organization in a tissue, restricted water diffusion in the tissue may be isotropic, e.g., in the brain gray matter and prostate gland, or anisotropic, e.g., in the brain white matter and kidney.

Accordingly, there have been developed two modes of d-MRI: (i) *diffusion-weighted imaging* (DWI) complemented by ADC *mapping* (calculated from DW images) for isotropic diffusion and (ii) *diffusion tensor imaging* (DTI) for anisotropic diffusion. While investigating diagnostic potentiality of d-MRI, we find it a truly biomarker of tissue microstructure in its prime applications of neuronal and oncologic imaging. We focus in this chapter only on oncologic imaging: detection, staging, and monitoring the response to therapy of cancer of central nervous system (CNS), breast, hepatic, pelvic, and other target organs.

Keywords

Self-diffusion · MRI · Diffusion MRI · Oncologic imaging

7.1 Introduction

This chapter gives a theoretical background to magnetic resonance imaging (MRI) starting with molecular diffusion, and an account of how it is utilized in cancer diagnosis.

The transport of matter in the absence of any bulk flow refers to diffusion. The chemical species diffuse down its chemical potential or concentration gradient. The phenomenon of diffusion being intrinsically a nonequilibrium one leads to an increase in the entropy of the system inducing it thereby to approach the state of equilibrium. Diffusion stops when chemical potential or concentration of the species at every point in the system becomes the same. Despite there being no net mass flux, self-diffusion of molecules keeps on as a manifestation of their random thermal motion.

Nuclear magnetic resonance (NMR) imaging is a powerful imaging modality in the clinical domain under the title *magnetic resonance imaging* popularly by its acronym: MRI. After achieving grand successes in applications to physics, chemistry, and molecular biology, NMR showed its promise as a vital tool in clinical medicine for investigating whole tissues in vitro and in vivo to trace out ongoing biological processes in our body.

The credit of advocating first the utilization of NMR for medical diagnosis, particularly for cancer detection, goes to Raymond Damadian who could demonstrate way back in 1971 that normal and malignant cells could be differentiated on the basis of relaxation rates of ^1H of water molecules between these cells. The experimental feasibility of macroscopic imaging by NMR was first demonstrated in 1972 by Paul Lauterbur by introducing magnetic field gradients for localizing NMR signal information in space. This gave birth to MRI. Hereafter, Richard Ernst's revolutionary contribution to the use of phase and frequency encoding and Fourier transform (FT) techniques for obtaining MR images raised the MRI instrument up to a commercial stage. Richard Ernst was awarded the Nobel Prize in 1991. This was followed by Peter Mansfield's spectacular development of echo-planer imaging (EPI) technique, which could provide images much more rapidly than the earlier methods. Paul Lauterbur and Peter Mansfield shared the Nobel Prize in 2003. Since then, MRI has been undergoing continuous developments to be a more and more powerful diagnostic modality.

MRI has the capability of exquisite imaging and mapping the spatial distribution of ^{1}H carried in molecules of H_2O in a given volume of interest (VOI) of a soft tissue in a target organ of ours and that too noninvasively without any exposure to hazardous radiation. It is, in fact, a well-proven safe modality for clinical diagnostics of different organs of patients barring those fitted with certain types of ferromagnetic implants and pacemakers.

^{1}H nuclei behave as mini spin magnets and emit radiofrequency signals in an NMR apparatus. These ^{1}H NMR signals are utilized for imaging our organs in MRI. The clues to the great success behind MR imaging of ^{1}H resonance of water molecules in soft tissues of various body organs are as follows: (i) astronomically high abundance $>10^{20}/\mu L$ of ^{1}H nuclei in our body fluids and (ii) the high sensitivity of a ^{1}H nucleus to NMR phenomenon due to its high gyromagnetic ratio γ (+267.522 × 10^6 radians per second per Tesla). A practical ^{1}HNMR sample like water molecules in tissues over the volume element of an organ is an ensemble of an extremely large number of isolated ^{1}H spins called a *spin packet*. The ^{1}H NMR signal of water from the volume element (voxel) for a given organ sample is represented as the intensity of a picture element (pixel) in the image of the organ. MRI has strikingly nonionizing, noninvasive, soft tissue-sensitive, quantitative, multicontrast, multipurpose, etc., characteristics, and the technique has been undergoing continuous improvement. MRI is a spectacularly fast and high-resolution diagnostic tool for deciphering details of our soft tissues within in a few seconds to minutes at a resolution even better than 100 μm. Truly, MRI is providing us images of tissue anatomy, pathology, metabolism, flow, diffusion, and functionality so excellently unmatched by other medical imaging techniques.

The initial oncologic applications were primarily confined to the field of neuro-oncology. The scope of *diffusion-weighted imaging-calculated apparent diffusion coefficient mapping* (DWI-ADC mapping) is keeping on broadening to include oncologic imaging of breast, liver, kidney, and prostate gland. Consensus has, in fact, been reached on the use of diffusion-weighted MRI (DW-MRI) as a cancer imaging biomarker. Without using any exogenous contrast agents or any hazardous radiation, DWI-ADC mapping has shown its utility for noninvasively and relatively rapidly assessing lesion aggressiveness and monitoring the response to tumor therapy quantitatively; e.g., when following a therapy, one observes the ADC value to increase significantly, and the therapy response is potentially positive, e.g., breast cancers and brain cancers. Clear trends have been observed between DWI-ADC parameters on one hand and tissue structure, cellularity, and necrosis on the other hand. In fact, DWI-ADC tool has exhibited its immense potentiality for diagnosis of a cancer patient throughout all stages of his treatment (Koh and Collins 2007; Padhani et al. 2009; White et al. 2014; Bozgeyik et al. 2013; Padhani et al. 2011; Malayeri et al. 2011).

A thumb rule for DWI-ADC mode of diffusion MRI to distinguish between malignant and benign lesions of a target organ is a malignant lesion as compared to its benign counterpart has significantly lower ADC. This has been observed without exception for focal hepatic lesions, breast lesions, etc. ADC is undoubtedly a potential biomarker of the response of a patient to the application of vascular disruptive drugs and to an apoptosis-inducing therapy.

For anisotropic diffusion of water in a tissue like brain white matter, one applies, as we know, the technique of *diffusion tensor imaging* (DTI) in place of DWI-ADC. Using DTI of a tissue where diffusion is anisotropic, one can estimate both the magnitude and the directionality of the diffusion coefficient of water there. This can be exploited to assess tumor infiltration in white matter tracts. For this purpose, a parameter called "*fractional anisotropy*" (FA) is introduced, which measures the extent of diffusion in a particular direction in a tissue. FA gets lowered when neuronal death, axonal loss, and irregular tumor cellular growth disorganize the architecture of primary brain tumors (Koh and Collins 2007; Padhani et al. 2009; Malayeri et al. 2011; Le Bihan 1991; Le Bihan et al. 1991; Mori and Barker 1999; Mukherjee et al. 2008; de Figueiredo et al. 2011; Le Bihan 2014).

Among clinical applications of DWI in neuroradiology are the evaluation of brain tumors and differentiation of brain abscesses from necrotic tumors besides the very early diagnosis of acute ischemia (Koh and Collins 2007; Malayeri et al. 2011).

This chapter discusses first the basic concepts of diffusion. It is then followed by discussion on NMR imaging, popularly known as MRI in the clinical domain. It then describes the role of diffusion process in MRI and discusses how *diffusion* MRI is useful in cancer diagnosis.

7.2 Diffusion

The transport of a species in a system in the absence of the bulk flow of the system refers to "diffusion." The flux $\mathbf{J}(\mathbf{r}, t)$/mol. cm^{-2}. s^{-1} of a species at a given concentration $c(\mathbf{r}, t)$/mol. cm^{-3} at a position vector $\mathbf{r}$ in a region of a system diffusing to another region of its lower concentration in the system is driven by the antigradient of its chemical potential or concentration $- \nabla c(\mathbf{r}, t)$/mol. cm^{-4} across the plane at $\mathbf{r}$ and at time t. The inherent nonequilibrium feature of diffusion induces an enhancement in the system's entropy with a gradual passage to a state of equilibrium when the species at every point in the system attains the same chemical potential or the concentration. Diffusion of a species will, of course, come to an end when the species at every point in the system attains the same chemical potential or the concentration.

7.2.1 Phenomenological Notion and Fick's Laws of Diffusion

Adolf Fick adopted a phenomenological approach to formulate a relation between $\mathbf{J}(\mathbf{r}, t)$ and $-\nabla c(\mathbf{r}, t)$ for a species in a bulk medium and obtained what is well known as Fick's first law (Eq. 7.1):

$$\mathbf{J}(\mathbf{r},t) = -\overline{\mathbf{D}} \nabla c(\mathbf{r},t) \tag{7.1}$$

where the proportionality constant $\overline{\mathbf{D}}$ is a symmetric tensor called "diffusion coefficient tensor" implying that the diffusion coefficient varies with direction, and diffusion is anisotropic. Further, $\overline{\mathbf{D}}$ being a real positive 3×3 symmetric matrix, D_{ij} equals D_{ji} (with $i, j, k = x, y, z$). However, when diffusion is isotropic, $\overline{\mathbf{D}}$ has no directional dependence and $\overline{\mathbf{D}}$ gets replaced by a positive "diffusion coefficient scalar" $D/\text{cm}^2 \cdot \text{s}^{-1}$. Equation 7.1 when substituted in the law of conservation of mass, $\partial c(\mathbf{r},t)/\partial t = -\nabla \cdot \mathbf{J}(\mathbf{r},t)$ leads to Fick's second law, the basic diffusion equation (Equation 7.2):

$$\frac{\partial c(\mathbf{r},t)}{\partial t} = \nabla \cdot [D \nabla c(\mathbf{r},t)] = D \nabla^2 c(\mathbf{r},t) \tag{7.2}$$

The solution of Eq. (7.2) depends upon the geometry of the system and the initial concentration distribution. However, for species initially located at the origin, and diffusing through an infinite medium later, the solution is the well-known Gaussian expression. The general solution in polar coordinates to the diffusion equation is thus written as Eq. (7.3):

$$c(\mathbf{r},t) = \frac{c_0}{(4\pi Dt)^{d/2}} e^{(-r^2/4Dt)} \tag{7.3}$$

where d denotes the dimensionality of the system (Graham 1850; Fick 1855; Jost 1960; Crank 1980; Philibert 2006; Mehre and Stolwijk 2009).

7.2.2 Molecular Backgrounds of Diffusion and Random Walk Model

The molecular backgrounds of Fick's diffusion were developed by Albert Einstein in terms of Brownian motion or random walks. The diffusing particles/molecules are self-propelled by internal thermal energy. The connection between Fickian diffusion coefficient D and the microscopic parameter, mean-squared displacement [MSD $= <r(t)^2>$] of the species in time t characterizing random walks of molecules, was established by Einstein: MSD $= 2dDt$ (where d, as stated, is the spatial dimension). This is the famous Einstein's diffusion equation or MSD equation. If a species is to spread out by diffusion over a distance of order L, an amount of time proportional to L^2 must be given. The length $\sqrt{2dDt}$ is referred as the "diffusion length." Obviously, Einstein's MSD equation as compared to Fick's laws provides a more fundamental approach for determining the coefficient of diffusion D (Graham 1850; Philibert 2006; Mehre and Stolwijk 2009; Einstein 1905; McQuarrie 2000).

7.2.3 Self-Diffusion

Einstein, while developing molecular backgrounds of diffusion, recognized further that diffusion could still occur when no concentration gradient is there. Diffusion would thus keep going as a result of thermally driven random motions of molecules, which continue to persist whether there is a concentration gradient or not. So, molecules in a system can keep mixing on their own even when there is no chemical potential or concentration gradient. This is what is referred as *self-diffusion*. Obviously, *self-diffusion* takes place under equilibrium. It is truly random in nature. The self-diffusivity is different from transport diffusivity/Fickian diffusivity. However, self-diffusion is also described by one of Fick's laws in which D is called the *coefficient of self-diffusion*. The coefficient of self-diffusion is truly a constant independent of $c(\mathbf{r})$ (Philibert 2006; Einstein 1905; McQuarrie 2000; De Schepper and Ernst 1979).

7.2.3.1 Self-Diffusion and Einstein's MSD Relation

Einstein showed that in an isotropic system, the mean-squared displacement (MSD) from the origin at time t is given by Einstein's equation in which the constant D is called the "self-diffusivity" in this context. It is worth mentioning here that molecules of water, which comprise 70% of our body weight, are moving randomly in our body tissues by self-diffusion much like in a glass of water where the motion of water molecules is completely random and is limited only by the boundaries of the container (McQuarrie 2000; De Schepper and Ernst 1979).

The self-diffusion coefficient of neat water is 2.299×10^{-5} cm. s^{-1} at 25 °C and 1.261×10^{-5} cm^2. s^{-1} at 4 °C. NMR is the method of choice to determine D from MSD of molecules and provides information about molecular displacements over diffusion distances, which could extend beyond the elementary molecular jump lengths, especially about translational displacements of water molecules in our body tissues. What is interesting is diffusion affords a highly sensitive tissue contrast mechanism with details of tissue microstructure useful for medical diagnosis. Diffusion MRI has become a pillar of modern clinical imaging. The superiority of diffusion MRI over conventional MRI is well known for early detection of brain stroke and of tumor in target organs. This is achieved by exploiting the magnetic moments of ^{1}H of H$_2$O molecules diffusing in our tissues (Le Bihan 1991; de Figueiredo et al. 2011; Le Bihan 2014; Nicolay et al. 2001; Luypaert et al. 2001; Winston 2012).

7.2.4 Anisotropic Self-Diffusion

The self-diffusion of water in our tissues may be isotropic or anisotropic depending on the tissue microstructure as seen in brain tissues it is isotropic in the gray matter but anisotropic in the white matter. In a tissue where self-diffusion is isotropic, D is a scalar. However, where self-diffusion is anisotropic the displacement of water molecules per unit time is not the same in all directions, as in white matter neural tissues, $D_{\parallel} \approx 2D_{\perp}$ showing that the self-diffusion coefficient is direction-dependent. This self-diffusion coefficient being greater along the axis parallel to the orientation of the nerve fiber but much less along the axis perpendicular to the nerve fiber must be a tensor $\overline{\mathbf{D}}$. Besides the white matter of brain, other tissues where water diffusion is anisotropic are the heart muscles, the articular cartilage, and the eye lens (Le Bihan 1991; Mukherjee et al. 2008; de Figueiredo et al. 2011; Le Bihan 2014; McQuarrie 2000; Evans et al. 1991).

In fact, for measuring and imaging in vivo self-diffusion of water noninvasively in tissues of our organs two MRI techniques based on whether water diffusion in a tissue is isotropic or anisotropic have been developed for imaging water diffusing in a tissue: DWI-ADC (diffusion-weighted imaging-apparent diffusion coefficient mapping) for isotropic cases and DTI (diffusion tensor imaging) for anisotropic cases. In either situation, the clue to their ability to produce exquisite image contrast between two tissues or healthy and diseased states of a tissue is the diffusion coefficient of the water molecules in the tissue(s) being imaged. A vivid illustration of the power of these techniques is the successful measurement of the structural integrity of brain white matter at different ages and in several diseases by DTI (Le Bihan 1991; Mukherjee et al. 2008; de Figueiredo et al. 2011; Le Bihan 2014).

7.3 MRI

MRI (magnetic resonance imaging) provides us images of our internal anatomy and ongoing physiological processes in our soft tissues. A strong magnetic field, a linearly oscillating RF (radiofrequency) magnetic field, and a set of very well-designed and computer-controlled magnetic field gradient coils are the primary components of a MRI unit for generating images of our body organs. It differs from a CT scan in the sense it is noninvasive: does not need ionizing X-rays used in CT. It has opened a large avenue for diagnosis and monitoring progress of a medical therapy.

7.3.1 NMR in the Context of MRI

In MRI of an organ in our body, ^{1}H NMR signals from $\sim 10^{20}$ ^{1}H nuclei/µL of H_2O molecules in tissues and cells of the organ emitted as the response to the stimuli of two external magnetic fields: one longitudinal $\mathbf{B}^0$ static and very strong and the other transverse $\mathbf{B}_{RF}(t)$ linearly oscillating at radiofrequency (RF) and very weak

($\sim 10^{-5} B^0$) are imaged spatially through imposing a set of magnetic field gradient coils $G_i(t)$ ($i = x, y, z$) on the static magnetic field $\mathbf{B}^0$. A nucleus has an intrinsic property spin interpreted as intrinsic angular momentum vector $\mathbf{I}$ of the nucleus. There is no such concept as the rotation of a nucleus or an electron around its own axis—there is only spin. $\mathbf{I}$ is completely distinct from $\mathbf{J}$ the rotational angular momentum vector of molecules. Quantum mechanics shows that $\mathbf{I}$ has the length of $\sqrt{I(I+1)}\hbar$, where $\hbar$ being Planck's constant h (6.634 × 10^{-34} J. s) divided by 2π ($\hbar = h/2\pi$) and has ($2I + 1$) components along the z-axis, i.e., I_z (defined by the direction of $\mathbf{B}^0$ in the "laboratory frame" of coordinates) with values $m_I \hbar$, where values of I (nuclear spin angular momentum quantum numbers) are restricted to 0, 1/2, 1, 3/2, 2, 5/2, 3, 7/2, etc., and those of m_I (magnetic quantum numbers) to $-I$, $-(I+1)$, ..., $(I-1)$, $+I$. Worth noting is the absence of half-integral quantum numbers for $\mathbf{J}$. Interesting is to note that at 0°K though all rotational motion ceases, an elementary particle like ^{1}H always has *spin*. The spin of a nucleus is designated by its I value, e.g., 0 for ^{12}C, ^{16}O, etc.; 1/2 for ^{1}H, ^{19}F, ^{31}P, ^{13}C, ^{15}N, etc.; 1 for ^{2}H, ^{14}N, etc.; 3/2 for ^{11}B, etc.; and 3 for ^{10}B, etc. Atomic nuclei besides their mass, charge, and spin may possess another intrinsic property magnetism. A result (Eq. 7.4) from quantum mechanics points to a direct link between $\mathbf{I}$ and $\boldsymbol{\mu}$ (spin magnetic moment vector) with the proportionality constant γ:

$$\boldsymbol{\mu} = \gamma \mathbf{I} \qquad (7.4)$$

Thus, a nuclear spin with $I \neq 0$ is a spin magnet and behaves as if a tiny bar magnet, which may interact with a static and with an oscillating magnetic field. Only nuclei with $I \neq 0$ can be NMR-active, and obviously, nuclei with $I = 0$ like ^{12}C and ^{16}O are not NMR-active. The constant γ is known as the magnetogyric or gyromagnetic ratio (rad. s^{-1}.Tesla^{-1}), and has its both magnitude and sign highly specific for a nucleus. The high value of γ of ^{1}H (+267.522 × 10^6 rad. s^{-1}. Tesla^{-1}) coupled with its astronomically high density ($\sim 10^{20}$/μL) in our body tissues has made MRI so successful for imaging our organs. A consequence of quantum mechanics is that the quantity $-\gamma B^0$ may be identified .as the famous "Larmor frequency" ω^0 (in rad. s^{-1}) (Eq. 7.5):

$$\omega^0 = -\gamma B^0 \qquad (7.5)$$

This is classically described as the angular velocity or frequency of revolution called *precession* of $\mathbf{I}$ or $\boldsymbol{\mu}$ around B^0 over a cone maintaining a constant angle between the spin polarization (the direction) and the field. The Larmor frequency ω^0 for ^{1}H being negative, the *precession* of its spin polarization around B^0 occurs in the clockwise direction. Further, for a ^{1}H, the spin up α state as compared to its spin down β state has lower energy. The very strong longitudinal field B^0 used is stable and provided by a superconducting magnet of strength 7 Tesla or above. The terms *Larmor frequency, precession frequency,* and *transition frequency* are often used interchangeably keeping in mind their signs (Abragam 1961; Merzbacher 1998; Keeler 2005; Levitt 2008; Chary and Govil 2008).

Before considering the interaction of a ^{1}H spin with the very weak transverse linearly oscillating RF magnetic field $\mathbf{B}_{RF}(t)$, let us look at the real situation of a tissue in an organ when placed in the static field $\mathbf{B}^0$. The tissue sample at thermal equilibrium is practically an ensemble of isolated ^{1}H spins wherein there is a slight excess of ^{1}H spins in the lower energy α state as per the Boltzmann law at a given temperature. The bulk magnetization of the tissue sample designated as **M** is directed parallel to $\mathbf{B}^0$, that is, along the z-axis. In the presence of $\mathbf{B}^0$ only, **M** is the equilibrium magnetization and symbolized as $M_0\mathbf{e}_z$, that is, $M_x = 0, M_y = 0$, and $M_z = M_0$ under this equilibrium condition.

Let now consider the interaction between **M** of a bulk sample of "isolated" ^{1}H nuclei and a pulse of the very weak (~$10^{-5}B^0$) RF field $\mathbf{B}_{RF}(t)$ along the +x - axis {$\mathbf{B}_{RF}(t) = B_{RF}\cos(\omega_{ref}t + \phi_p)\mathbf{e}_x$}, where ω_{ref}, the reference frequency (in rad.s^{-1}) of the spectrometer, is close or equal to ω^0 of the spins, and ϕ_p is the phase of the RF field. An RF pulse of duration (or length) t_p~10 – 20 μs followed by a period t of "free precession" (~50 ms—a few s, time of acquisition t_{acq} of NMR signals) "tilts" the vector $M_0\mathbf{e}_z$ away from the z-axis. The magnetization now begins to precess about the z-axis at ω^0. These superimposed motions because of their rapid and complicated time dependence are difficult to visualize. To solve such complicated issues, the trick of transforming the reference coordinate system from the laboratory frame to a new frame rotating clockwise (i.e., in the same sense for ω^0) round the z-axis at $\omega_{rot.\ frame} = \omega_{ref}$ is applied. This makes us to view nuclear spins from the angle of this rotating frame of coordinates (Keeler 2005; Levitt 2008; Chary and Govil 2008; Hore and Jones 2000).

Further simplification is achieved by noting that the linearly oscillating $\mathbf{B}_{RF}(t)$ field is resolvable into two counter-rotating circularly polarized components: one rotating clockwise (in the same sense as ω^0) at ω_{ref} and the other rotating anticlockwise (opposite to ω^0) at ω_{ref}, of which only the clockwise rotating one, the "resonant component" $\mathbf{B}_{res}^{RF}(t)$, is obviously the "effective" one.

Thus, once an x-pulse is turned on, the equilibrium magnetization $M_0\mathbf{e}_z$ "sees" an apparently static magnetic field $\mathbf{B}_{res}^{RF}$ and precesses more precisely "nutates" about the x-axis till it is turned off with the result: $M_0\mathbf{e}_z$ starts precessing anticlockwise during an x-pulse at ω_{nut} for a period t_p (called the duration or length of the pulse) and flips through an angle $\omega_{nut}t_p$ called the "flip angle" β_p. If t_p is so chosen that β_p is $\frac{\pi}{2}$, the x-pulse is referred as a $\left(\frac{\pi}{2}\right)_x$ pulse. Thus, the longitudinal magnetization $M_0\mathbf{e}_z$ aligned fully along the z-axis at equilibrium (under the influence of a very strong $\mathbf{B}^0$) is easily "tilted" or "knocked out" by a short $\left(\frac{\pi}{2}\right)_x$ pulse of a ~10^{-5} times weaker $\mathbf{B}_{res}^{RF}$ and on turning off the pulse at the end of t_p gets rotated anticlockwise into $-M_0\mathbf{e}_y$, a transverse one along the −y - axis and $M_0\mathbf{e}_z$ itself is destroyed. By altering ϕ_p in the "laboratory frame," one can shift the direction of the "pulse axis" in the rotating frame, e.g., by changing the phase ϕ_p from 0 to π/2, one instead of a $\left(\frac{\pi}{2}\right)_x$ pulse applies a $\left(\frac{\pi}{2}\right)_y$ pulse, which would rotate $M_0\mathbf{e}_z$ anticlockwise into $M_0\mathbf{e}_x$ at the end of the pulse, another transverse magnetization along the +x - axis. A $\left(\frac{\pi}{2}\right)_x$ or a $\left(\frac{\pi}{2}\right)_y$ pulse at the same time equalizes the population between the α and β spin states and excites coherences between the two states: The spins precess coherently or in phase generating an *x*-or a *y*-component of **M** (Abragam 1961; Keeler 2005; Levitt 2008; Chary and Govil 2008; Hore and Jones 2000).

The rotation of the longitudinal magnetization under such a strong field completely into the transverse *xy*-plane so easily by a much weaker field is a glare instance of the power of the idea of resonance or near resonance between ω^0 and ω_{ref}. It needs stressing here that unless the longitudinal bulk magnetization at equilibrium $M_0\mathbf{e}_z$ (which cannot emit any NMR signal) is "tilted" to the transverse *xy*-plane, no one could observe any NMR signal, which could be emitted only when the transverse bulk magnetization $M_0\mathbf{e}_y$ and/or $M_0\mathbf{e}_x$ starts decaying after turning off the RF pulse due to a process called transverse (T_2) relaxation, which is being described shortly.

To understand how a single signal ^{1}HNMR spectrum of water is obtained, we describe a basic "pulse-acquire" experiment with the exactly on-resonance condition: $\omega_{ref} = \omega^0$. Now, a $\left(\frac{\pi}{2}\right)_x$ pulse rotates **M** from the $+z$ - axis completely to the $-y$ - axis. Clearly, **M** does not then start to precess about the *z*-axis as switching over from the "laboratory frame" to the "rotating frame" has, under the exact on-resonance condition, completely removed the static field $\mathbf{B}^0$. However, if ω^0 is not equal to ω_{ref} there will be a "resonance offset" Ω^0 defined as the difference between ω^0 and ω_{ref}. Ω^0 may obviously be positive or negative. So, as soon as the pulse is turned off, **M** starts precessing about the $+z$ - axis at Ω^0 in the sense of its sign. In general, a "hard" ($\omega_{nut} >> \Omega^0$) virtually an "on-resonance" $(\pi/2)_x$ pulse is applied. So, at the end of this pulse, it may appear that **M** produced by this pulse would undergo "free precession" in the *xy*-plane indefinitely. This, however, does not happen since after turning off the pulse a new effect called the "relaxation effect" comes into force during the signal acquisition period *t*. The "relaxation effect" induces decay of coherences and transverse magnetization acquired during the $\left(\frac{\pi}{2}\right)_x$ pulse length. It is this decaying transverse magnetization that leads to the emission of the hard pulse excited NMR signals, which are acquired or recorded during the period *t* and accordingly *t* is often referred as t_{acq}, the acquisition period (Keeler 2005; Levitt 2008).

7.3.1.1 Relaxation Effects

The "relaxation effect" on free precession of magnetization after switching off the RF pulse was introduced by Felix Bloch (shared Nobel Prize in 1952 with Edward M. Purcell) in terms of two relaxation time constants: T_1 and T_2 in his phenomenological approach on the dynamics of bulk magnetization **M**. T_1 represents *longitudinal relaxation time constant* or *spin–lattice relaxation time constant*. We know that a $\left(\frac{\pi}{2}\right)_x$ pulse applied to spin population distribution at thermal equilibrium ($N_\alpha > N_\beta$) at a given temperature makes the population between the two states equal ($N_\alpha = N_\beta$) immediately after turning off the pulse. T_1 characterizes the return of this population equalization to their Boltzmann thermal equilibrium distribution in a few seconds (although in some samples it may take microseconds, milliseconds, minutes, hours, or even years), in other words, the return of the longitudinal magnetization to its equilibrium value $M_0\mathbf{e}_z$. The T_1 relaxation process follows a first-order kinetics and is generally described as $1 - e^{-t/T_1}$ (where t stands for the period of "free precession"). T_1 relaxation involves exchange of energy between the spin system and its molecular surroundings called "lattice" in a bulk sample. T_1, thus, is the time required for the buildup of longitudinal magnetization M_z from 0 to about 63% of its equilibrium value M_0.

On the other hand, T_2 represents the *transverse relaxation time constant*, the *coherence dephasing time constant*, the *coherence decay time constant*, or often improperly termed *spin–spin relaxation time constant*. T_2 characterizes the decay of magnetization in the transverse (*xy*-) plane or the loss of phase coherence excited by a $\left(\frac{\pi}{2}\right)_x$ pulse, after switching off the pulse. The transverse magnetization or coherence also decays with time but in this case back to its equilibrium value of zero.

T_2, thus, represents the time required for M_x or M_y to decay down to 1/e (about 37%) of their initial maximum value M_0. It is worth noting that the coherence decays at a remarkably slow rate: Spins execute hundreds of millions of precession circuits before losing synchronization. The decay of coherences increases the "entropy" of the spin ensemble pointing to its "irreversible" nature (Le Bihan 1991; Abragam 1961; Merzbacher 1998). For most of the cases, $T_2 \leq T_1$ and that both T_1 and T_2 are sources of valuable information about dynamics of the molecules in a system.

7.3.1.2 The Bloch Equation and Time Evolution of Bulk Magnetization

The general framework for the dynamics of **M** under the influences of a static field $\mathbf{B}^0$, a pulsed RF field $\mathbf{B}_{RF}(t)$, and the two relaxation effects (in terms of T_1 and T_2) is phenomenologically expressed as the famous "Bloch equation" (Eq. 7.6) (Abragam 1961; Levitt 2008; Bloch 1946; Hinshaw and Lent 1983):

$$\frac{d\mathbf{M}}{dt} = (\mathbf{M} \times \gamma \mathbf{B}) - \frac{(M_x \mathbf{e}_x + M_y \mathbf{e}_y)}{T_2} - \frac{(M_z - M_0)\mathbf{e}_z}{T_1} \quad (7.6)$$

where $\mathbf{B} = B^0 \mathbf{e}_z + B_{RF} \cos(\omega_{ref} t + \phi_p)\mathbf{e}_x$

The Bloch equation has the same central importance in NMR as the Lambert–Beer law has in optical spectroscopy. For the sake of compactness, the expressions for the transverse components of $\mathbf{M}(t)$: $M_x(t)$ and $M_y(t)$ are combined into a complex term (Eq. 7.7) as follows:

$$M(t) = M_0 \exp\left[i\Omega^0 t - (t/T_2)\right] \quad (7.7)$$

It is the decay of $M(t)$ rotating at the resonance offset Ω^0 toward zero during the acquisition period (t or t_{acq}) is at the origin of the emission of an NMR signal $s(t)$. The signal $s(t)$ is, in fact, proportional to $M(t)$ and is usually termed as "free induction decay" FID. On the other hand, the longitudinal component of **M** (Eq. 7.8) is as follows:

$$M_z(t) = M_0 \left(1 - e^{-t/T_1}\right) \tag{7.8}$$

starts growing up from its initial value of zero toward its equilibrium value M_0. Further, the expressions for $M(t)$ and $M_z(t)$ show how after turning off the $\left(\frac{\pi}{2}\right)_x$ pulse, the tip of **M** for suitable values of T_1 and T_2 ($T_2 \leq T_1$) tracks out a trajectory in three-dimensional space: The transverse magnetization $M(t)$ in the xy-plane oscillates and dies out as the longitudinal magnetization $M_z(t)$ relaxes back toward the $+z$-axis, and shorter is T_2, relaxation of $M_z(t)$ back toward the $+z$ - axis is faster and more abrupt. In the laboratory frame, **M** takes a spiraling path back to its equilibrium orientation. It may be noted that if an π_x pulse in lieu of an $\left(\frac{\pi}{2}\right)_x$ one had been applied, after switching it off the populations slowly relax back to their thermal equilibrium values, and $-M_0\mathbf{e}_z$ recovers back to the equilibrium value $M_0\mathbf{e}_z$ at a rate determined by $\frac{1}{T_1}$ (Keeler 2005; Levitt 2008; Chary and Govil 2008).

It is worth mentioning that the Bloch equation can be used to describe the behavior of a magnetization vector **M** under any conditions. In fact, it provides a valuable reference for developing equations of motion of magnetization for MRI, diffusion MRI, and other phenomena in NMR imaging.

7.3.1.3 An Outline of the Basic Pulsed-NMR Experiment

The magnetic nuclear spins in a sample are detected through "free induction decay" (FID) signals, and it is carried out by:
Allowing them to reach thermal equilibrium in a large magnetic field,
Rotating the nuclear spin polarizations by a hard RF pulse,
Detecting and amplifying the weak-transient radiofrequency signal, which is emitted after turning off the pulse when the spins resume their *precessional motion* in the main magnetic field. At the same time, the precession relaxes gradually with decrease in the xy-component $M(t)$ along with simultaneous increase in the z-component $M_z(t)$ with time. The decay of transverse magnetization $M(t)$ produces a voltage signal, which we measure as the NMR signal $s(t)$ or free induction decay (FID) (Keeler 2005; Levitt 2008; Chary and Govil 2008).

7.3.1.4 Signal-to-Noise Ratio (SNR)

The NMR signal from a single pulse-acquire experiment is often weak due to mixing of the signal with "noise" (the uncontrolled random RF signals due to thermal motion of electrons in the receiver coil of the spectrometer). The "signal-to-noise ratio" (SNR) is often low enough for the signal to be clearly discernable from the noise. However, by performing "signal averaging" the SNR may be improved upon to "pull" the signal very clearly "out" of the noise. The signal averaging involves repeating the same experiment called also "transients," "scans," or "shots" N times and averaging the resulting signals and the noise separately in which while the contribution of each signal adds, that of each noise being random begins to cancel as N increases. This procedure improves the SNR by $\sqrt{N}$ times; for example, by running 100 scans, the SNR for a single scan is magnified 10 times (Keeler 2005; Levitt 2008).

7.3.1.5 FID Signal s(t) to NMR Spectrum S(Ω) and Fourier Transformation

To make the information content in the time-domain FID signal s(t) more accessible to the human eye, one carries out Fourier transformation (FT) of s(t) to convert s(t) into the frequency-domain NMR spectrum S(Ω). The use of short RF pulses and FT technique has saved dramatically the time for recording an NMR spectrum by as high as 10^4 times.

The observed HWHM (half-width at half-maximum) of an NMR spectrum is, however, considerably larger than "true" $\frac{1}{T_2}$ and is represented as $\frac{1}{T_2^*}\left(T_2^* \leq T_2\right)$. This additional line width called "inhomogeneous broadening" arises from the technical difficulty in generating a "perfectly homogeneous" B^0. This makes B^0 and ω^0 spatially dependent modifying, thereby the relation: $\omega^0 = -\gamma B^0$ (Eq. 7.5) for ^{1}H nuclei to (Eq. 7.9):

$$\omega^0(\mathbf{r}) = -\gamma B^0(\mathbf{r}) \tag{7.9}$$

7.3.1.6 Measurements of Relaxation Times

The measurements of T_1 and T_2 are of great interest in the perspective of MRI. More viscous is the medium, lower are T_1 and T_2, which along with proton density PD may differ dramatically from one soft tissue to another in an organ of our body. PD, T_1, and T_2, more particularly T_2, represent the major sources of contrast mechanism in a standard MRI.

7.3.1.7 Inverse Recovery (IR) Method for Measuring T_1

In this method, a pulse sequence: $\pi_x - \tau - \left(\frac{\pi}{2}\right)_x$ is applied to the equilibrium magnetization $M_0\mathbf{e}_z$ of ^{1}H spin ensemble, e.g., water that gives only a single NMR line.

The FID NMR signal $s(\tau, t)$ has the amplitude $a(\tau)$, which reflects the history of the longitudinal magnetization and equals $\left(1 - 2e^{-\tau/T_1}\right)$. So, as the system keeps relaxing by T_1 mechanism during the period τ, the spectral peak amplitude starts from a negative value, crosses zero or null value, the null signal, obviously at $\tau_{null} = T_1 \ln 2$, and then moves over to positive values. Thus, the observation of the null signal permits one to evaluate directly T_1. In fact, the inversion recovery RF pulse sequence method in a NMR/MRI experiment is well-practiced for measurements of the longitudinal relaxation time constant T_1. The value of T_1 found for a tissue in our body ranges over 250 – 2500 ms (Keeler 2005; Levitt 2008; Chary and Govil 2008).

7.3.1.8 Spin-Echo (SE) Method and Measurement of T_2

Erwin Hahn discovered an ingenious method "spin echo" (SE) in 1950 for successful measurement of "true" T_2 that is free from any effect due to magnetic field in-homogeneity. The discovery of "spin echo" (SE) (also referred as "Hahn echo") may be regarded as the birth of modern pulsed NMR.

An SE experiment uses a pulse sequence: $\left(\frac{\pi}{2}\right)_x - \frac{\tau}{2} - \pi_y - \frac{\tau}{2}$, where transverse magnetization created by the $\left(\frac{\pi}{2}\right)_x$ pulse along the $-y$ - axis is refocused by the π_y pulse along the $-y$ - axis at the end of the pulse sequence of duration τ when the signal is acquired. The peak amplitudes of the signal are expressible as follows: $a(\tau) = a(0)\, e^{-\tau/T_2}$, where $a(0)$ is the peak height at $\tau = 0$. One can thus easily estimate the transverse relaxation time constant T_2 from the slope $-\frac{1}{T_2}$ of the linear plot of $\ln a(\tau)$ vs. τ. This estimate of transverse relaxation time constant T_2 is "true" independent of the inhomogeneous broadening. A term *echo time* (TE) needs introducing here, which stands for the duration counted from the end of the $\left(\frac{\pi}{2}\right)_x$ pulse to the time of acquisition of echo signal maximum, i.e., τ implying too that $\left(\frac{TE}{2}\right)$ is the timing for switching on the π_y pulse.

A vital component of a vast majority of MRI experimentations is the "spin-echo" (SE) experiment. The values of transverse magnetization relaxation time constant T_2 as evaluated by SE experiments of a number of our tissues fall generally within 25–250 ms, which is one order lower than T_1 values of the tissues: 250–2500 ms (Keeler 2005; Levitt 2008; Chary and Govil 2008; Hahn 1950).

7.3.1.9 Effects of Diffusion on Spin-Echo Signal

Perfect echo formation requires that the in-homogenous magnetic fields do not change while the pulse sequence is going on, so that the precession angle during the second $\frac{\tau}{2}$ interval is exactly the same as the precession angle during the first $\frac{\tau}{2}$ interval. However, if the spins diffuse into a region with a different magnetic field, the precession after the π_y pulse does no longer cancel exactly the precession before the pulse. This disturbs the echo formation and results in the signal attenuation at the echo maximum. This property allows, in fact, "spin-echo" (SE) experiments to use often for quantification of "diffusion" of molecules in a liquid. Further, diffusion of ^{1}H spins in water molecules, which are undergoing self-diffusion in our body tissues, has been found to induce such an attenuation in signal at the echo maximum in a SE experiment: A finding that has been brilliantly exploited to develop an advanced modality of MRI "diffusion MRI" (d-MRI) for successful neuro-imaging and oncologic imaging of our target organs (Levitt 2008).

7.3.2 NMR Imaging: MRI

A conventional NMR does not provide any spatial information. However, for imaging of an NMR signal $s(t)$ or its FT spectrum $S(\Omega)$, one requires encoding spatially the NMR signals. In 1972, Paul Lauterber (who shared Nobel prize with Peter Mansfield in 2003) showed that incorporation of precisely "controlled" in-homogeneity in the static field $B^0 \mathbf{e}_z$ by well-defined "magnetic field gradients" makes the total magnetic field along the z-axis to vary linearly as a function of position of the sample under study, and consequently to encode the NMR signals spatially. Spatially encoded NMR signals would, thus, depend on the distribution of the magnetic spins in the sample, in other words its shape. The map or portrayal of the spatial distribution of the NMR signal intensity or other NMR parameter in a water sample, for example, tissues in our organ thus obtained is basis for the construction of its ^{1}HNMR image. The noninvasive and apparently hazard-free nature of NMR imaging techniques has in view of high NMR sensitivity (γ) of ^{1}H, and its huge density ($\sim 10^{20}$/μL) in our body tissues has inevitably led to their application in the diagnosis of health status of soft tissues in our different organs as an exquisite clinical modality under the acronym "*MRI*" *where what we image actually is the distribution of spin density of water protons in a tissue of our body organ.*

7.3.2.1 MRI of Our Body Organs

We extend the above principle for MR imaging of 3D organs of our body to gain spatial information about soft tissues of our organs by applying spatially varying and time-varying three orthogonal gradient field components $G_i(t)$ ($i = x, y,$ and z) activated by computer-controlled very carefully designed x-, y-, and z-gradient coils. While these $G_i(t)$ components are turned on, the total magnetic field becomes:

$$\mathbf{B}^{\text{total}}(\mathbf{r},t)\mathbf{e}_z = \{B^0 + G_x(t)x + G_y(t)y + G_z(t)z\}\mathbf{e}_z = \{B^0 + \mathbf{G}(t)\bullet\mathbf{r}\}\mathbf{e}_z \quad (7.10)$$

Thus, the precession frequency of ^{1}H spins in an infinitesimal volume element (voxel) at $\mathbf{r}$ of our 3D tissue sample irradiated at $\omega_{\text{ref}} \cong \omega^0$ takes the form in the frame rotating at ω_{ref}:

$$\omega(\mathbf{r},t) = -\gamma\{B^0 + \mathbf{G}(t)\bullet\mathbf{r}\} - \omega_{\text{ref}} = -\gamma\{\mathbf{G}(t)\bullet\mathbf{r}\} = \omega_G(\mathbf{r},t) \quad (7.11)$$

The direct one-to-one correspondence between the frequency $\omega(\mathbf{r},t)$ or $\omega_G(\mathbf{r},t)$ and the spatial position $\mathbf{r}$ is the basic principle behind all magnetic resonance imaging (MRI) experiments where such space-dependent information is translated into images. Thus, for a given $\mathbf{G}$, two points displaced to each other by a vector orthogonal to $\mathbf{G}$ have the same $\omega(\mathbf{r},t)$ or $\omega_G(\mathbf{r},t)$. However, by transmitting another signal with a different $\mathbf{G}$, one can differentiate such two points. In a similar way, one can distinguish "all" points in the sample by transmitting sufficient number of different stimulus signals. An MR image of a tissue or organ is presented in the form of a set of pixels each of intensity proportional to the NMR signal amplitude of the contents of the corresponding volume element called voxel of a slice selected out of the organ.

It needs pointing that x-, y-, and z-components of $\mathbf{G}$ can be switched on selectively, one can obtain "axial," "sagittal," or "coronal" images of a target organ in a subject. Further, by exciting all the three gradients simultaneously, one can obtain an image in an oblique plane. This enables one to image a slice plane with any desired orientation through an appropriate choice of the direction of field gradients and of the RF pulses without any necessity of altering the position of the subject. This is a great advantage of MRI over other imaging techniques like CT. For seeing a position in our target organ, a fast RF pulse sequence with a large "field of view (FOV)" across an image with a distance of centimeter order is employed. The "region of interest

(ROI)" can now be decided and a smaller FOV and appropriate scan planes can be chosen (de Figueiredo et al. 2011; Chary and Govil 2008; Hinshaw and Lent 1983).

MRI signals from a sample have been shown to originate from its nonzero transverse magnetization of a sample $M(\mathbf{r}, t)$, the solution of the Bloch equation "generalized" with the inclusion of "magnetic gradient field" $\{\mathbf{G}(t) \cdot \mathbf{r}\}\mathbf{e}_z$ term in $\mathbf{B}(\mathbf{r}, t)$ (Abragam 1961; Hinshaw and Lent 1983):

$$\frac{\partial \mathbf{M}(\mathbf{r},t)}{\partial t} = \{\mathbf{M}(\mathbf{r},t) \times \gamma\mathbf{B}(\mathbf{r},t)\} - \frac{(M_x\mathbf{e}_x + M_y\mathbf{e}_y)}{T_2(\mathbf{r})} - \frac{\{M_z - M_0(\mathbf{r})\}\mathbf{e}_z}{T_1(\mathbf{r})} \quad (7.12)$$

where $[\mathbf{B}(\mathbf{r}, t) = \{B^0 + \mathbf{G}(t) \cdot \mathbf{r}\}\mathbf{e}_z + B_{RF} \cos(\omega_{ref}t + \phi_p)\mathbf{e}_x]$.

The equation is indicative of the dependence of image properties of a sample on (i) $M_0(\mathbf{r}) \propto$ proton density PD at $\mathbf{r}$, (ii) $T_1(\mathbf{r})$, and (iii) $T_2(\mathbf{r})$. The solution $M(\mathbf{r}, t)$ as found is as follows:

$$\mathbf{M}(\mathbf{r},t) = M_0(\mathbf{r}) \exp\{i\omega(\mathbf{r},t) - t/T_2(\mathbf{r})\} \quad (7.13a)$$

7.3.2.2 The Gradient Pulses

An important component of MRI "toolkit" is turning on a gradient for an interval τ shorter enough than the relaxation times so that one may regard the gradient a constant during the pulse (Chary and Govil 2008; Hinshaw and Lent 1983), and accordingly, $M(\mathbf{r}, t)$ after the gradient pulse takes the form:

$$\mathbf{M}(\mathbf{r},t) = M_0(\mathbf{r})\exp\{i\omega(\mathbf{r},t)\tau\} \quad (7.13b)$$

7.3.2.3 The "Receiver or Detector System" of an MRI Unit

It consists of (i) an antenna, a coil of wire wound around the sample, (ii) a matching network, (iii) a pre-amplifier, (iv) a quadrature phase-sensitive detector for M_x and M_y simultaneously as a function of time, and (v) an analog-to-digital converter (ADC) (Chary and Govil 2008; Hinshaw and Lent 1983), and detects $M(\mathbf{r}, t)$ and generates the output signal $s(t)$ from the quadrature phase-sensitive detector as a complex (two-channel) analog one (Eq. 7.14):

$$s(t) = K \int \mathbf{M}(\mathbf{r},t) \exp(-i\omega_{ref}t) d\mathbf{r} \quad (7.14)$$

where K is a complex "arbitrary constant" (for details, Ref. (Hinshaw and Lent 1983)). The analog $s(t)$ is converted by the ADC into two strings ("ones" and "zeros") of digital numbers $\{s_n\}$ given by (Eq. 7.15):

$$s_n = K \int \mathbf{M}(\mathbf{r},t) \exp(-i\omega_{ref}n\Delta t) d\mathbf{r} \quad (7.15)$$

7.3.2.4 Image Contrast

The acceptance of MRI for clinical diagnostics relies heavily on whether this modality can unequivocally and in a distinctly superior way differentiate different tissues or different health status of a tissue, in other words as compared to the other tools, how much an improved "contrast" it can create in images of the tissues. The simplest MR contrasting properties of ^{1}H of water molecules in tissues are obviously, as stated, their spatially dependent characteristics *viz.* (i) $M_0(\mathbf{r}) \propto$ PD at **r**, (ii) $T_1(\mathbf{r})$, and (iii) $T_2(\mathbf{r})$ the value of each of which changes from one normal tissue to another one, and more importantly from one diseased tissue to a healthy one.

Let us describe how to achieve these PD-, T_1-, and T_2-weighted images. It may be noted that $M_0(\mathbf{r})$ or PD at **r** provides the overall image shape, whereas $T_1(\mathbf{r})$ and $T_2(\mathbf{r})$ provide significant local information. Although a standard MR image is a composite of $M_0(\mathbf{r})$, $T_1(\mathbf{r})$, and $T_2(\mathbf{r})$, there are mechanisms for controlling their proportions to produce strikingly diverse images. This is magnificently exhibited through a "spin-echo imaging scheme" based on the basic SE pulse sequence incorporated with slice selective (G_z or G_{slice}), phase encoding (G_y or G_{phase}), and frequency encoding (G_x or G_{freq}) MRI gradient pulses. This classic imaging gradient SE pulse sequence, an advanced form of the basic SE, is commonly referred as "conventional spin echo" (CSE).

Among the three tissue contrast mechanisms (PD, T_1, and T_2), owing to their dramatically different values for soft tissues T_1 and T_2 are particularly useful and used as the primary ones. The values of T_1, T_2, and PD are specific to a tissue or pathology. Typical values of T_1 plus T_2 (at 1.0 Tesla) and PD for various tissues have been cited as Appendix in Table 7.1. T_1 and T_2 keep changing significantly with the onset of a disease and with aging. For clinical use, the contrast (C) between a pair of tissues A and B is defined as follows: $C = s_A - s_B$, where s_A and s_B are MR signals emitted from the tissues A and B, respectively.

T_2-weighted images are often found better than T_1-weighted ones for detecting infarcts and tumors. It is the well-designed nonuniformity imposed on the original

Table 7.1 Typical values of T_1 plus T_2 (at 1.0 Tesla) and PD for various tissues found in their MR images

Tissue	PD (%)	T_1 (ms)	T_2 (ms)
Fat	90	180	90
Gray matter	69	520	90
White matter	61	390	90
CSF	100	2000	300
Edema	86	900	77
Liver	91	270	50
Water	100	2500	2500
Cardiac muscle	100	600	34
Skeletal muscle	100	600	40

steady magnetic field through appropriate field gradient coils that enables spatially locating ^{1}H in a tissue, constitutes the basis of MRI. No doubt, a spectacular advantage of NMR imaging technology is one can image any desired section of an organ by selecting simply the direction of the field gradients. While discussing "diffusion MRI," we shall come across a vastly improved MR image contrast mechanism: Diffusion-weighted images are based on the coefficient of diffusion of water molecules in our tissues (Le Bihan 2014; Chary and Govil 2008).

7.3.2.5 Basic Components of a Standard MRI System

MRI systems are very complex instruments capable of performing a wide range of sophisticated experiments. Here, we mention only the basic components that make up the system in its simplest form (Chary and Govil 2008; Hinshaw and Lent 1983):
1. A superconducting magnet of strength (0.5–3.0 Tesla) which could provide a uniform and a stable magnetic field B^0.
2. An RF transmitter to irradiate the sample and at the same time subjecting the sample to a linearly oscillating practically a clockwise rotating weak magnetic field $\frac{1}{2}B_{RF}(t)$ perpendicular to the steady field B^0.
3. A time-varying magnetic field gradient system: $G_x(t)$, $G_y(t)$, and $G_z(t)$ generally of strength $(10 - 100) \times 10^{-3}$ Tesla. m^{-1} capable of generating controlled spatial in-homogeneity in the steady field. The experimenter controls the magnetic field imposed on the sample by modulating four-component vector $\mathbf{P}(t)$ which consists of the gradient field controlled by three input gradient signals: $G_x(t)$, $G_y(t)$, and $G_z(t)$ and of a RF field governed by $\frac{1}{2}B_{RF}(t)$ signal.
4. The signal output $s(t)$ is emitted by a receiver or detection system.
5. The image reconstruction and display are carried by an imager system with a computer.

7.3.3 Diffusion MRI

We have noted that during a spin-echo (SE) experiment, when ^{1}H spins carried in water molecules in a sample or a tissue self-diffuses from its one region into another one where the strength of the magnetic field is different, one comes across a significant attenuation in the SE signal: A characteristic of immense value for applications is ranging from quantifying "diffusion" rather "self-diffusion" of the spin carrying molecules in a liquid or in a liquid-like phase to the development of an advanced MRI modality: "Diffusion MRI" (d-MRI) for neuro-imaging and onco-logic imaging of our target organs. Differential rates of water ^{1}H spin diffusion among our different tissues and also between healthy and diseased states of a tissue afford a diffusion-weighted tissue MR contrast mechanism with a clarity much superior to that of a standard PD-, T_1-, or T_2-weighted one. This has led to the emergence of a new source of MRI contrast mechanism based on the self-diffusion coefficient of the tissue water molecules. Indeed, d-MRI provides such exquisite images of in vivo water molecules self-diffusing in our tissues and cells that it has attained a "state-of-art" status in clinical diagnostics. Truly, d-MRI is now the choice for imaging brain tissue structure and mapping the circuitry of our brain network for normal and diseased brains, and also for detection and monitoring the response to therapy of cancers and its metastatic forms in breast, prostate, liver, etc. The incredible success of d-MRI for very early detection of the brain stroke has, in fact, saved life of many patients sparing them from severe disabilities (Le Bihan 1991; Le Bihan et al. 1991; Mori and Barker 1999; de Figueiredo et al. 2011; Le Bihan 2014; Nicolay et al. 2001).

"Self-diffusion" more popularly known as "Brownian motion" of water molecules is significantly faster in a container outside our body than in our tissues and cells. This is obviously caused by the barriers in our tissues and cells which could modulate the rate and also direction of water diffusion in the tissues and cells. The prospect of d-MRI to provide details of such internal features of our tissue and cells makes it a tool of early harbinger of biologic abnormality. It is indeed fortunate that diffusion effects on NMR have provided us such a remarkable imaging tool to investigate in a completely noninvasive way in vivo molecular displacements of water in the micron and submicron range, encompassing the size of most biological tissue structures.

The diffusion coefficient D is truly a tensor quantity represented as $\overline{\mathbf{D}}$. No doubt, the diffusive flow or D of water molecules to be imaged for a given tissue would be affected by the environment of the tissue to a significant extent; e.g., D of pure water in a glass at body temperature (37 °C) is reduced from $\sim 10^{-4}$ cm$^2 \cdot$s^{-1} to $\sim 10^{-5}$ cm$^2 \cdot$s^{-1} in the gray matter of our brain. Thus, in a span of 40 ms, water molecules in the gray matter undergo an RMS displacement in the x-direction, i.e., $\sqrt{x(t)^2}$ of ~ 9 µm. This compares very well to the average size of our body cells (~ 10 µm) exhibiting the potentiality of the diffusion-weighted MRI (DW-MRI or DWI) for monitoring the changes that go on at a cellular spatial scale (Le Bihan 1991; de Figueiredo et al. 2011).

On the other hand, had the tissue environment no such effect on water diffusion in the tissue, these water molecules would undergo RMS displacement of ~ 30 µm, which is greater than cell dimensions. The hindrance to diffusion of water in a tissue or a cell with consequent lowering in D of water by an order of magnitude is easily traceable to the hydrophobic nature of the membranes and macromolecules present in the cell. This is further augmented through exchange of water between intra- and extracellular compartments coupled with the tortuosity of the extracellular space caused by the cell's size, organization, and packing density.

In spite of the fact that d-MRI has a spatial resolution of the order of mm, the technique exhibits high sensitivity to changes in D measured at the µm level of the cellular scale. d-MRI, thus, has the wonderful capability of shedding light on the cellular architecture at the mm scale: a vivid example of which is to document the modulation of the direction of water diffusion in a tissue by the underlying cellular structure. These findings equip one to derive information about anisotropy in highly organized structures such as myelinated white matter fiber tracts and, in fact, a sophisticated diffusion technique: "Diffusion tensor imaging (DTI)" has evolved for in vivo determination of directionality and magnitude of water diffusion in highly organized tissues (Le Bihan 1991; Mukherjee et al. 2008; de Figueiredo et al. 2011; Winston 2012).

7.3.3.1 Diffusion-Weighted Imaging (DWI)

We have noted that DWI probes diffusion of ^{1}H spins in water molecules of our body at extra-, intra-, or transcellular spatial scale: not possible by using conventional MRI. The DWI signal of an imaged volume ("voxel") at a location in a given slice of a target organ is a reliable measure of the diffusive flow of water at that location. The noninvasive characteristic of DWI makes it a tool of choice to visualize diffusion of water in a tissue.

Diffusion is found slower in an intracellular compartment than is in an intercellular one. It has already been pointed out that diffusion of water in a tissue is highly sensitive to its cellular environment. In fact, early pathological changes in the brain have been brilliantly manifested in the DW image: During the diagnosis of a patient suffering from a brain stroke when a PD-, T_1-, and even T_2-weighted MRI could show no infarct inferring as if the brain is normal, a 30 s DWI scan shows out very clearly the infarct as bright areas after a few minutes without using any contrast agent. The bright areas in the DW image point to higher concentration of water in the affected cell areas owing to slower diffusion there as compared to the rate of water diffusion in normal ones. Truly, a vastly improved sensitivity index has emerged out.

DWI is, however, most suited for tissues where diffusion of water is predominantly isotropic, for example, gray matter in the cerebral cortex and major brain nuclei. Applications of in vivo measurements of molecular diffusion of water in our tissues range from the determination of molecular organization in a tissue to the emergency management of stroke patients and detection of cancers. The diagnostic suitability of DWI has been examined for intracranial lesions, acute cerebral infarcts, tumor, and demyelinating diseases (Padhani et al. 2009; Mukherjee et al. 2008; de Figueiredo et al. 2011).

For quantifying diffusion MRI signal intensities, we need the expression for the transverse component of the bulk magnetization $\mathbf{M}(\mathbf{r}, t)$ of our tissue sample to which our required d-MRI signal intensities are proportional. The transverse component of $\mathbf{M}(\mathbf{r}, t)$ is obtained by solving the "Bloch–Torrey equations": The "generalized Bloch equations" in MRI with the inclusion of the terms due to the transfer of magnetization induced by spin diffusion of ^{1}H nuclei of water molecules in our tissues (Eq. 7.16) wherein the field gradient $\mathbf{G}(t)$ contains now bipolar symmetric diffusion-encoding pulsed gradients $\mathbf{G}_{\text{diff}}$ in the z-direction in addition to the usual MR imaging gradients (Abragam 1961; Torrey 1956; Stejskal and Tanner 1965):

$$\frac{\partial \mathbf{M}(\mathbf{r},t)}{\partial t} = \{\mathbf{M}(\mathbf{r},t) \times \gamma \mathbf{B}(\mathbf{r},t)\} - \frac{(M_x \mathbf{e}_x + M_y \mathbf{e}_y)}{T_2(\mathbf{r})} - \frac{\{M_z - M_0(\mathbf{r})\}\mathbf{e}_z}{T_1(\mathbf{r})}$$
$$+ \nabla \cdot \overline{\mathbf{D}} \cdot \nabla \mathbf{M}(\mathbf{r},t) \quad (7.16)$$

where $\mathbf{B}(\mathbf{r},t) = \{B^0 + \mathbf{G}(t) \cdot \mathbf{r}\}\mathbf{e}_z + B_{RF}\cos(\omega_{ref}t + \phi_p)\mathbf{e}_x$ for anisotropic diffusion where the diffusion coefficient $\overline{\mathbf{D}}$, a tensor characterized by a positive definite 3 × 3 symmetrical matrix with elements describing molecular displacements in three dimensions. Since water diffusion in most of the tissues is isotropic, the "Bloch–Torrey equations" simplifies to (Eq. 7.17)

$$\frac{\partial \mathbf{M}(\mathbf{r},t)}{\partial t} = \{\mathbf{M}(\mathbf{r},t) \times \gamma \mathbf{B}(\mathbf{r},t)\} - \frac{(M_x \mathbf{e}_x + M_y \mathbf{e}_y)}{T_2(\mathbf{r})} - \frac{\{M_z - M_0(\mathbf{r})\}\mathbf{e}_z}{T_1(\mathbf{r})}$$
$$+ D\nabla^2 \mathbf{M}(\mathbf{r},t) \quad (7.17)$$

where $\mathbf{B}(\mathbf{r},t) = \{B^0 + \mathbf{G}(t) \cdot \mathbf{r}\}\mathbf{e}_z + B_{RF}\cos(\omega_{ref} + \phi_p)\mathbf{e}_x$, and the diffusion coefficient D is a scalar. The corresponding equation for the transverse component of $\mathbf{M}(\mathbf{r},t)$ in complex form $M(\mathbf{r},t)$ can be expressed as (Eq. 7.18):

$$\frac{\partial M(\mathbf{r},t)}{\partial t} = i\Omega^0 M(\mathbf{r},t) - i\gamma\{\mathbf{G}(t) \cdot \mathbf{r}\}M(\mathbf{r},t) - \frac{M(\mathbf{r},t)}{T_2(\mathbf{r})} + D\nabla^2 M(\mathbf{r},t) \quad (7.18)$$

where $\Omega^0 = \omega^0 - \omega_{ref}$

Since the detector of the NMR/MRI system operates generally under resonance or nearly resonance condition ($\omega^0 \cong \omega_{ref}$, i.e., $\Omega^0 \cong 0$), the above equation modifies to (Eq. 7.19):

$$\frac{\partial M(\mathbf{r},t)}{\partial t} = -i\gamma\{\mathbf{G}(t) \cdot \mathbf{r}\}M(\mathbf{r},t) - \frac{M(\mathbf{r},t)}{T_2(\mathbf{r})} + D\nabla^2 M(\mathbf{r},t) \quad (7.19)$$

Stejskal and Tanner solved successfully this equation by introducing a novel pulse sequence: pulsed-field gradient spin-echo (PFG-SE) sequence which can sensitize MR to self-diffusion of water molecules in a sample or a tissue (Stejskal and Tanner 1965).

7.3.3.2 The Stejskal and Tanner's Pulsed-Field Gradient Spin-Echo (PFG-SE) Sequence

Toward solving Eq. (7.19) for exclusively diffusion-induced magnetization dynamics, one can rewrite this equation to arrive at the master equation (Eq. 7.20) for the Stejskal and Tanner's (PFG-SE) treatment:

$$\frac{\partial M(\mathbf{r},t)}{\partial t} = -i\gamma\{\mathbf{G}_{diff}(t) \cdot \mathbf{r}\}M(\mathbf{r},t) + D\nabla^2 M(\mathbf{r},t) \quad (7.20)$$

It is basically a single-shot spin-echo (SE) T_2-weighted ($\frac{\pi}{2} - \tau - \pi$) sequence into which two symmetric diffusion-sensitizing bipolar rectangular field gradient pulses of amplitude G_{diff} each of duration δ, one on either side of the π refocusing pulse, are incorporated into the conventional SE MR imaging (CSE) scheme, the gradient pulses being separated by an time interval Δ, the diffusion time. The diffusion gradient pulses "phase tag" or label diffusing ^{1}H spin magnets carried by water molecules. In course of the application of the gradient pulse applied before the π refocusing pulse, protons experience the first gradient pulse at the location a_1. On the other hand, when the second gradient pulse after the π refocusing pulse is applied, a phase shift depending on another location a_2 where the protons experience the second gradient pulse is produced. This is called "spin un-tagging." As in-between the two gradient pulses, a π pulse had been applied, the "net phase shift" (ϕ_{net}) is (Eq. 7.21):

$$\phi_{net} = \gamma G_{diff} \delta (a_2 - a_1) \quad (7.21)$$

If the proton spins would not have moved or diffused away from the position a_1 i.e., $a_1 = a_2$, the pair of the gradient pulses could not have any effect on the phase. So, the condition for ϕ_{net} to be nonzero is that the spins must undergo diffusion during the time interval Δ between the gradient pulses. The magnitude of ϕ_{net} would, of course, depend on the history of the spins during the interval Δ. However, for extending PFG-SE NMR pulse sequence to "diffusion MRI," one needs applying diffusion-weighted SE MRI sequence with three imaging gradient pulses G_x (or G_{freq}), G_y (or G_{phase}), and G_z (or G_{slice}) of CSE as follows. The sequence consists of slice selection by ramping up the G_z (or G_{slice}) gradient both during the excitation by the $\frac{\pi}{2}$ and the refocusing by the π RF pulses. The $\frac{\pi}{2}$ RF excitation pulse is followed by phase encoding achieved immediately by the G_y (or G_{phase}) gradient. Following the π RF refocusing pulse, frequency encoding is performed during the acquisition of the echo signal (readout period) by the G_x (or G_{freq}) gradient, designated commonly as the "readout gradient" (G_{read}). A pair of "diffusion-encoding gradient" (G_{diff}) pulses (one before and another after the π RF pulse as described above) is, of course, added to the pulse sequence after an interval since the end of the G_{phase} pulse.

The solution of the Bloch–Torrey equation in terms of M(**r**, *t*) as arrived at on the basis of Stejskal and Tanner's PFG-SE pulse sequence scheme at the echo time (TE) for a position vector **r** in a voxel of a tissue is (Eq. 7.22):

$$M(TE) = M_0 \exp(-TE/T_2) \exp\left\{-(\gamma G_{diff}\delta)^2 (\Delta - \delta/3) D\right\} \quad (7.22)$$

where M(TE) is the "diffusion gradient-induced" transverse magnetization of the sample at TE with M_0 as M(0) where $t = 0$ is counted just immediately after the $\frac{\pi}{2}$ pulse (when any diffusion gradient is yet to be applied), and $(\gamma G_{diff}\delta)^2 (\Delta - \delta/3)$, the attenuation factor is the value of the integral: $\gamma^2 \int_0^{TE} \left[\int_0^t G_{diff}(t')dt'\right]^2 dt$ in which the "term" $(\Delta - \delta/3)$ represents trailing to leading-edge separation.

This attenuation factor has been abbreviated as the "*b*" factor by Dennis Le Bihan in his name ("B"ihan): $b = \gamma^2 G_{diff}^2 \delta^2 (\Delta - \delta/3)$, and M(TE) is commonly expressed as (Eq. 7.23):

$$M(TE) = M_0 \exp(-TE/T_2) \exp(-bD) \qquad (7.23)$$

Since the diffusion-weighted or diffusion-sensitized echo signal ($s_b \neq 0$ usually denoted as s_b) called DWI (diffusion-weighted imaging) signal is proportional to M (TE), and the non-diffusion-weighted echo signal ($s_b = 0$ usually denoted as s_0) is proportional to $M_0 \exp(-TE/T_2)$, we get the DWI signal expression (Eq. 7.24):

$$s_b = s_0 \exp(-bD) \qquad (7.24)$$

On re-expressing it in terms of diffusion-induced echo signal attenuation (*E*) (Eq. 7.25):

$$E = s_b/s_0 = \exp(-bD) \qquad (7.25)$$

The attenuation factor *b* varies in magnitude from $0 - 20 \times 10^5$ s. cm^{-2}, the $b = 0$ value implying to a simple SE sequence. The *b*-value is the index for diffusion weighting. The relative attenuation of the signal intensity observed on images obtained at different *b*-values enables one to characterize the healthy and diseased states of a tissue based on their differential water diffusion.

To get away with the effects of T_2 decay, what one does is to measure the diffusion-induced echo signal, the DWI signal at two distinct values of the attenuation factor *b*. This way, one gets exclusively the diffusion-weighted attenuation (*E*). When one acquires both a diffusion-weighted scan (i.e., $b \neq 0$) and a non-diffusion-weighted scan (i.e., $b = 0$) at the same TE, the logarithmic relation between s_b and s_0 (Eq. 7.26) becomes:

$$\ln(s_b/s_0) = -bD \quad \text{or} \quad \ln s_b = \ln s_0 - bD \qquad (7.26)$$

One can now evaluate *D* of water in each voxel of the image of a tissue or an organ simply by dividing $\ln(s_b/s_0)$ by $(-b)$. However, to obtain a more accurate estimate of *D*, one does linear regression analysis of $\ln s_b$ on *b* for a series of DWI signals at of different values of the *b*-factor (generally by varying the gradient amplitude G_{diff} without changing any other parameter in the *b*-factor) (Le Bihan 1991; Le Bihan et al. 1991; Mori and Barker 1999; Mukherjee et al. 2008; de Figueiredo et al. 2011; Nicolay et al. 2001; Luypaert et al. 2001; Winston 2012; Sotak 2004; Le Bihan 2013). This procedure is generally adopted in clinical d-MRI scanners for evaluating *D* of water molecules in our tissues, e.g., white matter of our brain. The result found $(0.60 \pm 0.01) \times 10^{-5}$ cm^2. s^{-1} (like that its gray matter) is an order of magnitude less than *D* of pure water ($\sim 10^{-4}$ cm^2.s^{-1}). This reflects how powerful the role the cell and tissue structure factors (mentioned earlier) play for impeding or restricting in vivo water diffusion in a tissue or a cell (Malayeri et al. 2011; Le Bihan 1991; de Figueiredo et al. 2011; Winston 2012). On this ground, Dennis Le Bihan introduced, for practical purposes, a new concept "apparent diffusion coefficient" (ADC) as a substitute for the true "diffusion coefficient" *D* in white matter and other tissues. As a matter of fact, the ADC parameter which one may calculate from a DW image has proved its immense worth as a very early and clear indicator of the onset of an ischemic stroke. We discuss more about ADC in the following.

It needs pointing that the Stejskal and Tanner's equation is valid strictly only for a tissue where the rate of diffusion of water molecules is virtually the same over all its directions, in other words, is predominantly isotropic irrespective of whether water diffusion in the tissue is restricted or not. The Stejskal and Tanner's method introduced in 1960 for quantifying DW-MRI was implemented in routine practice by Le Bihan et al. in 1986. Imaging of anisotropic diffusion in tissues like the white matter of a brain will be discussed later.

7.3.3.3 Apparent Diffusion Coefficient Mapping (ADC Mapping): Quantitative Analysis with DWI

We have just noted that D of water molecules in a tissue as measured by DWI may be significantly lower than that of pure water in a test tube. In general, a DWI signal does not practically suffer from any T_1 confounding effect. However, it may suffer significantly from the T_2 confounding effect whenever long T_2 components are present. This is what is referred as "T_2 shine-through effects." This misleads seriously interpreting a DW image as artificially "hyperintense." Strikingly, ADC maps are insensitive to "T_2 shine-through effects" and are successfully utilized for quantifying diffusion. So, whenever a bright lesion is encountered on a DW image, the ADC map must be inspected to look for whether this bright area has really a low-intensity ADC signal to verify whether diffusion there is restricted or not.

For quantifying restricted diffusion of water in a tissue, as mentioned under DWI and also for doing away with just described possible misinterpretations of DW images, one obtains quantitative images rather parametric mappings in terms of Le Bihan's "apparent diffusion coefficient" (ADC) concept in an attempt toward elucidating the underlying mechanisms of ongoing diffusion in our tissues. The attenuation (E) in d-MRI echo signal (Eq. 7.25) for restricted diffusion of water in a tissue will now be re-expressed as (Eq. 7.27):

$$E = s_b/s_0 = \exp\{-b(\text{ADC})\} \qquad (7.27)$$

Thus, for a b-value chosen as the reciprocal of the expected ADC in a tissue, the diffusion-weighted signal s_b gets reduced to about 37% of the non-diffusion-weighted signal s_0. What we now measure as the slope of the linear plot of $\ln(s_b/s_0)$ vs b for a tissue in vivo is the "apparent diffusion coefficient" (ADC): An average of diffusion coefficients D over many restricted random walks portraying complex diffusion of water molecules in a tissue and exhibits deviations from "true" D. Lower is the value of ADC for a tissue, stronger the barriers to diffusion in the tissue: ADC is, thus, an index of water diffusion in tissues and cells.

An ADC map or image for a tissue in an organ is a quantification of its DW image and an important diagnostic aid to DWI. ADC of cerebral water gets reduced following acute brain stroke enabling rapid diagnosis and emergency management for the treatment of the patient. The ADC concept, in fact, was first employed to detect early ischemic stroke. Its value has been calculated during postprocessing of two DW images with at least two different b-values: b_1 and b_2 (Eq. 7.28):

$$\mathrm{ADC} = \frac{\ln(s_{b_1}/s_{b_2})}{(b_2 - b_1)} \quad (7.28)$$

In clinical practice, multiple b-values for a good number of DW images of a tissue are employed for improved characterization of the tissue. The image of different tissues in the region of interest (ROI) of an organ is displayed as a set of calculated parametric maps referred as ADC or diffusion maps. One can now derive ADC value of a given tissue from these maps. ADCs are preferentially calculated for homogenous areas of a tissue structure. The values of ADC ($\times 10^{-5}$ cm^2·s^{-1}) thus found for different tissues of a normal brain are as follows: cerebrospinal fluid (CSF): 2.94, gray matter: 0.76, and white matter (corpus callosum: 0.22; axial fibers: 1.07; transverse fibers: 0.64). The significant difference in the last two values of ADC along mutually perpendicular directions: ADC$_\parallel$ and ADC$_\perp$ are clear evidences for anisotropic diffusion of water in the white matter.

The cellular swelling that accompanies depolarization of cell membranes decreases the net displacement of water molecules with a concomitant reduction in ADC of tissue water. Whereas DWI signal intensity is a function of self-diffusion of water molecules in a tissue, ADC is sensitive to changes in the biological milieu that accompany a number of pathophysiological processes. The examples are as follows: ischemic stroke, transient ischemic attack TIA, ischemic depolarization IDs, cortical spreading depression, status epilepticus, and hypoglycemia, that is, the cases of cellular swelling "cytotoxic edema," which is accompanied by lowering of ADC for tissue water molecules. A lesion with restricted diffusion (e.g., strokes, abscesses) and lower ADC is obviously manifested as "hypo-intense" (darker pixels) on its calculated ADC map ($b = 1000$ s.cm^{-2}) but would appear as "hyperintense" (brighter pixels) on the DW image. Conversely, tissues like cerebrospinal fluid CSF with unrestricted water diffusion and high ADCs appear bright in ADC and dark in DWI. So, the signal intensity of a DW image is inversely related to that of ADC. Thus, between the diseased and healthy states of a tissue at a given b, the local concentration of water being higher due to its considerably slower diffusion in the diseased one, it looks brighter in the DW image but darker in the ADC image. For appreciating diffusion properties of water at each pixel for a human brain, an ADC scan has certainly distinct superiority over the corresponding DWI one. ADC is a proven clinical marker (Bozgeyik et al. 2013; Le Bihan 1991; Le Bihan et al. 1991; Mori and Barker 1999; Mukherjee et al. 2008; de Figueiredo et al. 2011; Nicolay et al. 2001; Luypaert et al. 2001; Winston 2012; Sotak 2004; Le Bihan 2013).

7.3.3.4 High-Speed Diffusion Imaging: Use of Echo-Planar Imaging (EPI)

The PFG-SE sequence makes DWI not only highly sensitive to diffusion of water molecules in our tissues and cells but also to bulk motion such as pulsations related to the cardiac cycle, involuntary twitches, and CSF flow. However, the technique is very slow and requires long acquisition time.

R. Turner and D. Le Bihan utilizing Peter Mansfield's echo-planer imaging (EPI) sequences developed an MRI technique capable of producing DW images at video rates and overcame the difficulty of long acquisition time *vis-à-vis* motion artifacts. In this technique, all data points for reconstructing the image are sampled after a single $\frac{\pi}{2} - \pi$ pair pulses and could reduce image acquisition times to as short as 20–100 ms. Such a high speed has been achieved by obtaining all spatial-encoding information after a single RF excitation. This has eliminated practically all motion-related artifacts and has also enabled imaging of rapidly changing physiological processes in a human body. This single-shot DW– SE – EPI technique provides heavily diffusion-weighted reliable images of the human brain free from motion artifacts in a matter of seconds and has made spectacular strides in the clinical evaluation of stroke and in studies on functional imaging of the human brain. It is now a tool of immense value for applications to our brain, abdomen, and heart (Chary and Govil 2008).

7.3.3.5 Diffusion Tensor Imaging (DTI)

Many of our tissues have highly organized structure and display anisotropic organization. Neural "axons" of "white matter" in our brain or muscle fibers in our heart, the internal fibrous structure of which has in many ways close similarity to the anisotropy of certain crystals, are vivid examples of such tissues. We have already come across an example of "diffusion anisotropy" in the fiber tracts of the white matter of our brain: Axial $ADC_{\parallel}$ is $1.07 \times 10^{-5} cm^2 . s^{-1}$, whereas transverse $ADC_{\perp}$ is $0.64 \times 10^{-5} cm^2 . s^{-1}$ showing clearly that water diffuses at a rate distinctly higher along the direction of the nerve fiber in the internal structure of white matter than along the direction perpendicular to the nerve fiber. Anisotropic diffusion of water is observed not only in the tissues quoted above but also in other highly organized tissues of our body. No doubt, the ADC values of water in all such tissues are direction-dependent rather orientation-dependent, and we shall consider ADC or D no more a scalar but a tensor, which describes diffusion in 3D space. The technique that is employed for measurements of ADC or D of water molecules in such highly organized tissues is popularly known as "diffusion tensor imaging" (DTI).

What is done in DTI is fiber tracking where instead of measuring many values of the *b*-factor, one measures the diffusion coefficient of water along many directions. Thus, another remarkable feature d-MRI is its noninvasive investigative capability of throwing light on the connectivity of the structure of highly organized parts of an organ like brain, e.g., white matter, and macroscopic axonal organization and skeletal muscles. The potential contributors of anisotropy to diffusion of water in the white matter are as follows: the cytoskeleton (neurofilaments and microtubules), the axonal membranes, and the myelin sheath. What we need is how to connect image intensities quantitatively to relative diffusion rates of water molecules along different directions in a tissue. This is achieved through the technique of "diffusion tensor imaging" otherwise called "diffusion tract imaging" (DTI). In fact, DTI has become the tool of choice for studying disorders in white matter of the brain (Mukherjee et al. 2008; de Figueiredo et al. 2011).

The data for anisotropic diffusion are described in terms of the diffusion (ADC) tensor $\overline{\mathbf{D}}$, a 3×3 symmetric matrix reflecting diffusion rates along six directions. The three diagonal elements (D_{xx}, D_{yy}, D_{zz}) of $\overline{\mathbf{D}}$ matrix represent ADCs measured along the principal (x-, y-, and z-) laboratory axes. For the symmetric $\overline{\mathbf{D}}$ matrix, the mirror image off-diagonal elements being equal ($D_{xy} = D_{yx}, D_{yz} = D_{zy}, D_{zx} = D_{xz}$), there are effectively only three off-diagonal terms: D_{xy}, D_{yz}, D_{zx}, which reflect the correlation of random motions between each pair of principal directions. The frame of reference for $\overline{\mathbf{D}}$ can be rotated to an optimal coordinate system based upon the "*diffusion ellipsoid*" in which form $\overline{\mathbf{D}}$ referred as $\overline{\mathbf{D}}_{prin}$ or $\overline{\Lambda}$ has only three real and positive diagonal elements called eigenvalues: $\lambda_1, \lambda_2, \lambda_3$ and no off-diagonal elements. The major axis of length assigned λ_1 of the *diffusion ellipsoid* is parallel to the principal diffusion direction within a voxel. This major axis often corresponds to anatomic features such as white matter tracts or fascial planes. The lengths of the two minor axes of the diffusion ellipsoid are assigned the other two eigenvalues: λ_2 and λ_3, i.e., ($\lambda_1 > (\lambda_2 \geq \lambda_3)$, λ_1 being orthogonal to each of λ_2 and λ_3. Obviously, each of the three λ_i's is proportional to Einstein's RMS diffusion displacement in the corresponding direction: λ_1 is the value of the maximum, primary, longitudinal, parallel, or axial diffusivity, whereas λ_2 and λ_3 represent the magnitude of diffusion in the transverse plane or of the radial diffusivity $\{(\lambda_2 + \lambda_3)/2\}$ (Mukherjee et al. 2008; de Figueiredo et al. 2011).

7.3.3.6 Indices of Anisotropic Diffusion

Based on the values of $\lambda_1, \lambda_2, \lambda_3$ for an ADC tensor $\overline{\mathbf{D}}_{prin}$, two commonly indices for estimating degree of diffusion anisotropy, mean diffusivity (MD) and fractional anisotropy (FA), have been introduced. These indices are defined as follows:

$$\mathrm{MD} = \frac{\mathrm{Tr}\overline{\mathbf{D}}_{prin}}{3} = \frac{\lambda_1 + \lambda_2 + \lambda_3}{3} \quad (7.29)$$

and

$$\mathrm{FA} = \sqrt{\frac{3}{2}} \frac{\sqrt{(\lambda_1 - \mathrm{MD})^2 + (\lambda_2 - \mathrm{MD})^2 + (\lambda_3 - \mathrm{MD})^2}}{\sqrt{\lambda_1^2 + \lambda_2^2 + \lambda_3^2}} \quad (7.30)$$

These two parameters of DTI are found sensitive but nonspecific measures of altered tissue structure, say of brain white matter. In essence, the "orientationally averaged" mean diffusivity is represented by MD and its values are remarkably similar across gray and white matter, between different subjects and across mammalian species at $0.7 \times 10^{-5} \mathrm{cm}^2 \cdot \mathrm{s}^{-1}$. On the other hand, FA values range over 0 to 1: $\mathrm{FA} = 0$ corresponds to spherical or fully isotropic diffusion, whereas $\mathrm{FA} = 1$ corresponds to tubular or fully anisotropic diffusion. The value of FA grades the degree of anisotropy of diffusion of water in a tissue. The FA index, thus, aids to assess how the principal directions of the diffusion tensor could be utilized to shed light on the connectivity in the white matter of the brain, the "tractograph." FA index is also useful for assessing brain maturation in children and connection disorders, for example, dyslexia and some psychiatric disorders (Mukherjee et al. 2008; de Figueiredo et al. 2011).

7.3.3.7 Application Potential of DTI

Introduction of the technique of DTI is a spectacular achievement in neuro-imaging. It enables us to gain remarkable insights into the connectivity in the network of the brain of an individual. The increasing popularity of DTI day by day is the testimony to its magnificent clinical success on a substantially large number of patients suffering from various nerve disorders. DTI has proven its usefulness not only for tract-specific localization of white matter lesions like trauma and defining of severity of traumatic brain injury, but also for distinguishing Alzheimer's disease from other types of dementia, and in characterization of skeletal and cardiac muscle. Since understanding the sensitivity to fiber orientation is a vital aspect in the field of sports medicine, DTI may aid greatly in the treatment of injury of muscle and tendon by imaging the structure of the muscle and tendon of the injured sports persons (Mukherjee et al. 2008; de Figueiredo et al. 2011).

What we have most interestingly seen is that in spite of diffusion being not directly related to the MR phenomenon, the coefficient of self-diffusion of water molecules in a tissue of our body provides a far superior contrast mechanism as compared to that by the spin relaxation times T_1 or T_2 for detecting diseased status of a tissue, particularly for early brain ischemia by MRI. No less interesting is MRI is the only tool to determine the diffusion coefficient in vivo of tissue water molecules with good accuracy and spatial resolution (Le Bihan 1991; Mukherjee et al. 2008; de Figueiredo et al. 2011; Levitt 2008).

7.4 Cancer

Cancer is the second largest killer after cardiovascular disease. Every year, millions of new patients are diagnosed with cancer. For detection, studying efficacy of ongoing cancer treatment and staging of the disease many modalities like sonography, X-rays, CT scan, and MRI are used by doctors. These are of varying cost and useful to differing extents in cancers of different organs. Many of these are invasive, and many may not give very precise location and extent of the lesion. MRI, though costly, is a very effective noninvasive tool to find fine diseased structures of inner parts of our body organs and cerebral components, which may not be possible with other modalities. We discuss some of these in the following subsections.

7.4.1 Interpretation of DWI and Its Assessment

By observing the relative attenuation of signal intensity on the images obtained at different b-values, tissue characterization based on differences in water diffusion is possible. For example, in a heterogeneous tumor, the more cystic or necrotic area of the tumor shows more signal attenuation on high b-value images, because water diffusion is less restricted. By contrast, the more cellular solid tumor areas continue to show relatively high signal intensity. Visual assessment of the relative tissue signal attenuation at DWI is applied for tumor detection, characterization, and the evaluation of treatment response in patients with cancer.

One of the pitfalls of visual assessment of directional or index DW images is high signal at DWI; this may be mistaken for restricted diffusion due to the T_2 "shine-through" effect. This effect can sometimes be reduced by the choice of a short TE, and a large b-value, but it cannot be avoided.

Though the diffusion anisotropy is an aid to identify tumor invasion of adjacent structures, it has no effective role for malignant cells due to their disorganized growth. However, the potential of DTI in organs with discernible structural organization such as the prostate and kidneys is quite high.

It has been mentioned earlier that the areas of restricted diffusion in highly cellular areas show low ADC values compared to the less cellular areas that exhibit higher ADC values. The areas of restricted diffusion appear to be higher in signal intensity on the directional or index DW images. These areas appear as low signal intensity areas (i.e., opposite to DW images) on the ADC map (Koh and Collins 2007; Malayeri et al. 2011; Bozgeyik et al. 2013). It may be added that a color overlay image of therapeutic-induced ADC change within the tumor called "functional diffusion map" (fDM) has been developed. The fDM provides the ability to objectively segment tumor into color regions based on the magnitude and direction of ADC change.

7.4.2 Clinical DWI in the Body: Some Technical Issues

In the implementation of DWI in the body, two main technical issues are as follows: *breath-hold* and *non-breath-hold* imaging strategies. Breath-hold imaging allows rapid assessment of a target volume (e.g., liver, kidney, and elsewhere in the abdomen). The images thus obtained retain good anatomic detail and are usually not degraded by respiratory motion or volume averaging. Small lesions may be better perceived, and the quantification of ADC is theoretically more accurate than with a non-breath-hold technique. One example of such technique is the breath-hold single-shot spin-echo EPI combined with parallel imaging (e.g., sensitivity encoding) and fat suppression. The image acquisition time at each breath-hold is 20–30 s, and imaging is typically completed in a few breath-holds. The disadvantages of breath-hold imaging include a limited number of b-value images that can be acquired over the duration of a breath-hold, poorer signal-to-noise ratio compared with multiple averaging methods, and greater sensitivity to pulsatile and susceptibility artifacts.

On the other hand, non-breath-hold spin-echo EPI combined with fat suppression is a versatile technique that can be used as a general purpose DWI sequence in the body imaging. Multiple slice excitation and signal averaging over a longer duration improve the signal-to-noise and contrast-to-noise ratios. However, the image acquisition time using this technique is longer compared with the breath-hold imaging, typically 3–6 min (*c.f.* 20–30 s) depending on the coverage required and the number of b (>5) values used. The evaluation of tumor heterogeneity may be compromised by the degree of motion and volume averaging (Koh and Collins 2007).

It needs pointing that while performing DWI in a target of our body, one should use meticulous techniques for minimizing artifacts due to bulk motion of the target. Otherwise, significant degradation of the quality of the DW image could occur. Further, the acquisition time for capturing the signal should be kept as short as possible. Depending on the tissue being investigated, the TR (pulse repetition time) should be long enough to minimize T_1 saturation effects (Malayeri et al. 2011).

7.4.3 Applications of DWI-ADC Mapping of Target Organs

The emerging applications of DWI-ADC are tumor detection, characterization, distinguishing tumor tissue from nontumor tissue, and monitoring and predicting treatment response.

DWI has been applied to the evaluation of *intracranial diseases, such as cerebrovascular accidents, trauma, epilepsy, depression, dementia, and neurotoxicity*. DWI has shown its capability of early detection or subtle changes within the brain before any visible abnormality can be seen on conventional morphologic imaging. The range of its applications got further widened to extracranial sites, including the abdomen and pelvis. The developments of echo-planar imaging (EPI), high-gradient amplitudes, multichannel coils, and parallel imaging have been instrumental in making DWI-ADC mapping so successful. The parallel imaging led to reduction in the TE (the echo time or echo-train length), thereby substantially diminishing motion artifact at image acquisition and enabling high-quality DW images of the targets in our body (Koh and Collins 2007; Malayeri et al. 2011; Padhani et al. 2011).

7.4.3.1 Detection of Tumors

DWI is superior to T_2-weighted imaging in the detection of lesions in organs. Tumors being frequently more cellular result in more restricted diffusion of water molecules as compared to their original normal tissues and show relatively more brightness on their DWI scans: Metastases appear as high signal intensity foci at DW images. *The criteria for tumor visibility on DWI are focal hyperintensities and the corresponding hypo-intensities on the ADC map relative to the rest of the target organ.* It is of interest to note that the sensitivity and the specificity for detection of metastases are considerably higher in DWI than in super-paramagnetic iron oxide

(SPIO)-enhanced MRI, the values observed for "liver" metastases being 82% and 94% in DWI as compared to 66% and 90%, respectively, in SPIO-enhanced MRI. Similarly, for "colorectal hepatic" metastases high values of the sensitivity and the specificity *viz.* 86% and 94%, respectively, have been observed in DWI. Further, there is increasing evidence that DWI improves sensitivity and specificity in the detection of prostate cancer. For the peripheral zone and the transition zone of the prostate gland, the sensitivities of 94% and 91%, respectively, and specificities of 90% and 84%, respectively, were found for cancer diagnosis of the gland (Koh and Collins 2007; Bozgeyik et al. 2013; Padhani et al. 2011; Malayeri et al. 2011).

7.4.3.2 Characterization of Tumors

Differences in cellularity of different tumors may reflect their histologic composition and biologic aggressiveness.

1. Central nervous system (CNS)
   (a) The first application of DWI for tumor characterization has been demonstrated in central nervous system (CNS) for "brain" tumors. The most common extra-axial brain tumors are *meningioma*. The differentiation between atypical/malignant and typical meningioma could be evaluated by DWI, and ADC values are found lower in atypical/malignant meningioma than in typical meningioma (Bozgeyik et al. 2013).
   (b) The most common intracranial tumor in adults is metastasis. For patients with intracerebral metastasis, the mean ADC value in metastasis found was 0.72×10^{-5} cm^2. s^{-1} showing restricted diffusion (Koh and Collins 2007; Bozgeyik et al. 2013).
   (c) For extracranial tumors such as *thyroid, orbit, head, and neck tumors*, mean ADC values of malignant and benign thyroid nodules found were 0.96×10^{-5} cm^2. s^{-1} and 3.06×10^{-5} cm^2. s^{-1}, respectively, demonstrating the utility of DWI-ADC in differentiating malignant and benign thyroid nodules (Bozgeyik et al. 2013).
2. Breast lesions
   As early as in 2002, oncologic imaging of "breast" by DWI-ADC mapping showed clearly that the mean ADC value of normal tissue and benign lesions is significantly higher than that of malignant lesions. ADC values exhibit an inverse relationship with tumor grade: The mean ADC value for less aggressive tumors (grade 1 and in situ lesions) is 1.19×10^{-5} cm. s^{-1} as compared to 0.96×10^{-5} cm^2. s^{-1} for more aggressive ones (grade 2 and grade 3 lesions).

   For discriminating between benign and malignant lesions, the ADC value of a breast lesion normalized to that of "ipsilateral remote glandular" tissue is often used: The normalized ADC values of benign and malignant breast lesions are 1.10 and 0.55, respectively (Koh and Collins 2007; Bozgeyik et al. 2013; Malayeri et al. 2011; Giannelli et al. 2014).
3. Hepatic lesions
   The observed mean ADC values are as follows: 2.5×10^{-5} cm^2. s^{-1} for benign "liver" lesions and 1.52×10^{-5} cm^2. s^{-1} for malignant ones, respectively. Further, benign "liver" lesions such as *cysts and hemangiomas* (where water

diffusion is much less restricted) exhibit higher mean ADC value: 2.45×10^{-5}cm^2. s^{-1} as compared to the ADC value of 1.08×10^{-5}cm^2. s^{-1} for malignant lesions such as metastases and hepatocellular carcinoma (where diffusion is very much restricted). In another study on the diagnostic role of DWI on "liver" lesions, it was shown that the mean ADC values of hepatoma and metastases were 0.90×10^{-5}cm^2. s^{-1} and 0.79×10^{-5}cm^2. s^{-1}, respectively.

On the basis of ADC values, abscesses (which have low ADC values) can be distinguished from cystic and necrotic metastases (which have higher ADC values). While drawing such inferences, one must keep in mind that although the ADC values differ in summary statistics between benign and malignant lesions, making use of an individual ADC value to prospectively characterize lesions might be risky in some instances. This risk could arise from considerable overlap in the ADC values of benign and malignant abnormalities.

4. Renal tumors
   (a) Accurate characterization of renal masses is essential to ensure appropriate staging and prognosis. The pattern of increased brightness at DWI and decreased ADC values in *renal cell carcinoma* (RCC) is similar to that seen in solid malignant lesions of other organs. In fact, as compared to simple or mildly complex cysts and oncocytomas, solid RCCs showed significantly lower ADCs. Further, the ADC value in RCC (2.71×10^{-5}cm^2. s^{-1}) was found higher than in transitional cell carcinoma (1.61×10^{-5}cm^2. s^{-1}). Lower ADC values of malignant tumors than those of benign ones are caused by restricted diffusion in renal neoplasms cell membrane integrity and tissue cellularity. A study on renal lesions by DWI-ADC mapping, the mean ADC value found for RCC (1.41×10^{-5}cm^2. s^{-1}) was significantly lower than *oncocytomas* (1.91×10^{-5}cm^2. s^{-1}) and still lower than benign lesions (2.23×10^{-5}cm^2. s^{-1}) (Koh and Collins 2007; Bozgeyik et al. 2013).
   (b) DWI is also used in detecting colorectal carcinoma and cystic lesions of the pancreas and ovaries (Koh and Collins 2007; Bozgeyik et al. 2013). We now cite an example of diagnostic utility of DWI-ADC for pancreatic tumors. Like other malignant tumors, it shows bright signal on DWIs and has lower ADC value compared to normal pancreatic tissue. ADC values of pancreatic cancer are in wide range with overlapping normal pancreatic tissue values. Restricted diffusion in the pancreatic cancer might be related to increased cellularity and fibrosis. In contrast to other tumors, fibrosis might contribute to diminished ADC values and cellularity (Koh and Collins 2007; Bozgeyik et al. 2013).
   (c) When DWI is used to characterize focal renal lesions, water diffusion in the normal kidney is anisotropic due to the organization of the renal tubules. DWI could readily distinguish between cystic from solid renal lesions. To confidently distinguish malignant from benign renal neoplasms on the basis of ADC measurements (unlike the case for breast lesions where the ADCs of malignant breast lesions were found lower than the ADCs associated with benign diseases), DWI was able to distinguish cystic *soft tissue sarcomas* from solid types (Koh and Collins 2007; Bozgeyik et al. 2013).

5. Pelvic lesions
   (a) For male pelvis, DWI-ADC provides a method for the assessment of the prostate gland in the characterization of prostate cancer and tumor aggressiveness with improved sensitivity and specificity. Prostate cancer is the most common genitourinary malignancy in men. Prostate cancer like other cancers has higher cellular density and an excess of intra- and intercellular membranes compared with normal glandular tissue. In prostate cancer, normal tissue of the gland is replaced by *adenocarcinoma*. The tumor is built up of high-density malignant epithelial cells, which results in decrease in ADC values. DWI-ADC has shown its potential on tumor staging and treatment response. DWI-ADC has demonstrated its use as a biomarker for local recurrence of prostate cancer. It needs pointing, in particular, the utility of ADC maps in this context because of the T_2 "shine-through" from the normal high signal intensity peripheral zone seen on DW images. The lower ADC values of prostate cancer correlate with higher cellularity at histologic analysis.

   It needs emphasizing that although DWI-ADC mapping for the diagnosis of prostate cancer shows improved sensitivity and specificity, it is difficult for this technique to detect tumors smaller than 5 mm. Further, tumor ADC values may help identify those patients with low risk, localized prostate cancer who may get benefit from radical treatment (Koh and Collins 2007; Bozgeyik et al. 2013; Malayeri et al. 2011).

   (b) DWI-ADC mapping is also widely used in the evaluation of uterine cancers: The ADC values of upper urinary tract cancer with grade 3 (median, $0.91 \times 10^{-5} cm^2 \cdot s^{-1}$) are significantly lower than those of upper urinary tract cancer with grades 2 and 1 (median, $1.22 \times 10^{-5} cm^2 \cdot s^{-1}$), whereas there was no significant difference in ADC value according to pathologic stage (pT2 or lower: median, $1.09 \times 10^{-5} cm^2 \cdot s^{-1}$; and pT3 or higher: median $0.94 \times 10^{-5} cm^2 \cdot s^{-1}$) (Bozgeyik et al. 2013; Malayeri et al. 2011; Yoshida et al. 2011).

   (c) DWI plays a significant role in the diagnosis of urinary bladder (UB) tumors based on the calculated ADC values. In fact, the bladder cancers are clearly detected in DW images. The mean ADC value in the UB carcinoma group was found significantly lower than that in the benign group and in the control group: $1.0684 \times 10^{-5} cm^2 \cdot s^{-1}$, $1.8030 \times 10^{-5} cm^2 \cdot s^{-1}$, and $2.010 \times 10^{-5} cm^2 \cdot s^{-1}$, respectively. There was also a significant difference among the mean ADC values of different grades of malignant tumors, corresponding to $0.9185 \times 10^{-5} cm^2 \cdot s^{-1}$ and $1.2815 \times 10^{-5} cm^2 \cdot s^{-1}$ in high-grade and low-grade malignant UB carcinomas, respectively. The sensitivity, specificity, and accuracy of DWI in the diagnosis of malignant UB lesions were 100%, 76.5%, and 93.65%, respectively. Thus, DWI-ADC mapping is a beneficial tool to differentiate between benign and malignant UB lesions, as well as between high-grade and low-grade UB carcinomas (Bozgeyik et al. 2013; Malayeri et al. 2011; Avcu et al. 2011).

(d) It was further found that the mean ADC value of cervical cancer (1.09×10^{-5}cm^2. s^{-1}) was found lower than that of normal cervical tissue (2.09×10^{-5}cm^2. s^{-1}) reflecting again the potential ability of DWI-ADC measurements to differentiate between cancerous and normal tissues (Bozgeyik et al. 2013; Malayeri et al. 2011).
(e) The ADC value of cancerous cervical tissue (1.09×10^{-5}cm^2. s^{-1} $\pm$ 0.20) is found significantly lower than that of normal cervical tissue (1.79×10^{-5}cm^2. s^{-1} $\pm$ 0.24). Moreover, DWI has shown promise for the staging of cervical cancer by allowing more accurate determination of extent of invasion and status of nodal involvement. In a 3.0 Tesla MR imaging study, it was demonstrated that *metastatic lymph nodes* had significantly lower ADC values (0.06×10^{-5}cm^2. s^{-1}) than that of intact lymph nodes (0.21×10^{-5}cm^2. s^{-1}) (Malayeri et al. 2011).

7.4.4 Distinguishing Tumors from Non-Tumors

In prostate cancer, although differentiating a tumor from other causes of a low-signal-intensity lesion in the prostate gland is quite difficult on conventional T$_2$-weighted MRI, DWI has demonstrated its potential for tumor identification. The normal central gland of the prostate has a lower ADC than the peripheral zone. Prostate cancers, which appear as low signal intensity foci on ADC maps, typically show lower ADC values than the peripheral zone, the transitional zone, and the central gland. There is, however, a difficulty owing to a significant overlap in the ADC values of prostate cancer and benign prostate changes. *A recent interesting observation is that low ADC in the central gland and the transitional zone accompanied by the loss of glandular anisotropy is more suggestive of tumor than glandular hyperplasia.*

In the spine, the ability to distinguish a malignant from a nonmalignant cause of vertebral collapse is highly challenging: Malignant vertebral infiltration and fracture frequently appear as high signal intensity areas on DWI compared to nonmalignant causes. A sensitivity of 42–100% and specificity of 92–94% were reported using visual qualitative assessment, although one study reported DWI to be unhelpful. However, quantitative analyses show that malignant vertebral body infiltration and vertebral body fracture return lower ADC values than benign causes. A clear separation of ADC measurements between malignant and benign groups was observed, though significant overlap was also reported. Another difficulty *one may face is that inflammatory conditions like tuberculosis infection and osteomyelitis can mimic malignant disease on DWI.* So, caution should be exercised while adopting these criteria for clinical practice (Koh and Collins 2007; Malayeri et al. 2011).

7.4.4.1 Monitoring Treatment Response
Anticancer treatment, if effective, should result in tumor lysis, loss of cell membrane integrity, increased extracellular space, and therefore an increase in water diffusion.

It is expected that after the initiation of this treatment, one would observe an increased value of ADC for the tumor. Since this scheme does not consider the contribution of intravascular perfusion to the diffusion measurement, which may be substantial in a tumor, therapies that are targeted against tumor vasculature may also result in a reduction in the ADC.

Since both cellular death and vascular changes in response to a treatment can precede changes in lesion size, changes observed in DWI scans can be an effective early biomarker for the treatment outcome both for vascular disruptive drugs and for therapies that induce apoptosis during monitoring the response to the treatment. *In fact, in most malignant tumors such as "breast" cancers, primary and metastatic cancers to the "liver," primary sarcomas of "bone," and in "brain" malignancies, the successes in the therapy are manifested by increases in ADC values.*

It has been observed in human studies of vertebral metastases and brain tumors that there is a reduction in high signal intensity at DWI as the response to the treatment. However, such visual appraisal may be confounded by T_2 "shine-through" effects. *Studies on hepatocellular carcinoma, cerebral gliomas, and soft tissue sarcoma using ADC measurements have shown that individuals responding to treatment show a significant rise in the ADC values after the therapy.* Also, in a study of colorectal hepatic metastases, an increase in ADC was observed in patients responding at least partially to the treatment, but such an ADC increase was not observed in the nonresponders.

DWI can play a significant qualitative and quantitative role in assessing posttreatment tumor response for evaluating tumor behavior, in planning future therapies, and in assessing tumor relapse. Qualitative analysis is based on visually assessing changes in DWI-ADC scans: an increase in ADC signal in lesions that have responded to treatment. Response to therapy often manifests as an increase in ADC values relative to a low ADC value of the tumor at baseline. Posttreatment DWI shows different signal intensity behaviors depending on the tissue component and the type of therapy used. After transarterial chemo- or radio-embolization, the ADC values of a hepatocellular carcinoma may show an early decrease followed by consistent increases resembling the case with cystic or necrotic changes. The ADC values are found significantly lower in patients with a higher degree of fibrosis than those in patients with less or no liver fibrosis. A correlation between ADC value and degree of fibrosis has been demonstrated in multiple studies.

It may be mentioned that different subtypes of renal cell carcinoma (RCC) respond differently to molecularly targeted therapies. ADC values have been shown to be a promising biomarker for monitoring the response of patients with prostate cancer to radiation therapy; *the mean ADC value of tumor was found increased after radiation therapy.*

The significant reduction in the diffusion properties of water protons in prostate cancer and the resulting reduction in the measured ADC value of prostate cancer relative to the normal prostate tissue caused by the restricted movements of water proton are due to a combination of the extracellular space and the complex intracellular environment resulting in a low ADC value.

ADC values have been shown to be a promising imaging biomarker for monitoring patients treated with radiation therapy. In a study, the mean ADC value of tumor was shown to increase after radiation therapy, whereas the mean posttherapy ADC values of benign peripheral and transition zones were lower than their pretherapy values. In patients with biochemical relapse after undergoing radiation therapy, the use of diffusion-weighted imaging together with T_2-weighted imaging showed greater sensitivity for the detection of recurrence than did T_2-weighted imaging alone. In patients with metastatic bone disease, an increase in tumor ADC value correlated with decreasing prostate-specific antigen levels in those who were treated with androgen deprivation therapy. DWI-ADC is also widely used for monitoring the effectiveness of the therapy for uterine cancers (Koh and Collins 2007; Malayeri et al. 2011).

7.4.4.2 Predicting Treatment Response

A most intriguing finding of the use of DWI-ADC measurements in cancer patients is predicting the response of tumor to chemotherapy or radiation therapy. *This has been demonstrated for the response of rectal carcinoma, cerebral gliomas, or colorectal hepatic metastases in patients to chemotherapy or radiation therapy with high sensitivity, if the cellular tumor has low baseline pretreatment ADC values.* The sensitivity of the response to the treatment, however, gets considerably diminished for tumors with high pretreatment ADC values.

Similarly, for cerebral gliomas and breast carcinoma, an early increase in the ADC value after commencing the treatment is predictive of a better therapy outcome. The increase in the ADC value preceded any reduction in tumor size.

The body of evidence obtained suggests that ADC measurement is a potentially useful tool providing unique prognostic information, and such measurements deserve wider investigations in large clinical studies (Koh and Collins 2007; Malayeri et al. 2011).

7.5 Conclusions

Nuclear magnetic resonance (NMR) is an incredible tool with very successful wings of applications spreading over staggering disciplines encompassing solid-state physics, materials science, quantum computing, chemistry, molecular biology, geology, agriculture, pharmaceuticals, medicine, etc. Based on the principle and instrumentation of this very powerful and highly informative tool, a remarkably powerful clinical imaging modality magnetic resonance imaging (MRI) has been developed with well-designed modifications necessary for imaging our body organs. In MRI, what we generally measure is ^{1}H NMR emission signals from water molecules in tissues and cells of our target organs as the basic primary step. For the purpose of imaging, the additions made to a NMR system are as follows: a gradient system that produces time-varying magnetic fields of well-controlled spatial nonuniformity; and an imager system that reconstructs and displays the images.

To obtain an image contrast between healthy and diseased tissues, one makes use of a property of water, which can provide means to introduce contrast in images of the tissues. Among such means, the most common are proton density (PD), longitudinal relaxation time T_1, and transverse relaxation times T_2 and T_2^*. This "standard MRI" is described as follows: PD-weighted, T_1-weighted, T_2-weighted, or T_2^*-weighted MRI. Of these, T_2-weighted or T_2^*-weighted MRI has proved to be the most useful diagnostic tool. Introduction of MRI has, no doubt, opened up a spectacular noninvasive clinical tool to investigate different soft tissues of our body without any exposure to ionizing radiation for imaging the target organ. It is well known but remarkably not widely that NMR offers a method for quantifying diffusion of molecules in a liquid or liquid-like phase. This fact has been brilliantly exploited for accurately determining the self-diffusion coefficient of water molecules in tissues and cells of our organs in MRI technology with the addition of a pair of bipolar field gradient (diffusion gradient) pulses in the basic spin-echo (SE) pulse sequence employed; a spectacular advancement in MRI has emerged under the title "diffusion MRI (d-MRI)," which has unparalleled sensitivity to water movements within the architecture of the tissues without requiring contrast agents, or chemical tracers. d-MRI is presented as "diffusion-weighted imaging (DWI)" with its complementary calculated "apparent diffusion coefficient (ADC) mapping" for isotropic diffusion in not well-organized or disorganized tissues, the most common cases, e.g., gray matter, CSF of the brain, tumors in breast, liver, and prostate gland; and as "diffusion tensor imaging (DTI)" for anisotropic diffusion in cases where the tissue is well-organized, e.g., along the white matter tracts of the internal capsule in the brain. It may be noted that unlike standard T_1 and T_2 mapping approaches, which are dependent on the magnetic field strength, diffusion-derived measures should be comparable between centers. Diffusion MRI is an incredible tool for neuro-imaging and oncologic imaging of our target organs. It has achieved spectacular success in early detection of brain stroke and other neurologic diseases, and in differentiating benign from malignant lesions, monitoring treatment response after chemotherapy or radiotherapy, differentiating post-therapeutic changes from residual active tumor and detecting recurrent cancer.

A magnificent development in MRI is functional MRI (fMRI), termed more appropriately as blood–oxygenation level-dependent functional MRI (BOLD fMRI), a very powerful technique for understanding the functioning of our brain. This technology has excited psychologists for studying mental disorders and other neurodegenerative diseases.

DWI-ADC mapping provides information about the functional environment of water molecules in tissues, thereby augmenting the morphologic information provided by conventional MRI. It is only when DWI signal intensity information is combined with calculated ADCs that DWI-ADC findings become a powerful tool to provide unique information related to tumor cellularity and the integrity of the cellular membrane. The DWI-ADC measurements for different parts in our body, in fact, provide information not only about cellularity and cell membrane architecture of tumoral tissues, but also about differentiation between malignant and benign tumoral tissues (despite overlapping ADC values existing in benign and malignant tumors), and assessment of the response of tumors to chemotherapy or radiation therapy in various target parts of our body. Challenges such as standardization of data acquisition and analysis are certainly there before the widespread adoption of DWI-ADC. However, DWI-ADC modality is a technique whose potential has been well exemplified through a number of case studies. Further, the technique is noninvasive, can be quickly performed, and deserves incorporation into standard clinical protocols after necessary evaluation.

Acknowledgement The author expresses his immense pleasure for the voluntary assistance rendered by Dr. Kheyanath Mitra, Ms. Rajshree Singh, Mr. Sambhav Vishwakarma, Mr. Sourav Mondal, and Mr. Satish Tiwary in various forms.

Appendix

We quote the typical values of T_1 plus T_2 (at 1.0 Tesla) and PD for various tissues found in their MR images in Table 7.1 (Chary and Govil 2008).

References

Abragam A (1961) The principles of nuclear magnetism (No. 32). Oxford University Press, Oxford

Avcu S, Koseoglu MN, Ceylan K, Dbulutand M, Unal O (2011) The value of diffusion-weighted MRI in the diagnosis of malignant and benign urinary bladder lesions. Br J Radiol 84(1006): 875–882

Bloch F (1946) Nuclear induction. Phys Rev 70(7–8):460–474

Bozgeyik Z, Onur MR, Poyraz AK (2013) The role of diffusion weighted magnetic resonance imaging in oncologic settings. Quant Imaging Med Surg 3(5):269–278

Chary KVR, Govil G (2008) NMR in biological systems. Springer, Dordrecht

Crank J (1980) The mathematics of diffusion. Oxford University Press, Oxford

de Figueiredo EH, Borgonovi AF, Doring TM (2011) Basic concepts of MR imaging, diffusion MR imaging, and diffusion tensor imaging. Magn Reson Imaging Clin N Am 19(1):1–22

de Schepper IM, Ernst MH (1979) Self-diffusion beyond Fick's law. Physica 98(1–2):189–214

Einstein A (1905) Über die von der molekularkinetischen Theorie der Wärme geforderte Bewegung von in ruhenden Flüssigkeiten suspendierten Teilchen. Ann Phys 322(8):549–560

Evans WAB, Powles JG, Dornford-Smith A (1991) Characterization of diffusive motion in anisotropic liquid systems. J Phys Condens Matter 3(11):1637–1648

Fick A (1855) Über diffusion von. Poggendroff's Annalen der Physik und Chemie 94(1):59–86

Giannelli M, Sghedoni R, Iacconi C, Iori M, Traino AC, Guerrisi M, Mascalchi M, Toschi N, Diciotti S (2014) MR scanner systems should be adequately characterized in diffusion-MRI of the breast. PLoS One 9(1):e86280

Graham T (1850) The Bakerian lecture-on the diffusion of liquids. Abstracts of the Papers Communicated to the Royal Society of London (No. 5, pp. 897–900). London: The Royal Society

Hahn EL (1950) Spin echoes. Phys Rev 80(4):580–594

Hinshaw WS, Lent AH (1983) An introduction to NMR imaging: from the Bloch equation to the imaging equation. Proc IEEE 71(3):338–350

Hore PJ, Jones JA, Wimperis S (2000) NMR: the toolkit. Oxford University Press, New York
Jost W (1960) Diffusion in solids, liquids, gases: 3d print., with addendum. Academic Press
Keeler J (2005) Understanding NMR spectroscopy. John Wiley & Sons, Chichester
Koh DM, Collins DJ (2007) Diffusion-weighted MRI in the body: applications and challenges in oncology. Am J Roentgenol 188(6):1622–1635
Le Bihan D (1991) Molecular diffusion nuclear magnetic resonance imaging. Magn Reson Q 7(1):1–30
Le Bihan D (2013) Apparent diffusion coefficient and beyond: what diffusion MR imaging can tell us about tissue structure. Radiology 268(2):318–322
Le Bihan D (2014) Diffusion MRI: what water tells us about the brain. EMBO Mol Med 6(5):569–573
Le Bihan D, Turner R, Moonen CT, Pekar J (1991) Imaging of diffusion and microcirculation with gradient sensitization: design, strategy, and significance. J Magn Reson Imaging 1(1):7–28
Levitt MH (2008) Spin dynamics, 2nd edn. John Wiley & Sons, Chichester
Luypaert R, Boujraf S, Sourbron S, Osteaux M (2001) Diffusion and perfusion MRI: basic physics. Eur J Radiol 38(1):19–27
Malayeri AA, El Khouli RH, Zaheer A, Jacobs MA, Corona-Villalobos CP, Kamel IR, Macura KJ (2011) Principles and applications of diffusion-weighted imaging in cancer detection, staging, and treatment follow-up. Radiographics 31(6):1773–1791
McQuarrie DA (2000) Statistical mechanics. Harper and Row, San Francisco
Mehre H, Stolwijk NA (2009) Heroes and highlights in the history of diffusion. Diffusion Fundam 11:1–32
Merzbacher E (1998) Quantum mechanics, 3rd edn. Wiley, New York
Mori S, Barker PB (1999) Diffusion magnetic resonance imaging: its principle and applications. Anat Rec 257(3):102–109
Mukherjee P, Berman JI, Chung SW, Hess CP, Henry RG (2008) Diffusion tensor MR imaging and fiber tractography: theoretic underpinnings. Am J Neuroradiol 29(4):632–641
Nicolay K, Braun KP, de Graaf RA, Dijkhuizen RM, Kruiskamp MJ (2001) Diffusion NMR spectroscopy. NMR Biomed 14(2):94–111
Padhani AR, Liu G, Mu-Koh D, Chenevert TL, Thoeny HC, Takahara T, Dzik-Jurasz A, Ross BD, Van Cauteren M, Collins D, Hammoud DA, Rustin GJS, Taouli B, Choyke PL (2009) Diffusion-weighted magnetic resonance imaging as a cancer biomarker: consensus and recommendations. Neoplasia 11(2):102–125
Padhani AR, Koh DM, Collins DJ (2011) Whole-body diffusion-weighted MR imaging in cancer: current status and research directions. Radiology 261(3):700–718
Philibert J (2006) One and a half century of diffusion: Fick, Einstein before and beyond. Diffus Fundam 4:6.1–6.19
Sotak CH (2004) Nuclear magnetic resonance (NMR) measurement of the apparent diffusion coefficient (ADC) of tissue water and its relationship to cell volume changes in pathological states. Neurochem Int 45(4):569–582
Stejskal EO, Tanner JE (1965) Spin diffusion measurements: spin echoes in the presence of a time-dependent field gradient. J Chem Phys 42(1):288–292
Torrey HC (1956) Bloch equations with diffusion terms. Phys Rev 104(3):563–605
White NS, McDonald CR, Farid N, Kuperman J, Karow D, Schenker-Ahmed NM, Bartsch H, Rakow-Penner R, Holland D, Shabaik A, Bjørnerud A, Hope T, Hattangadi-Gluth J, Liss M, Parsons JK, Chen CC, Raman S, Margolis D, Reiter RE, Marks L, Kesari S, Mundt AJ, Kane CJ, Carter BS, Bradley WG, Dale AM (2014) Diffusion-weighted imaging in cancer: physical foundations and applications of restriction spectrum imaging. Cancer Res 74(17):4638–4652
Winston GP (2012) The physical and biological basis of quantitative parameters derived from diffusion MRI. Quant Imaging Med Surg 2(4):254–265
Yoshida S, Masuda H, Ishii C, Tanaka H, Fujii Y, Kawakami S, Kihara K (2011) Usefulness of diffusion-weighted MRI in diagnosis of upper urinary tract cancer. Am J Roentgenol 196(1):110–116

Oncology: Radiation Oncologist's View

8

Satyajit Pradhan, Ashutosh Mukherjee, Vinay Saini, Govardhan HB, Ankita Rungta Kapoor, Abhishek Shinghal, Lincoln Pujari, and Sambit Swarup Nanda

Abstract

Radiation Oncology is the branch of medical science that deals with the use of ionising radiation, such as photons, X-rays or particulate radiation for the treatment of cancers. This branch of medicine is a fairly young branch, born just over a hundred years ago after the discovery of X-rays in 1895 and radioactivity in 1896. Since then, it has made rapid progress, with the development of protocols of multidisciplinary use of methods. Radiation has been combined with surgery, chemotherapy, hormone therapy, and immunocytic and novel biomolecules. With the development of linear accelerators and computers and with increased digitalisation, new methods of radiation delivery and quality control have been developed. The radiation oncologist now can accurately delineate his target volume that would include the gross tumour, areas of potential microscopic disease and potential areas of spread in adjoining areas, which precisely treat that volume to the highest dose possible in the shortest time possible and even take into account movement of that target due to physiological processes or tumour shrinkage. Radiation oncology has seen its development from the times of single doses and superficial X-rays to the development of cobalt therapy machines, linear accelerators and further on to the development of image guide radiation, proton beam radiation and MR-guided brachytherapy.

S. Pradhan (✉) · A. Mukherjee · V. Saini · A. R. Kapoor · A. Shinghal · L. Pujari · S. S. Nanda
Department of Radiation Oncology, Mahamana Pandit Madan Mohan Malaviya Cancer Centre, Banaras Hindu University, Varanasi, India
e-mail: satyajit.pr@gmail.com; satyajit@mpmmcc.tmc.gov.in

G. HB
Department of Radiation Oncology, Kidwai Memorial Institute of Oncology, Bangalore, India

© The Author(s), under exclusive license to Springer Nature Singapore Pte Ltd. 2022
S. K. Basu et al. (eds.), *Cancer Diagnostics and Therapeutics*,
https://doi.org/10.1007/978-981-16-4752-9_8

Keywords

Radiation Oncology · Radiation therapy techniques · Radiation Protection

8.1 Introduction

Radiation oncology is the branch of medicine that involves study, generating literature and research related to the use of ionising radiation in the treatment of cancer (malignancy). Radiotherapy (RT), also called radiation therapy, is the use of ionising radiation for the treatment of cancer (malignancy) and some nonmalignant conditions. Radiation therapy uses high-energy beams to damage tumour cells and stop their function and/or division and in turn their growth. RT, used alone or in association with other treatment modalities, has been an effective treatment of cancer for more than a century, and more than 60% of cancer patients are expected to receive RT at some point of time during their treatment protocol (Connell and Hellman 2009). Radiation therapy can be delivered either as teletherapy and/or brachytherapy. *Teletherapy* is described as the process of giving RT when the source of radiation is at a distance from the patient, while in *brachytherapy*, the source is either on or within the body of the patient in close proximity to the tumour area or through the tumour area.

8.1.1 Historical Perspectives

Radiation oncology is an integral part of present multimodality complex therapeutic approach for the effective treatment of cancer. Importance of radiation oncology in cancer treatment is increasing because of newer techniques and devices, which follow the basic principle of achieving maximum tumour control and minimal normal tissue toxicity. The origins of RT began with the experiments and findings of the German physicist Wilhelm Conrad Röentgen, who, in November 1895, discovered these unnamed rays, which could pass through tissues and produce images of them showing soft tissues and bone having different contrasts and named these '*X-rays*'. The discovery of the phenomenon of 'radioactivity' in uranium or pitch blend, as it was then called, by Antoine Henri Becquerel in 1896 and the isolation of polonium and radium by Pierre and Marie Curie provided the raw materials for the development of this new branch. Within a decade, radium was used to treat cancers like skin cancers and later cancers of the uterine cervix (Connell and Hellman 2009). Since then, high-energy 'X-rays' have been used for therapeutic purposes and low-energy 'X-rays' have been used for diagnosis.

By 1920, following experiments by Coutard, Bergonie and Tribendeau, physicians were in a position to study and understand the effects of radiation on living tissues (www.icrp.org). The specialty of radiobiology was established, and fractionated dose delivery (multiple smaller daily doses instead of single large dose to reduce toxicity) was enunciated following Coutard's famous experiments in 1928. A big step in the field of radiation protection was the establishment of ICRP

(International Commission on Radiation Protection) for the safety of radiation workers. Invention of ionisation chambers for accurate measurement of radiation dose in 1932 helped monitor exposure to radiation (www.icrp.org).

The *orthovoltage* era (1930–1950) saw the use of deep X-ray machines for teletherapy and the establishment of radium based interstitial brachytherapy. Brachytherapy allowed treatment of tumour deep inside body cavities or in accessible areas without access to external beam source and simultaneously limiting the side effects on unaffected tissue (Cameron 1993). The introduction of electron beam therapy and super voltage X-ray (50–200KV) therapy facilitated the treatment of deep-seated tumours, but these were still related to increased skin and superficial dose and hence increased sequelae. The *Megavoltage era* started after the Second World War with the development of the first linear accelerator and the first cobalt therapy machines. These were developed in Canada in 1952 and 1961, respectively. These machines helped deliver megavoltage X-rays to treat deeper tumours with an additional property of '*skin sparing*'. By the 1970s and the early 1980s, the advent of CT scans and its assimilation of the computer-aided digitalisation led to a revolution in the field of imaging. This helped in bringing about the concept of 3D conformal radiotherapy (**3DCRT**) in which beams and fields could be shaped to match the required target volume (Connell and Hellman 2009; Case 1953; Hirsch and Holzknecht 1926), while simultaneously sparing, as much as possible, the adjoining normal tissues. By the 1990s, the commercial availability of 3D treatment-planning systems led to widespread adoption of 3D planning and CRT as standard of practice.

Towards the end of the twentieth century and in the last 3 decades, computerisation and development of more powerful computing software have revolutionised radiation therapy. In essence, these developments have allowed the radiation oncologist to *see a moving target in 'real time', prescribe and deliver differential doses to different targets or even parts of different targets and nearly completely spare even 'next-door' structures*. The target need not be a regular shape. Any shape can be targeted, and the use of computer-aided imaging and beam modulation technology has allowed unheard and unthought of dose computations, which can be safely delivered.

In the present era, intensity-modulated radiation therapy (**IMRT**) has become the standard practice for many tumours. In this technique, the aim is to apply a planned unequal dose distribution across a beam to achieve the desired dose configuration. This is done by 'modulating' a beam, i.e., the dose intensity across a given unit area is varied across thousands of beamlets within a beam in a pattern, which is computer-generated and offers the best possible dose distribution (Brady et al. 2013). The use of **IMRT** has also introduced multiple other concepts such as modulated arc therapy, in which intensity-modulated treatment is delivered across a continuous arc with changing dose rates and beam angles; image-guided radiotherapy (**IGRT**) in which treatment delivery is done under real-time or near real-time image guidance to verify any change in tumour position following normal physiological movements and thus reduce margins required, reduce errors in delivery and thus treatment sequelae, as lesser amount of normal tissue is included in the treatment field. Other concepts introduced by **IMRT** include **Adaptive Radiotherapy**, which is a revision of the treatment plan with the help of repeat imaging when there has been a reduction in tumour volume and organ motion during the course of radiation treatment (Castelli et al. 2018; Murthy et al. 2011) with image guidance such as in head and neck cancer after change in contour of patient or organ or tumour displacement after weight loss or concept of plan of day in bladder cancer. Simultaneous integrated boost or **SIB** involves simultaneously treating large and smaller inner fields to reduce treatment time. **Stereotactic radiosurgery/therapy** is a new concept, where the well-localised target volume is defined in three planes (x-, y- and z-axes) and a sharply defined pencil beam is used to deliver very high doses over a short time period to ablate the tumour. This has been made possible by image guidance, tumour motion management and development of micro-multileaf collimators (mMLCs) and mini MLCs.

Strategies to address the organ motion during image guidance in radiotherapy, immobilisation devices, respiratory gating and surface-guided imaging techniques have been used. **Respiratory gating** or tracking is an advanced technology used in cases where radiation is delivered in either specific phases of respiration or radiation is delivered in breath-hold position such as the *deep inspiration breath-hold technique* (**DIBH**). *Intraoperative radiotherapy (IORT)* that delivers local radiation to the postoperative bed at the time of surgery with electrons or low-energy photons to prevent recurrence with very minimal doses to adjoining normal tissue doses has also come into practice. Whereas, chemical modifiers such as *radiosensitisers* make the tumour more sensitive without affecting the adjacent normal tissue, *radioprotectors* aim at preferentially protecting the normal tissues compared to tumours from the deleterious effects of radiation.

8.2 Mechanism of Action of Radiation and the Biological Basis of Radiation Therapy

In the human body, different types of cells have different reproducing capability and these cells replicate according to the need of the body. A tumour is basically a mass of cells, which has arisen from a common pool of cells and which, due to alteration in internal genetics and a host of external influences, has got into a phase of uncontrolled division resulting in destruction of surrounding tissues and/or spread to distant areas. RT kills tumour cells by enhancing the programmed cell death (apoptosis), stops their multiplication or passage through the cell cycle (mitotic death) by directly disrupting the DNA or other important cellular metabolic processes or indirectly by producing free radicals (atoms or molecules having an unpaired electron in the outer shell but are electrically neutral), which produce DNA damage. Unfortunately, radiation also acts on normal cells and damages them.

Radiobiology is the science which studies the biological effect of ionising radiations on tumour cells and normal cells. Tissues are made up of cells, which contain cytoplasm and nucleus in which DNA is the site of action of the radiation. When radiation hits on tissues, it undergoes processes that can be broadly divided into three phases:

1. *Physical phase*: Once the radiation hits the atom, in the time interval between 10^{-18} s and 10^{-14} s, some of the electrons in its orbit will be ejected out with

higher energy in a process, which is known as ionisation. These absorbed energies will lead to the chemical change.
2. *Chemical phase*: Ejected free electrons and water molecules help in the formation of *free radicals* and *ion radicals* (have an unpaired electron in the outer shell but are electrically charged). Free radicals have a short lifespan and are much more reactive compared to Ion radicals. These free radicals are highly reactive, and they eventually lead to the restoration of electronic charge equilibrium.
3. *Biological effects*: Within minutes after irradiation, signal transduction pathways mediated by *protein kinase C* and *tyrosine kinase* are stimulated. Genes and enzymes involved in genetic control of radiation damage repair are activated, stress genes are induced, and growth factors and cytokines are activated. This subsequently activates other genes, including those for fibroblast growth factor, tumour necrosis factor and transforming growth factor, tissue plasminogen activator, which are synthesised and then the cell invokes a series of activities in an attempt to tide over and survive the potentially deleterious dose of radiation. However, it has been seen that 37% of tissue exposed to the radiation will fail in that attempt and will die (Douglass 2018).

8.2.1 DNA Damage

Of all the radiation-induced events taking place in the cells, DNA damage is the most important and critical with regard to death of the living cell, both physical and mitotic death (the inability of the cell to go through the dividing or mitotic cycle). The level and type of DNA damage are established within seconds of irradiation, so that at this point cell signalling pathways are triggered and the biological repair pathways can recognise the problem they have to deal with. Of the various types of DNA damages that take place, **DSB** (double-strand break) are much more critical than **SSB** (single-strand break).

With regard to lethality of damage, there are three types of damages:

Lethal damage: This type of damage is irreversible and irreparable and results in cell death.
Potentially lethal damage (PLD): **PLD** can be influenced by postirradiation environmental conditions. The cell may survive if it does not attempt to divide immediately after the radiation exposure until the environment becomes conducive for cell division.
Sublethal damage: Under normal circumstances, **SLD** can be repaired in hours unless another **SLD** occurs in the cell leading to a lethal damage (Douglass 2018).

With regard to the mechanism of damage, there are two types of DNA damages:

Direct DNA damage: The atoms of the target itself may be ionised or excited, thus initiating a chain of events that leads to a biological change and finally death (e.g. double-stranded DNA break). This happens most commonly in irradiation with high **LET** (linear energy transfer) beams such as neutrons and alpha particles (Douglass 2018).

Indirect DNA damage: Here, radiation may interact with other atoms or molecules in the cells particularly water to produce free radicals. These free radicals can diffuse far enough to critically damage the targets (Douglass 2018).

These damages finally lead to cell death by one of the methods described below:

Mitotic catastrophe: During replication in the next mitotic cycle, the cells with damaged DNA lose their replicative potential and this leads to reproductive death of cell (mitotic death).

Apoptosis/programmed cell death: This is a physiological process that involves a series of genetically controlled steps including chromatin condensation and segmentation, fragmentation of the nucleus into apoptotic bodies, cell shrinkage and loss of cellular contact with neighbouring cells. This leads to these cells being engulfed by macrophages without a simultaneous inflammatory response. Radiation can also enhance the process of apoptosis.

Autophagy: A self-digestive process using the lysosomal enzymes causes degradation of long-lived proteins and organelles to maintain cellular homoeostasis.

Senescence: Programmed cellular stress response induced by radiation leads to permanent cell cycle arrest by activation of p53 and Rb proteins causing chromatin modifications that result in inhibition of genes necessary for transition from G1 to S phase of cell cycle.

Bystander death: If a large fraction of cells die because of radiation, the neighbouring cells are killed because of the toxins released nearby and by the action of inflammatory cells (Douglass 2018).

8.2.2 Biological Basis for Radiation Therapy

Radiation kills cells primarily by causing ionisation, which leads to **SSB** or **DSB** DNA damage. If the DNA damage is irreparable by cellular DNA repair mechanisms, the injury leads to cell death. Photons cause sparser ionising damage to cells along the path as compared to particles, which have higher **LET**, which result in denser ionisation along the tract. The basic assumption in radiotherapy is that there are critical targets in each cell, and when ionising radiation hits these targets, loss of cellular clonogenicity may result. This is called target hypothesis. **Therapeutic ratio (therapeutic index)** is defined as the ratio of tumour response for a particular level of normal tissue toxicity at a given radiation dose.

The ultimate outcome of radiation-induced cell damage is also dependant on the intrinsic ability of cells to repair damage. Cell survival after exposure can be expressed in terms of logarithmic curve of surviving fraction versus dose. The curve forms an initial shoulder followed by a logarithmic decline in survival, which varies with the dose. Cell survival is also related to the phase in which the cell is in the mitotic cycle, dose rate and oxygen concentration. The effect on the tumour cell and the cells of the adjoining normal tissues are dependent on the factors known as the **5R's of radiobiology**, i.e., repair, reoxygenation, redistribution, repopulation and also inherent radiosensitivity (Douglass 2018).

8.3 Principles of Radiotherapy

The practice of RT is based on the following principles: accurately define target volumes, define intent and hence technique of treatment, decide required dose and fractionation and deliver the required dose with minimal error. The aim is to deliver the required dose to the target volume with minimal spillage into surrounding normal tissue and with sparing of all critical normal structures in the shortest possible time.

8.4 Types of Radiation

Radiation is the emission and propagation of energy travelling in the form of waves or as high-speed particles, in space or through a material medium. Particulate radiations involve particles resulting from disintegration of an unstable (or radioactive) atom or artificially produced particles like accelerated electrons from a heated cathode. Electromagnetic radiation is the method of energy propagation in the form of light waves, heat waves, microwaves, radio waves, ultraviolet rays, gamma rays and X-rays that travel in form of waves and do not have mass.

Electromagnetic radiation propagates in a straight line, travels at the speed of light (nearly 300,000 km/s) and transfers energy to the medium through which it passes, the amount correlating directly with the frequency and inversely with the wavelength of the radiation. The energy of radiation decreases as it passes through a material, due to absorption and scattering, and is in inverse relation with the square of the distance travelled through the material. Electromagnetic radiation includes a spectrum ranging from radio waves to gamma rays. With radio waves at one end of the spectrum, we move through microwaves, infrared radiation (heat), visible light, ultraviolet and X-rays to gamma rays. As we move from left to right in the spectrum, the wavelength decreases, but frequency and energy increase. X-rays and gamma rays have the highest frequencies and energies.

Electromagnetic radiation is classified into ionising and nonionising radiations. Nonionising radiations have wavelengths of 10^{-7} m. Nonionising radiations have energies of <12 electron volts (eV); 12 eV is considered to be the lowest energy that an ionising radiation can possess. The different nonionising electromagnetic radiations are radio waves, microwaves, infrared light, visible light and ultraviolet light (Khan and Gibbons 2014).

8.4.1 Ionising Radiation

Ionising (high-energy) radiation (Fig. 8.1) can remove electrons from atoms creating charge imbalance and thus ionising them. Ionising radiations used clinically are both electromagnetic (photons) radiations and particulate radiations (electrons, protons, heavy ions).

8.4.2 Ionising Electromagnetic Radiation

Ionising electromagnetic radiations interact with matter resulting in the ejection of electrons from the orbits of the constituent atoms and molecules. These electrons hit other orbital electrons with decreasing energy after each hit to produce secondary electrons while traversing through the material. A mean energy of 33.85 eV is transferred during the ionisation process. When high-energy photons are used clinically, the resulting secondary electrons, which have an average energy of 60 eV per destructive event, are transferred to cellular molecules (Sowa et al. 2012; Khan and Gibbons 2014).

8.4.3 X-Rays

The German physicist Wilhelm Conrad Roentgen discovered X-rays in 1895 while experimenting with electricity passed through a hot cathode Roentgen tube, which had much of the air removed and with a much-reduced pressure (up to 10^{-3} mmHg)

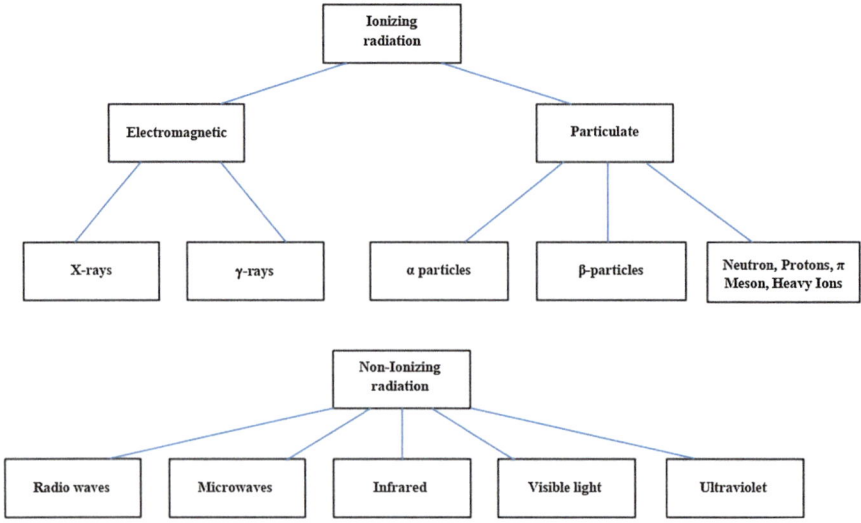

Fig. 8.1 Ionising and nonionising radiations

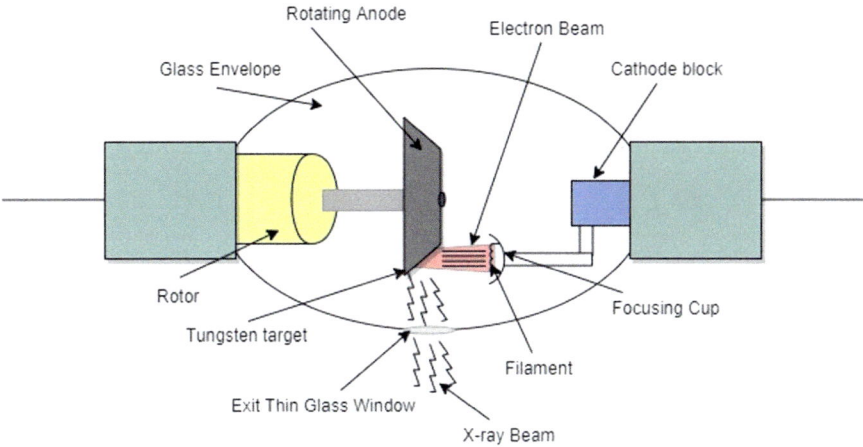

Fig. 8.2 Production of X-rays in diagnostic X-ray unit

in a glass tube containing anode (positive terminal) and cathode (negative terminal) ends between which a high-energy (10^6–10^8 V) potential was applied. Electrons are produced by a process called thermionic emission in the cathode, which gets accelerated towards the anode by the potential difference. These electrons at high speed hit the anode, which is made up of a metal with high melting temperature. The sudden deceleration of these electrons due to interactions with nuclei of the element in the anode (*bremsstrahlung interaction*) results in production of X-rays. The energy and the wavelength of the X-rays depend on the velocity and the kinetic energy of the electrons and the atomic number of the target (anode) metal. This process is used to produce radiation in diagnostic X-ray units (Fig. 8.2) and linear accelerators (Linacs) (Khan and Gibbons 2014). These X-rays are emitted by extranuclear events (i.e. interaction with electrons, which are seen outside the nucleus).

X-rays produced by X-ray tubes are of 2 types (Fig. 8.3).

(i) *Bremsstrahlung X-rays:* X-rays are produced as a result of sudden deceleration of high-speed incident electrons due to Coulomb interaction with the nucleus of an atom (bremsstrahlung interaction). These types of interactions generate X-ray photons with a continuous spectrum of energy.
(ii) *Characteristic X-rays:* It occurs when an electron in an inner atomic orbital is knocked out by an incoming electron, and the resulting space in orbit is filled by another electron that moves from an outer atomic orbit. The electron moving from the outer orbit must shed energy to move to the inner orbit, and the energy released in this process is radiated as characteristic X-rays (Fig. 8.3). They are characteristic due to the fact that the energy of the X-ray produced depends on the target metal onto which the electrons are bombarded. The energy of released radiation will be equal to the difference in the energy levels of the orbits of the

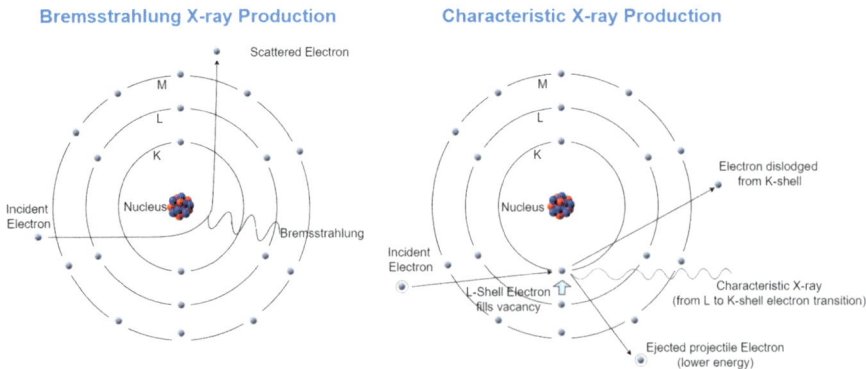

Fig. 8.3 Production of bremsstrahlung X-rays and characteristic X-rays

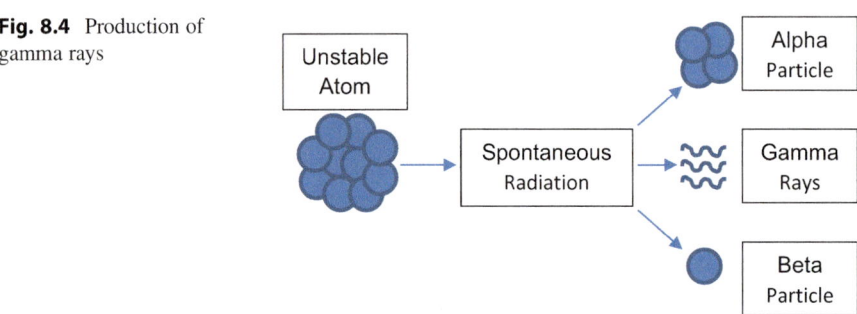

Fig. 8.4 Production of gamma rays

given target material between which the electron transition take place (Kramers 1923; Khan and Gibbons 2014).

8.4.4 Gamma (γ) Rays

Gamma rays are similar to X-rays, but are produced by intranuclear events. An unstable atomic nucleus gives away excess charge or energy to reach a stable state, which most commonly can be either as an intranuclear electron (e^- or beta particle-'ß') or as a doubly positively charged helium nucleus (an alpha particle-'α') (Fig. 8.4). If it still possesses excess energy, gamma rays are emitted in order to reach a steady state (Khan and Gibbons 2014).

8.4.5 Ionising Particulate Radiation

These are basically charged particles such as electrons, protons, alpha particles, neutrons, pi-mesons and heavy ions, which have been ejected from their nuclear or orbital locations by incident radiation. Electrons are by far the most commonly used

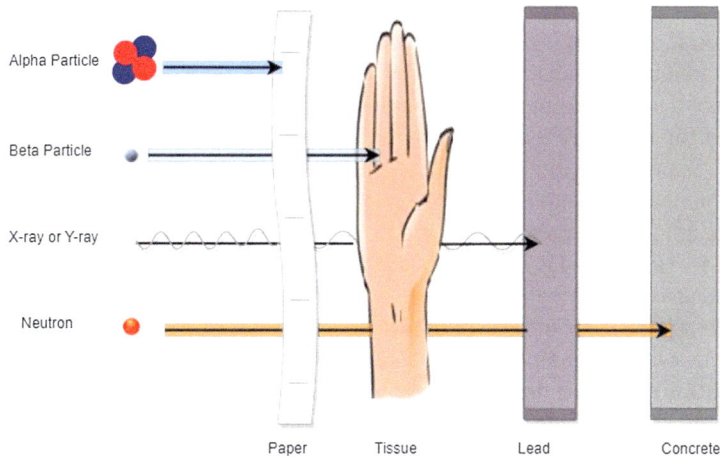

Fig. 8.5 Penetrating power of different types of radiation

particles and by virtue of their negative charge and low mass, and can be accelerated to high energies in *linear accelerators* or *betatrons*. In an atom, electrons are normally 'bound' by electrostatic attraction between itself and positive charge on the nucleus. It can become 'free' if it receives enough energy to overcome this binding energy. Nuclear decay processes can produce free electrons, which are called beta particles. Neutrons are charge-neutral particles forming stable large atomic nuclei by decreasing repulsion between positively charged protons in the nucleus. They have high penetration as they have no charge and hence indirect ionisation reactions in matter. Protons are like neutrons but are positively charged particles. An alpha particle is identical to a helium-4 nucleus (with 2 electrons missing). It has 2 protons and 2 neutrons in its nucleus, making it a heavily charged particle with twice the positive charge and thereby reducing its penetration power in a medium. It travels 2–5 cm in air and can be stopped by a piece of paper, or the top layer of the skin (Fig. 8.5) (Khan and Gibbons 2014).

8.5 Units Used in Radiation Oncology

Different terms are used for quantification of radiation depending on whether we are considering radiation coming from a radioactive source, amount of radiation travelling through the air, the amount of radiation absorbed by a person or any object, or the risk that a person will suffer health (biological) effects from the absorption of radiation.

When the amount of radiation being released from a radioactive atom is discussed, whether it emits alpha or beta particles, gamma rays, X-rays or neutrons, the **unit of measurement of radioactivity** is the conventional unit **Curie (Ci)** or the SI unit **Becquerel (Bq)**. This quantity represents as to how many atoms of the material decay in a given time period. **One Ci** is equal to **3.7 × 10^{10} disintegrations per second,** and **1 Bq** is equal to **1 disintegration per second**.

The amount of radiation travelling through the air is described in terms of exposure. It is a measure of the ionisation of air due to gamma-ray and X-rays. The **unit for exposure** is the **Roentgen (R)**, and the SI unit is **Coulomb/kilogram (C/kg)**. Several radiation monitoring devices (such as gamma zone monitors, ionisation chambers and GM-based survey metres) used in the radiation oncology department for radiation safety purposes can measure the radiation level in the air in units of **R**. However, **1R** equals **0.000258 C/kg**.

When an object or living tissue is exposed to radiation, some amount of radiation energy is absorbed by the object or tissue through which it passes. This amount of energy deposited per unit weight of an object or tissue is called the **absorbed dose**. The **unit for absorbed dose measurement** is **rad,** and the **SI unit** is **Gy**. However, **1 Gy** is equal to **100 rads**. It is a physical quantity that can be directly measured for which certain standards exist for measurement. In a radiation oncology department, this unit is used to prescribe the dose for patient treatment (Podgorsak 2005; Cember and Johnson 2009).

The combined effect of the amount of absorbed radiation and its medical effects (i.e., a person will suffer biological risk from exposure to radiation) is measured in terms of an equivalent dose (or effective dose) with **conventional unit** called **rem** and **SI unit Sievert (Sv)**, with **1 Sv** equivalent to **100 rem**. For beta and gamma radiations, the equivalent dose is the same as the absorbed dose. However, the equivalent dose is larger than the absorbed dose for alpha and neutron radiation, because these types of radiation are more damaging to the living tissues. These units are basically used for the evaluation of radiation protection for humans. It basically talks about the amount of energy that a radioactive source deposits in living tissue (Khan and Gibbons 2014; Podgorsak 2005; Cember and Johnson 2009).

8.6 Basics of Radiation Protection

International X-ray and Radium Committee was founded in 1928 and renamed as International Commission on Radiological Protection (ICRP) in 1950. The ICRP provides recommendations and guidance on protection against the risks associated with ionising radiation, from naturally occurring sources and from artificial sources, which are widely used in medicine, general industry and nuclear enterprises. The recommendation focused on preventing tissue reaction (or *deterministic effects*) and limiting *stochastic effects* to acceptable levels. According to ICRP, a system of radiological protection is based on the following basic principles:

Justification of Practice: This is the first principle, which emphasises on the justification of applications of radiation in a practice. No practice shall be adopted unless it produces a net positive benefit. It is judged from the total harm from the proposed practice involving radiation exposure that should be less than the expected benefit. Based on this application radiation in medicine, industry, agriculture and certain research can be justified.

Optimisation of protection: All exposure shall be kept **as l**ow **as r**easonably **a**chievable (**ALARA**), economic and social factors taken into account. The greater the level of protection, the higher the degree of safety achieved. A greater level of protection would involve expenditure, which may reduce the ultimate value of the practice. Therefore, it is recommended that protective measures should be optimised considering socio-economic factors to keep radiation exposure as low as reasonably achievable.

Dose limitation: In order to reduce the magnitude of the risks associated with a practice, limits on individual doses are established to prevent the occurrence of *deterministic effect* and minimise the likelihood of *stochastic effects* (cancer and hereditary effects). These dose limits take into account the doses received from external sources and by intake of the radionuclides into the body. However, while deriving these dose limits, no exposure due to natural background radiation and medical exposure are included in the limits. These dose limits are applicable to both workers and to members of the general public.

ICRP recommends the dose limit for radiation workers of whole body (effective dose) of 20 mSv/year averaged over a period of 5 years with not more than 50 mSv in a single year. In this dose limit, ICRP considers the 5% risk per Sv to be nominal and acceptable for the lifetime of radiation workers. A 20 mSv per year effective dose was set by ICRP for radiation workers. ICRP further recommends equivalent dose limit to prevent the deterministic effect for the lens of the eye and skin and extremities, with an annual dose limit of 20 mSv and 500 mSv, respectively (ICRP-60 1991; ICRP-103 2007; Charles 2008; ICRP-103 2011).

ICRP recommends an effective dose limit of 1 mSv in a year for the general public. However, in special circumstances, a higher value of effective dose could be allowed in a single year provided the average over 5 years does not exceed 1 mSv per year. Also for preventing deterministic effects, in the lens of the eye and skin, annual dose limit of 15 mSv and 50 mSv, respectively, is given for the public. This dose limit for the public is lowered because the risk for the whole population is more as it contains developing children who have a long life ahead and possess more risk as compared to radiation workers. For women radiation workers, the commission recommends no special dose limit in general. However, once pregnancy is declared, the embryo/foetus should be protected and the dose limit for the remainder of the pregnancy period is 1 mSv/year (ICRP-60 1991; ICRP-103 2007; Charles 2008; ICRP-103 2011).

To ensure the safety of radiation workers, members of the public and the environment at all levels, a suitable planned radiation workplace is designed in which all radiation safety standards is complied with along with area monitoring and personal monitoring facility in accordance with regulatory safety guidelines. This monitoring basically helps in assessment of workplace conditions and individual exposures. Area monitoring is carried out by installed monitor (area monitor), portable survey monitors and sample measurement. It provides the status of radiological conditions at various workplaces in a facility, and the results can be used for arriving at personal exposure. Personnel monitoring is carried out by personal monitoring dosimeter devices (e.g. TLDs, pocket dosimeter, OSLDs, RPLs and CR-39 polycarbonate films). The personnel monitoring dosimeter is commonly used by radiation workers because they work in a controlled area, in which the occupational exposure of personnel to radiation is higher than that of normal people. The personnel monitoring details are generally used for (a) an assessment of the effective dose value of a radiation worker in compliance with regulatory requirements, (b) in case of accidental overexposure provide valuable information for the initiation and support of any appropriate health surveillance and treatment, (c) demonstration of the adequacy of operation and design of facilities, (d) evaluation and development of radiation practices by means of collected data, (e) the motivation of workers to reduce their exposure as a result of information provided to them and (f) assessing high levels of exposure that may occur in a radiation accident. Thus, the protection of radiation workers and members of the public is ensured at all levels by following the regulatory safety requirements (Khan and Gibbons 2014; Podgorsak 2005; Cember and Johnson 2009).

8.7 Techniques of Radiotherapy

Different techniques of radiotherapy are discussed in this section.

8.7.1 Conventional Radiotherapy (2D Technique)

This technique is the simplest form of radiotherapy that is practised and does not use image guidance for beam placements, field design or treatment delivery. Target volumes are designated according to clinical imaging data, and simple beam arrangements are made with respect to anatomical landmarks such as parallel opposed, single enface, box fields and oblique arrangements. The aim of this method is to prescribe the dose to the required depth and obtain the required coverage.

In this method, since anatomical landmarks are used to surrogate imaging data, there is uncertainty over target volume, and hence, larger margins are required. The margins must account for extent of microscopic disease, voluntary and involuntary movements, routine variation in patient set-ups and errors in treatment delivery. Larger field volumes and little sparing of normal tissues can result in higher treatment-related sequelae. Field centre is fixed at maximum tumour dimension (e.g. oesophagus—maximum narrowing point). Standard shape fields with large margins with or without beam-modifying devices such as wedge filters and blocks are used. Large volume of irradiation leads to less tolerance, high integral dose and high probability of late morbidity. Dose prescription may not be adequate for satisfactory tumour control. Dose distribution will not be conformal in this method. However, this technique is still used wherever simple techniques are required and where the required dose can be safely delivered and in palliative care.

8.7.2 Three-Dimensional Conformal Radiation Therapy (3DCRT)

Conformal treatment means that the field designs match the target volumes. The aim of 3DCRT is to reduce normal tissue irradiation and increase tumour control probability, thereby increasing therapeutic ratio (increase tumour control probability (TCP) for an accepted level of normal tissue complication probability (NTCP)). In this technique, multiple beams from different directions are used to achieve a dose distribution that conforms to the target volume. Imaging systems like CT/MRI/PET-CT are used in order to acquire 3D anatomic information. Conformal radiotherapy is practised using serial sectional images obtained from CT scans done in the treatment position. Often, these images are fused with MRI images or even PET images for better assessment of the tumour volume. With the help of this information, the field shapes are designed. Use of computer-generated imaging and marking helps field shaping, field projection and development of beam's eye view as well as setting blocks and compensators. To improve the conformality of the dose distribution, conventional beam modifiers (e.g., wedge filters, partial transmission blocks and/or compensating filters) can be used. Either cerrobend blocks or MLC (multileaf collimator) is used for field shaping.

Organs at risk (**OAR**) are defined as *radiosensitive organs in or near the target volume whose presence influences treatment planning and/or prescribed dose.* The target volume definitions are divided into three distinct volumes: (a) *gross tumour volume* (**GTV**)—visible tumour assessed either clinically or radiologically; (b) *clinical target volume* (**CTV**)—this volume accounts for the suspected microscopic tumour spread; and (c) *planning target volume* (**PTV**)—it accounts for uncertainties in treatment delivery including set-up errors on a daily basis. Figure 8.6 illustrates the target volume concept in RT.

The success of 3DCRT depends on the accuracy of **CTV** drawn around **GTV**. The possible limitations are patient rotation, movement of tumour volumes, movements of **OARs** and movements of external fiducial markers placed on the patient body surface during imaging, simulation or treatment. **PTV** must take into account these systematic and random errors. The field size (**PTV + margin**) must be adequate enough to contain **PTV** in the isodose range of 95–107% of prescribed dose. The dose to **PTV** is affected by factors such as cross-beam profile, penumbra and lateral scattering (scattering from central axis) as a function of depth, radial distance and tissue density (ICRU Rep. 50, ICRU, Bethesda, MD 1993; ICRU Rep. 62, Bethesda, MD 1999; ICRU Rep. 83, Bethesda, MD 2010).

8.7.2.1 The Planning Process of 3DCRT
(i) *Patient treatment position, immobilisation and planning imaging:* Patient is positioned in proposed treatment position and immobilised using one or more of the many immobilisation devices that are available. Radiopaque markers are placed on the body surfaces to be covered within the treatment field and positioning lines on patient, and immobilisation devices are marked. Tomograms are obtained to check patient alignment. A volumetric CT scan of patient in treatment position is done and illustrative photographs taken for

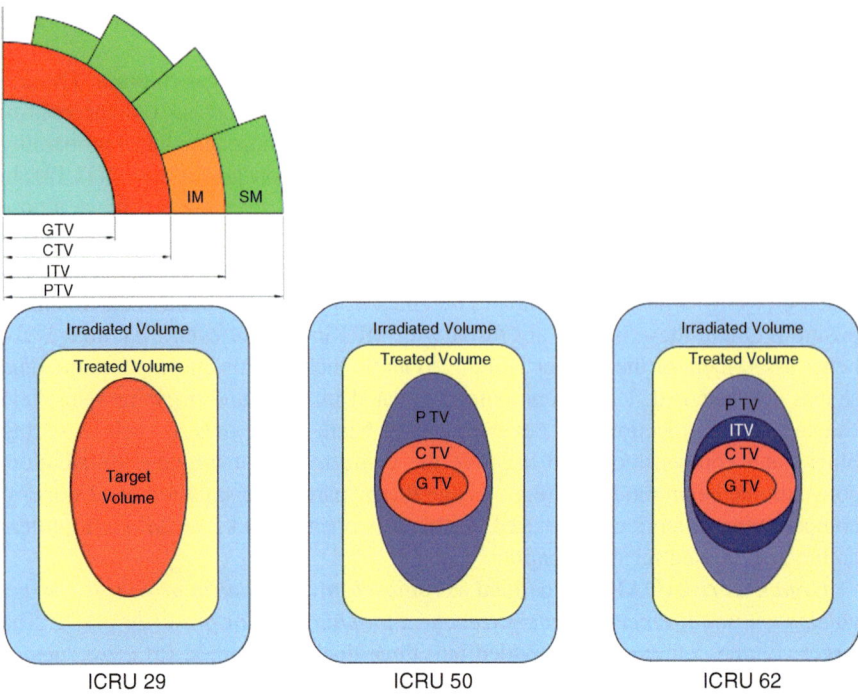

Fig. 8.6 Target volume concept in radiotherapy

ease of repositioning of the patient on treatment couch during treatment delivery. CT images are transferred to three-dimensional treatment-planning system (**3DTPS**) computers. Required serial sectional images from imaging studies, which have already been done or further required (e.g., magnetic resonance imaging and positron emission tomography/CT), are also transferred to **3DTPS**.

(ii) *Delineation of tumour/target volumes and organs at risk*: Physician contours the target volume(s) and **OAR**.

(iii) *Dose prescription*: Radiation oncologists prescribe radiation dose for (**PTV**) and dose–volume constraints for **OAR**.

(iv) *Planning*: *Forward planning* is used for **3DCRT**. In this technique, it is the oncologist and physicist who decide the beam configuration, number, direction, shapes, beam modifiers and beam weights. The system calculates the dose distribution based on these arrangements and shows the dose distribution, which may or may not be accepted by the oncologist. The team makes incremental changes based on the requirements.

(v) *Plan evaluation and improvement*: A plan is basically a sum total of beam placements, beam weightages and beam modifiers, which leads to a certain dose distribution. This is checked based on dose–volume histograms, planar isodose displays, 3D isodose displays and various measures as conformity, homogeneity indices, dose means and maximums, hot spots or cold spots, coverage and **OAR** doses. The accepted plan is mentioned in the patient chart (electronic medical record) and the treatment machine record and verification (**R & V**) system.

(vi) *Plan implementation and treatment verification*: The physicist performs second check of treatment plan and transfer of data to **R & V** system. On table, the patient position and treatment isocenters are verified using orthogonal electronic portal imaging devices (**EPID**s) and compared to the digitally reconstructed radiograph (**DRR**) from TPS or using onboard CT image with planning CT. In **3DCRT**, field shapes are checked by comparing treatment field **DRR**s with treatment beam **EPID**s. The new position settings are then fed into the **R & V** system. In the process of **3DCRT** plan verification, the treatment plan calculated by **TPS** is checked with the independent secondary calculation of monitor units (**MU**)/treatment time. Periodic imaging verification checks during treatment are done (e.g. orthogonal **EPID**s/**DRR**s or beam **EPID**s/**DRR**s, and cone-beam CT/planning CT) (ICRU Rep. 50, ICRU, Bethesda, MD 1993; ICRU Rep. 62, Bethesda, MD 1999; ICRU Rep. 83, Bethesda, MD 2010).

8.7.3 IMRT: Intensity-Modulated Radiation Therapy (IMRT)

The limitation of **3DCRT** beam is its inability to produce conformal isodose for complex shapes of targets. **IMRT** can shape the dose distribution to the required area while limiting dose to nearby vital structures. Because of its inherent ability to shape dose distributions in irregular or extended fields, the planner can consider prescribing higher doses to the target volume with tighter margins to save normal tissues. Higher doses may help in better tumour control and produce lower sequelae due to better sparing of **OAR**s. This is done by modulating or varying the fluence (number of electrons per unit area), and this means delivery of a nonuniform intensity of radiation within the field. In **IMRT**, a field is divided into many small beamlets, which are prescribed nonuniform intensities (i.e. weights). This manipulation of the intensities of individual beamlets within each beam results in greatly increased control over the radiation fluence, enabling custom design of desired dose distributions. **IMRT** is a process used to spare **OAR**s, treat irregular target shape, simultaneously treat multiple targets and dose escalation, which is not possible with **3DCRT**. In the case of **IMRT** plan verification, calculated by **TPS**, patient-specific Quality Assurance (**QA**-dosimetric verification of patient plan) is performed for each patient on either phantom or diode array-based commissioned **EPID**.

There are 2 types of **IMRT**: **static IMRT** and **dynamic IMRT**.

Static IMRT: Segments are created within the fields, which require identical doses. During treatment, MLCs take new position and form a shape, but MLCs do not move. After completing one segment, beam goes off. MLCs take new position and form a shape with beamlets requiring another set of identical doses, and the process continues till total field is completed. **Static IMRT** is more time-consuming, but good for long term stability of machine.

Dynamic IMRT: In this process, the beam will be continuously on and all the MLCs will be in movement with different speeds as dictated by the software. To deliver more doses at any beamlet, leading MLC moves fast and following MLCs moves slow. To reduce the dose, the leading MLCs move slow and following MLCs will move fast. More **MU** and lesser treatment time are used. Clinical results in literature for static and **dynamic IMRT** are identical. In **dynamic IMRT**, wear and tear are more; hence, maintenance is more. **IMRT** creates superior Dose Volume Histogram (**DVH**) compared to **3DCRT**.

8.7.3.1 Planning System of IMRT

Forward planning IMRT: Beam direction is selected in such a way as to just avoid **OAR**s between beam entry and target centre. Angular spacing between fields must be preferably uniform to avoid more overlapping areas. By trial and error methods changing beam weights and beam angles, best dose distributions can be obtained.

Inverse planning IMRT: In this type of planning after the completion of contouring work, segmentation, separation of target and **OAR** overlapping regions, the constraint table is filled and priorities are assigned for both **OAR**s and targets. In this technique, the oncologist provides the dose constraints and dose coverage requirements. These are converted by the physicist into a table of priority values and beam configurations and weights. It is the system, which then decides everything including beam weights, placements, modifiers and dose distribution. If the distribution is not acceptable, it again recalculates based on the input provided.

Uncertainties of various types (e.g. those related to daily or interfraction patient positioning; displacement and distortions of internal anatomy; intrafraction motion; and changes in physical and radiobiological characteristics of tumours and normal tissues during the course of treatment) may limit the applicability and efficacy of **IMRT**. Dosimetry characteristics of a delivery device, such as radiation scattering and transmission through the MLC leaves, introduce some limitations in the accuracy of delivery of desired **IMRT** fluence distributions (ICRU Rep. 83, Bethesda, MD 2010).

8.7.4 Image-Guided Radiation Therapy (IGRT)

The American College of Radiology and American Society of Radiation Oncology practice guideline defines **IGRT** as 'a procedure that refines the delivery of therapeutic radiation by applying image-based target relocalisation which helps in proper patient repositioning to ensure accurate treatment delivery by reducing the uncertainty in positioning and thereby reduces the need of extra margin required for set up errors, thus minimizing the volume of normal tissue exposed to ionizing radiation'.

8.7.4.1 Rationale for Image-Guided Radiation Therapy

Increasing the accuracy and precision of radiotherapy delivery has always been a therapeutic goal. *Inaccuracy* means systematic errors that, on average, bias the treatment delivery with respect to the true target location. Purpose of **IGRT** is the correction of inter/intrafraction variation in set-up and anatomy, variation on surface and volume of target and **OAR**s (Khan and Gibbons 2014).

8.7.5 Stereotactic Radiosurgery and Radiotherapy (SRS and SRT)

Stereotactic radiosurgery and radiotherapy are techniques to deliver high-dose radiation to a target with precise conformity to the target to create a desired radiobiological response and with minimal radiation dose to surrounding normal tissue. In this technique, the well-localised target volume is defined in three planes (x-, y- and z-axes), surrounding margins are minimal and very high doses are given over a short time period to ablate the tumour. The complication risks are reduced by decreasing or eliminating the margin of normal tissue from the tumour. When the total dose is delivered stereotactically in single session, it is called stereotactic radiosurgery (**SRS**), and if total dose of radiation is administered in more than one fraction, it is called stereotactic radiotherapy (**SRT**).

To precisely localise and irradiate the tumour, these techniques use frames or immobilisation devices, which have reference markers in three axes. **SRS** can be delivered with Gamma Knife (Elekta Inc., Norcross, GA), modified LINAC radiosurgery systems (including CyberKnife and image-guided radiotherapy systems), tomotherapy or proton beam systems. In **SRT**, typically noninvasive positioning techniques are used. **SRS** allowed clinicians to administer high single doses of radiation to intracranial targets with relative safety (Khan and Gibbons 2014; Benedict et al. 2010). The idea is to give sufficiently high dose to directly ablate or literally burn out the tumour using very high spatial precision to avoid normal tissue even in the margins and to account for motion management in such delivery.

8.7.6 Volumetric Modulated Arc Therapy (VMAT)

Volumetric modulated arc therapy (**VMAT**) is a new radiation therapy technique in which the radiation dose is delivered continuously as the treatment gantry rotates around the patient. **VMAT** works similarly to **IMRT** in the way the radiation dose is varied throughout treatment, but the radiation is delivered from all angles contrary to the **IMRT** where there are fixed gantry angles for radiation dose delivery and varying dose rates during the rotation of the gantry. **VMAT** also has the specific advantage of reduced treatment delivery time as compared to conventional static field **IMRT** (Khan and Gibbons 2014).

8.7.7 Surface-Guided Radiation Therapy (SGRT)

Surface-guided radiation therapy (**SGRT**) is a new technique, which uses stereovision technology to track patient's surface in 3D, for both set-up and motion management during radiation treatment. There are a number of publications supporting the use of **SGRT** in the treatment of breast, brain, head and neck cancer, sarcoma and many other conditions. **SGRT** can help in the following ways, which lead to many application-specific benefits:

Set-up: **SGRT** is less time-consuming and helps in improving the accuracy of set-up.
Treatment: **SGRT** can minimise the need for immobilisation while helping to ensure correct patient positioning for treatment in all 6° of freedom.
4D CT: **SGRT** enables contactless, noninvasive reconstruction of 4D CT data.

8.7.8 Brachytherapy

This is the technique of delivering RT in which the radioactive source is inside or on the patient's body near or through the tumour tissue. This technique of RT has developed slowly with the development of intracavitary cervical brachytherapy and skin brachytherapy. Over time, rules were framed for further uses in head and neck cancers, sarcomas, breast, gynaecological cancers, skin cancers, eye tumours, endobronchial tumours, biliary tract cancers and oesophageal cancers. In brachytherapy, the radioactive sources are housed in sealed compartments and may be available in the form of needles, wires, seeds, pellets, ribbons, plaques, etc., depending on the radioactive source and site where it is to be used. The application of catheters or applicators is done beforehand, and the sources are afterloaded remotely to prevent radiation exposure of the treating staff.

Exposure to the patient and target volume and hence the total dose given is calculated on the basis of the time the source spends at various points in each catheter or applicator before being withdrawn back into its sealed chamber. Here again, the sources are selected, which do not produce very high-energy beams and whose intensity falls off sharply with increase in distance. Thus, in brachytherapy the inverse square law is very important for radiation protection and dose calculation. As per the inverse square law, the radiation dose from a radioactive source at any point in the target area falls off inversely as the square of the distance from the source. Hence, the dose adjoining the source is very high in the tumour area and falls off rapidly in the surrounding area, and hence, the surrounding normal tissues are spared to a large extent from the deleterious effects of radiation.

Brachytherapy is typically performed in cases where the tumour is accessible, is well defined and not very deep-seated. In many other cases, brachytherapy can be given as boost dose after external beam radiotherapy to avoid dosing nearby structures, as brachytherapy because of its sharp dose fall-off gives very little dose to surrounding structures. However, performing this technique requires adequate training and skill learned over a period of time. It is important to remember to keep the dose hot spots in these applications inside the applicator and not in the tissue. Otherwise serious necrosis can happen, which may require reconstructive surgery.

With regard to the history of development of brachytherapy, the technique can be classified as (i) preloaded brachytherapy—in earlier times, the radioactive sources were manually placed directly into the tumour. On verification, if the application was not proper it had to be rectified. This resulted in considerable radiation doses to staffs involved in the procedure. (ii) Manual afterloading (**MAL**) brachytherapy—subsequently, this practice came into vogue. Applicators and catheters were placed, and after verifying that the applicators have been properly placed, the sources were manually loaded. This practice resulted in considerable reduction in dose to the staff. (iii) Remote afterloading (**RAL**) brachytherapy—the applicators and catheters are placed in position and after verifying the proper placement the staff comes out of the room and the sources are remotely loaded into the applicators by means of computer-controlled equipment.

The various types of brachytherapy are as follows:

Intracavitary Brachytherapy: In this method, an applicator is introduced inside a body cavity and the dose is prescribed at a fixed distance either from the central axis, at a fixed anatomical point, or from the applicator surface, e.g. cervix, vaginal vault and uterus.

Intraluminal Brachytherapy: It is similar to intracavitary brachytherapy in principle and practice. The applicator or catheters are introduced into the lumen of a hollow organ, e.g. nasopharynx, bronchus, oesophagus and blood vessels.

Interstitial Brachytherapy: In this technique, earlier on in the preloaded brachytherapy era, radioactive sources in form of needles/wires, etc., were applied through the tissue being treated with the distance between sources maintained as per rules of implantation being followed. A number of sources are thus placed through a tissue (in single or double or more planes), and the entire sleeve of tissue involved, in which these needles are placed, is treated. In present times, hollow tubes or catheters are placed into the area to be treated with brachytherapy as per rules of application and then afterloaded with the radioactive sources. This type of brachytherapy is usually done for sites such as breast, head and neck, and soft tissue sarcomas.

Template Interstitial Brachytherapy: This is similar to interstitial technique, but the needles are applied in a definitive pattern with the help of a rigid template, which is opposed to the tissue surface, e.g. pelvic template for gynaecology, prostate and anal canal, and breast template brachytherapy.

Surface Mould Brachytherapy: It is used in superficial tumours on the skin and mucosal surface, which are not more than 1 cm thick. The catheters are mounted on a mould, which is opposed to the treating surface. There are rules of dose prescription for mould brachytherapy. This type of brachytherapy can be done for superficial skin tumours, tumours of the lip, ear pinna and even in the postoperative situation in the cavity left after surgery for a maxillary tumour.

8.8 Fractionation in Radiotherapy

In the early days of RT, most protocols described delivering the calculated dose in single large fraction. This resulted in significant toxicities and limited the dose that could be given. Dividing or fractionating the dose into a number of smaller dose fractions at intervals was seen to help in tumour control for a given level of normal tissue toxicity than a single large dose. Dividing the total dose into several small fractions helps the normal tissues heal due to interfraction 'repair' of sublethal damage and 'repopulation' of cells if the overall treatment time is sufficiently long. Fractionated RT again increases the probability of greater damage to tumour cells because of 'reoxygenation' of previously hypoxic areas and 'reassortment' or 'redistribution' of cells into radiosensitive phases of ongoing cell cycle, between dose fractions.

8.8.1 Types of Fractionation in Radiotherapy

A number of dose fractionation schemes of radiation treatment are used. These are conventional fractionation, hyperfractionation, accelerated fractionation, split course radiotherapy and hypofractionation.

Conventional fractionation: Most common fractionation for curative radiotherapy is 1.8–2 Gy doses given daily for 5 days a week from Monday to Friday for 5–7 weeks. This was developed more as a convention than after research and helped by giving the machines rest on weekends. Fractionation helped by enabling delivery of high doses without exceeding either acute or chronic normal tissue tolerance.

Hyperfractionation: In hyperfractionation, doses smaller than conventional, typically 1.1–1.6 Gy, are given 2–3 times daily so as to achieve a higher total tumour dose in the same time period. This increases the therapeutic ratio (tumour control probability for an acceptable level of normal tissue complication probability), while lower fractional dose decreases late effects. This type of fractionation provides greatest benefit for head and neck cancers. An example of hyperfractionation is a schedule of 80.5 Gy delivered in 70 fractions, 1.15 Gy twice per day over a period of 7 weeks as compared to conventional fractionation of 70 Gy delivered in 35 fractions of 2 Gy each over a period of 7 weeks (Brady et al. 2013; Douglass 2018).

Accelerated hyperfractionation: Pure accelerated treatment delivers the same total dose in less overall treatment time by delivering two or more fractions each day. This, however, results in very severe acute side effects, which become a limiting factor. In impure accelerated treatment, dose is reduced or rest period is interposed in the middle of treatment, e.g. 72 Gy in 45 fractions (1.6 Gy three times per day) over a total time of 5 weeks with a rest period of 2 weeks in middle. This schedule can improve locoregional control but not survival. Increased acute side effects should be expected, and there may be an increase in late effects including lethal complications (Brady et al. 2013; Douglass 2018).

Split course radiation therapy: This type of radiation therapy is a form of periodic treatment in which treatment is divided into two or more phases with intervening rest periods. This method is based on the sound theoretical basis of different cell population kinetics of normal and malignant tissues. The recovery of normal proliferating cell of body is much faster than recovery of the tumour cells during resting phase. This variation of radiation therapy gives time for normal tissues such as the skin and mucosa to recover. Total dose is delivered in two parts with a gap in between, with interval of 4 weeks. Split course radiation therapy is applied to elderly patients, especially with bladder, prostate and lung cancer who are being treated with a radical intent (Brady et al. 2013; Douglass 2018).

Hypofractionation: Hypofractionation delivers higher daily doses over a shorter period of time (fewer days or weeks) than standard radiation therapy. It takes less time to complete the course of radiation as compared to the time taken by conventional fractionation. This is especially preferred in tissues like breast and prostate where the tissue has slower repair coefficient. It is also preferred in elderly people who may not be compliant or tolerant to full radiation course schedules (Brady et al. 2013; Douglass 2018). This type of fractionation is also used in brachytherapy.

8.9 Side Effects of Radiation

Early (acute) side effects are observed during or shortly after a course of RT up to 90 days. Late (chronic) side effects become clinically manifest after 90 days since onset of radiotherapy (from months to many years) (Cox 1995). The commonly followed toxicity grading is the RTOG (radiation therapy oncology group) grade of radiation reactions.

8.9.1 Early Effects (Acute Effects)

Skin: The skin consists of layers of keratinising epithelial cells and has an unlimited capacity for proliferation. Cells in the basal layer multiply, mature and differentiate as they keep on migrating towards the surface but retaining some proliferative potential. Hence, they are more prone to early effects of radiation. Damage to the epidermis results in an early reaction, usually observed 10 days after a dose of 10–12 Gy. There could be redness of skin, loss of sweating, loss of hair, dry desquamation, moist desquamation and, in the severe forms ulceration, bleeding and necrosis. Severe forms are usually in areas with moist skin folds.

Oral mucosa: The cellular organisation of the mucosa is similar to that of the skin. Cells multiply in the basal layer, and as differentiation takes place, they migrate towards the surface. Following radiation exposure to maximum tolerance dose, desquamation of oral cavity occurs by about day 12, after a dose of 20 Gy, with recovery in 2–3 weeks.

1st week: Asymptomatic to slight redness and oedema. Sensitivity to alcohol or tobacco use, chemotherapy, infection (oral candidiasis, herpes simplex virus) or immunosuppression (HIV).

2nd week: Increasing pain and lack of desire to eat. There is altered taste sensation with bitter and acid flavours most affected and salty and sweet tastes least affected. Erythema and oedema increase, and there is patchy mucositis.

3rd week: Mucositis and swelling with decrease in gland secretions leading to difficulty in swallowing. Confluent mucositis and plaques.

4th week: Progression of signs. Confluent mucositis sloughs. Mucosa becomes covered by fibrin.

5th week: Maximum radiation damage. Extreme sensitivity to touch, temperature and grainy food.

Post therapy: The basal cells migrate into the area and proliferate. In 2–4 weeks, complete resolution is observed.

Oesophagus: The mucosa consists of rapidly dividing cells. After radiation, the oesophagus displays an acute mucosal response of oesophagitis and increased thickness of the squamous layer. Symptoms appear that include substernal burning with pain on swallowing at about 10–12 days after the start of therapy, at a tolerance dose of 40–45 Gy. There is return to normal within a week of the end of therapy. In higher grades of toxicities, there could be severe difficulty or pain while swallowing with dehydration or weight loss >15%. Ultimately, there could be complete obstruction, ulceration, perforation and fistula.

Bowel: Acute mucositis frequently occurs, with symptoms such as diarrhoea or gastritis, depending on the treatment field. This occurs within 10–12 days of treatment at a tolerance dose of 10–20 Gy. In higher grades of toxicities, there could be diarrhoea requiring intravenous fluid administration, severe mucous or blood discharge and abdominal distension. There could be gastrointestinal bleeding requiring transfusion, acute or subacute obstruction and even fistula or perforation.

8.9.2 Late Effects

Skin: Late reactions occur months later, mediated through damage to the dermis, principally to the vasculature. Tolerance dose for these late reactions is 40 Gy. There could be darkening over the irradiated area, loss of hair and dryness in the skin which may feel little indurated.

Oral mucosa: Late reactions involving the oral mucositis include dryness of the mouth due to loss of salivary function. The tolerance dose for these tissues is 25 Gy. There could be telangiectasia, slight atrophy and dryness to marked atrophy and complete dryness and sometimes ulceration.

Oesophagus: Late effects are related to the muscle layer; they include necrosis and a thickening of the epithelium. Tolerance dose for whole volume is 55 Gy. There could be mild fibrosis and slight difficulty in swallowing, which could also be severe atrophy and severe difficulty in swallowing needing dilatation. In severe forms, there could be necrosis ulceration and perforation.

Bowel: Late bowel reactions involve all tissue layers and are caused by atrophy of the mucosa caused by vascular injury, with subsequent breakdown resulting from mechanical irritation and bacterial infection, which leads to an acute inflammatory response. Tolerance dose for whole volume is 40–45 Gy. There could be mild diarrhoea and cramping and excessive rectal mucous and intermittent bleeding. In severe forms, there could be obstruction or bleeding requiring surgery. There could be occasions where there could be necrosis, perforation and fistula formation.

8.10 Conclusions

Radiation oncology forms an important cog in the wheel in the multimodality management of cancer. Since inception, radiation oncology has come a long way. The understanding of radiation physics, biology and its utility in treatment of cancer has increased by leaps and bounds. Rapid advances in imaging technology, onboard imaging, computational methods and computer-controlled technology have enabled better tumour localisation, tumour delineation and advanced dose calculation and dose delivery. With all the technological advancement, modern radiation therapy has become very precise, which in turn has made accurate target delineation and precise quality-controlled treatment delivery possible. All these advances have made radiation treatments more tolerable, effective with better survival and lesser side effects. As with all specialty, radiation oncology also looks to the future with research ongoing in dose escalation practices to tumour, immunotherapy and biotherapy and the use of particle beam radiation.

References

Benedict SH, Yenice KM, Followill D, Galvin JM, Hinson W, Kavanagh B, Keall P, Lovelock M, Meeks S, Papiez L, Purdie T (2010) Stereotactic body radiation therapy: the report of AAPM task group 101. Med Phys 37(8):4078–4101

Brady LW, Wazer DE, Perez CA (2013) Perez & Brady's Principles and Practice of Radiation Oncology. Lippincott Williams & Wilkins

Cameron J (1993) Radiological oncologists: the unfolding of a medical specialty, by Juan. del Regato. Radiology Centennial, Reston, VA

Case JT (1953) Some early experiences in therapeutic radiology; formation of the American radium society. Am J Roentgenol Radium Therapy, Nucl Med 70(3):487–491

Castelli J, Simon A, Lafond C, Perichon N, Rigaud B, Chajon E, De Bari B, Ozsahin M, Bourhis J, de Crevoisier R (2018) Adaptive radiotherapy for head and neck cancer. Acta Oncol 57(10):1284–1292

Cember H, Johnson TE (2009) Introduction to health physics

Charles MW (2008) ICRP publication 103: recommendations of the ICRP. Radiat Prot Dosim 129(4):500–507

Connell PP, Hellman S (2009) Advances in radiotherapy and implications for the next century: a historical perspective. Cancer Res 69(2):383–392

Cox JD (1995) Toxicity criteria of the radiation therapy oncology group (RTOG) and the European organization for research and treatment of cancer (EORTC). Int J Radiat Oncol Biol Phys 31: 1341–1346

Douglass M (2018) Eric J. Hall and Amato J. Giaccia: Radiobiology for the Radiologist

Hirsch, J.S. and Holzknecht, G., 1926. The principles and practice of roentgen therapy

International Commission on Radiation Units and Measurements, Prescribing, Recording, and Reporting Photon Beam Therapy, ICRU Rep. 50, ICRU, Bethesda, MD, 1993

International Commission on Radiation Units and Measurements, Prescribing, Recording, and Reporting Photon Beam Therapy (Supplement to ICRU Rep. 50), ICRU Rep. 62, Bethesda, MD, 1999

International Commission on Radiation Units and Measurements, Prescribing, Recording, and Reporting Intensity-modulated Photon-beam Therapy (IMRT), ICRU Rep. 83, Bethesda, MD, 2010

International Commission on Radiological Protection (ICRP-103), statement on tissue reactions, ICRP 4825-3093–1464,Vienna 2011

Khan FM, Gibbons JP (2014) Khan's the physics of radiation therapy. Lippincott Williams & Wilkins

Kramers HA (1923) XCIII. On the theory of X-ray absorption and of the continuous X-ray spectrum. London Edinb Dublin Philos Mag J Sci 46(275):836–871

Murthy V, Master Z, Adurkar P, Mallick I, Mahantshetty U, Bakshi G, Tongaonkar H, Shrivastava S (2011) 'Plan of the day' adaptive radiotherapy for bladder cancer using helical tomotherapy. Radiother Oncol 99(1):55–60

Podgorsak EB (2005) Radiation oncology physics. IAEA, Vienna

Recommendations of the International Commission on Radiological Protection. Publication 60, Annals of the ICRP, Vol. 21, 1991

Recommendations of the International Commission on Radiological Protection, publication 103, Annals of the ICRP, Vol. 37, 2007

Sowa P, Rutkowska-Talipska J, Sulkowska U, Rutkowski K, Rutkowski R (2012) Ionizing and non-ionizing electromagnetic radiation in modern medicine. Polish Ann Med 19(2):134–138

Oncology: Biochemists' Perspective

Debolina Pal and Chinmay Kumar Panda

Abstract

In this chapter, the stepwise development of carcinogenesis has been addressed along with biochemical aspects of cancer cells. Different physical, chemical, viral, and other genetic causes of cancer and deregulation of different molecular pathways associated with cancer are also discussed. It was found that alterations of molecular pathways like cell fate, cell survival, and genome maintenance are important for the development of cancer and their key regulatory genes have been identified as the molecular targets to diagnose or understand the treatment procedure for cancer. However, it was found that different natural compounds could prevent the process of carcinogenesis. So precisely, it can be suggested that healthy lifestyle and food habit till date could help an individual to stay away from this dreadful disease. However, there has been reasonable advancement in the molecular targeted treatment procedure to treat the patients at different stages of cancer.

Keywords

Hallmark of cancer · Cancer genes · Carcinogenesis · Molecular pathways

9.1 Introduction

Cancer has now become one of the most dreadful diseases and the main cause of human death after heart diseases. It is hard to find any family without an incidence of cancer. This chapter is concerned with how a normal cell or a normal stem cell is acquiring some properties to behave abnormally and irreversibly to become a cancer

D. Pal · C. K. Panda (✉)
Department of Oncogene Regulation, Chittaranjan National Cancer Institute, Kolkata, India

© The Author(s), under exclusive license to Springer Nature Singapore Pte Ltd. 2022
S. K. Basu et al. (eds.), *Cancer Diagnostics and Therapeutics*,
https://doi.org/10.1007/978-981-16-4752-9_9

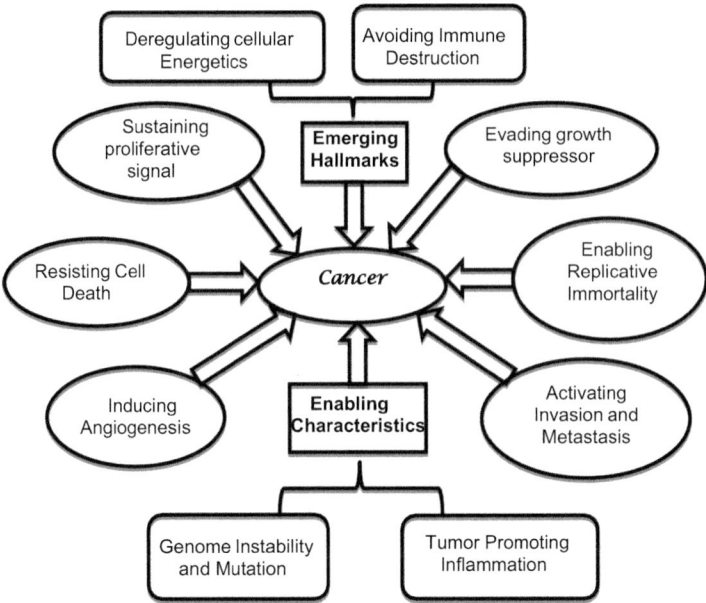

Fig. 9.1 Hallmark features of cancer (adapted and modified from Ref. (Hanahan and Weinberg 2011))

stem cell. The biochemical cause, natural prevention, and molecular therapies of cancer have also been discussed in this chapter.

In general, the characteristic features of human cancer cell are dependent upon the acquisition of the following capabilities: persistent cell division, escape from growth suppression, restriction of cell death, unlimited replicative potential, induction of angiogenesis, alteration of cellular metabolism, and escape from immune destruction (Hanahan and Weinberg 2011) (Fig. 9.1). Most of the above features are either inherited or acquired somatically by stepwise accumulation of alterations in the genes associated with cancer such as oncogenes, tumor suppressor, and stability genes (Table 9.1), each conferring a specific growth advantage to the cell, which leads to gradual conversion of normal cell into cancer cell (Romero-Garcia et al. 2011; Karakosta et al. 2005; Hanahan and Weinberg 2000).

Among these different hallmarks, the deregulation of cellular metabolism is the most important biochemical change during carcinogenesis. The main characteristic features of cancer cells compared with normal cells are to use aerobic glycolysis (Warburg effect), i.e., to use glycolysis pathway even in the abundance of oxygen. Healthy cells use anaerobic glycolysis (Phan et al. 2014). In cancer cells, glucose uptake also increased due to upregulation of glucose transporters mainly Glut1, Glut2, Glut3, and Glut4 (DeBerardinis and Cheng 2010). c-Myc and HIF-1α are important among oncogenes to play critical role in induction of some key glycolytic enzymes like HK2, PFK1, TPI1, and LDHA in tumors due to the presence of consensus Myc and HIF-1α-binding motifs in their promoter region (Vander Heiden

Table 9.1 List of genes associated with cancer

Genes associated with cancer	Definition	Activation/ inactivation	Example	References
Oncogenes	Initially, these genes were identified as genes those were carried by viruses that cause transformation of their target cells. These genes termed as "proto-oncogene" are essential for normal cellular functions like controlling cellular proliferation, differentiation, and apoptosis	Mutation, chromosomal translocation, gene amplification, and viral insertions altered their normal function to gain-of-function phenotype leads to constitutive activation	*Transcription factors* [e.g., MYC, FOS, JUN gene family], *chromatin remodelers* [e.g., SWI/SNF Complex], *growth factors* [e.g., FGF, PDGF, EGF], *growth factor receptors* [e.g., EGFR, PDGFR, VEGFR], *signal transducers* [e.g., ABL, SRC, RAF, RAS gene family, PI3K], and *regulators of cell cycle and cell death* [e.g., cyclin family, MDM2, BCL-2, BCLXL]	Croce (2008), Vogelstein and Kinzler (2004)
Tumor suppressor genes (TSGs)	These genes are required to suppress tumor formation.	Inactivated by (a) loss-of-function mutations; (b) Complete or part deletion of these genes; (c) Reduced expression due to promoter hypermethylation; (d) Deregulation of imprinting; and (e) alternate splicing	TSGs known so far are involved in *all major cellular physiological processes*, e.g., cell cycle, DNA damage repair pathways, cell signaling pathways [TP53, RB, LIMD1, RBSP3, APC, ATM, ATR, MSH2, MLH1, BRCA1, BRCA2, BUB1, SMAD4, PTEN, phospholipase A2, etc.]	Sharp et al. (2004), Kashuba et al. (2009), Berger et al. (2011)

(continued)

Table 9.1 (continued)

Genes associated with cancer	Definition	Activation/ inactivation	Example	References
Cancer susceptibility genes	The extent of susceptibility to cancer is often determined by the degree of penetrance of the genes, i.e., high penetrance and low penetrance. The alterations of the genes can be inherited via germline or can be acquired somatically in sporadic malignancies. High penetrance genes often result in multiple cases of cancer among first- and second-degree relatives, generally at young age. Nonhereditary sporadic cancers can also develop in genetically predisposed individuals resulting from alterations of several low penetrance genes	Inactivated by (a) loss-of-function mutations; (b) Complete or part deletion of these genes; (c) Reduced expression due to promoter hypermethylation	High penetrance genes: BRCA1, BRCA2, etc. Low penetrance genes: PTEN, TWIST1, etc.	Vogelstein and Kinzler (2004)
Replication fidelity genes	There is a definite life span for each type of eukaryotic cells, which is determined by the number of telomeric repeats on end of chromosome. After successive rounds of replication, the	In majority of human tumors, activation of telomerase resulting in acquisition of replicative immortality is an essential step	Telomerase is an example of this class of gene	Hanahan and Weinberg (2011)

(continued)

Table 9.1 (continued)

Genes associated with cancer	Definition	Activation/ inactivation	Example	References
	telomere repeats shorten, ultimately triggering senescence. Thus, telomere controls life span of a cell by replication			

Note: *EGF* epidermal growth factor, *EGFR* epidermal growth factor receptor, *FGF* fibroblast growth factor, *PDGF* platelet-derived growth factor, *PDGFR* platelet-derived growth factor receptor, *VEGFR* vascular endothelial growth factor receptor

et al. 2009; DeBerardinis et al. 2008). HIF-1α and c-Myc are mainly functional in hypoxia and normoxia, respectively (Dang 2007). This synchronization is quite important for continuous energy supply for cell proliferation and biosynthesis for glycolysis.

Moreover, pyruvate, the end product of glycolysis, is converted into lactate in tumor cells instead of acetyl-CoA in normal cells due to overexpressed lactate dehydrogenase. The role of lactate is very important for creating tumor microenvironment (Vander Heiden et al. 2009). Lactate restricts intracellular oxidative stress by reducing reactive oxygen species (ROS) in cancer cells and induces tumors survival. The pH of extracellular microenvironment is reduced by lactate accumulation, and it helps metalloproteases to induce invasion and metastasis by breaking down extracellular matrix (Phan et al. 2014).

Glycolytic metabolites, e.g., glucose-6-phosphate and dihydroxyacetone phosphate, could be used up in different other metabolic pathways like nucleotide and lipid biosynthesis pathway. Thus, glycolysis plays an important role in cell proliferation and tumor growth promotion.

9.2 Causes of Cancer

Multiple factors influence carcinogenesis. Among them, environmental factors, genetic constitution of an individual, diet, and lifestyle all share equal importance for causation of cancer. Broadly, environmental factors can be divided into physical, viral (Carrillo-Infante et al. 2007; Martin and Gutkind 2008), and chemical factors (Parsa 2012). Genetic constitution of an individual also determines the effectiveness of the environmental factors for pathogenesis of cancer. The genetic agents that influence carcinogenesis are shown in Table 9.2.

Table 9.2 List of agents influencing carcinogenesis

Compounds	Main source	Type of cancer
Physical carcinogens		
Ultraviolet (UV) radiation (UVA, UVB, UVC rays)	Sun light, tannin, lamps	Skin
Ionizing radiation	Cosmic ray, radioactive decay	Leukemia, skin
Viral carcinogens		
Epstein–Barr virus	Oral transfer of saliva, genital secretions	Burkitt's lymphoma, Hodgkin's lymphoma, nasopharyngeal carcinoma
Helicobacter pylori	Stomach	Stomach
Hepatitis B, hepatitis C	Blood transfusion, body fluid	Liver
Human immune deficiency virus (HIV-1)	Blood transfusion, body fluid	Cervical cancer, Kaposi sarcoma, non-Hodgkin lymphoma, etc.
Human papillomavirus (HPV 16, HPV 18, HPV 31, HPV 33, etc.)	Sexually transmitted infection	Oral cancer, cervical cancer, etc.
Chemical carcinogens		
Polycyclic aromatic hydrocarbon (PAH)		
7,12-dimethylbenz[a]-anthracene (DMBA)	Environmental pollutant, vehicles exhaust	Skin, lung, stomach
Benzo(a)pyrene (BaP)	Tobacco smoke, grilled meat, coal tar, smoke from the burning of fossil fuels, coal tar	Lung, skin
Benzo(g)chrysene (BgC)	Coal tar	Skin
3-methylcholanthrene (MCA)	Burning organic compounds	Prostate cancer, sarcoma
20-methylcholanthrene (MCA)	Research tool	Sarcoma, transformation of fibroblast
N-Nitroso compounds		
4-(methyl nitroso amino)-1-(3-pyridyl)-1-butanone (NNK)	Cigarette smoke, fried, foodstuffs, meat, beer, fish, latex product	Lung, nasal cavity, liver, oral, and pancreas
N'-nitroso nornicotine (NNN)		Lung, oral, esophagus
4-(methylnitrosamino)-1-(3-pyridyl)-1-butanol (NNAL)		Lung, nasal cavity, liver, and pancreas
N-Nitrosodimethylamine (NDMA)		Liver, gastric, esophagus
N-Nitrosodiethylamine (NDEA)		Liver, gastric, esophagus
N-methyl-N-nitrosourea (MNU)	Not used	Bone, brain, pancreas, blood

(continued)

Table 9.2 (continued)

Compounds	Main source	Type of cancer
Natural carcinogen		
Aflatoxin B1	Mycotoxin from *Aspergillus flavus* (found in contaminated peanut, grains)	Liver
Asbestos	Thermal insulation	Lung, mesothelioma, gastrointestinal, colorectal
Metals		
Arsenic (As)	Natural ores, alloys, groundwater	Skin, lung, liver
Cadmium (Cd)	Natural ores, batteries, pigment, ceramics	Lung, prostate, kidney
Chromium (Cr) (hexavalent)	Groundwater, tap water	Lung
Lead (Pb)	Battery, smelter, metal products, paint	Lung, bladder
Nickel (Ni)	Natural ores, electrodes	Lung, nasal cavity
Different dyes	Pigment, coloring oil, textiles, paints, printing inks, paper, and pharmaceuticals	Liver, lung, bladder, stomach, kidney, oral, larynx, esophagus, liver, gallbladder, pancreas
Ethanol	Alcoholic beverages	Liver, colon, oral, breast
Others		
Acetaldehyde	Alcoholic beverages	Liver, colorectal
Ethylene oxide	Textile, detergent, industry, cosmetics, sterilant for food	Leukemia, stomach, pancreas
Formaldehyde	Cigarette smoke, air pollution, fungicide, germicide, etc.	Lung, leukemia, brain cancer, etc.
Ortho-toluidine	Synthetic chemical used too	Urinary bladder cancer, liver
Vinyl chloride	Petroleum-derived chemicals	Liver cancer

9.3 Development of Cancer

A series of genetic and epigenetic alterations progressively convert the normal cell to a premalignant state and finally a cancerous state (Yokota and Sugimura 1993). Different proposed models are there to understand the whole process (Fig. 9.2).

Thus, to understand the genetic and molecular mechanism of cancer development, Fould (1957) was the first to propose a progression model stating that cancer was due to phenotypic manifestation of several genetic damages (Fig. 9.2) followed by Nowell's *clonal evolution model* of neoplastic progression, which was continuous appearance of genetically variant cells within a tumor of monoclonal expansion that could compete with each other on the basis of highest cellular growth rate (Fig. 9.2) (Nowell 1976). Then, it was proposed by Ref. (Bodmer 1997; Farber and Cameron 1980; Scrable et al. 1990) that normal cell progressing to fully malignant phenotype might be due to a nested set of aberrations (Fig. 9.2). Then, Ilyas et al.

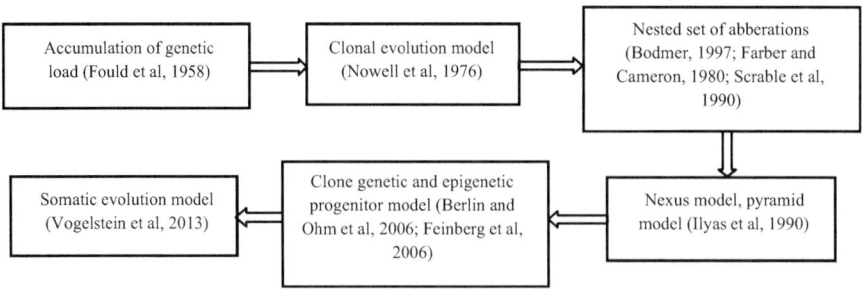

Fig. 9.2 Development of different predicted models explaining stage-wise progression of cancer

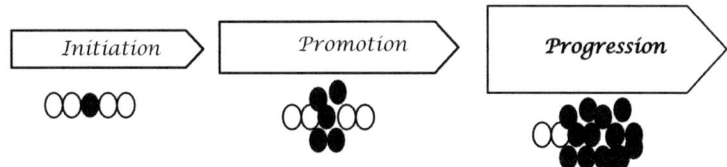

Fig. 9.3 Schematic diagram showing stepwise progression of cancer

(1999) explained tumor progression through *nexus model* (Fig. 9.2), which explained that tumor developed by nexus of interconnecting mutations and selection pressures applies at each point. They also proposed an *inverted pyramid model* (Fig. 9.2), where one mutation could predict the selection of next mutation and their interaction ensure optimal activity of both. Further studies lead to the development of *clonal genetic model and epigenetic progenitor model* of cancer progression (Baylin and Ohm 2006) (Fig. 9.2). Clonal genetic model was supported by induction of oncogenes and inactivation of tumor suppressor. Epigenetic model stated that cancer developed through three steps: an epigenetic alteration of progenitor cells; an initiating mutation along with genetic and epigenetic plasticity; and finally, the advanced *somatic evolution model* was proposed by Vogelstein et al. (2013). This model stated that mutation of certain gene associated for the development of carcinoma from normal cell was due to somatic evolution (Fig. 9.2).

Clinically, human tumors can be divided into three groups: *premalignant lesions, primary tumors, and metastases* (Yokota and Sugimura 1993). The process of carcinogenesis passes through three major sequential steps: (a) initiation, (b) promotion, and (c) progression (Fig. 9.3).

9.3.1 Initiation

The first step involves irreversible changes incorporated into the cellular genetic material. If cellular repair mechanism fails to detect damaged DNA, the base sequence would be modified (insertion/deletion/ modification) in the next round of replication. Ultimately, transcription and translation of this modified template would synthesize a modified protein with altered function. The initiated cell has proliferative stimulus, enough to generate clones of modified cell, though this is unable to generate malignant cell population (Yokota and Sugimura 1993; Feinberg et al. 2006). However, repeated exposure to initiating agent with certain frequency might lead to clonal expansion of mutated cells.

9.3.2 Promotion

The initiated cells have limited proliferative potential, which is not sufficient for continuation of the carcinogenesis process. Initiation followed by promotion provides an impact for further progress of the carcinogenic process. Tumor promotion involves genetic activation or inactivation to stimulate the proliferative potential of initiated cells, leading to the development of multiple *benign tumors* or *hyperplastic lesions*. Promoting activity can be achieved by alteration in normal signal transduction pathway with enhanced rate of transcription and translation of genes responsible for cellular proliferation.

9.3.3 Progression

This final step of carcinogenesis involves conversion of benign tumors to malignant neoplasms that is able to invade adjacent tissues resulting in metastasis (Slaga 1983; DiGiovanni 1992). Specific characteristics of metastatic cells like increasing cellular proliferation, reprogramming cellular metabolism, alteration in hormonal response, and loss of cellular differentiation, decreased antigenicity and acquisition of drug resistance provide selective growth advantage for tumor cell population (Nowell 1986). Multiple host tissue factors, for example, proteolytic enzymes, activators of plasminogen, tumor angiogenic factor, platelet-agglutinating capacity, and different membrane molecules including laminin, fibronectin, and major histocompatibility complex gene products, play important role in tumor progression (Welch et al. 1984).

9.4 Molecular Pathways Associated with Cancer

Stepwise accumulation of genetic and epigenetic changes leads to neoplastic conversion of a cell (Yokota and Sugimura 1993). Collections of cellular pathways are altered during the process of carcinogenesis. A brief description of a few pathways is given in Fig. 9.4. The pathways of the groups seem to have cooperativity to have selective growth advantage at each stage of tumor development.

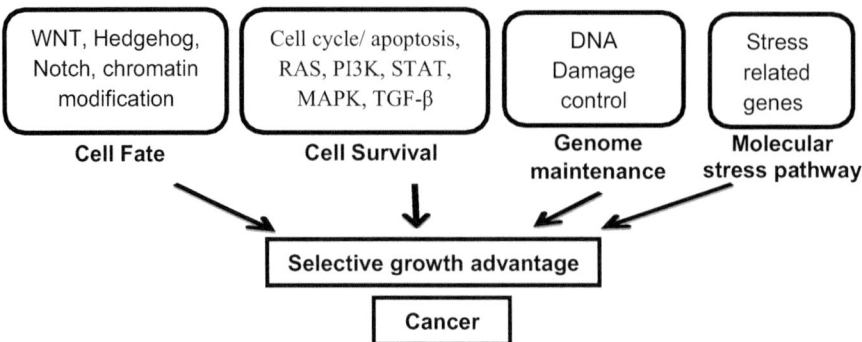

Fig. 9.4 Cellular pathways associated with cancer development (Edited and modified from Ref. (Vogelstein et al. 2013))

9.4.1 Cell Fate

Alteration in cell division and differentiation due to acquired growth advantage leads to tumor progression. These self-renewal signal transduction pathways, namely Wnt, Hedgehog Notch, and Bmi-1, are important for the determination of fate of the stem cells (Lei et al. 2017; Perrimon et al. 2012; Hoffmann 2012). Most of these pathways are frequently deregulated in several cancers and most importantly within the cancers that possess stem-like properties (Curtin and Lorenzi 2010). Also, the key genes involved in chromatin modification and transcription pathways belong to this category (Fig. 9.4).

9.4.1.1 Stem Cell Self-Renewal Pathway

WNT Pathway
This pathway is a well-known self-renewal pathway regulating embryonic development and tissue homeostasis and contributes to control the cell proliferation, differentiation, and epithelial–mesenchymal transition (Sarkar et al. 2010). β-catenin is the main effector molecule of this pathway. In the absence of active Wnt ligands, β-catenin forms complex with scaffold proteins axin and adenomatosis polyposis coli (APC). Then, β-catenin is phosphorylated by casein kinase Iα (CKIα) and glycogen synthase kinase (GSK-3β) at N-terminal serine/threonine residues sequentially followed by ubiquitination and proteasomal degradation. In the presence of Wnt ligand, Wnt binds to frizzled receptor and LRP co-receptor leading to the inhibition of β-catenin–axin–APC degradation complex formation resulting in release of β-catenin from the complex. The cytoplasmic-free β-catenin then either binds to E-cadherin at membrane or phosphorylated at tyrosine-654 residue by activated receptor tyrosine kinases (like EGFR) followed by phosphorylation at serine 675 by protein kinase A and translocate to the nucleus. In the nucleus, β-catenin complexes with T-cell/lymphoid enhancer transcription factors (TCFs/ LEF) and activates the transcription of Wnt target genes such as c-Myc, cyclin D1,

CD44, and EGFR. Several secreted or intracellular proteins like secreted frizzled receptor proteins (sFRPs) negatively regulate Wnt signaling by inhibiting Wnt ligands binding to receptor (Sarkar et al. 2010; van Veelen et al. 2011).

Aberrant activation of the Wnt/β-catenin pathway due to mutation of β-catenin gene (*CTNNB1*) at exon 3 and/or inactivation of APC, axin, or WNT antagonists SFRP1 and SFRP2 by mutation/deletion/ promoter hypermethylation could lead to nuclear β-catenin accumulation and transcriptional activation of WNT target genes like c-Myc and cyclin D1 as seen in different cancers (Sarkar et al. 2010).

Hedgehog (Hh) Pathway

The Hedgehog self-renewal pathway is an important regulator of cell proliferation, differentiation, and polarity. Alteration in this pathway leads to numerous human diseases including cancer (Sarkar et al. 2010; Chen and Jiang 2013). The effector molecule of this pathway is Gli. When Hh ligand is absent, PTCH receptor inhibits the transmembrane receptor-like protein smoothened (SMO) and Gli2/3 cytoplasmic form complex with Costal2 (Cos2), Fused (Fu) and Suppressor of fu (Sufu) leading to sequentially phosphorylation by PKA, CK1, and GSK-3β at several serine/threonine sites of Gli to form truncated repressor (Gli-r) (Sarkar et al. 2010; Chen and Jiang 2013). In the presence of Hh ligand, PTCH receptor activates SMO to engage COS2/Fu complex resulting in accumulation of activated full-length Gli that could enter into nucleus and transcribe several Hh target genes like Cyclin D1, c-Myc, Bcl2, Gli1, and PTCH (Chen and Jiang 2013; Sarkar et al. 2010).

Reduced expression of antagonists of the pathways like PTCH, HHIP, and SUFU due to deletion/mutation/promoter methylation and high expression of SHh, SMO, and Gli1 resulting in increased expression of target genes are reported in several cancers (Moeini et al. 2012; Chen and Wang 2015).

Notch Pathway

This signal is triggered by binding of ligand on the membrane of one cell (delta/delta-like/jagged/serrate) to a receptor (NOTCH1/2/3/4) on the membrane of the contacting cell leading to proteolytic cleavage of NOTCH receptors to release the cytoplasmic tail of NOTCH, i.e., intracellular domain of NOTCH (NICD). NICD translocates to the nucleus and associates with transcription factors p300, mastermind protein (MAM), and recombination signal-binding protein for κ-immunoglobulin kappa J region (RBPJκ) in mammals to turn on transcription of target genes [hairy/enhancer of split (HES) family of transcription factors] (Sikandar et al. 2010).

Alteration of NOTCH signaling was reported to be associated with several cancers like mutations in NOTCH1 in non-small cell lung cancer/oral cancer and upregulation of NOTCH2 in colorectal cancer (Andersson et al. 2011).

BMI Pathway

Self-renewal of hematopoietic stem cells takes place by this pathway. BMI1 (B-cell-specific Moloney murine leukemia virus integration site 1) is a polycomb ring finger oncogene. It promotes cell proliferation by transcriptional inhibition of cyclin-dependent kinase inhibitor INK4A (p16) and p19ARF (p14) (Park et al. 2003). Overexpression of BMI-1 has been previously reported in several cancers including gastric, ovarian, breast, head and neck, pancreatic, lung, liver, and endometrial carcinoma (Wang et al. 2015).

9.4.1.2 Chromatin Modification
Chromatin modification takes place mainly by DNA methyltransferases (DNMTs: DNMT1, DNMT3a, DNMT3b), histone acetyltransferase (HAT), histone deacetylases (HDACs), and histone methylase (HMT).

DNMT1 is a predominant methyltransferase for CpG methylation in hemimethylated DNA (Lopez-Serra et al. 2006). Overexpression of DNMT1 was reported in several cancers, for example, pancreatic cancer, pediatric gastric cancer, and retinoblastoma (Li et al. 2011; Ma et al. 2017; Qu et al. 2010). Among HDACs, HDAC 1, HDAC 2, HDAC 5, and HDAC 7 could play important roles in carcinogenesis (Miller et al. 2011; Urbich et al. 2009; Lei et al. 2017). Altered expressions of different HDACs have been reported in various cancers. HDAC 1 and HDAC 2 were found to be upregulated in colon cancer and gastric cancer, respectively (Miller et al. 2011). HDACs and histone acetyltransferases could bind to DNA indirectly through multiprotein complexes like co-repressors and co-activators (Sengupta and Seto 2004). Histone methylases (HMTs) have important role in cancer development (Albert and Helin 2010). HMTs could modify histones at specific Lys and Arg residues to alter their functions (Albert and Helin 2010). In addition, upregulation of histone demethylases has been seen in different cancers and suggested to be associated with chemoresistance (Yang et al. 2017).

9.4.2 Cell Survival

Cell survival is mainly controlled by several signaling proteins like EGFR, HER2, FGFR2, PDGFR, TGFbR2, MET, KIT, RAS, RAF, PIK3CA, and PTEN through cell cycle and apoptosis (Vogelstein et al. 2013) (Fig. 9.4). Progression through the cell cycle can be directly controlled by driver genes that directly regulate the cell cycle or apoptosis, such as P16, MYC, and BCL2, which are very frequently mutated in cancers. Inactivating mutations in VHL gene could enhance cell survival and stimulate angiogenesis through secretion of vascular endothelial growth factor (VEGF).

9.4.2.1 Alteration of Cell Cycle
Cell cycle is a highly ordered series of events, responsible for cellular duplication. Different extracellular signals, for example, growth factor binding, hormonal responses, cytokines, supply of nutrients, and anchorage attachments, stimulate a cell to divide (Michalides 1999). The process of cell cycle is highly regulated by sequential activation and degradation of the cyclins (cyclin D, cyclin E, cyclin A, and cyclin B), the cyclin-dependent kinases (CDKs: serine/threonine kinases; CDK1, CDK2, CDK4, and CDK6), and their inhibitory proteins known as cyclin-dependent kinase inhibitors [CKIs: INK (p16, p15) and KIP (p21, p27, and p57)

family members] (Michalides 1999). Each of these cyclin–CDK complexes, together with CKIs, is responsible for controlling cell cycle progression through checkpoints. Induction of DNA damage results in activation of cell cycle checkpoint proteins to arrest cell cycle and make necessary DNA repair or elimination of damaged cells by apoptosis (Hartwell and Weinert 1989). The eukaryotic cell cycle is secured by 4 checkpoints at G1/S phase; S phase; G2/M phase; and M phase (Tyson and Novak 2008). Deregulations of these checkpoints are important events during carcinogenesis.

9.4.2.2 Alteration of Apoptosis Pathway

Programmed cell death or apoptosis is one of the mostly altered pathways during carcinogenesis. This is an essential cellular event during embryonic development, immune system function, and control of tissue homeostasis (Vaux and Korsmeyer 1999). Programmed cell death follows two alternative pathways depending on death-inducing cellular response: (1) extrinsic pathway and (2) intrinsic pathway (Gupta 2003).

Extrinsic pathway is activated by binding of ligands to death receptors [tumor necrosis factor (TNF) receptor superfamily, including Fas/CD95, TNFR1, DR3, DR4, and DR5] on the cell surface. Upon ligation, this receptor recruits adaptor molecule (FADD, TRADD) by its cytoplasmic death domain (DD). The death effector domain (DED) in adaptor further recruits procaspase-8. Procaspase-8 (cysteinyl-aspartate-specific proteases) cleaves to form active caspase-8, which further activates effector caspase-3 to execute apoptosis process.

Intrinsic pathway is mitochondria-mediated pathway and is initiated by cellular stress (UV radiation, cytotoxic drug application) that alters the mitochondrial membrane potential. Mitochondrial membrane permeability is controlled by Bcl-2 family proteins (Bcl-2, Bax, Bad, Bak, Bcl-xL, Bid) (Walensky 2006). Cellular stress response mediated by p53 or c-Myc activates pro-apoptotic protein bax, which is translocated from cytosol to mitochondrial membrane to form dimer. During apoptosis, Bad is dephosphorylated and translocated to the outer membrane of mitochondria. Otherwise, the phosphorylated form of Bad is sequestered within cytoplasm. On outer membrane, Bad heterodimerizes with Bcl-xL to block its anti-apoptotic function (Walensky 2006). Bak also loosely associate with outer membrane. Bak forms homo-oligomers within mitochondrial membrane resulting in release of cytochrome-c in cytosol and binds with APAF1 (apoptotic protease-activating factor-1) to form apoptosome complex. Apoptosome activates procaspase-9 (initiator caspase), and subsequent caspase cascades to precede apoptosis (Pollack and Leeuwenburgh 2001; Okada and Mak 2004). Several other proteins were also released from mitochondria. For example, Smac (second mitochondria-derived activator of caspases) and DIABLO bind to IAPs (inhibitors of apoptosis proteins) and AIF (apoptosis-inducing factor). Translocation of AIF to nucleus induces chromatin condensation and DNA fragmentation. Deregulation in apoptosis pathway is a hallmark feature of carcinogenesis (Hanahan and Weinberg 2011).

9.4.3 Genome Maintenance

Alterations in the genes that control DNA damage response pathway (Vogelstein et al. 2013) (Fig. 9.4) allow cells to undergo chromosomal alterations like translocation, inversion, and duplication to survive and divide.

Direct reversal: This is the most simple DNA repair pathways in human that directly reverse the O6-methylguanine (O6-mG) (frequently mutated by alkylation) by the product of the MGMT gene (O6-methylguanine DNA methyltransferase) (Margison and Santibanez-Koref 2002). Cellular metabolism also produces low levels of O6-mG lesion in guanine residues of DNA molecule (Sedgwick 1997).

Base excision repair (BER): If the DNA bases are damaged by several cellular processes like oxidation, methylation, and deamination, this multistep pathway plays active role to detect and remove the damaged bases. This pathway is of two types "short patch" and "long patch." The former involves replacement only of the damaged base, whereas the later replaces a stretch of about 2–10 nucleotides including the damaged base (Memisoglu and Samson 2000).

Nucleotide excision repair (NER): This repair system helps to pyrimidine dimers caused by the UV component of sunlight, bulky chemical adducts, DNA intrastrand cross-links, and some forms of oxidative damage (Hess et al. 1997).

Double-strand break repair (DSB): This type of repair pathway is the most important to detect the problem of central dogma of a cell, i.e., replication and transcription (Mehta and Haber 2014). Several factors like ionizing radiation, exposure of genotoxic chemicals, and any mechanical stress on chromosomes can induce DSB. There are two pathways for the repair of DSBs viz homologous recombination (HR) and nonhomologous end joining (NHEJ) (Lieber 2010). Which pathway will be selected by the cell is unpredictable; the cell cycle stage at that time, however, plays important role for this decision (Jackson 2002).

Mismatch repair (MMR): This pathway corrects replication errors such as base–base mismatches and insertion/deletion loops (IDLs) that result from DNA polymerase misincorporation of nucleotides and template slippage, respectively (Fukui 2010). Mispairing generated by the spontaneous deamination of 5-methylcytosine and heteroduplexes formed following genetic recombination is also corrected via MMR. A defective MMR pathway leads to "mutator phenotype" characterized by increased frequencies of spontaneous mutations and microsatellite instability (MSI), which are the hallmarks of cancer (Loeb et al. 2008).

9.4.4 Molecular Stress Pathway

The hypoxic stress in the intratumor microenvironment augments molecular stress and provides the required stimulus for expression of the pro-angiogenic factors (VEGF), which mediates intricate interplay between various extracellular signaling pathways viz Notch and Hedgehog (Foxler et al. 2012). Hypoxia-inducible factor (HIF-1α) is a crucial player of tumor angiogenesis. Under normoxic condition, the oxygen-sensing prolyl hydroxylase (PHD) catalyzes hydroxylation of Pro-564 residue of HIF-1α. The TSG VHL binds and ubiquitinates this hydroxylated HIF-1α, subjecting it to proteasomal degradation, thereby shutting down the expression of hypoxia-specific genes (Fedele et al. 2002). Another candidate TSG, LIMD1, acts as a molecular scaffold to interact with PHDs and VHL to efficiently degrade HIF-1α during normoxia (Foxler et al. 2012). However, the intratumoral hypoxic condition inhibits the aforesaid hydroxylation of HIF-1α, thereby preventing VHL-mediated ubiquitination and degradation of HIF-1α, leading to its stabilization (Fedele et al. 2002). Under these hypoxic circumstances, the oxygen-dependent asparaginyl hydroxylase (FIH) fails to hydroxylate Asn-803 residue of HIF-1α, thereby enabling binding of the co-activators CBP/p300 and HIF-1β to HIF-1α leading to its transcriptional activation and expression of hypoxia-responsive genes viz VEGF and D-ll4 (Diez et al. 2007). VHL is inactivated in various malignancies, especially in kidney cancer, facilitating stabilization of HIF-1α and consequent tumor angiogenesis, even under normoxia (Banks et al. 2006).

9.5 Cancer Biomarkers

In response to an abnormal or disease conditions like cancer, our body produces some biological molecule that present in body fluids or tissues, according to National Cancer Institute (NCI) (Henry and Hayes 2012). According to World Health Organization (WHO), a biomarker can be any substance, structure, or process that can be detected, quantified, and influence or predict the incidence or outcome of disease (Sturgeon et al. 2010); for example, a cancer biomarker measures the risk of cancer development or measures the cancer progression risk or potential response to therapy. The widely used cancer biomarkers are mainly proteins (e.g., an enzyme or receptor) (Table 9.3). In addition, there are also other types of cancer biomarkers, for example, nucleic acid (e.g., microRNA or other noncoding RNA; microsatellite DNA markers), antibodies, and peptides (Sturgeon et al. 2008). A biomarker can be synthesized as a result of alterations of different metabolic and biosynthetic pathways. Biomarkers are usually detected in noninvasive procedures like collection of samples (blood, serum, plasma, stool, urine, sputum). Sometimes, it requires special imaging for evaluation or biopsy sampling for tissue-based analysis. Genetic biomarkers can be detected as DNA base sequence variations in germline DNA isolated from whole blood and sputum. Cancer biomarkers can be classified into the following categories:

Predictive biomarkers: With the help of this marker, a response against specific therapy such as response against certain chemotherapeutic drugs for specific cancers could be assessed (Cramer et al. 2011). Like in colorectal cancer, cetuximab treatment will be in vain if patient has KRAS-activating mutations. So, KRAS mutation status is a predictive biomarker for this case (Diamandis 2010) (Table 9.3).

Prognostic biomarker: With the help of these markers, disease recurrence or disease progression can be predicted; i.e., it helps to detect the clinical outcomes of the disease. An example of a prognostic cancer biomarker is the 21-gene recurrence

Table 9.3 Cancer biomarkers (edited and modified from Ref. (Goossens et al. 2015; Dawood et al. 2014; Clevers 2011))

Class of biomarker	Specific biomarker	Deregulation in cancer	Reported in cancer	Associated drug
Hormone receptor	Estrogen receptor (ER)/progesterone therapy (PR)	Receptor expression related to prognosis of the disease	Breast cancer	Tamoxifen
	Aromatase inhibitors	Decrease amount / The estrogen		Anastrozole, letrozole
Growth factor receptor	HER2	Overexpression	Breast cancer, esophagogastric adenocarcinoma	Tucatinib, trastuzumab, pertuzumab, ado-trastuzumab emtansine
	EGFR (HER1)	Mutations in tyrosine kinase domain lead to constitutive activation	Non-small cell cancer (NSCLC) — Lung	Receptor kinase erlotinib, / Tyrosine inhibitor: Gefitinib, afatanib, lapatinib
	EGFR (HER1)	Resistance to EGFR therapy if KRAS is mutated	Colorectal cancer	Receptor tyrosine kinase inhibitor: cetuximab, panitumumab
	BCR-ABL	BCR-ABL translocation Philadelphia chromosome (9; 22) leads to the formation of a constitutively active tyrosine kinase	Chronic myeloid leukemia (CML)	Imatinib
		Imatinib-resistant CML		Dasatinib, nilotinib
	PML–RARa	t(15;17)(q24;q21) translocation is associated with favorable prognosis	Acute myeloid leukemia (AML)	All-trans-retinoic acid
	Anaplastic lymphoma receptor tyrosine kinase gene (ALK)	Inversion in chromosome 2 leads to EML4-ALK fusion oncogene	Non-small cell lung cancer (NSCLC)	Crizotinib, ceritinib

Other cellular pathway markers	MEK in RAS-RAF-MEK pathway	Resistance to EGFR therapy	Pediatric gliomas, neuroblastoma, AML renal cell carcinoma (RCC), and NSCLC	AZD6244
	mTOR pathway			Ridaforolimus
	MET		Melanoma, colorectal cancer	Tivantinib, everolimus
	B-RAF V600	B-RAF enzyme inhibitor	Skin melanoma	Vemurafenib, dabrafenib, trametinib, binimetinib
	B-RAF	Small molecule B-RAF inhibitor that targets key enzymes in the MAPK signaling pathway	Melanoma	LGX818
	PI3Kα	Potent and selective PI3Kα inhibitor		BYL719
	BRCA1/2		Breast cancer	Olaparib, iniparib
	TOP2A	In subjects with HER2 overexpression		Anthracycline-based neoadjuvant chemotherapy
	RAS	Mutation type	Colorectal	FOLFOXIRI and bevacizumab
	HER	HER inhibitor	Head and neck squamous cell carcinoma; breast cancer	HM781-36B
	GATA-3		Peripheral T-cell lymphomas	MLN9708
Stem cell pathway	CD44	Broadly on many tissues	Breast/liver/head and neck/pancreas cancer	

(continued)

Table 9.3 (continued)

Class of biomarker	Specific biomarker	Deregulation in cancer	Reported in cancer	Associated drug
markers	CD90	T cells	Neurons Liver cancer	
	CD133	Proliferative cells in multiple organs	Brain/colorectal/lung/liver cancer	
	EpCAM	Panepithelial marker	Colorectal cancer, pancreatic cancer	
	CD19	Broadly on B lymphocytes	B-cell malignancies	
	CD20	Broadly on B lymphocytes	Melanoma	
	CD24	Broadly on B cells	Neuroblasts Pancreas/lung cancer, negative on breast cancer	
	CD34	Hematopoietic and endothelial progenitors	Hematopoietic malignancies	
	CD38	Multiple stages of B and T cells	Negative on AML	
	ABCB5	Keratinocyte progenitors	Melanoma	

score, which was predictive of breast cancer recurrence and overall survival in node-negative, tamoxifen-treated breast cancer (Diamandis 2010; Goossens et al. 2015).

Diagnostic biomarker: These markers help to diagnose the disease, i.e., the condition of a patient in specific disease (Diamandis 2010; Goossens et al. 2015).

Cancer stem cell biomarkers: In our normal system, a particular cell type with self-renewal ability is called stem cells, which give rise to all the cell lineages in the corresponding tissues. These cells undergo either asymmetric division generating one stem cell (S) and another differentiating cell (D) or symmetric division generating two stem cells. These stem cells have highest potential for proliferation and have longer life span so they have much greater tendency to acquire necessary number of transformation-associated genetic/epigenetic changes by exposure of inflammation, radiation, chemicals, or infection to become a *cancer stem cell* (CSC) (Moore and Lyle 2011). In a primary tumor or cancer cell lines, a very small percentage of the cells are CSC. The prevalence of CSCs is not the same in each tumor type and varies from tissue to tissue. Cancer stem cells (CSCs) express different surface markers. Based on the surface markers (Table 9.3 modified and edited from Ref. (Dawood et al. 2014; Clevers 2011)), e.g., CD44, CD90, CD133, and EpCAM, different cancers like head and neck, liver, breast, and lung can be evaluated.

9.6 Molecular Therapy of Cancer

A molecular therapeutic target for cancer can be identified with the following criteria; for example, it should be an important key regulatory protein or pathway without which cellular proliferation will be restricted. It should be upregulated in cancer tissue and not in normal tissue. However, drug should be available for the particular target. MYC, KRAS, and TP53 are the most common driver genes in human cancers but are reported to be resistant to therapeutic intervention (Kessler et al. 2012; Luo et al. 2009). On the other hand, approaches to inhibit kinases are well developed (Zhang et al. 2009).

To date, several studies identified large number of candidate targets and their anticancer therapies are now under development. However, only very few pathway-based targeted therapies have got place in clinical practice (Goossens et al. 2015). Some of these are listed in Table 9.3. One such example is treatment against chronic myeloid leukemia (CML) (Nowell and Hungerford 1960).

In breast cancer, amplification of the HER2 gene defines a subset of disease that is typically highly aggressive. Imatinib and trastuzumab were used as effective targeted therapy for patients with HER2-enriched breast cancer (Slamon et al. 2001). BRCA1/BRCA2 is also targeted for treating ovarian cancer (Table 9.3). Olaparib and iniparib are under clinical trial in patients with BRCA-driven breast cancer after being passed the preclinical test (Fong et al. 2009; O'Shaughnessy et al. 2011). A major difficulty in cancer treatment is the intracellular pathways that are interlinked. Thus, for the development of better molecular targeted therapy of cancer the discrete analysis of cellular pathways associated with tumor development should be analyzed to find out the key regulatory step(s). Then, respective drugs may be designed accordingly.

9.7 Cancer Chemoprevention

The term chemoprevention was first coined by Michael Sporn (Sporn 1976). Chemoprevention of cancer is to use natural, synthetic, or biological chemical agents to reverse, suppress, or prevent carcinogenic progression to invasive cancer (Sporn 1976). It is one of the important areas of current cancer research. Wattenberg L.W. pioneered the research by chemopreventive approach using animal model systems (Wattenberg 1993). Epidemiological studies and experiments in *in vitro* and *in vivo* models indicated that several dietary items/products viz. vitamins; beverages; and food components have cancer-preventive property (Table 9.4). The

Table 9.4 List of few chemopreventive agents along with their sources (Edited and modified from Ref. (Wang et al. 2012; Pal et al. 2012; Sur et al. 2016)

Food components	Name	Active compound
Beverages	Green tea	Epigallocatechin-3-gallate
	Black tea	Theaflavins
Fruits	Grapes	Resveratrol
	Berries	Resveratrol
Vegetables	Broccoli	Sulforaphane
	Cabbage	Indole-3-carbinol
	Carrot	Beta-caroteneCapsaicin
	Chili peppers	Genistein
	Soybean	Lycopene
	Tomato	
Spices	Bay leaves	Eugenol
	Cardamom	Do
	Cinamon	Do
	Clove	Do
	Coriander	Apigenin
	Cumin	Thymoquinone
	Garlic	Diallyl disulfide
	Ginger	Gingerol
	Mustard	Ferulic acid
	Parsley	Apigenin
	Pepper	Piperin
	Sesame seed	Ferulic acid
	Turmeric	Curcumin
Others	Honey	Caffeic acid phenethyl ester
	Peanut	Resveratrol
	Mushroom	Vitamin D2
	Sunflower oil	Vitamin E
Medicinal plant	Chirata	Amarogentin
	Karanja	Pongapin and Karanjin

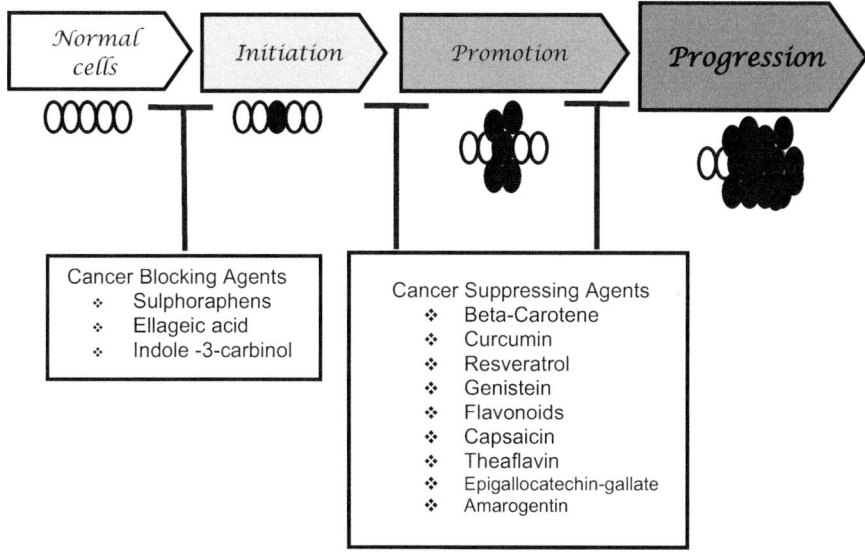

Fig. 9.5 Dietary phytochemicals that block or suppress different stages of carcinogenesis. Blocking agents block metabolic activation of pro-carcinogens and restrict the step of initiation. Cancer-suppressing agents can suppress either promotion or progression step (edited and modified from Ref. (Kotecha et al. 2016))

potential cancer chemopreventive compounds belong to different structural and functional chemical classes.

Chemopreventive agents can be classified according to their mechanism of action and must have the following properties (Fig. 9.5):

- Prevent absorption or metabolism of carcinogens (blocks *initiation*).
- Prevent carcinogens to react with specific cellular targets (blocks *initiation*).
- Suppress the expression of neoplasia in cells exposed to carcinogens (blocks *promotion*).
- Delay or prevent the conversion of initiated cells to preneoplastic cells and ultimately to neoplastic cells (blocks *promotion/progression*).
- Inhibit tumor progression by inhibiting cell proliferation and blocking metastasis (blocks *progression*). List of few such important chemopreventive compounds found in several daily used fruits, vegetables, and spices is listed in Table 9.4.

9.8 Conclusion

There are numerous causes of cancer. Altering molecular pathways like cell fate, cell survival, and genome maintenance are important for the development of cancer. Their key regulatory genes have been identified as the molecular targets to diagnose or understand the treatment procedure for cancer. Different natural compounds could prevent the process of carcinogenesis. Healthy lifestyle and food habit may help keep this dreadful disease away.

Acknowledgements The authors are grateful to the Director, Chittaranjan National Cancer Institute, Kolkata, for his kind support for the work. Financial supports for this work were provided by grant 09/030(0074)/2014 EMR-I from CSIR, New Delhi, to Dr. Debolina Pal and NASI Senior Scientist Platinum Jubilee Fellowship (2019) to Dr. C. K. Panda. The authors have no conflict of interests.

Appendix

Angiogenesis: In this physiological process, new blood vessels are formed from preexisting blood vessel. This process is guided mainly by VEGF (vascular endothelial growth factor)-mediated pathway. Other factors are also involved like VEGFR, FGF, PDGF, and PDGFR. This process is mainly found in case of invasive tumor.

Hypoxia and normoxia: Normoxia is the condition in which normal oxygen level in a cell remains between 10 and 21%, whereas in hypoxic condition it is reduced to less than 5%. Hypoxia is quite evident in the core area of a tumor due to lack of vascularization (McKeown 2014).

MicroRNA: This is popularly known as miRNA. This is one type of noncoding RNA, i.e., small noncoding RNA containing ~ 22 nucleotide length, which helps in gene silencing and alteration of gene expression.

Microsatellite DNA markers: It is a stretch of DNA motif (containing 1–6 nucleotides or more) repeated almost 5–50 times in a genome of an organism. It has higher mutation rate than other areas of the genome. For example, ATATATATAT is a dinucleotide microsatellite; GCTGCTGCTGCTGCT is a trinucleotide microsatellite. These microsatellites are located throughout the human genome at an average of approximately 30-Kb interval.

Noncoding RNA: These RNAs are not transcribed into proteins, but these have important role in gene transcription regulation. There are several types of noncoding RNAs such as transfer RNAs (tRNAs) and ribosomal RNAs (rRNAs), as well as small RNAs such as microRNAs, siRNAs, snRNAs, and the long noncoding RNAs.

Polycomb ring finger oncogene: This is BMI1 protein. It has one ring finger domain. It plays important role in self-renewal pathway, chromatin remodeling, and DNA repair pathway. Its overexpression is reported in several cancers, e.g., hematological malignancies, breast, ovarian, bladder, prostate, colorectal, and skin.

Stem cells: These are cells with self-renewal property and are capable to differentiate into other cell types. For example, hematopoietic stem cells are present mainly in bone marrow and can differentiate into different types of blood cells like red blood cells (RBC), white blood cells (WBC), and platelets.

Ubiquitination: In eukaryotes, a small (8.6 kDa) regulatory protein ubiquitin is found that helps in the process of ubiquitination. In this process, ubiquitin protein binds to the target protein and that protein is degraded via proteasome-mediated pathway or changes its cellular location or prevents interaction with other proteins.

References

Albert M, Helin K (2010) Histone methyltransferases in cancer. Semin Cell Dev Biol 21:209–220
Andersson ER, Sandberg R, Lendahl U (2011) Notch signaling: simplicity in design, versatility in function. Development 138:3593–3612
Banks RE, Tirukonda P, Hornigold TC, Astuti D, Cohen D, Maher ER, Stanley AJ, Harnden P, Joyce A, Knowles M, Selby PJ (2006) Genetic and epigenetic analysis of von Hippel-Lindau (VHL) gene alterations and relationship with clinical variables in sporadic renal cancer. Cancer Res 66:2000–2011
Baylin SB, Ohm JE (2006) Epigenetic gene silencing in cancer - a mechanism for early oncogenic pathway addiction? Nat Rev Cancer 6:107–116
Berger A, Knudson A, Pandolfi PA (2011) A continuum model for tumour suppression. Nature 476:163–169. https://doi.org/10.1038/nature10275
Bodmer W (1997) The somatic evolution of cancer. The Harveian oration of 1996. J R Coll Physicians Lond 31:82–89
Croce CM (2008) Oncognes and cancer
Carrillo-Infante CI, Abbadessa G, Bagella L, Giordano A (2007) Viral infections as a cause of cancer (review). Int J Oncol 30:1521–1528
Chen Y, Jiang J (2013) Decoding the phosphorylation code in hedgehog signal transduction. Cell Res 23:186–200
Chen C, Wang G (2015) Mechanisms of hepatocellular carcinoma and challenges and opportunities for molecular targeted therapy. World J Hepatol 7:1964–1970
Clevers H (2011) The cancer stem cell: premises, promises and challenges. Nat Med 17:313–319
Cramer DW, Bast RC Jr, Berg CD, Diamandis EP, Godwin AK, Hartge P, Lokshin AE, Lu KH, McIntosh MW, Mor G, Patriotis C, Pinsky PF, Thornquist MD, Scholler N, Skates SJ, Sluss PM, Srivastava S, Ward DC, Zhang Z, Zhu CS, Urban N (2011) Ovarian cancer biomarker performance in prostate, lung, colorectal, and ovarian cancer screening trial specimens. Cancer Prev Res (Phila) 4:365–374
Curtin JC, Lorenzi MV (2010) Drug discovery approaches to target Wnt signaling in cancer stem cells. Oncotarget 1:563–577
Dang CV (2007) The interplay between MYC and HIF in the Warburg effect. Ernst Schering Found Symp Proc 4:535–553
Dawood S, Austin L, Cristofanilli M (2014) Cancer stem cells: implications for cancer therapy. Oncology (Williston Park) 28:1101–1107
DeBerardinis RJ, Cheng T (2010) Q's next: the diverse functions of glutamine in metabolism, cell biology and cancer. Oncogene *29*(3):313–324
DeBerardinis RJ, Lum JJ, Hatzivassiliou G, Thompson CB (2008) The biology of cancer: metabolic reprogramming fuels cell growth and proliferation. Cell Metab 7:11–20
Diamandis EP (2010) Cancer biomarkers: can we turn recent failures into success? J Natl Cancer Inst 102:1462–1467
Diez H, Fischer A, Winkler A, Cheng JH, Hatzopoulos AK, Gessler M (2007) Hypoxia-mediated activation of Dll4-Notch-Hey2 signaling in endothelial progenitor cells and adoption of arterial cell fate. Exp Cell Res 313:1–9
DiGiovanni J (1992) Multistage carcinogenesis in mouse skin. Pharmacol Ther 54:63–128
Farber E, Cameron R (1980) The sequential analysis of cancer development. Adv Cancer Res 31:125–226
Fedele AO, Whitelaw ML, Peet DJ (2002) Regulation of gene expression by the hypoxia-inducible factors. Mol Interv 2:229–243
Feinberg AP, Ohlsson R, Henikoff S (2006) The epigenetic progenitor origin of human cancer. Nat Rev Genet 7:21–33

Fong PC, Boss DS, Yap TA, Tutt A, Wu P, Mergui-Roelvink M, Mortimer P, Swaisland H, Lau A, O'Connor MJ, Ashworth A, Carmichael J, Kaye SB, Schellens JH, de Bono JS (2009) Inhibition of poly(ADP-ribose) polymerase in tumors from BRCA mutation carriers. N Engl J Med 361: 123–134

Fould L (1957) The natural history of cancer. J Chronic Dis 8:2–37

Foxler D, Bridge K, James V, Webb TM, Mee M, Wong SCK, Feng Y, Constantin-Teodosiu D, Petursdottir TE, Bjornsson J, Ingvarsson S, Ratcliffe PJ, Longmore GD, Sharp TV (2012) The LIMD1 protein bridges an association between the prolyl hydroxylases and VHL to repress HIF-1 activity. Nat Cell Biol 14:201–208

Fukui K (2010) DNA mismatch repair in eukaryotes and bacteria. J Nucl Acids 2010:1–16

Goossens N, Nakagawa S, Sun X, Hoshida Y (2015) Cancer biomarker discovery and validation. Transl Cancer Res 4:256–269

Gupta S (2003) Molecular signaling in death receptor and mitochondrial pathways of apoptosis (review). Int J Oncol 22:15–20

Hanahan D, Weinberg RA (2000) The hallmarks of cancer. Cell 100:57–70

Hanahan D, Weinberg RA (2011) Hallmarks of cancer: the next generation. Cell 144:646–674

Hartwell LH, Weinert TA (1989) Checkpoints: controls that ensure the order of cell cycle events. Science 246:629–634

Henry NL, Hayes DF (2012) Cancer biomarkers. Mol Oncol 6:140–146

Hess MT, Schwitter U, Petretta M, Giese B, Naegeli H (1997) Bipartite substrate discrimination by human nucleotide excision repair. Proc Natl Acad Sci U S A 94:6664–6669

Hoffmann W (2012) Stem cells, self-renewal and cancer of the gastric epithelium. Curr Med Chem 19:5975–5983

Ilyas M, Straub J, Tomlinson IP, Tomlinson IP, Bodmer WF (1999) Genetic pathways in colorectal and other cancers. Eur J Cancer 35:335–351

Lei Y, Liu L, Zhang S, Guo S, Li X, Wang J, Su B, Fang Y, Chen X, Ke H, Tao W (2017) Hdac7 promotes lung tumorigenesis by inhibiting Stat3 activation. Mol Cancer 16:170–182

Li W, Liu H, Cheng ZJ, Su YH, Han HN, Zhang Y, Zhang XS (2011) DNA methylation and histone modifications regulate de novo shoot regeneration in Arabidopsis by modulating WUSCHEL expression and auxin signaling. PLoS Genet 7:e1002243

Jackson SP (2002) Sensing and repairing DNA double-strand breaks. Carcinogenesis 23:687–696

Karakosta A, Ch G, Charalabopoulos A, Peschos D, Batistau A, Charalabopoulos K (2005) Genetic models of human cancer as a multistep process. Paradigm models of colorectal cancer, breast cancer and chronic myelogenous and acute lymphoblastic leukaemia. J Exp Clin Cancer Res 24:05–14

Kashuba VI, Pavlova TV, Grigorieva EV, Kutsenko A, Yenamandra SP, Li J, Wang F, Protopopov AI, Zabarovska VI, Senchenko V, Haraldson K, Eshchenko T, Kobliakova J, Vorontsova O, Kuzmin I, Braga E, Blinov VM, Kisselev LL, Zeng YX, Ernberg I et al (2009) High mutability of the tumor suppressor genes RASSF1 and RBSP3 (CTDSPL) in cancer. PLoS One 4(5): e5231. https://doi.org/10.1371/journal.pone.0005231

Kessler JD, Kahle KT, Sun T, Meerbrey KL, Schlabach MR, Schmitt EM, Skinner SO, Xu Q, Li MZ, Hartman ZC, Rao M, Yu P, Dominguez-Vidana R, Liang AC, Solimini NL, Bernardi RJ, Yu B, Hsu T, Golding I, Luo J, Osborne CK, Creighton CJ, Hilsenbeck SG, Schiff R, Shaw CA, Elledge SJ, Westbrook TF (2012) A SUMOylation-dependent transcriptional subprogram is required for Myc-driven tumorigenesis. Science 335:348–353

Kotecha R, Takami A, Espinoza JL (2016) Dietary phytochemicals and cancer chemoprevention: a review of the clinical evidence. Oncotarget 7:52517–52529

Lei Y, Liu L, Zhang S, Guo S, Li X, Wang J, Su B, Fang Y, Chen X, Ke H, Tao W (2017) Hdac7 promotes lung tumorigenesis by inhibiting Stat3 activation. Mol Cancer 16:170

Lieber MR (2010) The mechanism of double-strand DNA break repair by the nonhomologous DNA end joining pathway. Annu Rev Biochem 79:181–211

Loeb LA, Bielas JH, Beckman RA (2008) Cancers exhibit a mutator phenotype: clinical implications. Cancer Res 68:3551–3557

Lopez-Serra L, Ballestar E, Fraga MF, Alaminos M, Setien F, Esteller MA (2006) Profile of methyl-CpG binding domain protein occupancy of hypermethylated promoter CpG islands of tumor suppressor genes in human cancer. Cancer Res 66:8342–8346

Luo J, Emanuele MJ, Li D, Creighton CJ, Schlabach MR, Westbrook TF, Wong KK, Elledge SJ (2009) A genome- wide RNAi screen identifies multiple synthetic lethal interactions with the Ras oncogene. Cell 137:835–848

Ma T, Li H, Sun M, Yuan Y, Sun LP (2017) DNMT1 overexpression predicting gastric carcinogenesis, subsequent progression and prognosis: a meta and bioinformatic analysis. Oncotarget 8:96396–96408

Margison GP, Santibáñez-Koref MF (2002) O6-alkylguanine-DNA alkyltransferase: role in carcinogenesis and chemotherapy. BioEssays 24(3):255–266

Martin D, Gutkind JS (2008) Human tumor-associated viruses and new insights into the molecular mechanisms of cancer. Oncogene 2:S31–S42

McKeown SR (2014) Defining normoxia, physoxia and hypoxia in tumors-implications for treatment response. Br J Radiol 87(1035):20130676

Mehta A, Haber JE (2014) Sources of DNA double-strand breaks and models of recombinational DNA repair. Cold Spring Harb Perspect Biol 6:a016428

Memisoglu A, Samson L (2000) Base excision repair in yeast and mammals. Mutation Res 451:39–51

Michalides RJ (1999) Cell cycle regulators: mechanisms and their role in aetiology, prognosis, and treatment of cancer. J Clin Pathol 52:555–568

Miller KM, Tjeertes JV, Coates J, Legube G, Polo SE, Britton S, Jackson SP (2011) Human HDAC1 and HDAC2 function in the DNA-damage response to promote DNA nonhomologous end-joining. Nat Struct Mol Biol 17:1144–1151

Moeini A, Cornellà H, Villanueva A (2012) Emerging signaling pathways in hepatocellular carcinoma. Liver Cancer 1:83–93

Moore N, Lyle S (2011) Quiescent, slow-cycling stem cell populations in cancer: a review of the evidence and discussion of significance. J Oncol 2011

Nowell PC (1976) The clonal evolution of tumor cell populations. Science 194:23–28

Nowell PC, Hungerford DA (1960) Chromosome studies on normal and leukemic human leukocytes. J Natl Cancer Inst 25:85–109

O'Shaughnessy J, Osborne C, Pippen JE, Yoffe M, Patt D, Rocha C, Koo IC, Sherman BM, Bradley C (2011) Iniparib plus chemotherapy in metastatic triple-negative breast cancer. N Engl J Med 364:205–214

Okada H, Mak TW (2004) Pathways of apoptotic and non-apoptotic death in tumor cells. Nat Rev Cancer 4:592–603

Pal D, Sur S, Mandal S, Das A, Roy A, Das S, Panda CK (2012) Prevention of liver carcinogenesis by amarogentin through modulation of G1/S cell cycle check point and induction of apoptosis. Carcinogenesis 33:2424–2431

Park IK, Qian D, Kiel M, Becker MW, Pihalja M, Weissman IL, Morrison SJ, Clarke MF (2003) Bmi-1 is required for maintenance of adult self-renewing haematopoietic stem cells. Nature 423:302–305

Parsa N (2012) Environmental Factors Inducing Human Cancers. Iran J Publ Health 41:1–9

Perrimon N, Pitsouli C, Shilo BZ (2012) Signaling mechanisms controlling cell fate and embryonic patterning. Cold Spring Harb Perspect Biol 4:a005975

Phan LM, Yeung SC, Lee MH (2014) Cancer metabolic reprogramming: importance, main features, and potentials for precise targeted anti-cancer therapies. Cancer Biol Med 11:11–19

Pollack M, Leeuwenburgh C (2001) Apoptosis and aging: role of the mitochondria. J Gerontol A Biol Sci Med Sci 56:B475–B482

Qu Y, Mu G, Wu Y, Dai X, Zhou F, Xu X, Wang Y, Wei F (2010) Overexpression of DNA methyltransferases 1, 3a, and 3b significantly correlates with retinoblastoma tumorigenesis. Am J Clin Pathol 134:826–834

Romero-Garcia S, Lopez-Gonzalez JS, Báez-Viveros JL, Aguilar-Cazares D, Prado-Garcia H (2011) Tumor cell metabolism: an integral view. Cancer Biol Ther 12:939–948

Sarkar FH, Li Y, Wang Z, Kong D (2010) The role of nutraceuticals in the regulation of Wnt and Hedgehog signaling in cancer. Cancer Metastasis Rev 29:383–394

Scrable HJ, Sapienza C, Cavenee WK (1990) Genetic and epigenetic losses of heterozygosity in cancer predisposition and progression. Adv Cancer Res 54:25–62

Sharp TV, Munoz F, Bourboulia D, Presneau N, Darai E, Wang HW, Cannon M, Butcher DN, Nicholson AG, Klein G, Imreh S, Boshoff C (2004) LIM domains-containing protein 1 (LIMD1), a tumor suppressor encoded at chromosome 3p21.3, binds pRB and represses E2F-driven transcription. Proc Natl Acad Sci U S A 101(47):16531–16536. https://doi.org/10.1073/pnas.0407123101

Sedgwick B (1997) Nitrosated peptides and polyamines as endogenous mutagens in O6-alkylguanine-DNA alkyltransferase deficient cells. Carcinogenesis 18:1561

Sengupta N, Seto E (2004) Regulation of histone deacetylase activities. J Cell Biochem 93:57–67

Sikandar SS, Pate KT, Anderson S, Dizon D, Edwards RA, Waterman ML, Lipkin SM (2010) NOTCH signaling is required for formation and self-renewal of tumor-initiating cells and for repression of secretory cell differentiation in colon cancer. Cancer Res 70:1469–1478

Slaga TJ (1983) Mechanisms of tumor promotion. Tumor promotion in internal organs, vol 1. CRC Pres, Boca Raton, FL

Slamon DJ, Leyland-Jones B, Shak S, Fuchs H, Paton V, Bajamonde A, Fleming T, Eiermann W, Wolter J, Pegram M, Baselga J, Norton L (2001) Use of chemotherapy plus a monoclonal antibody against HER2 for metastatic breast cancer that overexpresses HER2.N. Engl J Med 344:783–792

Sporn MB (1976) Approaches to prevention of epithelial cancer during the preneoplastic period. Cancer Res 36:2699–2702

Sturgeon CM, Duffy MJ, Stenman UH, Lilja H, Brünner N, Chan DW, Babaian R, Bast RC Jr, Dowell B, Esteva FJ, Haglund C, Harbeck N, Hayes DF, Holten-Andersen M, Klee GG, Lamerz R, Looijenga LH, Molina R, Nielsen HJ, Rittenhouse H, Semjonow A, Shih IM, Sibley P, Sölétormos G, Stephan C, Sokoll L, Hoffman BR, Diamandis EP (2008) National Academy of Clinical Biochemistry laboratory medicine practice guidelines for use of tumor markers in testicular, prostate, colorectal, breast, and ovarian cancers. Clin Chem 54:e11–e79

Sturgeon CM, Duffy MJ, Hofmann BR, Lamerz R, Fritsche HA, Gaarenstroom K, Bonfrer J, Ecke TH, Grossman HB, Hayes P, Hoffmann RT, Lerner SP, Löhe F, Louhimo J, Sawczuk I, Taketa K, Diamandis EP (2010) National Academy of Clinical Biochemistry. National Academy of Clinical Biochemistry Laboratory Medicine Practice Guidelines for use of tumor markers in liver, bladder, cervical, and gastric cancers. Clin Chem 56:e1–e48

Sur S, Pal D, Mandal S, Roy A, Panda CK (2016) Tea polyphenols epigallocatechin gallete and theaflavin restrict mouse liver carcinogenesis through modulation of self-renewal Wnt and hedgehog pathways. J Nutr Biochem 27:32–42

Tyson JJ, Novak B (2008) Temporal organization of the cell cycle. Curr Biol. 18:R759–R768

Urbich C, Rössig L, Kaluza D, Potente M, Boeckel JN, Knau A, Diehl F, Geng JG, Hofmann WK, Zeiher AM, Dimmeler S (2009) Angiogenesis and determines the angiogenic gene expression pattern of endothelial cells. Blood 113:5669–5679

van Veelen, W., Le, N.H., Helvensteijn, W., Blonden, L., Theeuwes, M., Bakker, E.R., Franken, P.F., van Gurp, L., Meijlink, F., van der Valk, M.A. and Kuipers, E.J., 2011. β-catenin tyrosine 654 phosphorylation increases Wnt signalling and intestinal tumorigenesis. Gut, 60(9), pp.1204-1212

Vander Heiden MG, Cantley LC, Thompson CB (2009) Understanding the Warburg effect: the metabolic requirements of cell proliferation. Science 324:1029–1033

Vaux DL, Korsmeyer SJ (1999) Cell death in development. Cell 96:245–254

Vogelstein B, Kinzler K (2004) Cancer genes and the pathways they control. Nat Med 10:789–799. https://doi.org/10.1038/nm1087

Vogelstein B, Papadopoulos N, Velculescu VE, Zhou S, Diaz LA, Kinzler KW (2013) Cancer genome landscapes. Science 339(6127):1546–1558

Walensky LD (2006) BCL-2 in the crosshairs: tipping the balance of life and death. Cell Death Differ 13(8):1339–1350

Wang H, Oo Khor T, Shu L, Su ZY, Fuentes F, Lee JH, Tony Kong AN (2012) Plants vs. cancer: a review on natural phytochemicals in preventing and treating cancers and their druggability. Anticancer Agents Med Chem 12(10):1281–1305

Wang MC, Li CL, Cui J, Jiao M, Jing LI, Nan KJ (2015) BMI-1, a promising therapeutic target for human cancer. Oncol Lett 10:583–588

Wattenberg LW (1993) Prevention therapy basic science and the resolution of the cancer problem. Cancer Res 53:5890–5896

Welch DR, Evans DP, Tomasovic SP, Milas L, Nicolson GL (1984) Multiple phenotypic divergence of mammary adenocarcinoma cell clones. II. Sensitivity to radiation, hyperthermia and FUdR. Clin Exp Metastasis 2:357–371

Yang C, Wang W, Liang JX, Li G, Vellaisamy K, Wong CY, Ma DL, Leung CH (2017) A rhodium (III)-based inhibitor of lysine-specific histone demethylase 1 as an epigenetic modulator in prostate cancer cells. J Med Chem 60:2597–2603

Yokota J, Sugimura T (1993) Multiple steps in carcinogenesis involving alterations of multiple tumor suppressor genes. Faseb J 7:920–925

Zhang J, Yang PL, Gray NS (2009) Targeting cancer with small molecule kinase inhibitors. Nat Rev Cancer 9:28–39

Oncology: Pathologist's View

Anup Kumar Roy

Abstract

Cancer is a leading cause of death in the present century and despite spectacular advances in diagnosis and treatment and a considerable up-gradation in the survival graph, we are still bothered about it. Risk of developing cancer increases with age and is related to certain risk factors, which vary according to the site of cancer, but the real demon is the disorder of gene expression, and epigenetic factors are mainly responsible for demonizing certain proteins that are related to our normal growth and development. We cannot regulate our genes directly, but their control through epigenetic factors has their stakeholders in environment, food, and lifestyle, which influence carcinogenesis, the process of transformation of a normal cell to a cancerous one. These transformations occur through mutational changes in certain genes, which give the cell some survival advantage over others, which is again an evolutionary process. It is interesting that during this transformation, which is generally a gradual and multistep process, body defense mechanisms play a dubious role of preventing the cancer cell to survive and proliferate on the one hand, while actively helping them to evade the defense surveillance system on the other. Previously, we were of opinion that cancer cells in a particular tumor are monoclonal, i.e., arise from a single transformed cell and gradually progress through obtaining different mutations that give them definite survival advantage over their neighbors and ultimately become autonomous. But recent progress in molecular study tools like genome-wide association studies (GWAS) reveals that cancer cells are heterogeneous and the different subclones exist with different characters from the very beginning of the tumor and only those subclones survive that win a fight against normal cells by their

---

A. K. Roy (✉)
Department of Pathology, Nil Ratan Sircar Medical College, Kolkata, India

© The Author(s), under exclusive license to Springer Nature Singapore Pte Ltd. 2022
S. K. Basu et al. (eds.), *Cancer Diagnostics and Therapeutics*,
https://doi.org/10.1007/978-981-16-4752-9_10

advantageous mutations. If mutations in a particular subclone could not confer the survival advantage, that particular subclone will extinct.

Keywords

Neoplasia · Cell cycle regulators · Carcinogenesis · Stem cell · Pathology

10.1 Introduction

The human society has faced death through violence, war, accidents, natural disasters, and fearsome array of infectious diseases over time immemorial. The leading causes of death in the twentieth century were pneumonia, influenza, and tuberculosis as are evident in the world literature. In the present century, the leading causes of death are heart diseases, cerebral strokes, and cancer. Antibiotics and other modern medicines have reduced morbidity from infectious diseases to a large extent; a conscious effort to change the lifestyle has reduced fatal heart ailments and strokes to some extent. But we are still bothered with mortality from cancer. A lot of progress has been achieved in the field of diagnostics and treatment, and consequently a considerable up-gradation in 5-year and 10-year survival graphs, but the "final diagnosis" still rests on the seemingly fragile shoulders of pathologists, the mysterious people sitting always with a microscope and writing reports in a language full of confounding words and dubious logic. Whenever the clinician thinks for a second opinion, the confusions often get exaggerated and the decision making becomes tougher. The irony is that hardly these two persons, that is, the clinician and the pathologist, meet and discuss, which might almost reduce the entire impasse.

This is the most talked-about perspective of a pathologist; they expect to receive a bit more information regarding the illness on the so-called requisition slip and sometimes a little bit of interaction with the clinician to arrive at an unequivocal tissue diagnosis with proper staging and grading that can save valuable time, which in turn may prove life-saving for the patient. The modern medicine as a subject experienced tremendous advancement in the last two to three decades as a consequence of human genome project and its aftermath, particularly in the areas of human genetics, immunology, therapeutics, and field of radiation oncology. Even the supportive fields like terminal care and rehabilitation have progressed a lot. Still, it is the early diagnosis, proper categorization, and prognostic evaluations that are vital for the choice of treatment from the available options.

10.2 Oncology, Neoplasia, and Cancer

Oncology is an integrated discipline of medicine dealing with the prevention, diagnosis, and treatment of cancer. A medical professional who practices oncology is designated as an oncologist, a comparatively new speciality not heard even a few decades ago. Oncology specialists include medical oncologists, surgical oncologists, radiation oncologists, and oncopathologists in their respective field of activities. But for a pathologist, the term "onco" does not always mean cancer.

Neoplasia literally means "new growth," which obviously encompasses both "benign" or good tumors and "malignant" or bad tumors, that is, cancer. The term "tumor" means any swelling that we often encounter in our daily living, even after a blunt injury or ankle sprain or following an allergic inflammation. So, neither tumor nor neoplasia literally means cancer. Therefore, a pathologist has to write his diagnosis in a proper and acceptable bunch of words that may appear Greek even to a literate common folk but carry meaningful information for the clinicians. Moreover, the "biopsy" report is not an expression of quantity as in the case of reporting blood glucose or lipid profile. The utmost concern of a "histopathologist" (the pathologist trained for and designated with reporting a biopsy sample) is to provide maximum information evident in a sample of tissue for the clinician so that he/she can make a proper choice of therapy for the best treatment outcome of the patient in the given situation.

So, cancer, presumably a malignant tumor or "bad tumor," can be defined as n growth of tissue that is autonomous, has escaped normal checkpoints of cell proliferation, and exhibits various degrees of similarity to their precursors (Strayer and Kluwer 2015). People often consider cancer as a disease of modern age and a curse of civilization, but the fact is that it is a disease of ancient times. Evidence of bone tumors has been found in prehistoric remains, and the disease is mentioned in some forms in the ancient literary works from India, Egypt, Babylonia, and Greece. Hippocrates is reported to have introduced the term *karkinos,* from which the term carcinoma is derived (Strayer and Kluwer 2015).

10.3 Risk Group for Developing Cancer

The incidence of cancer, in general, increases with age. Therefore, with gradually rising life expectancy, an increasing proportion of older population are falling prey to the claws of this horrifying demon. But the demon is actually a disorder in gene expression, may it be structural or functional deviation from so-called normal genome. The claws of the demon were blurred to our vision before the completion of Human Genome Project, but after that these are becoming clearer day by day. Now, we know the major proteins involved in development of carcinoma, its invasion and metastasis, and the genes coding these proteins. We are also exploring the factors that are involved in errant behavior of these proteins, factors that are known as "epigenetic" factors (epi = outside, of genes). These epigenetic factors that are mainly responsible for demonization of these benign proteins (which are mostly involved in normal growth of our cells and smooth running of cell cycles with a watchful policing of the whole complex events) are largely associated with our environment and lifestyle. Those cancers where the passwords (causative factors) are yet to be cracked will fall in line with others with further advance in cancer research, the cancer biologists presume.

The environmental factors that lead to carcinogenesis are a common threat for all of us, and we have to solve this threat ourselves. Regarding lifestyle, we may think of reverting to our younger days when we were less greedy, physically active, and socially more accessible to our friends and neighbor, be satisfied with minor gains, and used to live a simpler and happier life. But it is not possible to reverse the time line. Therefore, we have to innovate the ways of our living, which will be healthier, yet acceptable and feasible to all of us. For this, we should clearly understand the happenings during the development of cancer.

10.4 Carcinogenesis

Every living entity is composed of cell(s), which we can consider as the unit of life. It is basically the structural, functional, and biological unit of all known living organisms. The cell can replicate independently. It consists of cytoplasm enclosed within a membrane, which contains many biomolecules such as proteins and nucleic acids. Humans contain more than 10 trillion (10^{13}) cells. The cell was discovered by Botanist Robert Hooke, who coined the term "cell" for their resemblance to the cells inhabited by the Christian monks in the monastery. Human cells along with those of plants, animals, fungi, and protozoa belong to eukaryotic cells, although prokaryotic cells were the first form of life on earth. In eukaryotic cells, there is cell membrane, cytoplasm, different organelles, and vacuoles, but the most important among these is the cell nucleus that contains the cell's DNA or the genome. Human genome contains roughly 3.2 billion DNA base pairs. Within the genome, there are only around 20,000 protein-encoding genes, comprising less than 1.5% of the genome. The rest 98.5% of the human genome does not encode proteins, but here the DNA lies in long stretches, separating the protein-coding genes, almost like a buffer zone. The recently concluded ENCODE (Encyclopaedia of DNA Elements) shows that 80% of this noncoding DNA is involved in regulating gene expression (Kumar et al. 2014).

Now, the question is how the protein-encoding genes, or the genome, are implicated in cancer? Cancer or in the broad sense, neoplasia (new growth) is defined by British oncologist Willis as "an abnormal mass of tissue, the growth of which exceeds and is uncoordinated with that of the normal tissues and persists in the same excessive manner after cessation of the stimuli which evoked the change." Therefore, basically a cancer alias tumor alias neoplasia is a disorder of cell growth that is triggered by a series of acquired mutations (i.e., permanent DNA alterations). These mutations affect the normal mechanism of cell growth by setting the growth cycle out of gear, thus attaining freedom from the growth control mechanisms. Even the behavior of a tumor, whether it will be a benign one like lipoma (a tumor consisting of fat cells) or it will transform to a malignant or cancerous tumor-like liposarcoma, also depends on which set or sets of genes are mutated and at which stage of cell growth.

With this background, we may proceed into the mystery of carcinogenesis, that is, the genesis of cancer. The main objective behind the transformation of a normal cell into cancer cell is the attainment of survival advantage. This means the transformed cell will survive longer than its neighboring nontransformed cells. These mutations (i.e., a change in the primary nucleotide structure of the DNA) that will provide a survival advantage invariably involve those genes that are somewhat related to cell growth and differentiation. Those mutations that provide a survival advantage to the mutated cell will empower the cell to survive longer than its nonmutated neighbors, hence have a fair chance to undergo another mutation, which might not be possible without survival advantage acquired by the first mutation. This second mutation may or may not provide any survival or growth advantage. If it does not provide such advantage to the cells with the 1st mutation, then the matter ends there and this population of cells will face their natural death. But if, by any chance, this 2nd mutation also provides another survival advantage, then the descendants of these cells will survive longer and be prepared for undergoing the 3rd mutation. These survival advantages may be in the form of survival in low oxygenated condition or empowerment to avoid growth-inhibiting proteins or to avoid apoptosis and similar other advantages. After getting a series of such adventitious mutations, the cell population will attain "immortality"; that is, they will be free off normal regulations of cell cycle. We call this mutated population as "transformed" cell population. This transformed cell population forms the nidus for cancer.

10.5 Cell Cycle Regulators and their Role in Cancer

The most important drive for any living organism is its struggle for survival and growth. Dialectic materialism, which was theorized by Karl Marx in nineteenth century and applied in the evolution of human society, also influenced Charles Darwin who developed the great concept of evolutionary drive and survival of the fittest in animal and in plant kingdom. The experience an organism gains during its struggle for existence has to be shared with its progeny for their survival interest. This is being done by exchange of chromosome material during gametogenesis, and the gametes carry this experience in the form of genetic code to future generations. Therefore, cell division is a crucial step and it should be regulated properly so that no error, whether intentional or unintentional, occurs during cell division. This is ensured by a series of checks and counter checks by a gamut of proteins.

All cells do not have the same potential for cell division. Some cells do not divide at all after birth, for example, brain and heart (permanent cells); others have only minimal proliferative activity but are capable of dividing when called for like liver, kidney, pancreas (stable cells), and the rest of the cells are continuously being lost and replaced by proliferation, for example, bone marrow cells, skin, oral mucosa, gastrointestinal tract, and endometrium (labile cells). This compartmentalization is not always rigid; even permanent cells can be pushed from its quiescence state to activity of cell cycle through genetic switching.

Whenever a cell undergoes cell division, it has to pass through certain stages. First, it has to double its DNA blueprint so that the progeny cells get a normal DNA copy that is compatible for its survival and further growth. This is known as S (synthesis) phase as DNA synthesis occurs in this phase when the whole genome is copied. As the 3.2 billion base pairs are to be copied in a very tight schedule, there are possibilities for occurrence of some errors during copying and errors do occur at an approximate rate of 1 error in every million base pair and some of these errors are important enough to jeopardize the functioning of the progeny cells. Therefore, these errors have to be detected and rectified before the cell enters into its final lap of cell division that is M (mitosis) phase. There are proteins dedicated for detecting (spellchecking) the errors generated during copying and for rectifying (mismatch repairing) those errors. But this process will obviously take some time, therefore necessitating a temporary halt or applying brake to the cell cycle. The errant cell is then isolated from the highway of the cell cycle, and the repair can be accomplished properly and in time in the roadside garage and again place it on the pathway of cell cycle after successful repair. To supervise the whole process of detection of defect, putting a brake on the cell cycle, making necessary repair, and then put them back to cell cycle, strong supervisors are necessary who will do their job meticulously. This supervision is again no individual's job and a set of proteins are there under the able guidance of their team leader, the p53 protein, which has also a nickname in his team—the guardian of the genome. The team has their own standard operating procedure (SOP), which has been updated many times during the evolutionary process and currently the SOP includes thorough on-road checking of all cells that have completed the S phase and then going to the final destination, that is, M phase, so that no cell can enter M phase with a DNA defect. The SOP also includes erecting roadblocks by a separate group of proteins (cyclins, cyclin-dependent kinases, and CDK inhibitors) so that all cells running in the cell cycle have to stop and only be allowed to proceed further if the inspectors and supervisors found it ok and give signal to remove the roadblock. All these checks and counterchecks are intelligently designed so that they can function in an integrated way to ensure a full-proof result.

These are all about normal control of cell cycle. Problem begins if the above-mentioned checks and counterchecks did not function properly. Then, the errors generated in S phase go unnoticed and the cells enter the M phase of cell cycle either unchecked or underchecked or with faulty check. Some of these errors may be due to changes (mutations) in DNA that occurred due to environmental or chemical factors (carcinogens), while others are spontaneous errors of copying. The combination of genetic errors due to mutations and copying may provide the survival advantages that are prerequisites for survival advantages and subsequent malignant transformations.

10.6 Carcinogenesis Model

Our access to molecular analysis of biological specimens is expanding day by day, and the cost is also coming down; cancer genetics are revealing newer and unexpected information. The previous concept of monoclonal origin of cancer cells is being replaced by cancer heterogeneity. Previously, it was conceived that a series of mutations ultimately transform a cell irreversibly to a cancer cell with the characteristics of autonomy and other features like dysplasia, invasion, and lack of cohesion that imparts the phenotypic characters of a malignant lesion. Hence, it was presumed that all the cells present in a cancerous growth bear the same set of mutations and therefore same phenotypic characters.

Genome-wide sequencing (GWS) studies of cancer cells revealed that as few as ten or so mutations are required for the necessary transformation. The classical prototype is colorectal cancer. This multistep clonal model also revealed that not all mutations have the ability of transforming cells. Those mutations that are vital for transformation are known as "driver" mutations, and the other mutations that are nonvital but help propagating the driver mutations are known as "passenger" mutations. This theory says that tumor progressively accumulates carcinogenic mutations. These mutations occur in stepwise fashion. During the progression of clonal evolution, "driver" lesions that have some selective advantages lead to acquisition of "hallmarks of cancer" (e.g., sustained proliferation, avoiding growth suppression mechanisms and cell death, attaining replicative immortality, angiogenesis, invasion, and metastasis) (Flanagan 2016).

Now the genome-wide association studies (GWAS) are revealing that all cancer cases do not follow this clonal carcinogenesis model. Instead, there are multiple subclones in a tumor mass having different driver mutations. This proves that *cancers are heterogeneous rather than homogenous as all the cells do not carry the same set of mutations* and also they vary in their behavioral phenotype. This intratumor heterogeneity is clearly evident in cancers of kidney, lung, breast, ovary, leukemia, etc. According to the recently hypothesized *"Big Bang"* model of carcinogenesis, subclones do not always expand deterministically within tumors. Chance plays a great role in deciding the spread and success of a clonal growth. Because of the lack of selection, it is the exact timing of mutation that determines clone size, and mutations arising early in the tumor will tend to form larger subclones, whereas late mutations will form clones of restricted size. Further, all sizeable subclones arise early during cancer expansion. Thereafter, these subclones have to compete with the resident cells for survival and eventually their growth rate will slow down. Late arising subclones are extremely unlikely to expand to the size detectable by sampling. In contrast, early-arising subclone mutations have the advantage of sweeping through the population as the population size is small enough for new clones to be established. Hence, the naming of Big Bang model—the clonal composition of a tumor—is determined early on and remains effectively static thereafter. This is analogous to cosmological model wherein perturbations during the initial expansion of the universe still dominate in the present day (Flanagan 2016).

In sequential model, each driver mutation bestows a large increase in fitness to the recipient clone. Phenotypic evolution proceeds in an incremental fashion. In Big Bang model, significant selection occurs at the outset of a cancer growth, and the subsequent evolutionary selection within the expanding population is of negligible magnitude and/or consequence. The Big Bang model is thus an example of punctuated equilibrium (genotype–phenotype differentiation) whereby large phenotypic leaps can suddenly occur in an otherwise phenotypically static population. Most preinvasive and premalignant lesions are in a state of evolutionary stasis, most will not progress to malignancy. The main driver mutations occur early in the neoplastic process—"born to be bad" (Flanagan 2016).

Do these observations mean the multistage carcinogenesis theory has been totally replaced by the Big Bang theory? The answer is not affirmative as in many cases multistage carcinogenesis is still valid.

10.7 Evolution and Cancer

Animals have evolved some potent tumor suppressor mechanisms to prevent cancer development. These mechanisms at the earlier stage of evolution were important for growth of multicellular organisms and large-body animals. Thus, the development of animals was evolutionarily constrained by the need to limit cancer. Cancer development within an individual is thus an evolutionary process, and it often mimics evolution of different species.

What are the hallmarks of species evolution? Species are evolved by mutation and by the process of selection acting on individuals in a population. Tumor cells also reveal these characteristics of evolving by mutation and selection in a tissue, same as that of evolution of individuals in a population. Cancer evolution leaves "information" in cellular genomes that evolutionary theory can decode. Species evolution also leaves such information in the genome; in cancer, such information is easily verifiable.

From a biological point of view, cancer is an evolutionary disease. In cancer, cells not only evolve morphologically, but also functionally. These new set of functions thus acquired by the cancer cells are beneficial for them, but ultimately lethal to their hosts. What new functions do the cancer cells gain by these evolutions? Apart from growing quickly, it also ignores signals to die (evade apoptosis), evades host immune defenses, grows blood vessels to obtain nutrients (angiogenesis), invades surrounding tissue, survives in bloodstream, and establishes new colonies throughout the body (metastasis), and they may even resist treatment (chemo- and radioresistant) (Swamidass 2017; Casás-Selves and DeGregori 2011).

All these gains in function are related to overcoming the growth control mechanisms in some way or other. Earlier, we have mentioned that the growth-controlling (or tumor suppressor) mechanisms are robust in comparison with the growth-promoting mechanisms, and this is an integral part of evolution of species. But whenever an individual cell has to attain autonomy in growth, which is also an evolutionary drive, it has to overcome the stubborn attitude of an array of tumor suppressor proteins, and this is achieved by a series of sequential mutations of genes encoding these tumor suppressor proteins.

10.8 Cancer Stem Cells

The concept of stem cell is not new, but the evidence indicating the presence of cancer stem cells (CSCs) is a recent development. We know that stem cells are those cells that have *pluripotency* to develop different types of mature cells. Let us give an example. When an embryo has formed after the successful fusion of a sperm and an ovum, it forms a tiny bubble-like aggregate of cells within a few days, which we call *blastocyst*. It undergoes cell division, and the cell number increases by doubling. Now, these cells are capable of creating an entire human embryo and then into a fetus and ultimately a full-formed newborn. To achieve this highly specialized form, all the cells in a blastocyst have the potential to create an entire baby. These are known as pluripotent embryonal stem cells. These are the most potent stem cells that can give rise to any type of mature cells needed for organogenesis. There are other types of stem cells, for example, multipotent stem cells (hematopoietic stem cells that form different blood cells), oligopotent stem cells (neural stem cells that are more restricted to neuron cell formation), and unipotent or monopotent stem cells (restricted for developing single type of cells). All these stem cells have some common characteristics. Stem cells are usually immortal; that is, they have an unlimited replication potential. When a stem cell divides, it is usually asymmetric; i.e., it always gives rise to a stem cell (to replenish stem cell population) and a differentiated cell (that will be a mature and functional cell). Stem cells can be induced in vitro to differentiate into any organ or tissue-specific cells by genetic reprogramming. These induced stem cells are known as induced pluripotent stem cells or iPSC (Nature Outlook: Cancer 2014).

A lot of research work is going on stem cells and its application in targeted therapy in different cancers. One may find frequent news items on some recent controversies related to stem cell research, particularly on ethical issues, religious controversies, federal funding, and other issues. Stem cells have already become an integral part of our survival. In different blood cancers and lymphomas, hematopoietic stem cell transplantation is the only hope for survival. Stem cell transplantation is also coming up for the treatment of different neurological diseases as well.

How did the scientists come to know about the existence of cancer stem cells? Obviously, this is not an accidental finding. As we mentioned earlier, almost all labile tissues in our body (those tissues which have a high cellular turnover, e.g., skin, and gastric and intestinal mucosa) do harbor adult stem cells, which are basically unipotent stem cells destined to maintain a high turnover of adult cell population. With the advancement on stem cell research, attention was focused also on these adult stem cells in cancer tissue. As these stem cells are immortal (in the sense that they always maintain a constant pool of themselves by pursuing asymmetric cell division), they have more chance of acquiring and propagating the mutations that are necessary for transforming them from adult stem cells into cancer stem cells. The evidences in favor of this theory came in the form of immune histochemical (IHC) marker study.

IHC is detection of expression of different cellular proteins by specific antibodies tagged with fluorescent dye. The proteins being antigenic in nature will be identified by these tagged antibodies, and after washing, those tagged fluorescent markers will be detected using a fluorescent microscope. Detection can also be done by nonfluorescent tagging, for example, horseradish peroxidase. These techniques have been utilized for detecting different cancer cells since the 1970s. IHC has been universally accepted as markers of cancer cells as they are definite evidences of the cell of origin, markers of prognostic factors that determine the survival rate, and, the most important one, the target molecules for therapeutic approach. Initially, only a few markers were used for cancer cell study, but with tremendous advancement of molecular biology more and more protein molecules are being identified for depicting the functional heterogeneity of mutated cancer cells. Some of these IHC markers were successfully developed for detection of stem cells also. These stem cell markers were also applicable for detection of stem cell population in different adult tissue to locate their residence (known as *niche*) and study their behavior and interaction with the microenvironment that dictates their activation and hibernation.

The first conclusive evidence for CSCs came in 1997 when Bonnet and Dick were succeeded to isolate a subpopulation of leukemia cells that expressed surface marker CD34, but not CD38. The authors established that the $CD34^+/CD38^-$ cell subpopulation is capable of tumorigenesis in mice that were histologically similar to the donor. The first evidence of a solid tumor containing cancer stem-like cell found in 2002 with the discovery of a clonogenic, sphere-forming cell isolated and characterized from human brain gliomas. More evidence comes from histology. Many tumors are heterogeneous, and they contain multiple cell types that are native to the host organ. Tumor heterogeneity is commonly expressed by tumor metastases. This suggests that the cell that produced them had the potentiality to generate multiple cell types, a classical hallmark of stem cells. IHC marker study also confirmed this property of "stemness." The existence of leukemia stem cells prompted research into other cancers. CSCs have recently been identified in several solid tumors (e.g., brain, breast, colon, ovary, pancreas, and prostate), melanoma, nonmelanoma skin cancers, and multiple myeloma.

10.9 Cancer Stem Cells and Growth of Cancer

Cancer stem cells possess the characteristics of normal stem cells, particularly self-renewal and differentiation into multiple cell types. CSCs are therefore tumorigenic cancer cells that acquire driver mutations, in contrast to other nontumorigenic cancer cells that form the bulk of the tumor by passenger mutations. CSCs persist in a tumor as a separate population and are responsible for cancer relapse and metastasis. Development of specific therapies targeted at CSCs holds hope for the improvement of survival and quality of life of cancer patients, especially for patients with metastatic disease. Therefore, the histopathologists are now more interested in studying CSC by IHC markers and a lot of publications are on the way. But there are some challenges also: Some of these markers are shared by non-CSCs present in the tumor mass, therefore compromising the specificity of these markers, Not all the cases of a particular tumor express these CSC markers, If we accept this CSC model of carcinogenesis, then the question is how we will explain cancer formation in stable or permanent tissue (where cell turnover is low or absent), which do not harbor adult stem cells that are subsequently converted into CSCs.

However, this CSC model is now widely accepted as a valid model of carcinogenesis in a good number of cancer growths.

10.10 Pathologist in Diagnosis of Cancer

Cancer is suspected by the clinician by carefully listening the complaints of the patient or the relatives, asking him/her a few leading questions, which include the family history, food habit, working conditions, personal habits and hygiene, addictions if any, history of medication, and then a thorough clinical examination. The clinician then advises some investigations to confirm or exclude his clinical suspicion. In this whole process, the most definite part of the diagnosis is rendered by the pathologist in the form of cytopathology or/and histopathology. Other diagnostic methods like radiodiagnosis and biochemistry have also an important ancillary role, but final diagnosis rests on pathologists. But there are some important issues that need to be solved.

Whenever cancer is suspected by the clinician, he/she should act promptly to confirm the diagnosis. Tumor load grows in geometric progression, and therefore, we should not miss the opportunity of offering early treatment to the patient. It can be readily calculated that the originally transformed cell must undergo at least 30 doublings (1 doubling = 1 cell division) to produce 10^9 cells weighing 1 gm., which is the smallest clinically detectable mass. In contrast, only 10 further doubling cycles are required to produce a tumor containing 10^{12} cells weighing approximately 1 kg, which is usually the maximal size compatible for life (Kumar et al. 2014).

For making a faster diagnosis, cytopathology is an important tool. Fine needle aspiration cytology (FNAC) is an approach, which is widely used. The procedure involves aspirating cells with a small-bore needle, followed by cytological examination of the stained smear. This method is used most commonly for assessing readily palpable tumors such as breast, thyroid, and lymph node. Modern imaging techniques like ultrasonography, CT, or MRI permit guided FNAC of deep-seated lesions and nonpalpable ones. It is an outpatient procedure and obviates surgical biopsy and its attendant risks.

Cytological smears provide yet another method for early detection of cancer and are widely used to screen for carcinoma of cervix. Detection of cancer cells can also be made in other body secretions and excretions like bronchial fluid and sputum in lung cancer, urine in bladder and prostate cancer, gastric fluid in stomach cancer, and for identification of cancer cells in abdominal, pleural, cerebrospinal, and joint fluids. The cancer cells that are exfoliated in different body fluids due to their lack of cohesion when they are transformed into malignant cells exhibit a range of

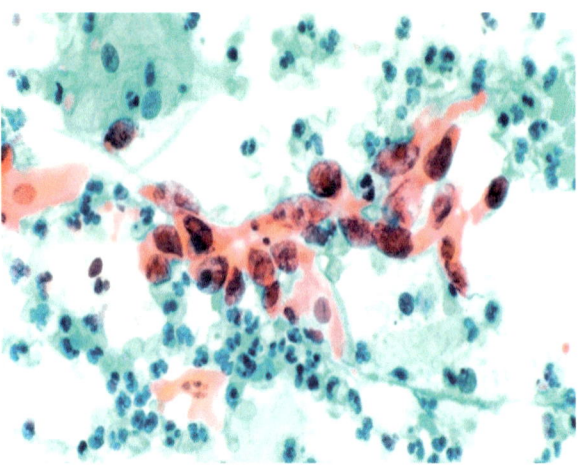

Fig. 10.1 An abnormal cervicovaginal smear showing numerous malignant cells that have pleomorphic, hyperchromatic nuclei

morphological changes encompassed by the term *anaplasia,* which literally means "to form backward," implying a reversal of differentiation to a more primitive level.

Lack of *differentiation* usually means morphological variations in a cell from its normal counterpart and can be detected microscopically by observing some characteristics: (i) variation in size and shape, which are termed as *pleomorphism*, (ii) abnormal nuclear morphology, which include nuclei being disproportionately large for the cell, and the nuclear-to-cytoplasm ratio is increased and may approach 1:1 instead of normal 1:4 or 1:6, (iii) the nuclear membrane is irregular and the chromatin is often coarsely clumped and distributed along the nuclear membrane, giving it a vesicular appearance, (iv) nucleus is more darkly stained than normal, which we call *hyperchromatic* nuclei, and (v) more frequent mitotic figures than normal and atypical mitosis, giving it a bizarre look. But cytological reporting of a cancer requires more experience as judgment must be rendered based on the features of individual cells or, at most, a clump of cells without the supporting evidence of loss of orientation of one cell to another (loss of polarity) and the evidence of invasion, which can be judged only in histology.

Figure 10.1 points to an abnormal cervicovaginal smear showing numerous malignant cells that have pleomorphic, hyperchromatic nuclei. Figure 10.2 shows anticytokeratin immune peroxidase stain (IHC) of a tumor of epithelial origin. Figure 10.3 shows papillary adenoma of colon, which is a benign tumor. Figure 10.4 shows beautiful transition from normal surface squamous epithelium to invasive carcinoma. Figure 10.5 shows well-differentiated squamous cell carcinoma showing prominent "pearl" formation. Figure 10.6 shows well-differentiated adenocarcinoma of colon. Note the cellular distinguishing it from Figure 10.3, *i.e.*, papillary adenoma. Figure 10.7 shows anaplastic large cell carcinoma of lung showing marked cellular pleomorphism. Figure 10.8 shows highly dysplastic epithelium with high N:C ratio, nuclear hyperchromatism, mitotic figures, and loss of polarity.

Histopathology or biopsy is the mainstay in cancer diagnosis, but it needs a very good histology laboratory with modern instruments and facilities but, more

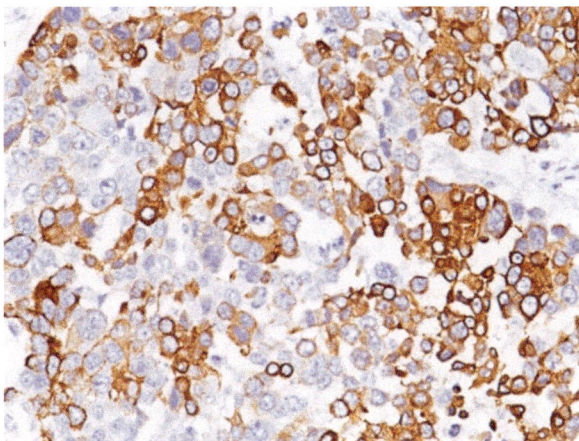

Fig. 10.2 Anticytokeratin immunoperoxidase stain (IHC) of a tumor of epithelial origin

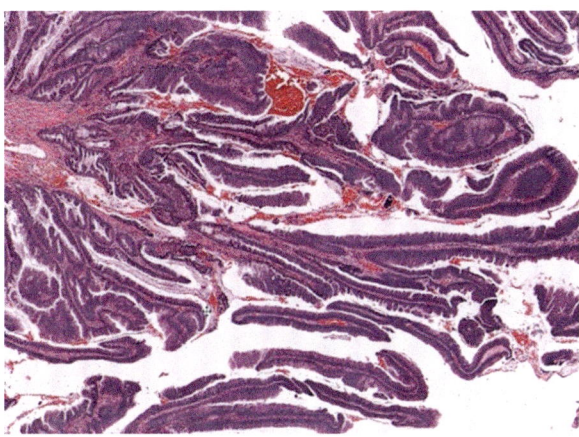

Fig. 10.3 Papillary adenoma of colon showing finger-like projections

importantly, trained manpower, which is often lacking. The pathologist, while reporting a histology slide under a microscope, will require a good quality thin section, appropriately stained, to ascertain the architectural and morphological characteristics and to arrive at a logical conclusion. He/she also needs a morphological description of the specimen from which the tissue section is taken and the relevant clinical information with other diagnostic parameters. Therefore, good coordinating information flow among clinician, pathologist, and laboratory staff is an essential prerequisite for proper diagnosis.

While reporting the pathologist should adhere to standard protocol of reporting, as provided by the nomenclature guidelines of WHO and cancer protocol templates by CAP (College of American Pathologists) or cancer staging manual of AJCC (American Joint Committee on Cancer), reporting based on standard guidelines and protocols carries pertinent information to clinicians to decide for the best choice of treatment, which has an immense value for cure and survival.

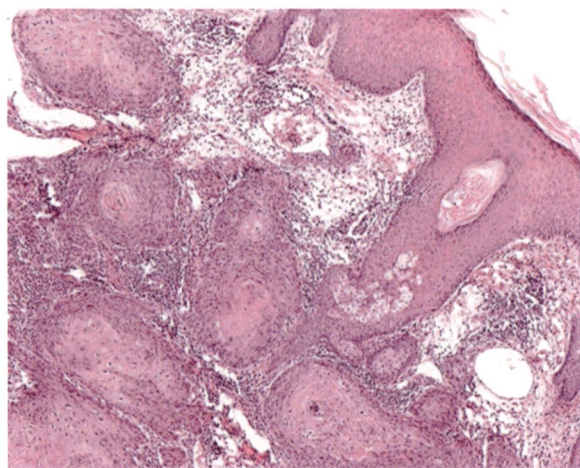

Fig. 10.4 Transition from normal squamous epithelium to invasive carcinoma showing invasion

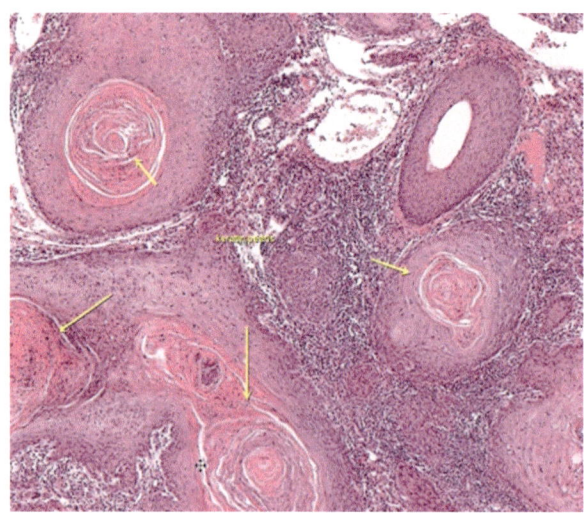

Fig. 10.5 Squamous cell carcinoma, well-differentiated, showing "pearls"

The availability of specific antibodies has greatly facilitated the identification of cell product by immunohistochemistry (IHC). But this facility requires judicious use of IHC markers in a cost-effective manner. Nowadays, most of the cancers are categorized by their protein expression, which is of immense importance for targeted therapy. Therefore, all laboratories reporting cancer must be equipped with IHC facility. This is particularly important for (i) categorization of undifferentiated tumors, that is, when the histology has failed to recognize the tumor properly, (ii) determination of site of origin of metastatic tumor, that is, when we could not ascertain the tissue of origin of a metastatic or far away deposit of a cancer, and (iii) detection of protein molecules expressed by the cancer cells that have prognostic (predicting the outcome) or therapeutic significance.

10 Oncology: Pathologist's View

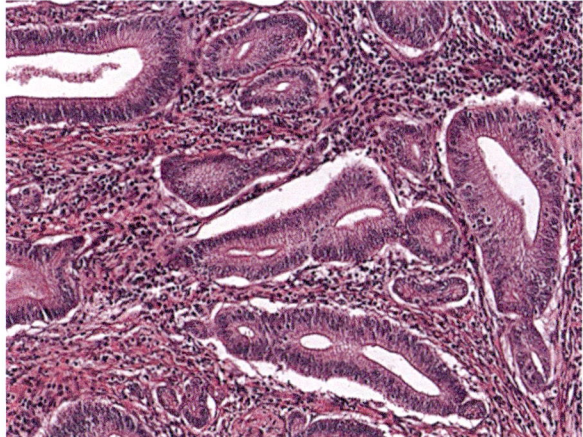

Fig. 10.6 Adenocarcinoma of colon. Glandular pattern is well-recognized

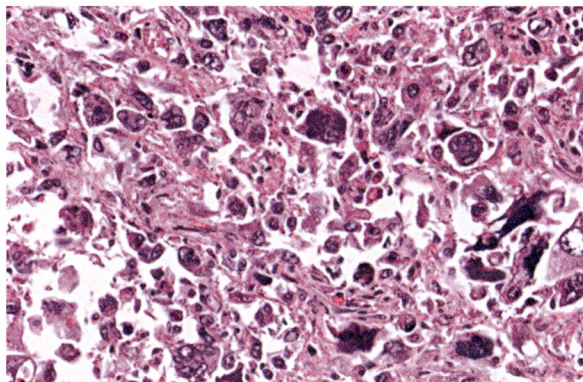

Fig. 10.7 Anaplastic large cell carcinoma of lung showing marked pleomorphism

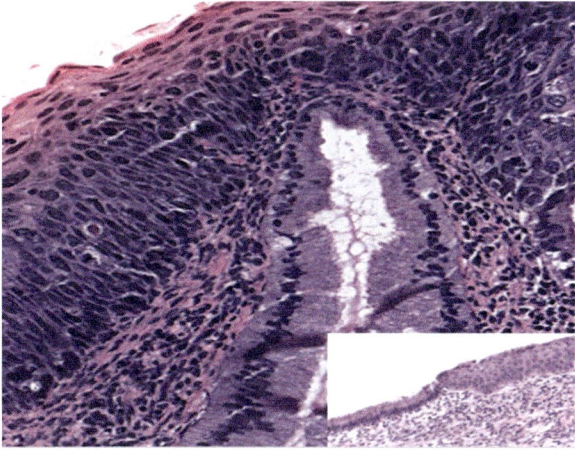

Fig. 10.8 Dysplastic epithelium showing high N:C ratio, nuclear hyperchromatism, mitotic activity, and loss of polarity

10.11 Histopathology Techniques in Biopsy Interpretation

Pathology literally means "study of diseases" and pathologists are a speciality of doctors trained for examining different body samples and providing their opinions, based on which the clinician will diagnose and treat the patient. This broad speciality has subsequently become compartmentalized into several subdisciplines: Histopathologists study morphological alterations in tissues and cells. Microbiologists study the microbes, biochemists study the alterations in biochemical substances, and so on. Cancer is basically an alteration in growth of tissue and therefore has implications on almost all specialities in medical sciences. A cancer patient may come to a clinician of any speciality for certain complaints related to the organ affected. Cancer may be suspected during investigation procedure for any other disease. The process can affect any age-group, although it is much more common in middle-aged or elderly patients, probably due to the time taken for a chain of mutations that need to occur for the malignant transformations. However, whenever a clinician suspects a cancer in his/her patient, the final diagnosis (this will remind us of the famous namesake novel by Arthur Hailey centered on the chief pathologist Joe Pearson) usually rests on the pathologists who have to take all the responsibilities to declare the presence of this dreadful disease.

The clinician usually sends a small tissue sample from the suspected organ which he collects either in an OT setup or in an outpatient procedure room which we called "biopsy." If the biopsy sample is collected by taking out a small bit of tissue from the tumor or growth, we call it incisional biopsy, and if it is collected after taking out the whole tumor, we call it excisional biopsy. Whether it is a small incisional biopsy or the whole specimen, it should be sent to the pathology department in an appropriate container containing formol saline, a fluid mixture of formalin and normal saline. This fluid maintains the tissue architecture and prevents degeneration of cells. This biopsy sample should be accompanied by a requisition slip that contains detailed information regarding the tissue, the clinician's impression about the growth accompanied by relevant history of current illness, relevant operative note, and few other technical information that are absolutely necessary for the pathologists to derive an evidence-based conclusion. Whenever the pathology department receives a biopsy sample, the sequential steps that follow are as follows:

A unique identification number is provided after entering into a register.
The tissue is kept in freshly prepared fixative solution overnight.
On the next day, the tissue piece or organ is taken out from the container, and if the tissue is small, as in the case of incisional biopsy or biopsy collected through endoscopy, the whole specimen is put in a cassette for histological processing. If the tissue is not a small one, its appearance, size, surface, etc., are noted, then it is dissected on a paraffin tray with the help of dissection instruments (forceps, knives, scissors, and others) following an established protocol and the details of description noted on the requisition form. This process is known as *grossing*. This is very much important for diagnosis of cancer as the cancerous tissue has some characteristic features that are revealed during this process and without which the

diagnosis of cancer will be incomplete. These gross findings are an important part of final report.

After grossing is done and small tissue pieces thus obtained are put into the cassettes, labeled with the identification number provided to it earlier during registration, and these cassettes are then put into an automated tissue processor or *histokinet* where these cassettes containing tissue pieces are dipped into different chemical solutions for a particular period of time (set by timer) to drive out the water portion of the tissue and make it suitable for paraffin embedding.

Paraffin embedding is a process by which the tissue pieces contained in the cassettes are immersed in paraffin bath (hot and liquefied paraffin wax) and then cooled down to make a paraffin-embedded block in which the tissue is impregnated inside the 3D paraffin blocks. Now, these tissues are called formalin-fixed paraffin-embedded (FFPE) tissues. This embedding is meant for hardening the tissue so that sections can be made for examination under microscope.

Now, these paraffin blocks, with the identification or accession number inscribed on one side of the block, are cut into thin sections or slices of about 3–5 micron thick in a microtome machine.

These sections are then taken on glass slides, deparaffinized, and stained to make the tissue and the cells in it to view under microscope.

10.12 Final Diagnosis by a Pathologist

The task of a histopathologist is quite difficult because, unlike the clinician, he is not examining the patient directly, and therefore, he has to depend solely on the clinical information provided in the requisition slip by his/her clinician friend. After examining the histology slides under microscope, he/she has to correlate his/her microscopic findings with that of the clinical and operative notes, biochemical reports, and the findings noted during grossing to arrive at a logical conclusion. When he corroborates the positive with the negative findings for a particular diagnosis, his analytic mind follows the dialectic pathway to arrive at the final diagnosis.

Why the process of deduction is dialectic in nature? Because the pathologist has to explore all possibilities with an open mind without any clinical bias on the one hand, while on the other hand trying to integrate the clinical findings with the histological findings so that the final diagnosis is a logical conclusion in that particular case. As biological science as a subject is full of variances, and multiple variants act on the disease process (e.g.. age, sex, geographical location, ethnicity, food habit, lifestyle, education, and culture), the same logic is not applicable in all similar cases, that is, the logic varies from case to case, depending on the variables involved in each case. As this logical conclusion is the highest level of cognitive function, error is not a rare thing and this might sometimes prove costlier for the patient. Several procedures may be followed by the pathologist to minimize errors: using proper and relevant clinical information along with the reports of biochemical and radiological investigations, interviewing the patient and/or the relatives for bridging the gap in information, regressing the specimen to corroborate with histology findings, recutting the section from the block to get a thinner and deeper section, and using special histochemical stains if needed. Ancillary methods like immunohistochemistry (stains tagged with antibody to detect antigens expressed in the tissue), cytogenetics (alterations in chromosomes that are characteristics of a particular tumor), and molecular genetic study may be used where indicated. Apart from these routine techniques for diagnosis of cancer and predicting the prognosis and guidance to therapy, there are certain ancillary diagnostic tools available at state-of-the-art laboratories to help the final diagnosis. These are as follows:

Flow cytometry can rapidly and quantitatively measure several cellular antigens, particularly in cases of leukemias and lymphomas. It has the advantage over IHC as multiple proteins (antigens) can be assessed simultaneously on individual cells.
Circulating tumor cells can be detected in blood as in cases of many carcinomas and melanomas, and their identity is confirmed by coating them with antibodies.
Molecular and cytogenetic techniques such as antigen receptor gene rearrangement detected by PCR-based evaluation in T- and B-cell neoplasms, detection of genetic translocation by routine cytogenetic analysis, or FISH (fluorescent in situ hybridization) in leukemias and lymphomas, NGS (Next Generation Sequencing) that can cover entire human genome to detect any genetic mutation anywhere in the genome, or SNP (single nucleotide polymorphism) chips which allow high-resolution mapping of nucleotides (either deletions or amplification). Have also profound implications in prognosis of malignant neoplasms, detection of minimal residual disease after treatment, diagnosis of hereditary predisposition to cancer, and guiding therapy with oncoprotein-directed drugs.

10.13 Molecular Profiling of Tumors: The Future of Cancer Diagnostics

Until recently, molecular studies of tumors involve the analysis of individual genes in selected cancer. However, the past decade, particularly after completion of Human Genome Project, has seen the introduction of revolutionary technologies that can rapidly sequence an entire human genome; assess epigenetic or around the gene modifications (the epigenome); quantify all the RNA expressed in a cell population (the transcriptome); measure many proteins simultaneously (the proteome); and take a snapshot of all of the cell's metabolites (the metabolome). DNA sequencing is technically simpler than RNA sequencing, permitting the development of massively parallel sequencing methods (so-called next-generation [NextGen] sequencing). The time taken for NextGen for an individual tumor today is 28 days, and the cost has fallen under $3,000.

These advances have enabled systemic sequencing and cataloging the genomic alterations in various human cancers under a consortium, The Cancer Genome Atlas (TCGA). The complexity of the genetic aberrations identified in these genome-wide studies has inspired biomedical informaticians to display the data in a creative way, known as circos plot, which provides a snapshot of all the genetic alterations present in a particular tumor. This information is required for a "personalized" approach if targeted therapy is to succeed.

Another molecular method that is moving rapidly in clinical practice is "DNA microarray" to identify changes in DNA copy number, such as amplifications and deletions. Other "omics" such as proteomics and epigenomics are currently being used mainly in the field of clinical research.

These developments in technology have led some scientists to predict that the end of histopathology is in sight. Though it sounds as a premature one, it can well be argued that the histopathologists will remain the key anchor for the show; we are probably in the midst of a paradigm shift in which the most important workup of a cancer specimen is the identification of molecular targets, rather than histopathological diagnosis. For example, the histologically distinct cancers all often harbor the same gain-in-function mutation of BRAF, a serine/threonine kinase, and a component of RAS signaling pathways. All these "BRAFomas" are candidates for treatment with BRAF inhibitors. But again for predicting the response to such therapy, histological subtypes are important, which obviously fall within the domain of histopathologists. Moreover, histopathological inspection of tumor will provide information about other important characteristics such as anaplasia, invasiveness, and tumor heterogeneity. Thus, what lies ahead is not the replacement of one set of techniques by another. On the contrary, the most accurate diagnosis and assessment of prognosis in cancer patients will be arrived by a combination of morphologic and molecular techniques.

Medical science has been immensely enriched by fundamental research in other scientific fields including basic sciences. The knowledge base in genetics and immunology is expanding in an exponential way, and it is impossible for an individual to keep track with these developments. Therefore, the perspectives of a pathologist are gradually drifting from an individual effort to a team effort. Moreover, integration with clinical and allied diagnostic departments is a must for an effective problem-solving approach to the patients. A pathologist will definitely sharpen his macroscopic and microscopic skills and pursue a dialectic approach of problem-solving, but he has to confirm his diagnosis also by judicious use of special stain, genetic markers, immune markers, and other ancillary techniques. Only a good morphological observation is not sufficient nowadays, due to advent of targeted therapy. Therefore, the perspectives are changing and a pathologist should remain updated and relevant in the diagnosis and management in oncology.

References

Casás-Selves M, DeGregori J (2011) How cancer shapes evolution, and how evolution shapes cancer. National Institute of Health Public Access Author Manuscript

Flanagan AM (2016) Recent advances in histopathology. Chapter 3. JP Medical Publishers, New Delhi, pp 29–40

Kumar V, Abbas AK, Fausto N, Aster JC (2014) Robbins and Cotran: pathological basis of disease, 9th edn. Saunders Elsevier, Philadelphia, PA
Nature Outlook Cancer (2014) 29 May / Vol 509 / Issue No. 7502
Strayer DS, Kluwer W (2015) Rubin's pathology: clinicopathologic foundation in medicine, 7th edn. Wolters Kluwer, Philadelphia, PA
Swamidass SJ (2017) Cancer and evolution. Biogloss

Part III
Cancer Therapeutics

Surgical Oncology: An Overview

11

Aseem Mishra and Vivekanand Sharma

Abstract

This chapter aims to share with the readers a glimpse of the evolution of the role of surgery in oncological practice, and provide them with an understanding of what role does surgery play in management of cancer from diagnosis, treatment, and prevention of cancer as per tumor biology and stage of the disease.

Keywords

Surgery · Screening · Diagnosis · Staging

11.1 Introduction

The treatment of cancer is complex and is governed by multitude of factors. This requires an integrated multidisciplinary team approach of various specialities like surgical oncologist, medical oncologist, and radiation oncologist working together for treatment planning and administration. With the advent of personalized care in modern oncology practice, the role of allied subspecialities, for example, pathology, diagnostic and interventional radiology, and rehabilitation, has widened considerably.

The aim of this chapter is to share with the reader a glimpse of the evolution of role of surgery in oncological practice and to provide them with an understanding of

A. Mishra (✉)
Department of Head and Neck Surgical Oncology, Homi Bhabha Cancer Hospital, Varanasi, India

V. Sharma
Department of Surgical Oncology, Homi Bhabha Cancer Hospital, Varanasi, India

what role does surgery plays in management of cancer from diagnosis, treatment, and prevention of cancer as per tumor biology and stage.

11.2 Evolution of Role of Surgery in Oncology

Historically, surgery was the sole method used for treating cancer and there are numerous texts documenting the same since thousands of years back. Leonidas of Alexandria is widely quoted as the first person to have devised an oncologically correct surgery in the first century AD for breast cancer (Retief and Cilliers 2011). However, in those days with the lack of understanding of antisepsis, anesthesia, blood transfusion, and disease biology, the outcome was usually poor and mortality with or without treatment remained high.

With advances in science and technology, a greater understanding for perioperative physiology, disease pathology, and treatment modalities, there has been a paradigm shift and many major cancer surgeries today are associated with less than 5% mortality rate (Ghaferi et al. 2009). This along with similar advances in sister specialities in medical and radiation oncology has allowed cancer to be treated as curable, when detected early.

11.3 Role of Surgical Oncologist

The roles of a surgical oncologist include cancer screening, diagnosis, and staging, which are discussed in this section.

11.3.1 Cancer Screening

As understanding of the development of cancer increased, some cancer types were recognized which could be identified early with simple diagnostic tests and if timely treated led to improved patient survival. Cancers of breast, colorectal, cervix, lung, and prostate are hence being increasingly identified in early stages, and the surgical oncologists are playing a key role in this process (Sankaranarayanan et al. 2005; White et al. 2020).

11.3.2 Diagnosis and Staging

Accurate diagnosis of cancer type is essential, since it not only dictates initiating any therapy, but also has implications in patient counseling and medico-legal issues as well. Identification of cancer is usually made by sampling a part of the suspected tissue, followed by pathological analysis. The method employed to obtain the tissue, called biopsy, may change depending upon the site of disease, presentation type, and individual patient.

1. Lesions over any accessible surface like on the skin or mucosa (e.g., in oral cavity, vagina, cervix, rectum, and anal canal) can be biopsied easily via an incisional or punch biopsy (Griffiths et al. 1964).
2. Lesions involving the mucosal surfaces in the gastrointestinal tract (GIT), hypopharynx, urinary bladder, etc., are usually biopsied with an aid of endoscopic access (colonoscopy/laryngoscopy/cystoscopy). The endoscopic vision helps in locating the lesion and targeting the appropriate lesion for biopsy. Certain scopy can be performed on out-patient basis which does not require any hospital admission or major preparation (Zalvan et al. 2013).
3. *Fine needle aspiration (FNA):* This can be performed in palpable but deeply located lesions or in nonpalpable lesions as in the case of breast lesion, lymph nodes, and thyroid and salivary gland malignancy (Lin et al. 1997; Zagorianakou et al. 2005). FNA is usually guided with an imaging modality like an ultrasonography, mammography, or a computed tomography scan. Rapid OnSite Evaluation (ROSE) permits the aspirate from the needle to be assessed immediately under microscope for adequacy and ensures successful result (O'Leary et al. 2012; Schmidt et al. 2013). FNA has a central role in diagnosing thyroid nodule, parotid gland, metastatic nodes, and borderline resectable pancreatic tumors. With recent advances and experience in immunohistochemistry markers, even cytological aspirate can be used in many diseases for an accurate diagnosis. The cytological examination is helpful in establishing a diagnosis, but it provides limited information regarding tumor architecture and hence biopsy still remains the gold standard for tissue diagnosis.
4. *Core needle biopsy (CNB):* Akin to FNA, core biopsy can be performed in out-patient setting with added advantage of providing additional information about the tumor architecture. CNB provides adequate tissue to make an appropriate diagnosis in most of the patients. It provides adequate tissue for pathologist to make a diagnosis and perform immunohistochemistry evaluation, whenever necessary. CNB is the cornerstone of diagnosis in all solid malignancy where the lesion is not accessible endoscopically or other less invasive techniques. CNB is superior as it is less invasive and can be performed quickly without any external scar and is less painful (Verkooijen et al. 2000; Yao et al. 2012).
5. *Open incisional biopsy:* Open biopsy is the gold standard for histopathological diagnosis in all cases where there is a diagnostic dilemma, but it should be reserved for those in which less-invasive procedures are inconclusive. It provides accurate diagnosis as the sampling error is minimized. However, care should be taken while planning an excisional biopsy so as to not compromise future treatment. Incision for biopsy must be planned in a way that the scar can be incorporated in the future excision, if needed. Improper incision placement leads to compromised surgical resection in future or may cause an unnecessary larger normal tissue resection with resultant loss of form or function or both.
6. *Open excisional biopsy:* Excisional biopsy involves excision of entire tumor with a margin of normal tissue. It offers the advantage of providing adequate tissue for diagnosis. However, selecting patient for excisional biopsy is of utmost importance. It should be reserved for small malignant or premalignant lesions where it

may be the definitive treatment. Similar to incisional biopsy planning, the final treatment and the future surgical plan should always be kept in mind before planning any excisional biopsy (Estourgie et al. 2007; Ghaem-Maghami et al. 2007).

In certain solid tumors of the pancreas, ovarian tumors, testicular cancer, and certain benign nasal mass like angiofibroma, routine biopsy is not recommended. Definitive treatment is planned in these cases without tissue diagnosis. Thus, clinical knowledge of tumor biology and characteristics is essential. The present era of technology has provided several lesser invasive investigations, which can be utilized to avoid a biopsy.

While the role of surgical oncologist remains central in most methods of biopsy, there is an increase in the role of endoscopist and interventional radiologists in an effort to explore less-invasive methods in challenging presentations.

11.4 Staging

Treatment of any cancer starts with accurate staging and is the most important component in the cancer treatment planning. The clinical history, physical examination, imaging, and biopsy are useful tools in clinical staging. These are essential component of clinical TNM staging. The pathological TNM staging is derived from the histopathological evaluation of the tissue from the resection of the primary tumor and regional lymph nodes. The Union for International Cancer Control (UICC) provides the staging system based on the clinical data and a similar staging system is also provided by American Joint Committee on the cancer AJCC - Cancer Staging Manual (2017).

Surgeries were a common method of staging in earlier times. With improved imaging methods like multi-detector computed tomography (MDCT), PET-CECT (PET-contrast enhanced computed tomography), and high-resolution magnetic resonance imaging (MRI), the need for invasive methods like open surgery for tumors of spleen and liver or pancreas has become obsolete. There is a paradigm shift in the last decade, and newer diagnostic methods are being used intraoperatively for diagnosing lymph node metastasis. One such technology is sentinel node biopsy (SNB) (Morton et al. 2006; Veronesi et al. 2003). This technique involves assessing the first echelon nodal station draining the primary tumor using a radiotracer dye. Primary tumor is injected with radiotracer dye (technetium labeled colloid sulfur or isosulfan blue), and a gamma camera is used to analyze radioactive uptake at the first echelon nodes. This helps in determining the possibility of lymphnodal metastasis at a very early stage and helps in surgical planning for the nodal clearance. Multiple scientific trials have proven this method to be effective in cancers of penis, breast, vulva, melanoma, and squamous head and neck cancers. However, en-bloc nodal dissection is almost certain in tumors like head and neck cancer and in en-bloc gastrectomy with extended lymph nodal dissection gastrointestinal cancer. The newer improved technology and diagnostic modalities have played a significant role in adequate treatment planning.

11.5 Surgery: From More, to Less, to None

Principle of oncological resection involves complete removal of tumor with tumor-free margins, and the whole specimen must come out en-bloc. In the earlier centuries, cancer surgeries entailed major, often mutilating surgeries. Advances in multimodality therapy have led to the development of conservative approaches toward cancer, with better preservation of organ and function without compromising oncological outcomes. Notable examples in this context are cancers of the anal canal, hypopharynx, larynx, and upper esophagus, where traditional mutilating surgeries have been replaced with chemoradiation for the majority of patients. In cancers of breast, similar advances led to better cosmetic outcomes with breast conservation and oncoplasty.

With the advent of adjuvant therapy, surgery for the primary tumor has become less extensive. Following the NSABP B-06 trial, breast cancer patients now undergo breast conservation surgery and postoperative irradiation. Following the onset of VA trial in laryngeal and hypopharyngeal cancer, the organ preservation protocol became the need of the day (Wolf et al. 1991). Previously, the patients of laryngeal cancer underwent total laryngectomy. This led to loss of organ and subsequent loss of voice. After the successful completion of the Veteran Affairs trial, the organ conservation protocol came forward in a very big way. Subsequent to this, organ preservation became the treatment of choice. Induction chemotherapy followed by radiotherapy became treatment of choice to avoid a laryngectomy. A better understanding of early disease treatment and modalities has also expanded treatment options for early mucosal cancers in larynx, esophagus, stomach, and rectum where endoscopic options like endoscopic mucosal resection and dissection allow organ preservation. In recent years, even wait and watch policies are being attempted in cancers of the rectum and esophagus.

11.6 Plastic and Reconstructive Surgery

The development of *plastic surgery* has led to a remarkable improvement of surgical oncology. Extensive resections which were earlier deemed unresectable due to associated morbidity and lack of reconstructive option can now be resected and reconstructed with a good cosmetic and functional outcome (Munhoz et al. 2013). Advances including oncoplastic surgery are associated with improved cosmetic outcomes with less morbid measures in breast cancer, better aesthetic and functional outcome in cancers of the head, neck, and skull base region, and greater and earlier restoration of function in musculoskeletal cancers. These outcomes encourage more patients to lead a good quality of life after surgery.

The extensive resections of the tumors of the oral cavity can now be reconstructed using a microvascular free flap like a free anterolateral thigh flap and free fibula osteocutaneous flap. These flaps have offered significant improvement in quality of life of hitherto inoperable patients.

11.7 Minimally Invasive Surgery (MIS)

Minimally invasive surgery involves providing a lesser traumatic access to a disease site or organ. This provides good visualization and exposure, and this makes the resection easy. The development of MIS is an important development in recent surgical developments. Surgeries which were hitherto deferred due to the morbidity associated with the access and large incisions can now be performed with lesser complications. This reduces the trauma associated with creating the exposure for surgery. After initial dilemma over the oncological safety of MIS over open approach, oncological efficacy is now recognized in colorectal, endometrial, renal cancers, anterior skull base, and paranasal sinus tumors with encouraging results in esophageal, lung, gastric, and even thyroid and pancreatic cancers in carefully selected patients (Gemmill and McCulloch 2007; Nagpal et al. 2010).

11.8 Role of Robotic Surgery

Robotic-assisted surgical procedures are performed with the use of robotic system. Robotic surgery avoids several limitations associated with MIS. It provides superior image resolution and enhances the dexterity of surgeon due to robotic arms. The operating surgeon controls the console of the robot, and the main robotic system with the arms is utilized to deliver instruments at the operating site. The surgeon performs the surgery by sitting on console. It provides the advantage of better optical visualization, camera stability, three-dimensional view, and filters the tremors. Robotic assistance offers enhanced degree of freedom in a limited space, making robotic system useful in urological and gynecological surgery.

Robot-assisted surgery offers shorter length of stay of a patient, fewer readmissions, and reduced intraoperative morbidity. Robotic surgery in head and neck is performed through transoral access and is popularly known as transoral robotic surgery (TORS). TORS offers the advantage of excellent approach to tumors of the oropharynx and hypopharynx (Genden et al. 2009; Weinstein et al. 2012). It is particularly useful for salvage surgery of the oropharyngeal tumors like a recurrent disease of the base of the tongue and tonsillar fossa. Other indications in head and neck have also been developed, and robotic arms have been used to perform surgeries like thyroidectomy and neck dissection. The approach to these areas is via a postauricular incision and sublingual access through midline (Benhidjeb et al. 2009; Dionigi et al. 2017).

The application of robotic surgery in oncology has broadened in the recent times with the use of robot-assisted surgery in almost all specialities like urology, gynecology, thoracic, gastrointestinal, and head and neck oncology. However, it is essential to have a careful case selection as in minimal access surgery. The indication of surgery has to be analyzed carefully. The option of converting to an open procedure should be discussed with the patient beforehand. Another important aspect which has to be highlighted is not to popularize the robot-assisted surgery as stigma or superior against open procedure. As a common myth, the patient perceives robotic surgery being performed by robot which needs to be dissuaded. Judicious use of this technology is essential, and it depends on the patient selection and should not be popularized as patients' choice.

A patient undergoing surgery for prostate cancer is concerned about the urinary and sexual function. Robotic surgery is now being performed in several centers in India for oncological and nononcological indications. While robotic surgeries have found acceptance in cancers of multiple sites, one must also be cognizant of its cost-effectiveness as against other modalities.

11.9 Impact of Emerging Technology in Surgical Oncology

Integration of augmented reality, artificial intelligence, and technological advances has found place in guided surgical systems, which have a role in not only safer surgeries but also structured surgical training- and distance-based learning. The use of image guided navigation system has reformed the learning and practice of complicated surgeries like endoscopic surgeries of the anterior skull base. The CECT images of the patient are reformatted and utilized for anatomical guidance during surgery. This avoids several complications as the major vessels and nerves can be visualized through the system and the surgeon is alarmed well before dissecting these key structures.

11.10 Expanding Indications of Treatment and Indian Context

With increased survival of cancer patients, the surgical oncologist is playing a greater role even in metastatic cancers, which were earlier deemed beyond the purview of cure. Excision of metastatic sites in liver, lung, and even disseminated abdominal disease can now be operated with procedures like metastatectomy, peritonectomy, and hyperthermic intraperitoneal and intra-thoracic chemotherapy. With emerging technologies, several recurrent tumors and extensive resections are now being performed. The branch of surgical oncology has developed immensely in the Indian subcontinent. The improvisation of surgical oncology is now progressing toward subspecializations within surgical oncology. The surgical oncology training is offered as a superspecialization course through a three-year training (MCh Surgical Oncology). Other than that, various subspecialities like superspeciality training in head and neck oncology, gynecological oncology, and breast surgical oncology have also developed. The various surgical specialities like gastrointestinal surgery, neurosurgery, and urology have also played an immense role in development of these cancer subspecialities. However, in majority of tier-II and tier III cities in the third world countries, the oncology treatment is still delivered by general surgeons, otolaryngologists, gynecologists, etc. The development of the training programs in India has led to significant improvement in the quality of care delivered to the patients. This has improved the survival outcome of these cancers and offered an excellent quality of life to cancer patients. With extensive training programs being structured, and formation of the national cancer grid, several institutes have been upgraded to provide adequate clinical care in oncology. The extension of the central institutes like Tata Memorial Hospital under the department of atomic energy to various towns like Sangrur, Varanasi, Mullanpur, Vizag, and others has also made affordable cancer available in these areas. The opening up of various institutes like All India Institute of Medical Sciences in various cities has helped in decentralizing not only cancer care but also has made available superspecialist clinical care available in various cities. These projects are going to change the phase of healthcare in India in the near future.

11.11 Conclusion

Surgery in oncology has developed rapidly in the last century. Surgery is curative in early solid malignancy and has superior survival outcome in head neck, breast, gastrointestinal, and gynecological cancers. While surgery remains the most effective single treatment modality, multidisciplinary team approach is essential for a better clinical outcome. While many tools are becoming available for surgery, their careful usage and patient selection are more important aspects of patient care.

References

AJCC - Cancer Staging Manual (2017) [WWW document]. https://cancerstaging.org/referencestools/deskreferences/pages/default.aspx. Accessed 27 Jan 20

Benhidjeb T, Wilhelm T, Harlaar J, Kleinrensink GJ, Schneider TA, Stark M (2009) Natural orifice surgery on thyroid gland: totally transoral video-assisted thyroidectomy (TOVAT): report of first experimental results of a new surgical method. Surg Endosc 23(5):1119–1120

Dionigi G, Lavazza M, Wu C-W, Sun H, Liu X, Tufano RP, Kim HY, Richmon JD, Anuwong A (2017) Transoral thyroidectomy: why is it needed? Gland Surg 6(3):272–276. https://doi.org/10.21037/gs.2017.03.21

Estourgie SH, Olmos RAV, Nieweg OE, Hoefnagel CA, Rutgers EJT, Kroon BBR (2007) Excision biopsy of breast lesions changes the pattern of lymphatic drainage. Br J Surg 94(9):1088–1091. https://doi.org/10.1002/bjs.5763

Gemmill EH, McCulloch P (2007) Systematic review of minimally invasive resection for gastrooesophageal cancer. Br J Surg 94(12):1461–1467. https://doi.org/10.1002/bjs.6015

Genden EM, Desai S, Sung C-K (2009) Transoral robotic surgery for the management of head and neck cancer: a preliminary experience. Head Neck 31(3):283–289. https://doi.org/10.1002/hed.20972

Ghaem-Maghami S, Sagi S, Majeed G, Soutter WP (2007) Incomplete excision of cervical intraepithelial neoplasia and risk of treatment failure: a meta-analysis. Lancet Oncol 8(11): 985–993. https://doi.org/10.1016/S1470-2045(07)70283-8

Ghaferi AA, Birkmeyer JD, Dimick JB (2009) Variation in hospital mortality associated with inpatient surgery. N Engl J Med 361(14):1368–1375. https://doi.org/10.1056/NEJMsa0903048

Griffiths CT, Austin JH, Younge PA (1964) Punch biopsy of the cervix. Am J Obstetr Gynecol 88(5):695–703. https://doi.org/10.1016/0002-9378(64)90900-7

Lin J-D, Huang B-Y, Weng H-F, Jeng L-B, Hsueh C (1997) Thyroid ultrasonography with fine-needle aspiration cytology for the diagnosis of thyroid cancer. J Clin Ultrasound 25(3):111–118. https://doi.org/10.1002/(SICI)1097-0096(199703)25:3<111::AID-JCU3>3.0.CO;2-J

Morton DL, Thompson JF, Cochran AJ, Mozzillo N, Elashoff R, Essner R, Nieweg OE, Roses DF, Hoekstra HJ, Karakousis CP, Reintgen DS (2006) Sentinel-node biopsy or nodal observation in melanoma. N Engl J Med 355(13):1307–1317. https://doi.org/10.1056/NEJMoa060992

Munhoz AM, Montag E, Gemperli R (2013) Oncoplastic breast surgery: indications, techniques and perspectives. Gland Surg 2(3):143

Nagpal K, Ahmed K, Vats A, Yakoub D, James D, Ashrafian H, Darzi A, Moorthy K, Athanasiou T (2010) Is minimally invasive surgery beneficial in the management of esophageal cancer? A meta-analysis. Surg Endosc 24(7):1621–1629. https://doi.org/10.1007/s00464-009-0822-7

O'Leary DP, O'Brien O, Relihan N, McCarthy J, Ryan M, Barry J, Kelly LM, Redmond HP (2012) Rapid on-site evaluation of axillary fine-needle aspiration cytology in breast cancer. J Br Surg 99(6):807–812. https://doi.org/10.1002/bjs.8738

Retief FP, Cilliers L (2011) Breast cancer in antiquity. SAMJ: South African Medical Journal 101(8):513–515

Sankaranarayanan R, Ramadas K, Thomas G, Muwonge R, Thara S, Mathew B, Rajan B, Trivandrum Oral Cancer Screening Study Group (2005) Effect of screening on oral cancer mortality in Kerala, India: a cluster-randomised controlled trial. Lancet 365(9475):1927–1933. https://doi.org/10.1016/S0140-6736(05)66658-5

Schmidt RL, Witt BL, Matynia AP, Barraza G, Layfield LJ, Adler DG (2013) Rapid on-site evaluation increases endoscopic ultrasound-guided fine-needle aspiration adequacy for pancreatic lesions. Dig Dis Sci 58(3):872–882. https://doi.org/10.1007/s10620-012-2411-1

Verkooijen HM, Peeters PHM, Buskens E, Koot VCM, Rinkes IB, Mali WTM, van Vroonhoven TJ (2000) Diagnostic accuracy of large-core needle biopsy for nonpalpable breast disease: a meta-analysis. Br J Cancer 82(5):1017–1021. https://doi.org/10.1054/bjoc.1999.1036

Veronesi U, Paganelli G, Viale G, Luini A, Zurrida S, Galimberti V, Intra M, Veronesi P, Robertson C, Maisonneuve P, Renne G (2003) A randomized comparison of sentinel-node biopsy with routine axillary dissection in breast cancer. N Engl J Med 349(6):546–553. https://doi.org/10.1056/NEJMoa012782

Weinstein GS, Quon H, Newman HJ, Chalian JA, Malloy K, Lin A, Desai A, Livolsi VA, Montone KT, Cohen KR, O'Malley BW (2012) Transoral robotic surgery alone for oropharyngeal cancer: an analysis of local control. Arch Otolaryngol Head Neck Surg 138(7):628–634

White MC, Kavanaugh-Lynch MMH, Davis-Patterson S, Buermeyer N (2020) An expanded agenda for the primary prevention of breast cancer: charting a course for the future. Int J Environ Res Public Health 17(3):714. https://doi.org/10.3390/ijerph17030714

Wolf GT, Fisher SG, Hong WK, Hillman R, Spaulding M, Laramore GE, Endicott JW, McClatchey K, Henderson WG (1991) Induction chemotherapy plus radiation compared with surgery plus radiation in patients with advanced laryngeal cancer. N Engl J Med 324(24): 1685–1690. https://doi.org/10.1056/NEJM199106133242402

Yao X, Gomes MM, Tsao MS, Allen CJ, Geddie W, Sekhon H (2012) Fine-needle aspiration biopsy versus core-needle biopsy in diagnosing lung cancer: a systematic review. Curr Oncol 19(1):e16–e27. https://doi.org/10.3747/co.19.871

Zagorianakou P, Fiaccavento S, Zagorianakou N, Makrydimas G, Stefanou D, Agnantis NJ (2005) FNAC: its role, limitations and perspective in the preoperative diagnosis of breast cancer. Eur J Gynaecol Oncol 26(2):143–149

Zalvan CH, Brown DJ, Oiseth SJ, Roark RM (2013) Comparison of trans-nasal laryngoscopic office based biopsy of laryngopharyngeal lesions with traditional operative biopsy. Eur Arch Otorhinolaryngol 270(9):2509–2513. https://doi.org/10.1007/s00405-013-2507-z

Medical Oncology in Cancer Treatment

12

Mandira Saha

Abstract

Malignancy is the second most common disease worldwide after heart disease. Cancer of any subsite is generally treated through one or more modalities of surgery, chemotherapy, and radiotherapy. Surgery and radiotherapy are local forms of treatments, but chemotherapy is systemic. The specialization in the field of systemic therapy is called medical oncology. The concept of medical oncology has come into the picture a couple of decades back. Chemotherapy, hormonal therapy, and targeted therapy have been established for the treatment of cancer. Immunotherapy and gene therapy are still in the investigational phase. Many new molecules are getting approval from the USFDA every year and changing the treatment protocol continuously.

Keywords

Chemotherapeutic agent · Cell cycle · Systemic therapy · Cancer cell

12.1 Introduction

Cancer is the second most common disease worldwide after cardiac diseases (Nagai and Kim 2017). The branch of medicine dedicated to diagnosing, treating and researching cancer is known as oncology. Cancer can be treated by surgery, systemic therapy or radiotherapy. The specialization in the field of systemic therapy is called medical oncology. Treatment of cancer depends on its location, stage and type. The chemical agents that are used to stop the uncontrolled growth of malignant cells,

M. Saha (✉)
Department of Radiation Oncology, HCG-EKO Cancer Centre, Kolkata, India

© The Author(s), under exclusive license to Springer Nature Singapore Pte Ltd. 2022
S. K. Basu et al. (eds.), *Cancer Diagnostics and Therapeutics*,
https://doi.org/10.1007/978-981-16-4752-9_12

either by killing the cells or by stopping them from cell division, are called chemotherapy (Chu and Sartorelli 2004). It can be administered through oral route, injections, infusion or skin, depending on the nature and type of chemotherapeutic agents. For many cancers like hematological malignancies, chemotherapy is the primary treatment of choice. Sometimes this can be given as an adjunct to other treatments, like surgery or radiotherapy, which is the primary mode of treatment. When chemotherapy is used before surgery/radiotherapy (primary treatment), it is called *neoadjuvant chemotherapy*. The aim is to shrink down the size of tumors. Secondly, when it is used during primary treatment like radiotherapy, it is called *concurrent chemotherapy*, which potentiates the action of primary treatment. It is also used after the completion of primary treatment as *adjuvant chemotherapy*, which has a purpose to eradicate the microscopic cancer cells from the body and prevent recurrence of cancer in future. Another name of chemotherapeutic agents is cytotoxic *medicine* (cyto = cells; toxic = harmful). The intent of treatment by systemic therapy could be either curative or palliative where the aim is to alleviate the symptoms only. The systemic therapy not only consists of cytotoxic medicines but also includes other agents like *hormonal therapy* and *biological therapy*. The treatment protocol for malignant disease is changing or modifying itself every year with the development of new drugs as well as new concepts about the existing drugs.

12.2 History (DeVita and Chu 2008)

The term *chemotherapy* was first used by the famous German chemist Paul Ehrlich in the early 1900, and he developed drugs to treat infections. He actually synthesized an organic arsenic compound to treat parasitic infection in rabbit. It was the first chemical agent against infection, and at that time it was named *salvarsan*, meaning *the savior of mankind*. In the earlier days, drug research was very difficult because of the lack of proper animal models and access to the chemical agents. So the initial few decades were invested to make appropriate animal models. Multiple trials failed either due to lack of responsiveness or due to unacceptable toxicities. Human trials were difficult due to scarcity of information of potentially toxic chemicals. One of the landmark achievements was the use of hormonal agents for cancer treatment in 1937. Another chemotherapeutic medicine, called *alkylating agent*, was developed as military weapon at the time of World War I. That chemical was a nitrogen mustered. There was a explosion at Naples harbor, accidentally exposing the US army to that chemical, which showed marked deletion of blood cells in bone marrow and in the lymph nodes at the time of World War II. They started research on that chemical at Yale university with the help of two famous pharmacologists, Louis Goodman and Alfred Gilman. They produced nitrogen mustards and applied them on the animal model of lymphoma to examine the therapeutic effects of these chemicals. Human trial was first done in 1943 in a lymphoma patient. As the result was promising, multiple related compounds were manufactured and the drugs were called alkylating agents. At the same time, they found out another compound from green leafy vegetables, called folic acid which had opposite effects of alkylating agent, like increased proliferation of bone marrow cells. From this agent, they got the concept of *folate antagonism* and synthesized antifolates. Dr. Sidney Farber, an American pediatric pathologist, first used antifolate to treat leukemia. He was the first person who identified its effect on blood cancer, and for his work in this field, he is regarded as the father of modern chemotherapy.

12.3 Chemotherapeutic Drug: Its Evolution and Development

After the discovery of *alkylating agent*, multiple trials were started on chemotherapeutic drugs development. Initially, single drug was used for treatment. Choriocarcinoma was cured by *methotrexate* and after that *Burkitt's lymphoma* was treated with an alkylating agent (cyclophosphamide). At that time, blood cancers were mostly treated with chemotherapy. But the problems arose with inadequate response to the treatment and early recurrence of the disease. So, the combination of chemotherapy was started during 1970s to overcome those difficulties. As the knowledge about cancer biology is gradually increasing, the evolution in chemotherapeutic drug development is supposedly touching the zenith.

Whenever a new molecule is discovered with anticancer property, it must go through multiple phases of drug development. If the results come out satisfactory, it gets approval for the treatment of cancer from the United States Food and Drug Administration (USFDA). The molecules are found by many ways, such as accidental exposure, separated from other chemicals like plant alkaloids or developed in the laboratory by chemical research according to biology of cancer cells. After indentifying the drug, it has to go through multiple trials which are broadly categorized into 3 main steps: preclinical, clinical, and postclinical trials. These are necessary to know the chemical structure of the drugs, mechanism of actions on cancer cells, side effects, etc.

Preclinical trials are done in the laboratory on animals to know whether it can be tried on human or not.

Clinical trials have 5 phases as follows (Sedgwick 2011):

Phase 0—It is done on a few people (10–15) with small dose to see how it acts in the human body.
Phase I—It is also done in small group of 15–30 patients to assess the safety dose of the medicine.
Phase II—This is to assess safety as well as efficacy and acute side effects of the drug. In this phase, the drugs are tested on specific cancer type. If the phase II results are satisfactory, the drugs are tested for the next phase.
Phase III—This is the most important trial when it is compared with the current standard of care for treatment. The patient is randomized to standard treatment arm as well as study drug arm. A large number of patients are needed (more than 100) in this phase. The trial is stopped immediately if the side effects are very severe or if the interim result shows much better result with any arm. Phase III trial result is needed before getting approval from U.S. FDA.

Phase IV or postclinical trial—It is done after FDA approval with a large number of patients and to find out the long-term and rare side effects of the drugs.

Therefore to develop a drug and to get approval from the USFDA, a prolonged time, involvement of thousands of people and a lot of resources are needed.

12.4 Indications of Chemotherapy

Chemotherapy is one of the most important weapons to fight cancer. It can be used either with curative or palliative intent. Curative intent means when we are able to completely eradicate the cancer cells from the body. Here chemotherapy is used as the primary modality of treatment. Hematological malignancies can be cured mostly by chemotherapy only. The list of diseases where chemotherapy is used as the primary treatment are as follows: acute leukemia, lymphoma, Hodgkin's disease choriocarcinoma, Wilm's tumor, testicular carcinoma, etc. (Chabner and Roberts 2005).

Sometimes, it can be added with the primary treatment which is either surgery or radiotherapy. It can be used either before or after the primary treatment which is called *neoadjuvant* and *adjuvant* chemotherapy respectively. The neoadjuvant/adjuvant chemotherapy are used in the following cancers—brain tumors (glioblastoma multiforme, anaplastic astrocytoma, medulloblastoma, etc.), retinoblastoma, sarcoma, head and neck cancers, esophageal cancer, lung cancer, breast cancer, gastrointestinal cancer (stomach, gall bladder, pancreas, colon, rectum, etc), ovarian cancer, endometrial carcinoma, renal cell carcinoma, carcinoma urinary bladder, etc. The aim of adjuvant chemotherapy is to prevent future recurrence of the disease and prolong life. It is used in those cancers which have the propensity to spread to the other organs of the body, eradicating the microscopic spread of the disease. In case of neoadjuvant chemotherapy, it helps to decrease the size of tumors and makes them amenable for either surgery or radiotherapy.

There are other situations where chemotherapy can be used as radiosensitizer; that is, chemotherapy enhances the action of radiation therapy. Here chemotherapy is used during radiotherapy and it potentiates the actions of radiotherapy, that is, increases radiation-induced cell killing. This is called *concurrent chemotherapy*. The examples are glioblastoma multiforme, medulloblastoma, nasopharyngeal carcinoma, head and neck cancers, esophageal carcinoma, lung carcinoma, carcinoma rectum and anal canal, cervical carcinoma, etc.

The other use of chemotherapy is to alleviate the symptoms of advanced stage of cancer. This is called *palliative chemotherapy*, where the aim of chemotherapy is only to decrease the distressing symptoms of the disease and prolong the quality of life for sometimes by slowing down the progression of the disease.

12.5 Cell Biology and Chemotherapy

Cancer was first scientifically investigated by Sir Percival Pott in 1775. He had noticed that scrotal cancer was frequent in chimney sweeps in the UK and the causative agent might be chimney soot (now known to be tar) (Tannock et al. 1987).

When normal cells of our body are converted to cancer cells, the specific features of cancer are always present. These comprise eight biologic features which are uncontrolled and purposeless cell division, inactivation of growth suppression, resisting cell death, enabling replicative immortality, new blood vessel formation, tendency to spread to other organs, bypassing the effect of body's own immunity and reprogramming energy metabolism. These are called the *hallmark of cancer* which makes its behavior complex. These changes in the cells are the target of chemotherapy (Govindan and DeVita 2009).

Normally human body is made up of approximately 40 trillion cells. Cells are made up of cell membrane (covering of cells), cytoskeleton and nucleus. So, dysfunction of any part of a cell causes either cell death or loss of cell's function. Using this concept, multiple chemotherapeutic agents have been developed. Cell membrane is made up of protein, lipid and multiple cell membrane receptors. These receptors get attached to various molecules like enzymes, hormones, growth factors, proteins, vitamins and minerals for their functions. The action of cytoskeleton is to organize and maintain cell shape and anchor the organelles. The multiple chemical and enzymatic processes also take place in the cytoplasm including cell division.

Inside the nucleus, two types of genetic material are present - deoxyribonucleic acid (DNA) and ribonucleic acid (RNA). The main information is hidden inside the DNA as the genetic code which is the main constituent of nucleus (Hartwell et al. 1999). Almost every function of cells is regulated by these genetic codes. So the primary target of cell destruction is DNA. It has two chains (strands) which are connected to each other by hydrogen bonds between the bases (adenine and thymine or cytosine and guanine) and they are twisted into a double helix. During cell division, one mother cell divides into two daughter cells which lead to growth of that organ/tumor. This is the most vulnerable state for DNA, when it can be broken very easily. There are multiple phases in cell cycles and chemotherapy can be either cell cycle specific or nonspecific.

12.6 Types of Systemic Therapy and Its Functions

It is broadly classified into three categories—*chemotherapy*, *hormonal therapy*, and *biological response modifier* (*targeted therapy*, *immunotherapy*, and *gene therapy*). Chemotherapy is mainly of two types:

Cell cycle nonspecific chemotherapy.
Cell cycle specific chemotherapy.

12.6.1 Cell Cycle Nonspecific Chemotherapy

It is the group of chemotherapeutic drugs that act as anticancer agents at all or any phase of the cell cycle including the resting phase (Mitchison 1979). Alkylating antineoplastic agents and antitumor antibiotics are two examples.

12.6.1.1 Alkylating Agents

Previously, it was best known as *sulfur mustard/mustard gas*, which was used as the chemical weapon in the World War I. This was the first chemical agent tried for cancer treatment. These compounds have highly reactive carbonium ion intermediates which transfer alkyl group to DNA strands. It is called alkylation which results in DNA cross-linking, abnormal base pairing and eventually cell death. Alkylating agents have cytotoxic and radiomimetic (like ionizing radiation) actions (Tripathi 1994). Though it is a cell cycle nonspecific agent, better result is seen in rapidly dividing cells like cancer cells and in normal tissue like bone marrow, hair follicles and mucosal lining. It acts by directly breaking the DNA strands through DNA cross-linking and abnormal base pairing. The examples of *alkylating agents* are as follows:

Mechlorethamine (Mustine HCl) (Hall and Giaccia 2006): First alkylating agent had been approved for multiple hematological malignancies like Hodgkin's disease and T-cell lymphoma. The main side effect is bone marrow suppression.

Cyclophosphamide: FDA approved this for many malignancies but is commonly used for breast carcinoma, non-Hodgkin lymphoma, ovarian carcinoma and testicular cancer. The main side effects are bone marrow suppression and hemorrhagic cystitis (inflammation of urinary bladder), etc.

Ifosfamide: It has molecular similarity with cyclophosphamide. It is used for many types of tumors, including adult sarcomas, lymphoma, Hodgkin disease, breast cancer and ovarian cancer. The side effects are similar to that of cyclophosphamide.

Chlorambucil: It is commonly used in chronic lymphocytic leukemia, low-grade lymphoma, Waldenström macroglobulinemia, multiple myeloma, hairy cell leukemia and rarely in some solid tumors. The main side effect is bone marrow suppression.

Melphalan: Primarily, it is used for multiple myeloma. But it can be used in special situation of malignant melanoma and sarcoma for *regional limb perfusion*. The side effects are like the other alkylating agents, mainly bone marrow suppression.

Busulfan: This is mainly used in hematological cancer like chronic myeloid leukemia, polycythemia vera and very high dose therapy for bone morrow transplant. Primary side effects are bone marrow suppression, pulmonary toxicity, etc.

Carmustine (BCNU): FDA approved it for primary brain tumors, multiple myeloma, Hodgkin disease, and other lymphomas. The main side effect is bone marrow suppression, especially thrombocytopenia (low platelet count).

Lomustine (CCNU): It is chemically similar to carmustine. It has been approved for primary brain tumors and Hodgkin's disease and occasionally used in multiple myeloma, other lymphomas and breast cancer. It has similar side effect like carmustine. These two drugs are used for primary brain tumors, because they are permeable to *blood–brain barrier*.

Dacarbazine (DTIC): It is approved for the treatment of malignant melanoma and Hodgkin disease and also used for adult sarcomas and neuroblastoma. The side effects are bone marrow suppression, severe nausea, vomiting, etc.

Temozolomide: It has molecular similarity with dacarbazine, but it is permeable to blood–brain barrier. The main indications are primary brain tumors like high-grade gliomas and metastatic malignant melanoma.

Bendamustine: It is used for low-grade lymphoma like follicular lymphoma and chronic lymphocytic leukemia. The primary side effect is myelosuppression.

Procarbazine: It has been approved for Hodgkin disease and might also be useful in non-Hodgkin lymphoma, multiple myeloma, brain tumors, melanoma and lung cancer. The side effects are the same as other alkylating agents.

Atypical alkylating agents are those that have similar mechanism of actions, but chemically different from the common alkylating agents. They have platinum molecule with the chemical structure as the active group instead of alkyl group like the other alkylating agents. The examples are cisplatin, carboplatin and oxaliplatin.

Cisplatin: It is used for almost every class of solid tumor and lymphoma. The main side effects are kidney damage, severe nausea, vomiting, etc.

Carboplatin: It is extensively used in testicular cancer, squamous cell cancers of the head and neck, cervical cancer, lung cancer, ovarian cancer, breast cancer, etc. The primary side effect is bone marrow suppression.

Oxaliplatin: It is commonly used in colorectal cancer, stomach and gall bladder cancer and pancreatic cancer. It is a neurotoxic drug.

12.6.1.2 Antitumor Antibiotics

They are also known as cycle nonspecific chemotherapy. These products are obtained from microorganisms and have prominent antitumor activity. Practically, all of them intercalate between DNA strands and interfere with its template function. The drugs under this are as follows:

Actinomycin D: It is used for Wilms tumor, Ewing sarcoma, rhabdomyosarcoma, uterine carcinoma, germ cell tumors, sarcoma botryoides and also for other sarcomas, melanoma, acute myeloid leukemia, ovarian cancer and trophoblastic neoplasms. Bone marrow suppression, nausea and vomiting are the commonly occurring side effects.

Bleomycin: It has been approved for germ cell tumors, Hodgkin's disease, melanoma, ovarian cancer and Kaposi sarcoma. It is also used as a sclerosing agent for malignant pleural or pericardial effusions. The important side effect is pulmonary toxicity.

Doxorubicin: It breaks DNA strands by activating the enzyme topoisomerase II and generates free radicals. The main indications are breast cancer, ovarian cancer, sarcomas, pediatric malignancies and lymphomas. It causes permanent damage to cardiac muscle (congestive heart failure). Other side effects are nausea, vomiting and bone marrow suppression. Radiation-induced skin toxicity increases when it is used concurrently or after radiotherapy. Daunorubicin and mitoxantrone are other analogues of doxorubicin.

Mitomycin C: It is approved for adenocarcinoma, stomach, pancreas and squamous cell carcinoma of anal canal. The other indications are recurrent breast cancer and lung cancer. Common side effects are myelosuppression, nausea, vomiting, etc.

12.6.2 Cell Cycle Specific Chemotherapy

These agents act on the specific phase of cell cycle. For tumors or growth of any organ, existing cells in a living body are to pass through successive cell divisions. Cell cycle is a series of events where cells are divided into daughter cells. When cells enter into cell cycle, it passes through multiple phases. G_0 phase is the resting phase of a cell. In cell cycle, the first phase is G1 which is a checkpoint where it is checked that the cell is ready for the next phase or not. The second phase is synthetic phase or the S phase when the cell contents are increased in amount. The S phase is relatively resistant phase of cell cycle in terms of chemo-sensitivity. The next phase is second checkpoint which is called the G2 phase. Here, the cells are checked for the final stage, that is, mitotic phase/cell division phase or not. The kind of errors are detected and those are removed in this phase before entering into cell division. The next phase is cell division phase or mitotic phase (M phase), when one mother cell is devided into two daughter cells. The G2-M phase is the most chemo-sensitive phase of cell cycle. When chemotherapeutic agents act in those phases of cell cycles and cause cell death, are called cell-cycle specific chemotherapy. Examples of agents under this category are as follows:

Antimetabolites: These drugs act in the S phase of cell cycles. It prevents the synthesis of DNA and arrests the cell cycle. It acts better in highly dividing cells. The examples are antagonists of purine, pyrimidine and folate.

Purine antagonists are 6-marcaptopurine, azathioprine and fludarabine. These groups of drugs are mainly used in hematological cancer and the common side effect is bone marrow suppression.

Pyrimidine antagonists are gemcitabine, 5-fluorouracil, capecitabine, etc. The indications are head and neck cancer, lung cancer, breast cancer, gastrointestinal cancers and ovarian cancer. The most common side effects are bone marrow suppression, diarrhea, mucositis, etc. Folinic acid is used here to potentiate the action of the chemotherapy.

Folate antagonists are methotrexate, pemetrexed and methotrexate. These are used in sarcoma, hematological malignancies, choriocarcinoma, etc. Bone marrow suppression and renal toxicity are the commonly occurring side effects. The pemetrexed is the drug of choice for mesothelioma and adenocarcinoma of lung. Intermittent administration of folic acid is necessary with this chemotherapy to prevent neurotoxicity.

Antimicrotubule agents: These are mitotic inhibitors and mainly act on the microtubule protein called *tubulin* during mitotic phase or M phase of cell cycle. By inhibiting the actions of microtubules, it arrests cell division and eventually results in cell death. There are two types of these agents—vinca alkaloids and taxanes.

Vinca alkaloids: The main action is to prevent its polymerization and assembly of microtubules causing disruption of mitotic spindle and interfere in cell division. Vincristine, vinblastine and vinorelbine are vinca alkaloids. Vincristine is commonly used in Ewing sarcoma, Wilm's tumor, lymphomas and acute lymphoblastic lymphoma, etc. Neurotoxicity is the main side effect. Vinblastine is commonly used in lymphoma and testicular tumors and vinorelbine is used in lung cancer, breast cancer, etc. Neurotoxicity and bone marrow suppression are the common side effects.

Taxanes: It also acts on microtubules, but the actions are opposite to vinca alkaloids though the final result is the same, that is, to arrest the cell division and cause cell death. Here, it causes excessive polymerization of microtubules and prevents its normal functions. The important examples are paclitaxel and docetaxel. The indications are head and neck cancers, esophageal cancer, lung cancer, breast cancer, stomach cancer, ovarian and cervical cancer, etc. The common side effects are bone marrow suppression, mucositis, hair loss, etc.

Topoisomerase inhibitors: It mainly acts in the G2 phase of cell cycle. The topoisomerase I inhibitors are topotecan and irinotecan. The indications are lung cancer, GI cancer, ovarian cancer, etc.

Etoposide: It is a kind of plant alkaloid, not a mitotic inhibitor, but causes arrest of cell cycle by inhibiting topoisomerase II which results in DNA strand breakage. The common indications are lung cancers, stomach cancer, germ cell tumors, and lymphomas. Bone marrow suppression is the main side effect.

There are many other chemotherapeutic agents. Mostly, the chemotherapeutic agents are used in combinations to increase its activity against cancer cells and also to prevent resistance. As multiple drugs act on multiple phases of cell cycles, there are increased chances of cell killing. The combination drugs are chosen in such a way to increase cell destructions without increasing the side effects. The dose is decided optimally to the cytotoxic range (safe maximal tolerable dose) with acceptable toxicities, and it is measured according to either body surface area or body weight of the patient. The tumor grows exponentially and chemotherapy-induced cell killing is proportional to the available dose of chemotherapeutic agents, which is regardless of tumor burden. So, the chance of cure is more when tumor burden is less. It is not possible to cure the disease with single-dose of chemotherapy and sometimes it is difficult with single agent also. But a few cancers with very rapidly growing cells like choriocarcinoma and Burkitt's lymphoma are curable with single-agent chemotherapy. For those tumors with slow to moderately growing cells, combination chemotherapy is useful to eradicate the tumors. Single-agent chemotherapy causes drug resistance and unable to cure the disease.

Goldie–Coldman model (1979) is followed for combination chemotherapy. The strategy is to use as many effective drugs as soon as possible when the tumor size is small. When drugs are used in combination, it is called chemotherapy regimen. It is always better if the chemotherapeutic agents are used in combinations or strict alternating regimen rather than sequential regimen. Because when a single agent is used constantly for sometimes, the cancer cells adapt themselves by doing some mutational changes in the cells and make them ineffective to that particular chemotherapeutic drug. This way chemo-resistance usually develops. These drugs are administered mostly through intravenous (into vein). The main advantage is chemotherapy medicine directly enters into bloodstream. Very few agents are available in oral tablet form and the main advantage is easy to administer and needs less monitoring. Among the intravenous form of chemotherapeutic agents, a few drugs are strongly vesicants; that is, they cause ulceration and fibrosis if they extravasate to surrounding soft tissue of the vein. Chemotherapy is always administered in fixed time interval, which is called *cycle*. This gap is given between cycles for the normal tissue to recover, mainly bone marrow. But the gap should not be too long for the tumor cells to grow again and the administered chemotherapy becomes ineffective. The good news is that during this time interval between chemotherapy cycles, normal cells recover from the damage caused by chemotherapeutic agents but the cancer cells cannot recover because they are already mutated and unable to repair the damage. Usually these intervals are optimum for a particular chemotherapeutic agent. The usual time interval between chemotherapy cycles is 3 weeks, but it could be given at weekly, biweekly or monthly intervals. A few drugs which are available in oral form, unlike intravenous drugs, can be given every day in small doses. As the chemotherapy agents get directly mixed in the bloodstream, it affects almost every organ but the common sites are bone marrow, heart, lung, liver, hair follicles and mucosal lining. The side effects are seen because of the involvement of the abovementioned organs.

To decrease the acute symptoms of chemotherapy, few supportive medicines are used in the form of "premedications" before chemotherapy. These medicines are antiemetics (prevent nausea and vomiting), antiallergic (prevent allergic reactions), diuretics (prevent extra fluid retention) and steroids.

12.6.2.1 Hormonal Agents

The hormones are the biochemical agents that are produced in the glands and act on different organs in the body. This chemical agents control the metabolism, growth of many organs, and reproduction. It has effect on growth and maturity of many organs, such as breast, prostate and uterus. The important hormones are *estrogen* and *progesterone* for breast cancer and uterine cancer (endometrial adenocarcinoma) and *androgens* (*testosterone*) for prostate cancer. There are hormonal receptors on the cells of that particular organ where the respective hormone gets attached; it starts multiple *signaling pathways*, which are responsible for its action. One of the common pathways is signaling for cell proliferation, which sometimes becomes uncontrolled and ends up in hormone sensitive tumor. The hormone receptor positivity is detected by immunohistochemistry (IHC) test of tumor cells, for example, breast cancer and endometrial cancer (Mcleod 2003) for estrogen and progesterone receptors. So there is no role of antihormone therapy, if the tumor cells are negative for hormone receptors. But in the case of prostate cancer, the scenario is completely different. Here, the hormone receptors are never checked on tumor cells, but antihormone therapy is almost always required in any stage of the disease (growth of the prostate gland is entirely regulated by androgens namely testosterone). The antihormonal agents are synthetic or semisynthetic products which are structurally similar to the hormones and bind to the hormonal receptors to stop the hormonal functions. This is a kind of systemic treatment for cancer, and it is also called *endocrine therapy*. The routes of administration for these agents are oral, subcutaneous and intramuscular. It is not a curative treatment when it is used as single modality therapy. In the case of curative treatment, it is always added with other treatment modalities to prevent or delay the recurrence.

It is also used as *palliative treatment in metastatic disease* to delay the disease progression. As it works on the hormonal receptors only, the side effects related to endocrine therapy are minimal. For example, estrogen (female sex hormone) and androgens (male sex hormone) both have similar effects on maintaining bone growth and bone mineral density. So, any antihormone therapy against estrogen or androgen receptors causes low bone mineral density resulting in osteoporosis. The other side effects are metabolic derangement, which is mainly related to lipid metabolism and its distribution causing obesity, sexual dysfunction, cardiac problems, mood swing, hot flashes, etc. These two important hormones are secreted from ovaries and testis respectively. They can be stopped at multiple levels from the start of secretion to the hormonal receptors on tumors/organs. The sex hormone secretion is controlled by another group of hormones secreted from hypothalamus and pituitary gland. The hormones are secreted from hypothalamus in pulsatile manner which stimulate pituitary gland to secret a group of hormones that act on ovaries or testicles to produce sex hormones. Commonly two types of antihormonal agents are used— one acts by stopping the secretion for hormones and another by blocking the hormonal receptor. The hormonal secretion can permanently be stopped by removing the organ that produces hormones, in males' removal of testicles and in females' removal of ovaries.

There is another group of hormones called *steroids* (*corticosteroids*) extensively used in cancer treatment. These agents do not have any anticancer property, but they are extremely useful for palliative treatment to relieve the symptoms and also help to combat chemotherapy-related toxicities.

12.6.2.2 Biological Response Modifiers

There are three main groups of drugs—targeted therapy, immunotherapy and gene therapy. These are called biological response modifiers because they destroy cancer cells by immunostimulation. It is defined by the National Cancer Institute as the *agents or approaches that modify the relationship between tumor and host by modifying the host's biological response to tumor cells with resultant therapeutic effects* (Bisht et al. 2010). These agents act in multiple ways to destroy the cancer cells, such as by blocking the growth factor receptors on cell surface and preventing proliferation of cells or blood vessels (*targeted therapy*). It also works by increasing body's own immunity against cancer cells (*immunotherapy*). Here the treatment is cancer cell-specific; hence, the side effects are minimal. These are the foundations of *precision medicines* in cancer treatment. The nature, behavior of cancer cells and response to treatment vary from patient to patient, even if it is the same form of malignancy because genetically every cancer is different from one another. It is important to understand the genetic and molecular basis of the cancer and decide treatment according to that. These are termed as a *personalized medicine* (Jain 2002).

12.6.2.3 Targeted Therapy

Nowadays, targeted therapy is one of the major modalities of systemic therapy and is an important part of *precision medicines* in cancer treatment. Cancer cells produce multiple molecules due to genetic alteration which have positive effect on uncontrolled cell proliferation and new blood vessels formation. These molecules express on the cell membrane or cytoplasm and regulate multiple signal transduction pathways of cell proliferation. The main targets are these molecules in targeted therapy which inhibit these signaling pathways to stop the tumor growth. So, it is very important to identify those targets on the cancer cells in order to decide the effective anti-neoplastic medicines against those molecules. There are many tests to detect the targets on the cells, such as *immunohistochemistry* (IHC), *fluorescence in situ hybridization* (FISH) and DNA sequencing. Unlike chemotherapy which acts on rapidly growing any cells in the body including normal healthy cells, targeted therapy specifically acts on the cells having particular targets. So the side effects of targeted therapy are much lesser than chemotherapy (Wu et al. 2006).

There are multiple types of targeted therapy; some of them enter inside the cells and act on the targets (e.g., TKIs—imatinib, erlotinib and sorafenib); others act on the cell surface receptors (e.g., monoclonal antibodies—cetuximab, bevacizumab and trastuzumab). The route of administration is either intravenous or oral. Intravenous administration scheduling is the same as that of chemotherapy; the cycles are given commonly at 3-weekly intervals, but weekly, biweekly, or monthly intervals are also the options. Oral drugs are usually taken regularly. Targeted therapy is still in the investigational phase and emerging day by day. Many clinical trials are on going for these drugs developement and few of them get approval for treatment in every year. Though it is very much cancer cell-specific treatment but not a curative one as a single modality treatment. Sometimes it is used as adjuvant/neoadjuvant treatment with other chemotherapy to delay the recurrence (e.g., trastuzumab therapy in HER2 positive breast cancer) or as a part of curative treatment. Many a time, it is used as palliative treatment to delay the progression of the disease and keep the patients free of symptoms (e.g., erlotinib in EGFR positive metastatic adenocarcinoma of lung). Usually the patients tolerate this therapy well and continue this till progression of the disease (in metastatic patients). The commonly occurring side effects are diarrhea, mucositis, rash, healing problem, bleeding, high blood pressure, occasionally cardio toxicity, liver problems, etc.

12.6.2.4 Immunotherapy

It is a type of biologic therapy that helps body's own immune system to fight cancer. Normally it is well known that immune system fights against infection or anything which is foreign to our body but does not attack its own cells. The immune system is made up of *white blood cells, lymphatics, organs and tissues, etc*. The immune system is the unsleeping defender of our body and protects every time from so many external assaults. But if it goes against body's own cells, it can be very harmful which is known as *autoimmune disease*. So there are multiple immune checkpoints which actually control the exacerbated immune reactions against its own cells and also make them familiar to its own cells. Tumors originate in the body from normal cells though they are genetically altered and different from normal cells. Immune system can find those abnormal cells to destroy them and eradicate them from the body. That is why sometimes immune reactions are seen in and around the tumors and these are called *tumor infiltrating lymphocytes* (TILs). It is a sign of the immune system trying to kill the tumor cells (Gonzalez et al. 2018). But most of the time immune system cannot eradicate the cancer cells completely, because malignant cells have some property to escape the immunological attack. They produce some abnormal proteins (called *antigen*) which are detected by immune cells. So cancer cells alter themselves genetically in such a way that those proteins become less visible to the immune cells or sometime immune cells become inactive when they come in contact with cancer cells. The aim of immunotherapy is to make it more powerful to destroy the cancer cells. There are multiple types of immunotherapy, such as immune checkpoint inhibitors (e.g., ipilimumab, pembrolizumab, and atezolizumab), T-cell transfer therapy (CAR T-cell therapy), monoclonal antibody (discussed in the previous paragraph as targeted therapy), cancer treatment vaccine (e.g., sipuleucel T for prostate cancer) and immune system modulator (e.g., cytokines, interleukins and BCG vaccine for urinary bladder). Immunotherapy alone is not a curative therapy. It is usually added with other modalities like surgery, radiotherapy and chemotherapy. As it has emerged recently in the cancer treatment, still it is in the investigational phase. Commonly they are used in palliative situations, because the long-term toxicities are still unknown. The common routes of administration are intravenous, oral, intravesical (into the urinary bladder), etc. There are multiple side effects in immunotherapy, mostly because of excessive immunological reactions, such as diarrhea, skin rash, flu-like syndrome (fever, chill, and weakness), fatigue, joint pain or inflammation of any organ.

12.6.2.5 Gene Therapy

It has been extensively studied for cancer treatment for the last few decades. It is well known that cancer cells result due to genetic alterations (mutation) and there are multiple mutations which control its behavior like uncontrolled proliferation of cells and blood vessels (angiogenesis), escaping natural death of cells (apoptosis) and immune reaction, tendency to spread to other organs (*metastasis*) (Wirth and Ylä-Herttuala 2014). The aim of gene therapy is to replace those abnormal parts of gene either by normal gene or by genes with counteracting the cancer cell's behavior, such as by increasing immunity against cancer cells, inhibiting the proliferation of blood vessels and activating cell death process. Though theoretically it is very much appealing, practically it is very difficult to apply for treatment. Multiple research projects are going on regarding gene therapy and some of them show few satisfactory results.

12.7 Drug Resistance

Cancer cells have sometimes inherent or acquired nature to escape the effect of cytotoxic drugs, that is called chemoresistance (Alfarouk et al. 2015). This is very dangerous situation when chemotherapeutic medicines cannot kill the malignant cells and the disease is progressed eventually. This resistance can arise due to multiple factors as well as multiple levels. The commonest cause is mutation in the tumor cells which become resistant to chemotherapy in many ways. The changes can occur at initial steps like absorption, distribution or metabolism of the cytotoxic drugs. Mutations in the tumor cells can cause changes in the drug receptors on cell membrane or the cells may pump out the drugs to decrease its activity. The drug resistance is more when single agent is used for the treatment. One of the initial efforts to overcome resistance was use of combination chemotherapy. Nowadays, many researches are going on to combat the drug resistance including development of new chemotherapeutic drugs.

12.8 Conclusion

The systemic therapy is one of the main weapons to fight cancer. The concept of medical oncology has come into picture many years ago. Chemotherapy, hormonal therapy and targeted therapy have been established for the treatment of cancer. Immunotherapy and gene therapy are still in the phase of investigation. Many new molecules are getting approval from the USFDA every year and changing the treatment protocol for cancer.

Medical Notes
Autoimmune disease: Body's own immune system attacks and destroys its own healthy tissue by mistake and causes dysfunction of that particular organ.
Burkitt's lymphoma: type of blood malignancy (lymphoma).
Ewing sarcoma: type of bone cancer (sarcoma).
Hodgkin's disease: type of blood malignancy (lymphoma).
Kaposi sarcoma: type of soft tissue malignancy occurred in immune-compromised (e.g., HIV-positive) patient. It is an AIDS-defining disease.
Wilm's tumor: type of childhood kidney tumor.
Waldenström macroglobulinemia: type of hematological malignancy causing excessive production of abnormal immuneglobulin.

References

Alfarouk KO, Stock CM, Taylor S, Walsh M, Muddathir AK, Verduzco D, Bashir AH, Mohammed OY, Elhassan GO, Harguindey S, Reshkin SJ (2015) Resistance to cancer chemotherapy: failure in drug response from ADME to P-gp. Cancer Cell Int 15(1):1–13

Bisht M, Bist SS, Dhasmana DC (2010) Biological response modifiers: current use and future prospects in cancer therapy. Indian J Cancer 47(4):443

Chabner BA, Roberts TG (2005) Chemotherapy and the war on cancer. Nat Rev Cancer 5(1):65–72

Chu E, Sartorelli AC (2004) Cancer chemotherapy. Basic Clin Pharmacol 9:898–930

DeVita VT, Chu E (2008) A history of cancer chemotherapy. Cancer Res 68(21):8643–8653

Goldie JH, Coldman AJ (1979) A mathematic model for relating the drug sensitivity of tumors to their spontaneous mutation rate. Cancer Treat Rep 63(11–12):1727–1733

Gonzalez H, Hagerling C, Werb Z (2018) Roles of the immune system in cancer: from tumor initiation to metastatic progression. Genes Dev 32(19–20):1267–1284

Govindan R, DeVita VT (eds) (2009) DeVita, Hellman, and Rosenberg's cancer: principles & practice of oncology review. Lippincott Williams & Wilkins

Hall EJ, Giaccia AJ (2006) Radiobiology for the radiologist (vol. 6)

Hartwell LH, Hopfield JJ, Leibler S, Murray AW (1999) From molecular to modular cell biology. Nature 402(6761):C47–C52

Jain KK (2002) Personalized medicine. Curr Opin Mol Ther 4(6):548–558

Mcleod DG (2003) Hormonal therapy: historical perspective to future directions. Urology 61(2):3–7. (Mcleod, 2003)

Mitchison DA (1979) Basic mechanisms of chemotherapy. Chest 76(6):771–780

Nagai H, Kim YH (2017) Cancer prevention from the perspective of global cancer burden patterns. J Thorac Dis 9(3):448

Sedgwick P (2011) Phases of clinical trials. BMJ 343

Tannock IF, Hill RP, Bristow RG, Harrington L (eds) (1987) The basic science of oncology. Pergamon Press, New York, p 398

Tripathi KD (1994) Essentials of medical pharmacology. Indian J Pharmacol 26(2):166

Wirth T, Ylä-Herttuala S (2014) Gene therapy used in cancer treatment. Biomedicine 2(2):149–162. https://doi.org/10.3390/biomedicines2020149

Wu HC, Chang DK, Huang CT (2006) Targeted therapy for cancer. J Cancer Mol 2(2):57–66

Chemotherapy Effects on Immune System

Debasish Hota and Amruta Tripathy

Abstract

The primary goal of conventional anticancer drugs always has been killing the tumor cells in the body. Cancer growth starts with certain irreversible changes within the immune system, and this in turn gave rise to the concept that tumor growth can be halted only by destroying the malignant cells. However, with the advent of targeted chemotherapeutic agents, drugs that specifically suppress cells promoting tumor growth and stimulate the immune system to act against the cancer cells, the management of cancers took a novel approach. Various conventional chemotherapeutic agents also have showed immune modulatory effects in recently conducted studies. Thus, targeting the immune system rather than the cancer cell seems to be a new and better perspective in cancer treatment.

Keywords

Chemotherapy · Immune system · Immune surveillance · Chemotherapeutic agents

13.1 Introduction

Cancers have been known to develop due to an imbalance of genes which promote development of tumor cells (proto-oncogenes) and genes which suppress the growth of tumor cells (tumor suppressor genes) (Hanahan and Weinberg 2000). In the physiological state, a delicate balance between these two prevents cancerous transformation of vulnerable cell in the body. There are many triggers that can lead to disruption of this balance ranging from sudden changes in the gene structure (genetic

D. Hota (✉) · A. Tripathy
Department of Pharmacology, All India Institute of Medical Sciences, Bhubaneswar, Odisha, India

© The Author(s), under exclusive license to Springer Nature Singapore Pte Ltd. 2022
S. K. Basu et al. (eds.), *Cancer Diagnostics and Therapeutics*,
https://doi.org/10.1007/978-981-16-4752-9_13

mutations) or certain infections like human papillomavirus causing cervical cancer or exposure of the body to certain substances or chemicals that can lead to subsequent development of cancers (carcinogens) like cigarette smoke.

From the belief that no child with leukemia has ever been cured (Walker 1964) to the use of cell lines and targeted drugs for curing cancers, cancer chemotherapy has traversed great leaps. Surgery and radiotherapy were considered the mainstay in treating cancers up to the 1960s. However, cure rates plateaued despite development of more and more localized surgeries (DeVita and Chu 2008). As knowledge about behavior and properties of cancer cells expanded, it became clear that no single treatment modality would be sufficient for treating cancers. Thus, the management of malignancies gradually started involving interplay of surgery, radiotherapy, and chemotherapy.

13.2 The Immune System and Cancer

As the human body is exposed to a large number of bacteria, viruses, fungi, parasites, and other antigens on a daily basis, the immune system is responsible for the overall protection of the body. Immune response is the reaction of the above system toward averting any infection that can be detrimental to the body (Chaplin 2010). The immune system is of paramount importance in cancer patients because:

The disease *per se* can affect the immune system by destroying the bone marrow responsible for production of immune cells in the body.

Agents aiding in cancer treatment can in turn destroy the bone marrow, thereby weakening the immune system of the body.

A stronger immune system can forge a better fight against any kind of cancer.

White blood cells (WBCs) are responsible for development of immunity in the body. They are produced in the bone marrow. Cancer: Once it affects the bone marrow, or cancer drugs used to treat the disease can suppress the normal functioning of the bone marrow, thereby inhibiting proper functioning of WBCs. However, bone marrow suppression caused by antitumor agents is reversible most of the times and gets corrected once the drug is discontinued.

There are two kinds of immunity as depicted Fig. 13.1.
Innate immunity—immunity present since birth;
Acquired immunity—immunity that develops during growth of the person.

Innate immunity can be provided by anything ranging from the skin, acid in the stomach, and hair to neutrophils (a specific kind of WBCs) (Chaplin 2010). The components of innate immunity are always ready to attack any intruder into the body. They are the first-line defense against any infection. Acquired immunity, on the other hand, is something the body learns and slowly develops. The learning happens either because of a previous infection or vaccination. Once a particular infection is introduced into the body, the body remembers to recognize it the next

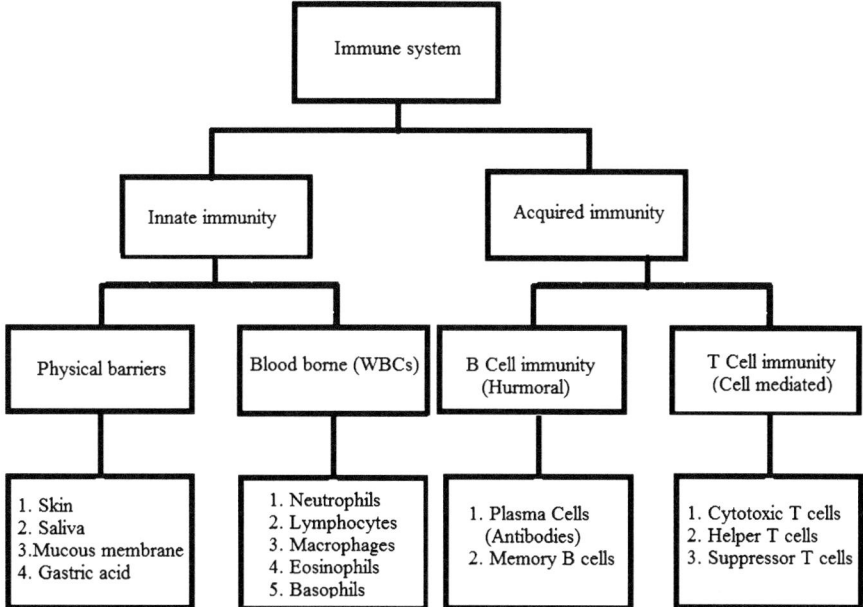

Fig. 13.1 Immune system

time a similar infection occurs and reacts in a timely manner to stop the infection from spreading.

Skin, tear, mucosa of the eyes, nose, and mouth form a shield against entry of possible germs or toxic substances into the body. Cancers or anticancer drugs can breach the innate immunity and thus make the body more susceptible to various infections. Once the physical barrier is breached and the germ or toxic substance enters into the body, the body deals with them through the WBCs, antibodies, and complement system present inside the body. They can either be already present within the body or are produced once the harmful substances enter the body.

Lymphatic system is the next line of defense that faces the infective organisms. It consists of lymph nodes present ubiquitously throughout the body. Lymph nodes secrete lymph, a clear fluid which essentially is blood plasma minus the red and white blood cells. The germs on entering the cells are filtered and drained by the lymphatic fluid, and they eventually make their way to the lymph nodes (Chaplin 2010). Swollen lymph nodes that can be felt during an infection or fever are thus a good sign of the body fighting the infectious organism. Conversely, swollen nodes in a cancer patient might indicate the disease progression to the next stage.

Organisms that escape the lymphatic system are then attacked by the next line of defense, the antibodies. Antibodies are produced in the WBCs and are highly specific; a particular kind of toxin or organism (antigen) has a specific type of antibody acting against it. They are also called immunoglobulins and are of 5 types—IgG, IgM, IgA, IgE, and IgD. Besides antibodies, the body produces a vast array of compounds like interleukins (ILs) and tumor necrosis factor (TNF) in the neutrophils (Chaplin 2010) and interferons that are part of the immune system and are circulated throughout the body in the eventuality of any kind of infection.

White blood cells or WBCs are produced in the bone marrow from stem cells and are one of the primary components of the immune system. There are three types of WBCs in the body—granulocytes (further divided into neutrophils, basophils, and eosinophils), lymphocytes (subdivided into B cells formed in the bone marrow and T cells formed in the thymus), and monocytes which can form macrophages.

Neutrophils: These are the most abundant type of WBCs seen in the circulation. They have a very short life span (few hours). In case of a cut, scratch, or aggregation of any kind of infective organisms within the body (like urinary or respiratory tract infections or skin infections), neutrophils are the first cells which are recruited in the concerned area. Once neutrophils infiltrate the wound or the infected tissue, they engulf the infective organism by a process called phagocytosis (Kennedy and DeLeo 2009) and release further enzymes to prevent the spread of infection into the surrounding tissues.

Eosinophils and basophils: They are much less abundant than neutrophils in the circulation. Eosinophils are predominantly associated with protecting the body against parasitic infections (Chaplin 2010). Basophils are associated with release of histamine, a substance released in the eventuality of an allergy of any kind.

Lymphocytes: They are responsible for protecting the body against majority of the bacterial and viral infections. Although production of lymphocytes starts in the stem cells as all other WBCs, the differentiation either completes in the bone marrow itself or spleen (B cells) or in the thymus (T cells). B cells further divide into plasma cells which then go on to produce antibodies that eliminate various foreign objects that enter into the circulation. Apart from helping in antibody production, B cells also have a memory of infective agents that have invaded the body previously so that future infections can be prevented. T cells are of three types—killer cells that kill various infected cells and prevent the spread of infection or tumor inside the body, helper cells that aid in the production of antibodies by the B cells, and suppressor cells that help in controlling the overall immune response by B cells and killer T cells.

Monocytes: Once formed in the bone marrow, they enter into the cells of the body and get converted into macrophages, which as the name suggests are the largest of all WBCs. They are omnipresent throughout the body and are responsible for clearing up the debris after neutrophils have phagocytosed the infective organisms. Macrophages persist for longer duration of time at the site of inflammation and thus play an important role in chronic inflammation (Chaplin 2010).

Major histocompatibility complex (MHC) (also known as the human leukocyte antigen (HLA)) are the molecules that help the B and T cells to distinguish between the normal and affected cells of the body (Chaplin 2010). The MHC molecules express a small component of the infective organism or the toxin or the tumor cell on the surface of the host cell, thereby prompting the immune cells to attack the affected cells of the body and kill them. Whenever there is a malfunction of this component of the immune system, it turns the B and T cells against the normal cells of the body leading to a wide variety of autoimmune diseases.

The immune system plays a major role both in disease progression and in treatment of various cancers. Progression of cancer in human body is attributed to failure of the immune system to check the process of multiplication of tumor cells from the initial stages. Treatment of cancer therefore, quite obviously, involves the use of certain medications which will activate the immune system of the body which will, in turn, mount a defense against the cancer cells and kill them. However, certain cancer chemotherapeutic agents also tend to suppress the immune system by destroying bone marrow which in turn blocks synthesis of various components of the immune system.

13.3 Anticancer Immune Surveillance

According to the immune surveillance theory, initiation and progress of any cancer eventually happen due to a failing immune system (Schreiber et al. 2011; Zitvogel et al. 2006). This makes boosting the immune system an important goal in treating cancers. The gradual increase in the use of specific agents that modulate the immune system, targeted agents, in cancer treatment has further strengthened the cause of reinstating a full-fledged immune system in a cancer patient. The baseline composition and function of various components of the immune system is a major prognostic marker in cancer patients (Fridman et al. 2012). However, studies in various animal models have also suggested immune system acts as a double-edged sword when it comes to cancer. Certain components of the immune system are protective, whereas there are few immunosuppressive components as well which prevent the killing of cancer cells. This immune surveillance is called immunoediting (Schreiber et al. 2011) and consists of three phases:

Elimination phase: Destruction of the cancer cells by the immune system.
Equilibrium phase: Production of cancer cells surpasses the destruction creating a pressure on the immune system. This also leads to genetic changes in the cancer cells (editing).
Escape phase: The edited cancer cells are resistant to the immune system and proliferate in the body leading to growth of the cancer.

The first two phases show an abundance of B and T cells in the cancer tissues, whereas in the third phase, there is a sudden paucity in numbers of these immune cells. This sudden shift shows the capability of cancer cells in not only generating their own immunosuppressive molecules but also attracting host immunosuppressive cells, thereby immobilizing the entire immune system.

13.4 Clinical Scenario for Use of Chemotherapeutic Agents

Drugs that have been used as conventional chemotherapeutic agents are those that destroy the tumor cells with or without any effect on the immune system. The reasons for use of chemotherapeutic agents may be treatment of blood cancers, treatment in advanced cases once the cancer has spread to other organs (metastasis), and for reducing the size of the tumor before exposing it for surgery or radiotherapy.

Primary chemotherapy: It refers to the use of chemotherapeutic agents as the main modality of treatment when no other alternative exists. This is commonly used in case of blood cancers or advanced stages of any cancer when the aim of treatment is to prolong the life of the patient. It is primarily used in leukemia and lymphomas, and childhood tumors like Wilm's tumor.

Neoadjuvant chemotherapy: It refers to use of chemotherapeutic agents in those cancers where alternative therapy like surgery is the mainstay of treatment but is less efficacious when used singly ("How Is Chemotherapy Used to Treat Cancer?" 2016). Initially, cancer chemotherapy is administered for a short period which helps in reducing the bulk of the tumor. This is subsequently followed by surgery or radiotherapy. After the definitive procedure, chemotherapy is again continued for another 3–4 cycles for better control. Neoadjuvant chemotherapy has been found to be highly effective in cancers like bladder cancer, breast cancer, nonsmall cell type of lung cancer, and laryngeal cancer.

Adjuvant chemotherapy: The main aim of chemotherapy in this setting is to prevent the recurrence of the tumor and improve the overall survival of the patient ("How Is Chemotherapy Used to Treat Cancer?" 2016). Adjuvant chemotherapy is usually administered after surgical removal of the tumor or following radiotherapy. It has been used with great success in breast cancer, colon cancer, and gastric cancer among others.

13.5 Conventional Cancer Chemotherapeutic Agents

The logical goal in treating cancers is destruction of the tumor cells. This has been traditionally achieved by the use of cytotoxic agents which, as the name suggests, are toxic to the rapidly progressing cancer cells. Based on their principal mechanism of action, conventional cancer chemotherapy drugs can be broadly subdivided into (Galluzzi et al. 2015):

Alkylating agents: They contain an alkyl group that attaches to the DNA of the malignant cells and prevents the replication of the same (e.g., cyclophosphamide).

Platinum analogs: Platinum complex binds to the DNA of the tumor cells, thereby preventing the replication of these cells and subsequently leading to the death of the same (e.g., cisplatin).

Antimetabolites: They act by interfering with the DNA and RNA synthesis in the cells, thereby restricting the growth and multiplication of cancer cells (e.g., 5-fluorouracil [5-FU]).

Topoisomerase inhibitors: Topoisomerases are enzymes that help in separating the double-stranded structure of DNA before replication and then rejoining of the replicated strands of DNA to form the double strand again. Inhibition of these enzymes leads to DNA damage, thereby causing death of the cancer cells (e.g., irinotecan).

Microtubular poisons: They bind to the cancer cells during mitosis or cell division and thereby prevent replication of the same (e.g., paclitaxel).

Cytotoxic antibiotics: They exert their action either by deranging DNA synthesis by preventing action of topoisomerase enzyme or by generating free radicals which further increase the killing of tumor cells (e.g., bleomycin).

Conventional chemotherapeutic agents, albeit predominantly kill the cancer cells, have been found to exert some action on the cells of the immune system as well. For example, taxane-based neoadjuvant therapy in breast cancer patients has been shown to increase the immune response in the body and thereby increase the overall survival of the patient (Issa-Nummer et al. 2013; Senovilla et al. 2012).

Increase in CD20 (a marker for showing B-cell activity) levels in biliary tract (Goeppert et al. 2013) and colorectal cancers (Kasajima et al. 2010), increase in CD4 (a marker for showing T-cell activity) levels in malignant melanoma (Mignot et al. 2014), increase in CD3 and CD8 (markers for showing T-cell activity) levels in ovarian cancer (Han et al. 2008), and increase in CD68 (a marker for showing macrophage activity) levels in gastric (Wang et al. 2011) and pancreatic cancers (Di Caro et al. 2016) have been associated with overall prolonged survival after cancer chemotherapy. High levels of B and T cells have been associated with poor prognosis as well. Increase in CD20 levels in malignant melanoma (Neagu et al. 2013), increase in CD68 levels in breast cancer (DeNardo et al. 2011), and increase in CD138 (a marker for plasma cell activity) levels in breast cancer (Mohammed et al. 2013) and malignant melanoma (Neagu et al. 2013) have been associated with low survival and bad prognosis.

13.6 Immunological Effects of Commonly Used Cancer Chemotherapeutic Drugs

Conventional anticancer drugs have been shown to help in mounting a robust immunological attack against cancer cells in both clinical and animal studies. The effects of these agents on immune system have been shown in Table 13.1.

However, not all anticancer agents produce a favorable immune response. Most of the anticancer drugs, when given at the highest dose that is safe, produce immunosuppression leading to destruction of bone marrow and reduced number of WBCs in the body. Sometimes, combination anticancer therapy has produced immune stimulatory responses. This might be attributed to the lower toxicity of the individual drug as the dose is reduced when given in a combination therapy with other anticancer agents.

Table 13.1 Immunological effects of conventional anticancer drugs

Drug	Effect on immune system
Bleomycin	Promotes immunological cell death of tumor cells and stimulates action of T helper cells (Bugaut et al. 2013)
Cyclophosphamide	Restores T-cell function (Gershan et al. 2015; Ghiringhelli et al. 2007) Increases macrophage function (Wu and Waxman 2015) Expands the production of stem cells in bone marrow (Ding et al. 2014)
Docetaxel	Favors immunosurveillance (Senovilla et al. 2012, Valent et al. 2013)
Doxorubicin	Promotes immunological cell death of tumor cells (Casares et al. 2005) Favors the expansion of stem cells in bone marrow (Ding et al. 2014)
5-FU	Increases infiltration of tumor cells by T lymphocytes (Lim et al. 2014) Increases expression of MHCs on cancer cells (Khallouf et al. 2012) Favors stem cell differentiation in bone marrow when given as a combination regimen in colorectal cancer (Kanterman et al. 2014)
Gemcitabine	Increases circulating monocytes (Soeda et al. 2009) Improves recruitment of natural killer cells (Liu et al. 2010; Xu et al. 2011) Enables action of targeted chemotherapeutic agents (Sawant et al. 2013)
Irinotecan	Increases stem cell production in the bone marrow when given as combination therapy in colorectal cancer (Kanterman et al. 2014)
Oxaliplatin	Promotes T-cell-dependent immune reaction to cancer cells (Shalapour et al. 2015) Promotes immunological death of cancer cells (Martins et al. 2011) Promotes activity of neutrophils and macrophages on cancer cells (Iida et al. 2013)
Paclitaxel	Promotes tumor infiltration by macrophages (DeNardo et al. 2011) Promotes tumor infiltration by T lymphocytes (Demaria et al. 2001)

13.6.1 Other Uses of Cancer Chemotherapeutic Agents

Considering the multitude effect of anticancer agents on the immune system, these drugs have been used in various other autoimmune disorders, wherein the immune system of the host produces antibodies against its own cells and destroys them. Rheumatoid arthritis is one such autoimmune condition where anticancer drugs have been used to halt the progression of the disease by inhibiting the action of the cells of the immune system on various joints. Cyclophosphamide, methotrexate, and thalidomide have been used routinely in the treatment of rheumatoid arthritis ("Chemotherapy Drugs Used to Treat Arthritis" 2016). Methotrexate has also been used in psoriasis, an autoimmune skin condition, with good results. Cyclophosphamide has been used in other autoimmune conditions like Wegener's granulomatosis and nephrotic syndrome. Thalidomide is one of the drug of choice in case of type I lepra reaction, an autoimmune complication seen in leprosy ("Thalidomide: Research advances in cancer and other conditions" 2016). Anticancer drugs have also been used as radiosensitizers. Radiosensitizers are agents that sensitize the tumor cells to radiation therapy, thereby maximizing the effects of radiation therapy in the treatment of various cancers (Raviraj et al. 2014). Anticancer agents that have been used as radiosensitizers include 5-FU, gemcitabine, fludarabine, paclitaxel, docetaxel, and irinotecan.

13.7 Toxicity of Conventional Anticancer Drugs on Immune System

Bone marrow suppression: This is one of the commonest side effects seen with antitumor agents. There is an overall reduction in the production of RBCs, WBCs, and platelets in the bone marrows. Bone marrow suppression manifests as anemia (reduction in the number of RBCs), increased susceptibility to infections (reduction in the number of WBCs), and increased bleeding tendencies (reduced number of platelets). This side effect does not manifest immediately because the already formed blood products (RBCs, WBCs, and platelets) have to be consumed first before the manifestations of bone marrow suppression are seen. WBCs are affected the most since they have a very short life span of 6–12 h and the reserves thus get used up very early. RBCs, with a life span of 120 days, are affected the least by bone marrow suppression. The full-blown effects of bone marrow suppression are normally evident by 10–14 days with most of the chemotherapeutic agents.

However, the onset is delayed with certain drugs like nitrosoureas and mitomycin (Medina and Fausel 2008). Onset of fever during the chemotherapy course is the commonest symptom suggestive of bone marrow suppression. The drugs might have to be temporarily stopped, or the dosage reduced till the blood counts come back to the normal range. However, in certain cases, reduction of the drug dosage might lead to bad prognosis of the disease. Hence, a judicious decision has to be taken. Blood transfusion used to be the most common treatment option for correcting anemia. However, the current treatment guidelines recommend the use of epoetin alfa or darbepoetin alfa, which are erythropoietic agents (increase the production of RBCs in the bone marrow). Iron supplementation is necessary for further optimizing RBC production. Since fever, which is a sign of infection, is the most common manifestation of reduction in the number of WBCs, treatment usually consists of antibiotics to resolve the underlying infection. However, in some cases, when the WBC number falls down to dangerously low levels, colony-stimulating factors (CSFs), which increase the production of WBCs in the bone marrow, can be used. Reduction in the number of platelets is managed by platelet transfusion to prevent any episode of bleeding.

Secondary malignancies: Drugs like alkylating agents, anthracyclines, and etoposide have been associated with causing this long-term serious complication. Blood cancer is the most common secondary malignancy seen in majority of cancer patients (Medina and Fausel 2008).

13.8 Targeted Cancer Chemotherapy

Conventional cancer chemotherapeutic agents acted as a double-edged sword. On the one hand, they killed the cancer cells in the body, whereas they also had a detrimental effect on the normal cells of the body. This led to the cascade of side effects seen with these drugs, immunosuppression, anemia, and bleeding episodes being the most common ones. As cancer research expanded, scientists came up with

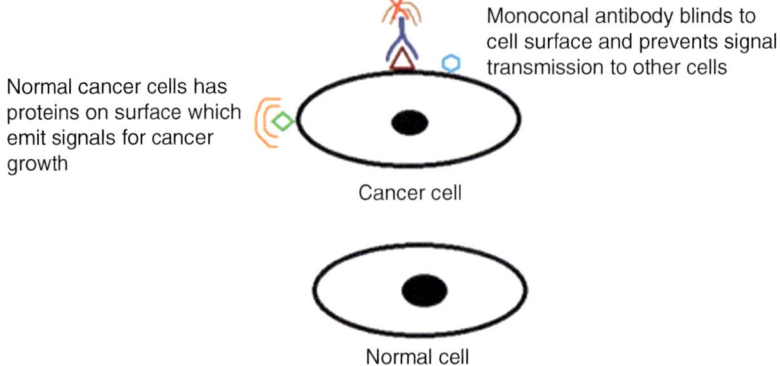

Fig. 13.2 Mechanism of action of monoclonal antibodies

more and more cellular features and metabolic pathways that were unique to the tumor cells. This helped in developing drugs which specifically targeted tumor cells only with minimal or no effect on the normal body cells. Targeted cancer chemotherapy is thus defined as a special type of cancer chemotherapy which takes advantage of the difference between cancerous cells and normal cells of the body ("What Is Targeted Cancer Therapy?" 2016). It is used along with other antitumor agents or as an adjunct to surgery or radiotherapy.

Cancer cells differ from the normal cells in certain perspectives:

They divide and multiply more rapidly than the normal cells of the body.

Certain genetic changes (mutations) help in converting a normal cell to a cancerous cell.

Tumor cells then generate certain signals that further aids in the rapid growth and multiplication of other cancerous cells.

As cancer cells are growing at a more than normal rate, their requirement for nutrients increases many folds. This is met by an increase in the blood vessels around the cancer cells which facilitate the provision of nutrients to the cancer cells.

Targeted cancer chemotherapeutic agents disrupt any of these processes, or they can trigger the natural defenses of the body to kill the cancerous cells. These targeted cancer chemotherapeutic agents are also known as monoclonal antibodies (Nelson et al. 2000). These drugs have features akin to the antibodies present naturally in the body. They have selective action against certain types of cancer cells or certain proteins which are expressed on the surface of cancer cells specifically. Monoclonal antibodies either kill the cancer cells directly or in most cases prevent generation of signals by cancer cells which is necessary for growth of cancer cells. This has been represented in Fig. 13.2.

Rituximab (a monoclonal antibody against CD20) was the first among these targeted therapies which was approved for use in cancer chemotherapy (Medina and Fausel 2008). It is a monoclonal antibody which activates the B cells, responsible for producing immunity in the body. Once these B cells are activated, they attack the tumor cells and kill them leading to reduction in tumor size. Tumors express a wide range of growth factors which are important for the growth and multiplication of tumor cells, notable among them being epithelial growth factor and vascular endothelial growth factor. Drugs like cetuximab, trastuzumab, erlotinib, and gefitinib inhibit various epithelial growth factors and therefore block signaling pathways associated with growth and development of cancer cells. The development of new blood vessels around tumor cells is of paramount importance for delivery of nutrients to the cancer cells which in turn helps in the rapid multiplication of these cells. Vascular endothelial growth factors aid in new blood vessel formation, and drugs like bevacizumab, sunitinib, and sorafenib inhibit these factors. In chronic myeloid leukemia (CML), a variant of blood cancer, there is a specific mutation that leads to breakage in two chromosomes and the resealing of the broken parts (BCR-ABL mutation) leads to the formation of Philadelphia chromosome. Imatinib is a targeted chemotherapeutic agent that specifically blocks the action of this Philadelphia chromosome and thus prevents the spread of CML (Medina and Fausel 2008). Another group of drugs, bortezomib and carfilzomib, inhibit the protein synthesis inside the tumor cells. Since proteins are required for the normal functioning of all cells, inhibition of protein synthesis leads to death of cancer cells.

Targeted cancer chemotherapeutic agents have been used in a wide variety of cancers like leukemia, solid organ tumors like colorectal cancer, lung cancer, breast cancer, head and neck cancers, and pancreatic cancer (Medina and Fausel 2008). These agents have become the first-line therapy as part of a combination therapy in various kinds of cancers albeit high cost of the drug still remains a major concern.

Although targeted cancer chemotherapy heralded a new era in the treatment of cancers, they are not without their share of adverse effects. The most common side effect seen with these agents is hypersensitivity seen during injecting the drug into the body. Chances of allergic reactions are highest with the first dose of the drug. Hence, before starting intravenous injections of these chemotherapeutic agents, it is preferred to administer antiallergic medications like diphenhydramine to all the patients. The injection should be given at a very slow rate, and the patient should be observed thereafter for any signs of allergy like fever, rash, itching, swelling of eyes, and lips. Medications inhibiting vascular endothelial growth factors can increase the blood pressure. Diarrhea is a common side effect seen with drugs like imatinib (Medina and Fausel 2008).

13.9 Immunological Effects of Targeted Anticancer Agents

Monoclonal antibodies have more precise and accurate effects on the immune system as compared to the conventional anticancer agents. Effects of these targeted chemotherapeutic drugs have been demonstrated in combination therapies only. These have been briefly described in Table 13.2.

Table 13.2 Immunological effects of certain targeted anticancer drugs

Drug	Effect
Bevacizumab	Reduces T cells within the tumor, thereby preventing growth of the tumor (Terme et al. 2013)
Erlotinib	Increases susceptibility of cancer cells toward natural killer cells leading to increased cell death (Kim et al. 2011)
Gefitinib	Increases susceptibility of cancer cells toward natural killer cells leading to increased cell death (He et al. 2013; Kim et al. 2011)
Imatinib	Promotes circulation of natural killer cells that kill the cancer cells in CML (Mizoguchi et al. 2013) Promotes the production of BCR-ABL1-specific cytotoxic T lymphocytes in CML (Riva et al. 2014) Increases production of interferon γ that kills cancer cells (Ménard et al. 2009) Enhances destruction of cancer cells by antibodies (Murray et al. 2014)
Sorafenib	Increases accumulation of natural killer cells in the cancer tissues (Romero et al. 2014)
Sunitinib	Increases the cytotoxic T-cell levels in tumor tissues (Roselli et al. 2013)
Vorinostat	Increases levels of interferon-γ-producing T lymphocytes and promotes killing of cancer cells by plasma cells (West et al. 2013)

13.10 Conclusion

Conventional cancer therapy always has been directed at killing the tumor cells, and this in turn has caused notable adverse effects to the normal cells of the body as well. Traditionally cancer chemotherapy has always been associated with a decrease in the quality of life of the cancer patients due to the inadvertent immunological side effects associated with the use of these drugs. Targeted cancer chemotherapeutic drugs thus proved to be an invaluable weapon in the treatment of various types of cancers with clinically significant results. However, as more proof gathers regarding the modulation of the immune system by both conventional and targeted anticancer agents, treatment modalities in various cancers are bound to go through a major change.

As more evidence regarding the immune stimulatory role of conventional anticancer agents like methotrexate, cyclophosphamide, bleomycin, and cisplatin comes to light, the combination of these agents with immune modulatory drugs seems a much safer and more effective treatment plan. Combination can either be based on the principle of: (a) counteracting the immunosuppressant action of anticancer drugs by targeted agents or (b) boosting the immune stimulatory action of conventional anticancer agents.

Overall survival in cancer has tremendously increased due to the emergence of various new treatment modalities. The development of the concept of modulating the immune system to control the growth rate of tumors instead of directly killing the cancer cells has thus paved the way for use of both the conventional and newer cancer chemotherapeutic agents as combination therapies. The optimal use of these combination therapies, their long-term benefits, and adverse effects on cancer patients, however, are few of the important questions that still remain unanswered.

References

Bugaut H, Bruchard M, Berger H, Derangère V, Odoul L, Euvrard R, Ladoire S, Chalmin F, Végran F, Rébé C, Apetoh L (2013) Bleomycin exerts ambivalent antitumor immune effect by triggering both immunogenic cell death and proliferation of regulatory T cells. PLoS One 8(6): e65181

Casares N, Pequignot MO, Tesniere A, Ghiringhelli F, Roux S, Chaput N, Schmitt E, Hamai A, Hervas-Stubbs S, Obeid M, Coutant F (2005) Caspase-dependent immunogenicity of doxorubicin-induced tumor cell death. J Exp Med 202(12):1691–1701

Chaplin DD (2010) Overview of the immune response. J Allergy Clin Immunol 125(2):S3–S23

Chemotherapy Drugs Used to Treat Arthritis (2016). Retrieved from: https://www.webmd.com/arthritis/chemotherapy-drugs#1

Demaria S, Volm MD, Shapiro RL, Yee HT, Oratz R, Formenti SC, Muggia F, Symmans WF (2001) Development of tumor-infiltrating lymphocytes in breast cancer after neoadjuvant paclitaxel chemotherapy. Clin Cancer Res 7(10):3025–3030

DeNardo DG, Brennan DJ, Rexhepaj E, Ruffell B, Shiao SL, Madden SF, Gallagher WM, Wadhwani N, Keil SD, Junaid SA, Rugo HS (2011) Leukocyte complexity predicts breast cancer survival and functionally regulates response to chemotherapy. Cancer Discov 1(1):54–67

DeVita VT, Chu E (2008) A history of cancer chemotherapy. Cancer Res 68(21):8643–8653

Di Caro G, Cortese N, Castino GF, Grizzi F, Gavazzi F, Ridolfi C, Capretti G, Mineri R, Todoric J, Zerbi A, Allavena P (2016) Dual prognostic significance of tumor-associated macrophages in human pancreatic adenocarcinoma treated or untreated with chemotherapy. Gut 65 (10):1710–1720

Ding ZC, Lu X, Yu M, Lemos H, Huang L, Chandler P, Liu K, Walters M, Krasinski A, Mack M, Blazar BR (2014) Immunosuppressive myeloid cells induced by chemotherapy attenuate antitumor CD4+ T-cell responses through the PD-1–PD-L1 axis. Cancer Res 74(13):3441–3453

Fridman WH, Pages F, Sautes-Fridman C, Galon J (2012) The immune contexture in human tumors: impact on clinical outcome. Nat Rev Cancer 12(4):298–306

Galluzzi L, Buque A, Kepp O, Zitvogel L, Kroemer G (2015) Immunological effects of conventional chemotherapy and targeted anticancer agents. Cancer Cell 28(6):690–714

Gershan JA, Barr KM, Weber JJ, Jing W, Johnson BD (2015) Immune modulating effects of cyclophosphamide and treatment with tumor lysate/CpG synergize to eliminate murine neuroblastoma. J Immunother Cancer 3(1):1–11

Ghiringhelli F, Menard C, Puig PE, Ladoire S, Roux S, Martin F, Solary E, Le Cesne A, Zitvogel L, Chauffert B (2007) Metronomic cyclophosphamide regimen selectively depletes CD4+ CD25+ regulatory T cells and restores T and NK effector functions in end stage cancer patients. Cancer Immunol Immunother 56(5):641–648

Goeppert B, Frauenschuh L, Zucknick M, Stenzinger A, Andrulis M, Klauschen F, Joehrens K, Warth A, Renner M, Mehrabi A, Hafezi M (2013) Prognostic impact of tumor-infiltrating immune cells on biliary tract cancer. Br J Cancer 109(10):2665–2674

Han LY, Fletcher MS, Urbauer DL, Mueller P, Landen CN, Kamat AA, Lin YG, Merritt WM, Spannuth WA, Deavers MT, De Geest K (2008) HLA class I antigen processing machinery component expression and intratumoral T-Cell infiltrate as independent prognostic markers in ovarian carcinoma. Clin Cancer Res 14(11):3372–3379

Hanahan D, Weinberg RA (2000) The hallmarks of cancer. Cell 100(1):57–70

He S, Yin T, Li D, Gao X, Wan Y, Ma X, Ye T, Guo F, Sun J, Lin Z, Wang Y (2013) Enhanced interaction between natural killer cells and lung cancer cells: involvement in gefitinib-mediated immunoregulation. J Transl Med 11(1):1–11

How Is Chemotherapy Used to Treat Cancer? (2016). Retrieved from: https://www.cancer.org/treatment/treatments-and-side-effects/treatment-types/chemotherapy/how-is-chemotherapy-used-to-treat-cancer.html

Iida N, Dzutsev A, Stewart CA, Smith L, Bouladoux N, Weingarten RA, Molina DA, Salcedo R, Back T, Cramer S, Dai RM (2013) Commensal bacteria control cancer response to therapy by modulating the tumor microenvironment. Science 342(6161):967–970

Issa-Nummer Y, Darb-Esfahani S, Loibl S, Kunz G, Nekljudova V, Schrader I, Sinn BV, Ulmer HU, Kronenwett R, Just M, Kühn T (2013) Prospective validation of immunological infiltrate for prediction of response to neoadjuvant chemotherapy in HER2-negative breast cancer–a substudy of the neoadjuvant GeparQuinto trial. PLoS One 8(12):e79775

Kanterman J, Sade-Feldman M, Biton M, Ish-Shalom E, Lasry A, Goldshtein A, Hubert A, Baniyash M (2014) Adverse immunoregulatory effects of 5FU and CPT11 chemotherapy on myeloid-derived suppressor cells and colorectal cancer outcomes. Cancer Res 74(21):6022–6035

Kasajima A, Sers C, Sasano H, Jöhrens K, Stenzinger A, Noske A, Buckendahl AC, Darb-Esfahani S, Müller BM, Budczies J, Lehman A (2010) Down-regulation of the antigen processing machinery is linked to a loss of inflammatory response in colorectal cancer. Hum Pathol 41(12):1758–1769

Kennedy AD, DeLeo FR (2009) Neutrophil apoptosis and the resolution of infection. Immunol Res 43(1–3):25–61

Khallouf H, Märten A, Serba S, Teichgräber V, Büchler MW, Jäger D, Schmidt J (2012) 5-Fluorouracil and interferon-α immunochemotherapy enhances immunogenicity of murine pancreatic cancer through upregulation of NKG2D ligands and MHC class I. J Immunother 35(3):245–253

Kim H, Kim SH, Kim MJ, Kim SJ, Park SJ, Chung JS, Bae JH, Kang CD (2011) EGFR inhibitors enhanced the susceptibility to NK cell-mediated lysis of lung cancer cells. J Immunother 34(4):372–381

Lim SH, Chua WEI, Cheng C, Descallar J, Ng W, Solomon M, Bokey L, Wong K, Lee MT, De Souza P, Shin JS (2014) Effect of neoadjuvant chemoradiation on tumor-infiltrating/associated lymphocytes in locally advanced rectal cancers. Anticancer Res 34(11):6505–6513

Liu WM, Fowler DW, Smith P, Dalgleish AG (2010) Pre-treatment with chemotherapy can enhance the antigenicity and immunogenicity of tumors by promoting adaptive immune responses. Br J Cancer 102(1):115–123

Martins I, Kepp O, Schlemmer F, Adjemian S, Tailler M, Shen S, Michaud M, Menger L, Gdoura A, Tajeddine N, Tesniere A (2011) Restoration of the immunogenicity of cisplatin-induced cancer cell death by endoplasmic reticulum stress. Oncogene 30(10):1147–1158

Medina PJ, Fausel C (2008) Pharmacotherapy, a pathophysiologic approach seventh edition: cancer treatment and chemotherapy. Edited by DiPiro JP, Talbert RL, Yee GC, Matzke GR, Wells BG, Posey LM

Ménard C, Blay JY, Borg C, Michiels S, Ghiringhelli F, Robert C, Nonn C, Chaput N, Taïeb J, Delahaye NF, Flament C (2009) Natural killer cell IFN-γ levels predict long-term survival with imatinib mesylate therapy in gastrointestinal stromal tumor–bearing patients. Cancer Res 69(8):3563–3569

Mignot G, Hervieu A, Vabres P, Dalac S, Jeudy G, Bel B, Apetoh L, Ghiringhelli F (2014) Prospective study of the evolution of blood lymphoid immune parameters during dacarbazine chemotherapy in metastatic and locally advanced melanoma patients. PLoS One 9(8):e105907

Mizoguchi I, Yoshimoto T, Katagiri S, Mizuguchi J, Tauchi T, Kimura Y, Inokuchi K, Ohyashiki JH, Ohyashiki K (2013) Sustained upregulation of effector natural killer cells in chronic myeloid leukemia after discontinuation of imatinib. Cancer Sci 104(9):1146–1153

Mohammed ZMA, Going JJ, Edwards J, Elsberger B, McMillan DC (2013) The relationship between lymphocyte subsets and clinico-pathological determinants of survival in patients with primary operable invasive ductal breast cancer. Br J Cancer 109(6):1676–1684

Murray JC, Aldeghaither D, Wang S, Nasto RE, Jablonski SA, Tang Y, Weiner LM (2014) c-Abl modulates tumor cell sensitivity to antibody-dependent cellular cytotoxicity. Cancer Immunol Res 2(12):1186–1198

Neagu M, Constantin C, Zurac S (2013) Immune parameters in the prognosis and therapy monitoring of cutaneous melanoma patients: experience, role, and limitations. BioMed Res Int 2013

Nelson PN, Reynolds GM, Waldron EE, Ward E, Giannopoulos K, Murray PG (2000) Demystified...: monoclonal antibodies. Mol Pathol 53(3):111

Raviraj J, Bokkasam VK, Kumar VS, Reddy US, Suman V (2014) Radiosensitizers, radioprotectors, and radiation mitigators. Indian J Dental Res 25(1):83

Riva G, Luppi M, Lagreca I, Barozzi P, Quadrelli C, Vallerini D, Zanetti E, Basso S, Forghieri F, Morselli M, Maccaferri M (2014) Long-term molecular remission with persistence of BCR-ABL1-specific cytotoxic T cells following imatinib withdrawal in an elderly patient with Philadelphia-positive ALL. Br J Haematol 164(2):299–302

Romero AI, Chaput N, Poirier-Colame V, Rusakiewicz S, Jacquelot N, Chaba K, Mortier E, Jacques Y, Caillat-Zucman S, Flament C, Caignard A (2014) Regulation of CD4+ NKG2D+ Th1 cells in patients with metastatic melanoma treated with sorafenib: role of IL-15Rα and NKG2D triggering. Cancer Res 74(1):68–80

Roselli M, Cereda V, di Bari MG, Formica V, Spila A, Jochems C, Farsaci B, Donahue R, Gulley JL, Schlom J, Guadagni F (2013) Effects of conventional therapeutic interventions on the number and function of regulatory T cells. Oncoimmunology 2(10):e27025

Sawant A, Schafer CC, Jin TH, Zmijewski J, Hubert MT, Roth J, Sun Z, Siegal GP, Thannickal VJ, Grant SC, Ponnazhagan S (2013) Enhancement of antitumor immunity in lung cancer by targeting myeloid-derived suppressor cell pathways. Cancer Res 73(22):6609–6620

Schreiber RD, Old LJ, Smyth MJ (2011) Cancer immunoediting: integrating immunity's roles in cancer suppression and promotion. Science 331(6024):1565–1570

Senovilla L, Vitale I, Martins I, Tailler M, Pailleret C, Michaud M, Galluzzi L, Adjemian S, Kepp O, Niso-Santano M, Shen S (2012) An immunosurveillance mechanism controls cancer cell ploidy. Science 337(6102):1678–1684

Shalapour S, Font-Burgada J, Di Caro G, Zhong Z, Sanchez-Lopez E, Dhar D, Willimsky G, Ammirante M, Strasner A, Hansel DE, Jamieson C (2015) Immunosuppressive plasma cells impede T-cell-dependent immunogenic chemotherapy. Nature 521(7550):94–98

Soeda A, Morita-Hoshi Y, Makiyama H, Morizane C, Ueno H, Ikeda M, Okusaka T, Yamagata S, Takahashi N, Hyodo I, Takaue Y (2009) Regular dose of gemcitabine induces an increase in CD14+ monocytes and CD11c+ dendritic cells in patients with advanced pancreatic cancer. Jpn J Clin Oncol 39(12):797–806

Terme M, Pernot S, Marcheteau E, Sandoval F, Benhamouda N, Colussi O, Dubreuil O, Carpentier AF, Tartour E, Taieb J (2013) VEGFA-VEGFR pathway blockade inhibits tumor-induced regulatory T-cell proliferation in colorectal cancer. Cancer Res 73(2):539–549

Thalidomide: Research advances in cancer and other conditions (2016). Retrieved from: https://www.mayoclinic.org/diseases-conditions/cancer/in-depth/thalidomide/art-20046534

Valent A, Penault-Llorca F, Cayre A, Kroemer G (2013) Change in HER2 (ERBB2) gene status after taxane-based chemotherapy for breast cancer: polyploidization can lead to diagnostic pitfalls with potential impact for clinical management. Cancer Genet 206(1-2):37–41

Walker EA Jr (1964) Management of the child with a fatal disease. Clin Pediatr 3:418–427

Wang B, Xu D, Yu X, Ding T, Rao H, Zhan Y, Zheng L, Li L (2011) Association of intra-tumoral infiltrating macrophages and regulatory T cells is an independent prognostic factor in gastric cancer after radical resection. Ann Surg Oncol 18(9):2585–2593

West AC, Mattarollo SR, Shortt J, Cluse LA, Christiansen AJ, Smyth MJ, Johnstone RW (2013) An intact immune system is required for the anticancer activities of histone deacetylase inhibitors. Cancer Res 73(24):7265–7276

What Is Targeted Cancer Therapy? (2016). Retrieved from: https://www.cancer.org/treatment/treatments-and-side-effects/treatment-types/targeted-therapy/what-is.html

Wu J, Waxman DJ (2015) Metronomic cyclophosphamide eradicates large implanted GL261 gliomas by activating antitumor Cd8+ T-cell responses and immune memory. Oncoimmunology 4(4):e1005521

Xu X, Rao GS, Groh V, Spies T, Gattuso P, Kaufman HL, Plate J, Prinz RA (2011) Major histocompatibility complex class I-related chain A/B (MICA/B) expression in tumor tissue and serum of pancreatic cancer: role of uric acid accumulation in gemcitabine-induced MICA/B expression. BMC Cancer 11(1):1–11

Zitvogel L, Tesniere A, Kroemer G (2006) Cancer despite immunosurveillance: immunoselection and immunosubversion. Nat Rev Immunol 6(10):715–727

Telomerase and Its Therapeutic Implications in Cancer

Raman Kumar, Nidhi Gupta, and Alpana Sharma

Abstract

Telomeres are specialized nucleoprotein structures localized at the ends of eukaryotic chromosomes. Telomere biology is frequently associated with human cancers where dysfunctional telomeres have been proved to participate in genetic instability. Due to end replication problem, there is a loss of nucleotides from telomere after every cell division. Thus, after limited divisions, normal somatic cells cease to divide and undergo senescence and/or apoptosis. This inability of normal cells to continually proliferate is bypassed in immortal or cancer cells via reactivation of telomerase. Telomerase is a highly specialized ribonucleoprotein molecule that promotes addition of telomeric repeats at the $3'$ end of the chromosome. Multiple disease-specific studies have unveiled involvement of various genetic and epigenetic mechanisms in regulation of telomerase expression. Further, selective expression of telomerase in cancer cells in comparison with normal somatic cells makes telomerase a suitable therapeutic target. Thus, large numbers of natural and artificial compounds targeting telomerase have been screened in multiple cancers. This chapter focuses on the telomerase structure, function, and its therapeutic implications in cancer.

Keywords

Telomere · Telomerase · Cancer · Therapeutics

R. Kumar · N. Gupta · A. Sharma (✉)
Department of Biochemistry, All India Institute of Medical Sciences, New Delhi, India

14.1 Introduction

In 1930s, first ever evidence of specialized structure presented at the chromosomal ends which prevented end-to-end fusion of chromosome was unveiled. Hermann Muller, a geneticist, termed this structure as "telomere" which was obtained from Greek words meaning "end" (telos) and "part" (meros). Muller in 1938 in a lecture entitled "The Remaking of Chromosomes" proposed that ....*the terminal gene must have a special function, that of sealing the ends of the chromosome, so to speak, and that for some reason, a chromosome cannot persist indefinitely without having its ends thusly "sealed." This gene may accordingly be distinguished by a special term, the "telomere"* (Muller 1938).

Simultaneously, Barbara McClintock while performing experiments on maize indicated that these naturally occurring chromosome ends or telomeres demonstrate a very special property which was necessary for chromosomal stability thus supported Muller's work (McClintock 1941). Later on, in 1970s, studies performed in *Tetrahymena* and yeast established that structure protecting the chromosome ends is tandem repeat of hexanucleotide blocks that could function across the diverse species and is structurally and functionally conserved (Blackburn and Gall 1978; Klobutcher et al. 1981). Then, in mid-1980s, Blackburn and Greider showed that cellular extract possesses enzymatic activities capable of adding tandem hexanucleotides repeats to natural chromosome ends that eventually leads to the discovery of telomerase for which Blackburn in 2009 shared the Nobel Prize in Physiology along with Greider and Szostak (Szostak and Blackburn 1982).

14.2 Telomere Structure and Function

In vertebrates, telomeres are very specialized nucleoprotein structures spanning about 10–50 kb in length and are composed of tandem repeats of the hexanucleotides, $5'$-(TTAGGG)$_n$-$3'$ (Fig. 14.1). According to species, types, and nature of cells, telomere length ranges from 10 kb to 50 kb in both human and *mus musculus* (de Lange et al. 1990; Kipling and Cooke 1990). It is a single–double-stranded hybrid structure in which approximately 10–15 kb (in human at birth) long is double-stranded nucleotide sequences and 75–200 single-stranded DNA sequences rich in guanine nucleotides are present at $3'$ chromosomal ends (Doksani et al. 2013). This G-rich single-stranded overhang plays crucial role in telomere capping and conformation stabilization by forming higher-order structure. Telomere protects chromosomal ends from genome surveillance machinery, nucleolytic attacks, and recombination proteins by burying, otherwise double-stranded breaks through generation of stabilized secondary structures. Among these structures, one is the formation of circular structure called a T loop in which the single-stranded part of telomere folds back on the double-stranded repeats and invades them by displacing Watson–Crick base pairing. Further, G-strand overhang by displacing normal Watson–Crick base pairing results in the generation of a stabilized structure called as displacement loop (D loop). Additionally, others in the category of secondary

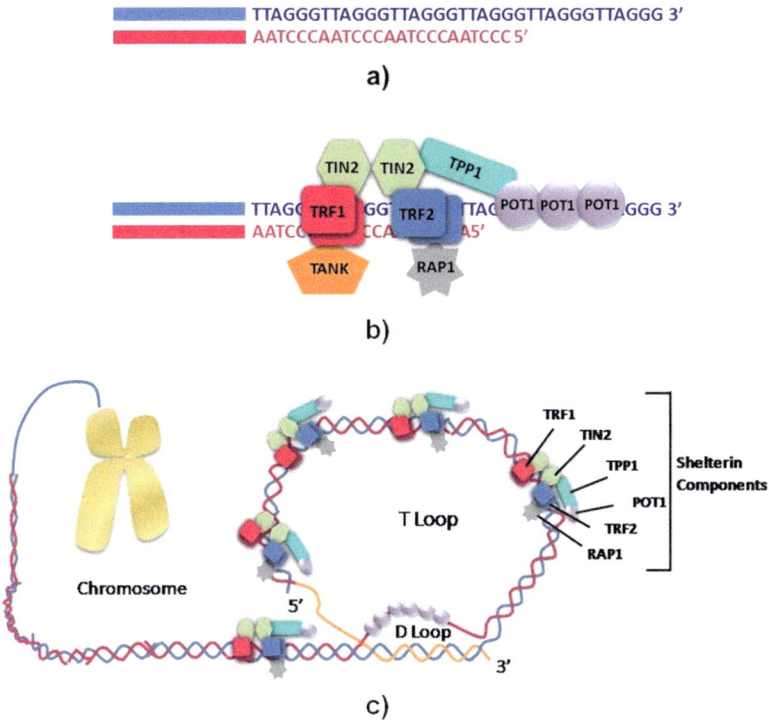

Fig. 14.1 Schematic representation of telomere sequence, structure, and telomeric proteins

structures are G-quadruplexes. In vitro experiments revealed that G-rich region of telomere DNA forms a four stranded stabilized structure known as G-quadruplex (or G-quartet). This motif forms a stable planar four-stranded structure by stacking upon each other and further trapped a cation by coordinating with carbonyl group of each guanine residue. This coordination and trapping of cation further stabilizes the telomere structure. Few experimental studies have indicated that interaction of G-quadruplex and telomerase RNA template might influence telomerase activity at chromosomal ends and hence provided evidences of being used as a suitable therapeutic target (Kim et al. 2003; Shin-ya et al. 2001).

14.3 Telomeric Proteins

Secondary structures of telomere maintain the chromosomal capping with the help of a number of proteins that directly or indirectly localized to telomere (Smith et al. 2020). These nucleoprotein structures together play crucial roles in evading DNA repair and recombination (DDR) mechanisms. In mammals, these proteins on the basis of their localization on telomere were divided into three different categories. The initial two categories were structural proteins, and together were called as shelterin proteins (Doksani et al. 2013). First category of proteins directly interact with telomeric DNA using specialized DNA-binding motifs present in their primary sequence. In this category, two proteins including TTAGGG telomere repeat binding factor 1 (TRF1) and telomere repeat binding factor 2 (TRF 2) attach directly to double-stranded telomeric DNA, whereas single-stranded structure is guarded by protection of telomere 1 (POT1) protein (Doksani et al. 2013). The TRF1 protein negatively regulates telomere length, whereas TRF2 after binding at chromosomal termini suppresses DDR pathways and prevents their recognition as double-stranded breaks. These three proteins perform special functions to ensure telomere stability. Such as, TRF2 promotes T-loop formation and prevents the activation of ataxia telangiectasia mutated (ATM)-dependent DDR pathways and nonhomologous end joining (NHEJ) (Arnoult and Karlseder 2015). TRF1 plays significant role in the replication of telomere DNA (Zimmermann et al. 2014). Further, POT1 suppresses ataxia telangiectasia and Rad3-related protein (ATR)-dependent DDR pathways by inhibiting access of replication protein A (RPA) at single-stranded telomeric DNA sequences (Denchi and de Lange 2007). Second category of proteins interact with above mentioned telomere binding proteins and form a multiprotein complex involved in telomere length regulation. Three unique proteins, namely POT1 and TIN2 organizing protein 1 (TPP1), TRF2 interacting protein 1 (RAP1) and TRF1 and TRF2 interacting nuclear protein 2 (TIN2) along with above proteins, associate with each other to form highly specialized structure known as Shelterin complex (de Lange 2005; Doksani et al. 2013). TIN2 promotes the integrity and stability of this complex by acting as a scaffold protein on which TRF1, TRF2, and TPP1/POT1 heterodimers are attached (Ye 2004). This attachment further stabilizes TRF1 and TRF2 interaction with telomeric DNA (Frescas and de Lange 2014a, b). RAP1 by attaching with TRF2 increases the efficiency of interaction between TRF2 and telomeric DNA (Janoušková et al. 2015). Additionally, few shelterin complex associated proteins that regulate the telomere structure and its length include Pin2/TRF1-interacting factor (PINXI) and tankyrases 1 and 2 (Smith 1998; Zhou and Lu 2001). PINXI is a negative regulator of telomere length, while tankyrases 1 and 2 both positively regulate the telomere length by recruiting telomerase on the telomere. Final class of proteins includes those which regulate the biological processes including DNA damage regulators. Similar to above-discussed second class, these proteins do not bind directly to the telomeric DNA but otherwise require telomere-associated factors to localize on the telomeres. It includes MRE11/RAD50/NBS1 (MRN) complex, Ku70/86 heterodimer, RecQ helicases, Werner (WRN), and Bloom (BLM); these proteins play crucial roles in maintaining telomere homeostasis. In addition, some proteins implicitly recognize telomere dysfunction such as ATM, ATR kinases, 53BP1 (p53-binding protein 1), γH2AX, and RAD17 (radiation sensitive 17) (Lillard-Wetherell et al. 2004; Opresko et al. 2002). These proteins, in response to critically shortened telomeres, are recruited to special location at telomeres known as telomere dysfunction induced foci (TIF) and induce senescence and/or apoptosis.

It has been observed that telomere length quantitatively associates with expression levels of telomere-associated proteins. Dysregulation of telomere is a very common observation seen in many cancers; it has been observed that with alteration in telomere length there is simultaneous alteration in expression levels of these proteins. Butler et al. reported that there were increased mRNA levels of TRF1, TRF2, POT1, and TIN2 mRNA, but not TERT mRNA with decrease telomere contents in the breast carcinoma patients (Butler et al. 2012). Similarly, in 2010, Hu et al. reported in gastric cancer patients that increased protein levels of TRF1, TRF2, TIN2, and TERT molecules correlate negatively with the telomere length (Hu et al. 2010).

In our research laboratory, we studied the association between expression levels of shelterin complex and its associated molecules and telomere length in multiple myeloma patients. We observed significantly decreased telomere length and increased telomerase activity in bone marrow samples of myeloma patients in comparison with age-matched controls (Fig. 14.2). Further, we demonstrated significantly increased mRNA and protein expression of TRF1, TRF2, POT1, RAP1, TPP1, TIN2, and TANK1 in myeloma patients while drastically decreased molecular expression of PINX1 in these patients was observed. At mRNA levels, expression of these molecules correlated significantly with each other and with patients' stages. Further, mRNA expression also correlated negatively with telomere length in myeloma patients (Kumar et al. 2018a, b). Thus, these results suggest that there is a strong association between telomere length and expression of shelterin complex molecules in most of cancers and projected these molecules as suitable theranostic markers.

14.4 Telomere and Mechanism of Cellular Mortality

Telomere length maintenance is of utmost importance for normal cells to grow, sustain, and divide. Telomere length shortening along with other potential oncogenic changes contributes to genomic instability which ultimately leads to initiation of early stage cancer. Literature reported that normal somatic cells divide for a limited number of cell divisions. This is because our classical replication machinery is incapable to replicate the end part of chromosome which is also called as "end-replication problem". Due to end replication problem after each round of cell division, a part of telomere (in humans 50–150 bp) breaks off, thereby leading to telomere shortening. Once the telomere becomes critically shortened, they are more prone to induce damage to the genome or thus will undergo senescence by activating DNA repair and recombination pathways. Thus, this capacity of limited cell division correlates well with the telomere length. Further, the average age of an individual depends weakly on the mean telomere length of their somatic cells. Nevertheless, these observations implied the role of telomere as a molecular clock which limits the cellular lifespan on the basis of number of cellular divisions.

Inside cell at molecular level, two crucial barriers which prevent immortalization and ultimately malignant transformation are replicative senescence (M1) and crisis (M2) (Wright et al. 1989). It has been seen that human somatic cells cultured in vitro for extended period of time ultimately ceases to divide by entering nonproliferative

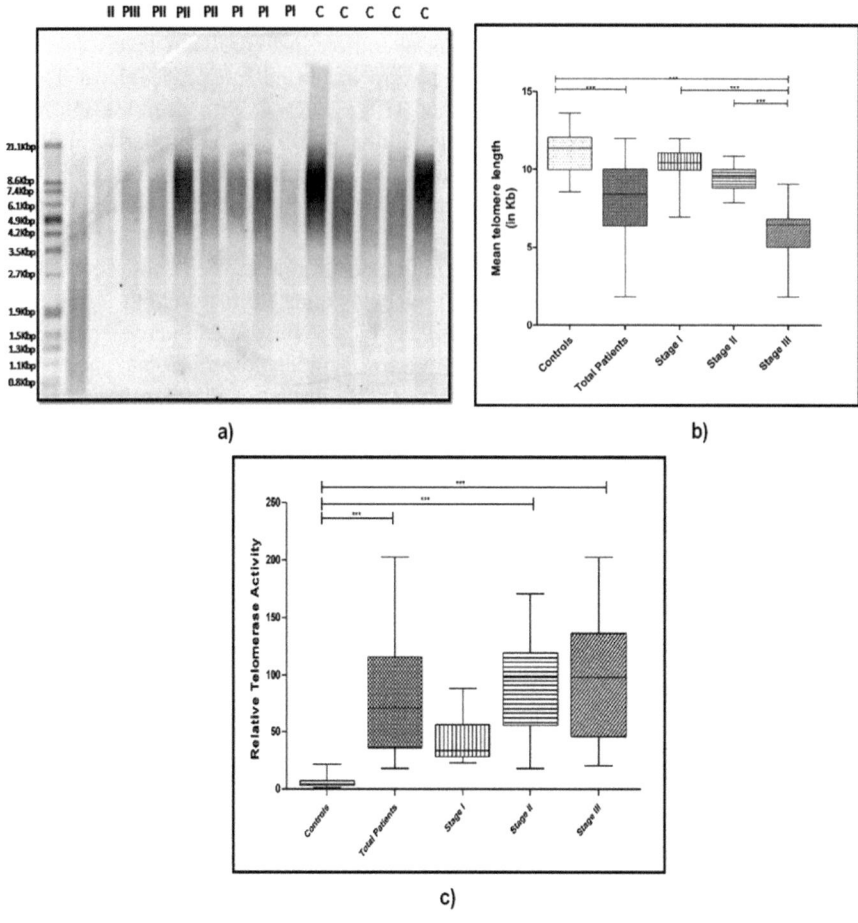

Fig. 14.2 (**a**) Representative image showing southern blot analyses in MM patients (all stages) and controls. Lane 1—ladder showing molecular weights on left side, lane 2–4—stage III MM patients, lane 5–7—stage II MM patients, lane 8–10—stage I MM patients, and lane 11–15—control subjects; (**b**) Box and Whisker plot showing mTL determined using the telotool software in southern blot experiment in controls and MM patients (total patients and all stages); (**c**) Box and Whisker plot showing RTA in total cell lysate of controls and MM patients (total patients and all stages): MM—multiple myeloma; mTL—mean telomere length; RTA—relative telomerase activity. "*" is denoted as $p < 0.05$ to determine the level of significance. (Adapted with permission from (Kumar et al. 2018a, b)

though metabolically active state known as mortality stage 1 (M1) or also known as replicative senescence (Fig. 14.3). This M1 stage involves critically shortened length of telomeres which ultimately leads to abrogation of cellular proliferation. Some studies suggested that this inhibition can be made by uncapping of single telomere, while majority suggested that telomere state rather than telomere length determines the fate of those cells. Majority of normal somatic cells stay either in the M1 state for

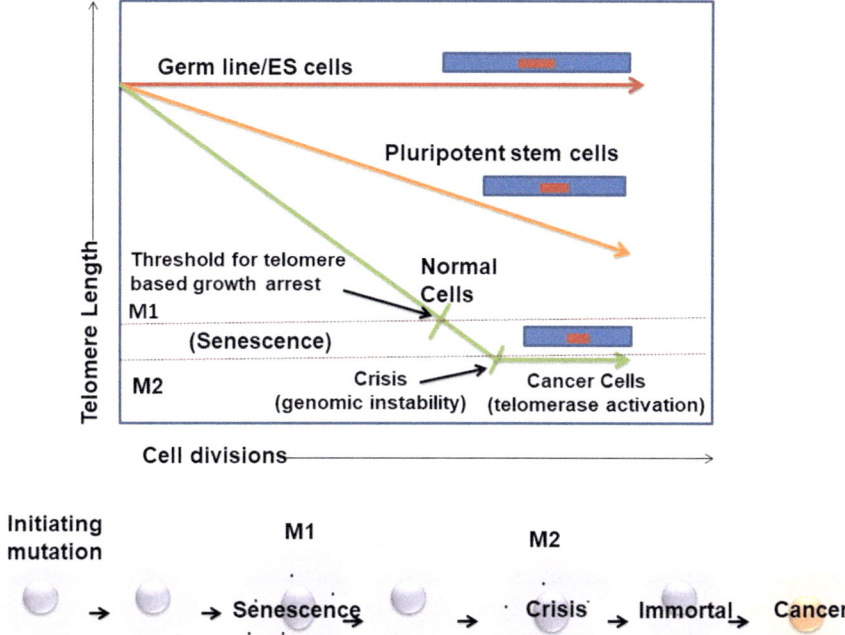

Fig. 14.3 Schematic representation of telomere length dynamics in different cell types. Embryonic cells and germ cells maintain their telomere length throughout their life, while pluripotent stem cells have reduced rate of telomere loss and therefore survive for longer duration. In contrast, most somatic cells lack telomerase activity and thereby divide for a limited number of cycles and then undergo senescence. Some cells are able to bypass senescence and reach crisis stage which involves wide-scale genomic instability and apoptosis in cells. One of ten million cells escapes the crisis stage by reactivating telomerase and become a cancerous cell with limitless replicative potential

their entire life or undergo apoptosis by activating apoptosis-inducing pathways. While in the existence of cancer-inducing changes, M1 stage can be by-passed by deregulating tumor suppressor pathways which predominantly involve inactivation of *p53* and *Rb* genes, thereby underlining their potential in maintaining these cells at senescence or M1 stage. Once the ablation of these genes or pathways occurs, cells continue to proliferate until they reach a new dysfunctional state termed as mortality stage 2 (M2 stage) or crisis stage. This stage is accompanied by critically shortened or naked telomeres leading to formation of multiple TIF characterized by breakage-fusion-bridge (BFB) formation. This stage also involves massive cellular apoptosis along with continued DNA replication, thereby indicating that cells entering M2 stage are no longer able to stop the cell cycle progression (Hayashi et al. 2015). However, a rare clone (1 in 10^7 cells) progresses toward immortalization (Castro-Vega et al. 2015). These surviving cells will maintain their telomere lengths by expressing detectable levels of telomerase or through a totally different lesser-known mechanism called as alternative lengthening of telomeres (ALT) (Shay and

Wright 2011). Though ALT method has been seen in only about 15% of malignancies, the rest 85% of human cancers involve reactivation of telomerase.

14.5 Structure and Function of Telomerase

Telomerase is a highly specialized ribonucleoprotein structure that helps in maintenance of telomere length and ultimately genome integrity by enzymatically repairing ends of the chromosome (Nguyen et al. 2019; Roake and Artandi 2020). In the eukaryotic evolution, two critical drawbacks arose because of linear chromosomes in the genomic material which could deprotect eukaryotic cells genomic integrity (de Lange 2015). First, cellular DNA polymerase is incapable to completely replicate their lagging strand of linear DNA known as "end-replication problem" (Watson 1972). These unreplicated ends can be recognized by DNA repair and recombination machinery as damaged DNA whose repair results in the loss of genetic information (Olovnikov 1973). Second, these unreplicated ends are always prone to be targeted by endonuclease activity which is known as "end-protection problem." Some special mechanisms were needed to protect these chromosomes termini from nuclease activity and to prevent their shortening. In the late 1980s, telomerase enzyme discovery unfurled the end-replication mystery. The ribonucleoprotein complex, telomerase, using RNA template adds repetitive DNA sequences and maintains the telomere length (Greider and Blackburn 1989). Then, telomeric proteins in a sequence-specific manner attach to these telomeric repeats and form a protective cap around the chromosome, thus solving the "end-protection problem" (de Lange 2009).

In 1985, the enzyme which can de novo synthesize and elongate telomeric DNA was discovered as telomerase (Greider and Blackburn 1985, 1989). Later on, in 1997, it was identified and characterized (Harrington et al. 1997). This enzyme is a large ribonucleoprotein complex which though error-prone progressively adds DNA repeats (TTAGGG) to the telomere. It is a RNA-dependent DNA polymerase which comprises two minimally functional subunits: an RNA component called as telomerase RNA (TR) and a functional protein known as telomerase reverse transcriptase (hTERT). In humans, hTR (human telomerase RNA) is synthesized from TERC gene localized on 3q26 region of the chromosome, whereas hTERT protein is encoded from TERT gene present on 5p15.33 region of the chromosome 5 (Cong 1999; Feng et al. 1995). Other proteins implicated in the generation of holoenzyme are Dyskerin, Gar1, Nhp2, Pontin, Reptin, Tcab1, and telomerase protein component (TEP1). These proteins play crucial roles in proper assembly and its recruitment to the telomere (Cohen et al. 2007; Saito et al. 1997; Venteicher et al. 2008). Additionally, Es1p and Es3p (Ku heterodimer) are required for proper assembly and maturation of telomerase protein complex (Liu et al. 2004).

Evolutionary, hTERT is highly conserved and transcriptionally regulated protein whose expression levels correlate significantly with telomerase activity, thereby presenting as rate limiting and important determinant for enzymatic activity (Cong et al. 2002). In contrast, hTR is ubiquitously and constitutively expressed mRNA which is required for de novo synthesis of telomeric DNA (Cong et al. 2002). Further, hTERT comprises four domains: telomerase "essential" N-terminal (TEN) domain, telomerase RNA-binding domain (TRBD), reverse transcriptase (RT) domain, and C-terminal extension (CTE) or thumb domain. The N-terminal TEN domain binds with single-stranded telomeric DNA and also interacts with hTR (O'Connor et al. 2005; Robart and Collins 2011). TRBD domain is made of all helical domains where α-helices arranged to form asymmetric halves which collectively forms TRBD RNA-binding pocket which can binds with both single-stranded and paired RNA (Rouda and Skordalakes 2007). RT domain is the most evolutionary conserved and characterized domain that forms the catalytic/enzymatic domain of TERT (Lue et al. 2003). The hTERT CTE domain is made up of three highly conserved regions required for formation of stable RNA–DNA duplex. Mutation in these regions can lead to multiple human diseases including aplastic anemia, dyskeratosis congenital and idiopathic pulmonary fibrosis as they lead to decreased telomerase activity and processivity by disrupting interaction between RT and DNA (Hoffman et al. 2017). hTR is a very versatile component that shows sequence and size divergence in wide variety of eukaryotic species. In vertebrates, the size of hTR ranges from ~310 to 560 nt (Xie et al. 2008), while in yeast the hTR's size ranges between ~780 and 1820 nt (Gunisova et al. 2009). At one time, ~1.5 repeats of the RNA template domain of hTR are copied and incorporated into the growing telomere DNA as a complementary sequence. The RNA template domain comprises two distinct segments: $3'$ region helps in pairing with the DNA sequence, while $5'$ region provides the template which is used by the hTERT for telomeric DNA synthesis (Greider 1991). The template region helps in telomerase processivity, regulating telomerase enzymatic activity and the template utilization. Initially, both hTERT and hTR are located in different nucleolar compartment, but with beginning of S-phase, hTERT is colocalized with hTR in the cajal bodies. Further, loading of human telomerase onto the telomere sequence requires TIN2 and TPP1 shelterin complex proteins already present on the double-stranded region through TRF1 and TRF2 (Abreu et al. 2010; de Lange 2010). Additionally, TPP1 interacts with single-stranded DNA-binding protein POT1, thereby forming a TPP1-POT1 complex essential for the telomerase activity (Broccoli et al. 1995).

14.6 Implications of Telomerase in Cancer

Due to stringent regulation of hTERT, telomerase activity ranges from very low to absent in normal somatic cells, while it is higher in stem cells, germline, and other rapidly renewing cells (Broccoli et al. 1995). However, some mitotically active cells such as endometrial tissue, hair follicles, and proliferative tissue of intestinal crypts demonstrate significantly increased telomerase activity (Brien et al. 1997; Broccoli et al. 1995; Ramirez et al. 1997). Telomere length and telomerase activity show significantly great variation between embryonic stem cells and normal somatic cells. Embryonic stem cells consistently maintained their telomere length by displaying considerably higher telomerase activity, whereas normal somatic cells show progressive decrease in telomere length due to lack of telomerase activity. This property of selective expression of hTERT has recognized its potential as suitable diagnostic and prognostic biomarker in various cancers including bladder, prostate, thyroid, breast, colon, cervical, gastric, lung, and myeloma cancer (Fernández-Marcelo et al. 2015; Glybochko et al. 2014; Kulić et al. 2016; Kumar et al. 2018a, b; Sharma et al. 2007; Tahara et al. 1995; Umbricht et al. 1997; Wu et al. 2000).

Cancer occurs when normal cells undergo genomic instability due to multiple genetic mutations and acquires the ability to proliferate indefinitely by bypassing the senescence via reactivation of telomerase or other less explored mechanisms such as ALT pathway (Shay and Wright 2011). Telomerase levels are upregulated in >85% of malignancies, while only 10–15% neoplasias follow mechanism like ALT for continued proliferation. In the last 20 years, exemplary research has been taking place in identifying different mechanisms regulating telomerase upregulation in cancer. These mechanisms include majorly hTERT promoter mutation (31%) and hTERT promoter methylation (53%), whereas minor pathways for upregulation were hTERT gene amplification (3%) and gene rearrangements (3%) (Elisabeth Naderlinger and Klaus Holzmann 2017) (Fig. 14.4).

14.6.1 Role of Gene Amplification and Rearrangements

During oncogenesis, gene amplification has been observed as one of the most important mechanisms leading to gain or loss of genetic material. Gene amplification involves increased gene copy number which leads to enhanced gene expression. Multiple models including replication error, telomere dysfunction, and occurrence of fragile sites in chromosome lead to increased expression (Albertson 2006). Clinically, poor disease outcome in multiple malignancies including breast and thyroid cancers significantly correlated with upregulated hTERT expression due to increased copy number (Piscuoglio et al. 2016; Wang et al. 2016). A study involving a cohort of multiple cancers demonstrated that only 3% of hTERT expressing tumors has gene amplification, thereby suggesting other important mechanisms in deregulating hTERT expression (Zhang et al. 2000). Another potential mechanism for hTERT upregulation is genomic rearrangements by placing activators and enhancers at the promoter region of hTERT gene locus (Peifer et al. 2015). In 2016, Kawashima et al. reported that hTERT upregulation through genomic rearrangements correlated with poor clinical outcomes in neuroblastoma patients (Kawashima et al. 2016).

14.6.2 Role of Promoter Mutations in hTERT Regulation

Genetic alterations such as mutation in TERT promoter (TERTpMut) are unique but frequent, leading to increased hTERT expression and telomerase activation. The 260 bp long core hTERT promoter comprises several transcription factors binding motifs involved in regulation of gene activation and transcription (Kyo et al. 2008).

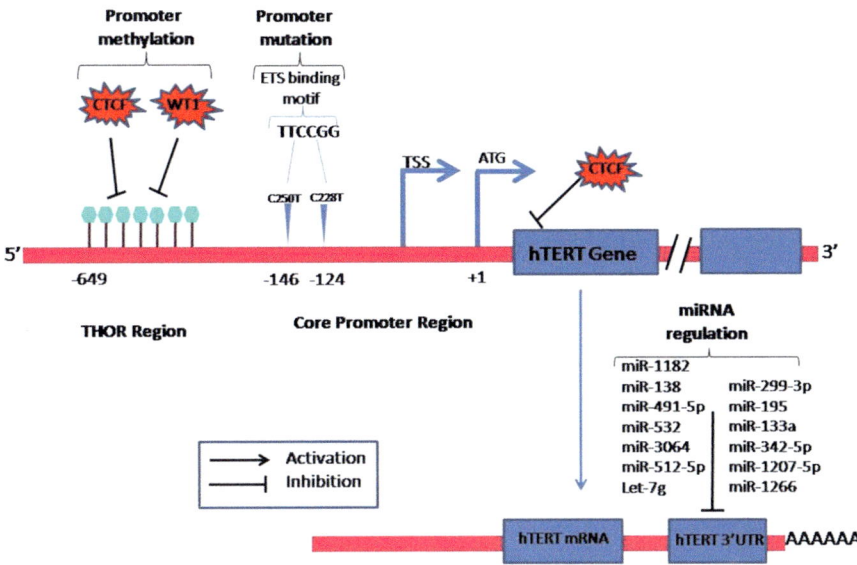

Fig. 14.4 Regulation of hTERT gene expression. Most common modes of hTERT gene expression regulators are promoter mutation, promoter methylation, and miRNA expression. Promoter methylation prevents the binding of transcriptional repressor, whereas promoter mutation generates novel binding sites for transcriptional activators. miRNA, however, regulates hTERT translation by binding to the 3′-UTR regions of the hTERT mRNA

Two landmarked studies in 2013 uncovered the potential of hTERT promoter mutation at two noncoding regions in both familial and sporadic melanomas (Horn et al. 2013; Huang et al. 2013). These mutations which were observed at −124 bp and −146 bp positions upstream to the transcription start site involved C > T transitions (at positions 1,295,228 (C228T) and 1,295,250 (C250T) on chromosome 5) and thereby generate a novel nucleotide stretch suitable for binding with E26 transformation-specific (ETS) transcriptional factor involved in both activation or repression of hTERT. Since then, studies have shown significant association of TERTpMut with distinct tumors including glioblastoma, thyroid cancer, bladder cancer, etc. (Vinagre et al. 2013). The multifarious transcriptional controlling of hTERT gene is evidenced by the fact that its promoter region contains binding motifs for both enhancers and suppressors transcription factors. Widespread occurrence of TERTpMut at difference stages and in grades of various carcinomas supported this as crucial early event during carcinogenesis (Kinde et al. 2013; Wang et al. 2014).

Clinically, biomarker utility of hTERT was observed when tumors carrying TERTpMut expressed high hTERT mRNA and proteins levels in comparison with tumors having wild-type promoter (Jin et al. 2018; Leão et al. 2018; Vinagre et al. 2013). In 2018, Spiegl-Kreinecker et al. reported the involvement of hTERT promoter mutation in poor prognosis and cellular immortalization in meningioma

(Spiegl-Kreinecker et al. 2018). A study reported by Wu et al. reported that simultaneous occurrence of mutation in TP/RB1 and TERTpMut might play significant role in the disease progression (Wu et al. 2014). Barczak et al. (2017) have emphasized that 36% of head and neck cancer patients had hTERT C250T promoter mutation in early stage tumors (Barczak et al. 2017). Further, in urothelial bladder carcinoma, the identification of TERTpMut in urine and tissue samples projected its role as non-invasive diagnostic and prognostic biomarker. Additionally, a report has shown that apart from disease prediction or outcome, TERTpMut was a better predictor of response to other adjuvant therapies including radiotherapy resistance (Gao et al. 2016). Yuan et al. (2016) in a meta-analysis reported the prognostic significance of TERTpMut along with clinical characteristics such as patient's age, gender, and distant metastasis in nonsmall cell lung carcinoma (NSCLC) (Yuan et al. 2016). Literature survey in adult gliomas reported that frequency of TERTpMut is highest in glioblastomas (70%), followed by oligodendrogliomas (60%) and then oligoastrocytomas (35%) (Killela et al. 2013). In urological malignancies, TERTpMut varies widely ranging from 85% in urothelial carcinoma of bladder to 9% in renal cell carcinoma to a complete absence in testicular and prostate cancers. Some cancers such as colorectal carcinoma, prostate cancer, and testicular carcinoma do not harbor TERTpMut but display telomerase activation and self-renewal properties, thereby suggesting other mechanisms of its activation.

14.6.3 Role of Promoter Methylation in hTERT Regulation

In contrast to above mechanisms, epigenetic process such as DNA methylation is a very stable and frequent mode of gene expression regulation. In whole genome, this occurs at CpG sites generally located in the noncoding regions. Methylation is performed by DNA methyltransferases which adds a methyl group on the 5-carbon of cytosine which is always preceded to a guanine base. Generally, CpG dinucleotide repeats are widely distributed throughout genome with some specific clustered regions known as CpG islands. Most of these methylated CpG islands are located in the intergenic region specifically in the promoter part upstream to the transcription initiation site. Thus, DNA methylation at the promoter region plays crucial role in gene expression regulation. In general, genes which are having hypermethylated promoters are transcriptionally silenced, while those genes having less methylations are transcriptionally active genes. It is because hypermethylated DNA in the promoter region interferes with the proper positioning of transcription activators or chromatin conformation (Baylin and Jones 2011). Research established that during cancer progression, there is an elevated hypermethylation of CpG islands in the promoter region of oncogenes, thereby proposing them as a suitable hallmark of oncogenesis (Bartlett et al. 2013). Some studies have demonstrated contradictory results where hypomethylation of CpG islands in the hTERT promoter suppresses hTERT expression, while others have reported that hTERT overexpressing cancer cells displays promoter hypermethylation (Guilleret et al. 2002; Shin et al. 2003). Later on, it was observed that the region involving hTERT core promoter was hypomethylated and promotes hTERT overexpression, whereas region upstream to the core promoter was hypermethylated which did not allow repressors to bind and decrease their expression (Zinn et al. 2007). A recent report by Tsujioka et al. evidenced that hTERT promoter methylation occurs mainly in the core promoter which generates binding sites for ETS family of transcription activating factors (Tsujioka et al. 2015). Further CTCF, a transcription repressive factor, interacts with core promoter and decreases hTERT transcription by compacting chromatin organization. Therefore, promoter DNA methylation prevents binding of CTCF, thereby enhancing the telomerase expression (Renaud 2005). In addition to that, c-myc (transcription activator) binds to the hTERT core promoter in a hypomethylation state and activates telomerase activity and ultimately promotes cellular proliferation and differentiation (Wu et al. 1999a, b).

Clinically, hypermethylation of region upstream to the hTERT core promoter has been observed in the multiple cancer types including brain, prostate, urothelium, colon, and blood. Svahn et al. in 2018 reported that methylation of hTERT promoter is significantly correlated with poor outcomes in adrenocortical carcinoma (Svahn et al. 2018). In a study done in pediatric gliomas, methylation of hTERT promoter is considered as a promising biomarker of tumor progression (Castelo-Branco et al. 2013). THOR stands for TERT hypermethylated oncological region, which is 100% specific and 96% sensitive in detecting malignant neoplasm positive for hTERT expression. The identification of THOR methylation has demonstrated an unmatched potential to be used as promising diagnostic and prognostic biomarker in multiple cancers including thyroid cancer, acute myeloid leukemia/myelodysplastic syndrome, esophageal carcinoma, meningioma. and hepatocellular carcinoma (Castelo-Branco et al. 2013; Deng et al. 2015; Fürtjes et al. 2016; Wang et al. 2016; Zhang et al. 2015; Zhao et al. 2016). In these cancers, hTERT methylation pattern was positively correlating with hTERT reactivation and its expression and in many cases associated significantly with worst clinical outcomes.

14.6.4 Role of MicroRNAs in hTERT Regulation

MicroRNAs (miRNAs) are small, endogenously synthesized 20–25 nucleotides long noncoding RNA molecules known to regulate gene expression. They also play critical roles in regulation of various pathophysiological processes including cellular proliferation, apoptosis, and differentiation in several diseases and are implicated in genome instability by acting as oncogenic and suppressor drivers (Li et al. 2015; Vincent et al. 2014). Functionally, they regulate the posttranscriptional gene silencing by mRNA degradation and repressing translation. During cancer occurrence, decreased expression of miRNAs in tumor tissue indicates their suppressive nature, as decreased levels contribute to tumorigenesis. Further, overexpression of miRNAs (oncomiRNAs) regulated tumor suppressor genes also leads to occurrence of tumors (Cho 2007; Eckburg et al. 2020). Therefore, their functions as tumor suppressors or oncogenes depend on targeted genes. Multiple miRNAs discovered so far demonstrate enormous potential to regulate hTERT translation in many cancers. miRNAs and hTERT expressions are inversely correlated, thus during tumorigenesis high hTERT expression will have downregulated levels of respective miRNAs (Cho 2007; Hrdličková et al. 2014). miRNAs regulate the hTERT expression in two ways: In the direct mode, they attach to $3'$ untranslated region ($3'$-UTR) of hTERT and inhibit its translation (Bai et al. 2017). Study reported in thyroid carcinoma cells that decreased expression of miR-138 was inversely correlated with hTERT overexpression; further induced expression of miR-138 significantly decreased hTERT protein levels (Mitomo et al. 2008). Additionally, another report demonstrated that interaction of let-7 g*, miR-133a, miR-342-5p, and miR-491-5p with the $3'$-UTR region of hTERT mRNA significantly downregulated telomerase activity and inhibited cellular proliferation (Hrdličková et al. 2014). Furthermore, miR-1182 was also found to be downregulated in tumor tissues and cell lines of bladder cancer and restoration of miR-1182 level inhibited cellular proliferation and its invasion (Zhou et al. 2016). Indirectly, miRNAs regulate hTERT transcription by regulating transcription factors required for hTERT gene expression such as c-myc. For example, in esophageal squamous cell carcinoma, miR-1294 downregulates c-myc levels (Liu et al. 2015a, b). Further, miR-34a, a well-known tumor suppressor, decreases telomerase activity and induces cellular senescence by targeting c-myc/FoxM1 pathway in hepatocellular cancer cells (Xu et al. 2015). Due to high stability in tissue and body fluids, biomarker and therapeutic potential of miRNAs have been highlighted in multiple studies (Weber et al. 2010). In bladder, gastric and ovarian cancers downregulated levels of miR-1182, miR-1207-5p, miR-1266, miR-532, and miR-3064 associated poorly with clinical outcomes. In bladder cancer, increased expression of miR-1182 sensitizes cancer cells to chemotherapeutic drug cisplatin, thereby leading to induce better patients' response during or after treatment (Zhou et al. 2016). Potential of miRNA targeting hTERT and other factors crucial for telomere pathway holds great prospects to be used as promising therapeutic approach to suppress telomerase activity and inhibits other cancer pathways in future (Rupaimoole and Slack 2017).

14.7 Therapeutic Implications of Telomerase in Cancer

Majority of human tumors (>85% of cancers) and tumor cell lines showed increased telomerase expression. Telomerase is critically involved in telomere length maintenance and limitless cancer cells proliferation. Further, in contrast to its expression in tumor cells, normal cells display minimal to complete absence of telomerase expression. Thus, this property of selective telomerase expression could be exploited as important biomarker and further generates the possibility of development of telomerase inhibitors as anticancer agents (Guterres and Villanueva 2020). Due to its structural and functional complexity, it can be targeted in many ways. Some telomerase inhibitors target its catalytic components, whereas other targets its RNA template part. These telomerase inhibitors range from natural compounds to synthetic products along with artificially created oligonucleotides. These compounds include terpenes, alkaloids, polyphenols, xanthones, and artificial products

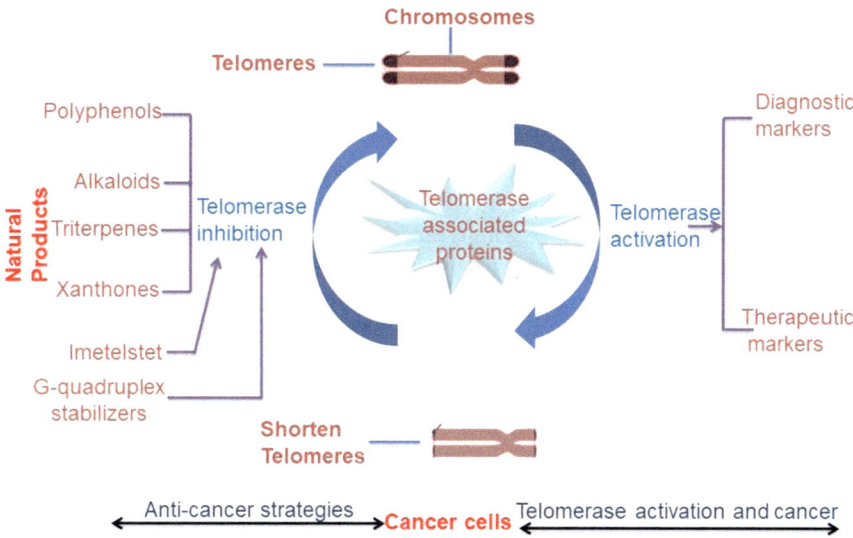

Fig. 14.5 Telomerase-related anticancer strategies by natural and chemically synthesized products

(Imetelstat, G-quadruplex stabilizers) which inhibit the telomerase expression and/or activity and thus restrained cellular proliferation (Fig. 14.5). Therapeutic implications of telomerase in various cancers by these products are discussed below.

14.7.1 Polyphenols

Curcumin, a major component of rhizome in turmeric (Curcuma longa L.), is a phenolic compound of medicinal importance. Literature reported that curcumin significantly induced apoptosis and showed anti-inflammatory, antioxidant, and neuroprotective activities (Griffiths et al. 2016). In 2006, Cui et al. reported that curcumin inhibited cells proliferation and decreased telomerase activity in a dose-dependent manner in multiple cancer cell types (Bel7402, HL60, and SGC7901) (Cui et al. 2006). Similarly, Ramachandran et al. reported that curcumin at 50–100 μM doses decreased telomerase expression and its activity in MCF-7 cancer cells in a c-myc-independent manner (Ramachandran et al. 2002). Further, Lee and Chung demonstrated that curcumin also induced cellular apoptosis by inhibiting the nuclear translocation of hTERT, thereby suppressing telomerase activity (Lee and Chung 2010).

Similarly, quercetin is a naturally occurring polyphenol found in most fruits, vegetables, green tea, and food grains. Multiple studies unearthed antiproliferative and pro-apoptotic properties of quercetin in various cancer cells (Lou et al. 2016; Ren et al. 2017). Further, multiple leukemic cells lines showed significantly decreased cell viability and telomerase activity after treatment with quercetin, thereby supporting its potential as suitable therapeutic target (Avci et al. 2011). Furthermore, several studies showed that quercetin dose dependently decreased the hTERT expression, lowered telomerase activity, and induced apoptosis in multiple cancer cells including lung (Wang et al. 2003), stomach (Wei et al. 2007), brain (Zamin et al. 2009), colon (Behjati et al. 2017), gastric cancer (Wei et al. 2007), and nasopharyngeal (Zheng and Chen 2017).

Resveratrol (3,5,40-trihydroxy-trans-stilbene) is a naturally occurring phenolic phytoalexin compound obtained from many plants and fruits. Chen, RJ et al. reported that resveratrol decreased cell viability in lung cancer cells by downregulating expression and activity of telomerase via modulating p53 gene expression (Chen et al. 2017). In another study, the effects of resveratrol on declining hTERT mRNA expression and telomerase activity in colorectal cancer cells were reported (Fuggetta et al. 2006). Similarly, Wang XY et al. reported that in colorectal cancer cells, resveratrol decreased cellular proliferation by inhibiting the hTERT promoter activity (Wang et al. 2010). Pterostilbene, a natural analog of resveratrol, induces apoptosis in cancer cells by binding and blocking the active site of telomerase (Tippani et al. 2014).

Tannic acid is naturally occurring polyphenol compound present in red wine, grapes, beans, tea, coffee, nuts, vegetables, and fruits. In a study published by Cosan et al. significant apoptosis and reduction in telomerase activity by tannic acid have been reported in breast and colon cancer cells (Turgut Cosan et al. 2011). Epigallocatechin-3-gallate (EGCG), one of the well-known tannic acids, demonstrates significant anticancer potential by inducing apoptosis through mitochondrial membrane potential, activating caspase-3 expression, and inhibiting telomerase activity in cancer cells (Gurung et al. 2015; Liu et al. 2017). Further, studies reported that it has also decreased hTERT and c-myc gene molecular expression (mRNA and protein) in various cancer cells (Liu et al. 2017; Zhang et al. 2014).

14.7.2 Alkaloids

Boldine (1,10-dimethoxy-2,9-dihydroxy aporphine) is a naturally occurring aporphine alkaloid synthesized in boldo tree (*Peumus boldus*) and also in lindera (*Lindera aggregata*). Boldine, dose, and time dependently induced cell death and exhibited antitumor effects against multiple cancer cell lines including bladder (T24), brain (U138-MG, U87-MG, and C6), and hepatocarcinoma (HepG-2) cells. Studies reported that induction of cell death is due to decreased hTERT gene expression along with lowered telomerase activity (Gerhardt et al. 2014; Kazemi Noureini and Wink 2015).

Similarly, berberine is an alkaloid extracted from the rhizome, stem barks, and roots of *Berberis vulgaris* (barberry) and many other plants. Wu et al. and Naasani et al. showed that berberine displayed time-dependent increase in apoptosis and decreased telomerase activity in human leukemic cancer cells (Naasani et al. 1999; Wu et al. 1999a, b). Further, berberine also increased the formation of G-quadruplexes at telomeres leading to cellular growth arrests and thus demonstrated anticancer potential (Franceschin et al. 2006; Ji et al. 2012).

14.7.3 Terpenes

Pristimerin is a triterpene extracted from Celastraceae and Hippocrateaceae families of plants and is known to display chemopreventive potential. This drug demonstrated significant antiproliferative potential against multiple human cancer cells (Cevatemre et al. 2018; Tiedemann et al. 2009). Liu et al. reported that in prostate cancer cells, pristimerin decreased telomerase activity by inhibiting expression of transcription factors involved in regulation of hTERT gene transcription (Liu et al. 2015a, b). Further, in another study performed on pancreatic duct adenocarcinoma cells, pristimerin substantially decreased cell proliferation by both arresting and inducing apoptosis along with decreased telomerase activity (Deeb et al. 2015). Similarly, oleanane (methyl-2-cyano-3,12-dioxooleana-1,9(11)-dien-28-oate) is a triterpene which is known to have substantial anti-inflammatory, pro-apoptotic activities and thus demonstrated anticancer potential against pancreatic ductal adenocarcinoma cancer cells (Deeb et al. 2013).

In our laboratory, we have studied anticancer potential of Tanshinone I (TanI), a diterpene, on myeloma cancer cell lines (RPMI 8226 and U266). We observed that Tanshinone I alone or in combination lenalidomide time and dose dependently induced apoptosis in both myeloma cells lines. Similarly, very low doses (2.0 μM) of TanI significantly decreased telomerase activity in both myeloma cancer cells at 24-h time point (Fig. 14.6). Further, TanI alone and in association with lenalidomide significantly decreased the molecular expression (mRNA and protein) of shelterin complex and its associated factors in myeloma cancer cells (Kumar et al. 2018a).

14.7.4 Xanthones

In this category, two important secondary metabolites are gambogic acid and gambogenic acid extracted from the resin of *Garcinia hanburyi* tree. Due to their unique color, they are generally used as coloring substances. Studies have reported that these xanthones induced cytotoxicity in cancer cells in both dose- and time-dependent manner (Pan et al. 2017; Zhao et al. 2017). Additionally, report suggested that these xanthones induced apoptosis and decreased telomerase expression and activity by inhibiting binding of transcriptional activators at the hTERT promoter region (Yu et al. 2006).

Though these natural compounds demonstrated significant anticancer potentials and drastically reduced telomerase levels in cancer cells, till today very few of them have been used as chemotherapeutic agents. Thus, synthetic drugs have provided significant hopes as anti-telomerase inhibitors against cancer cells.

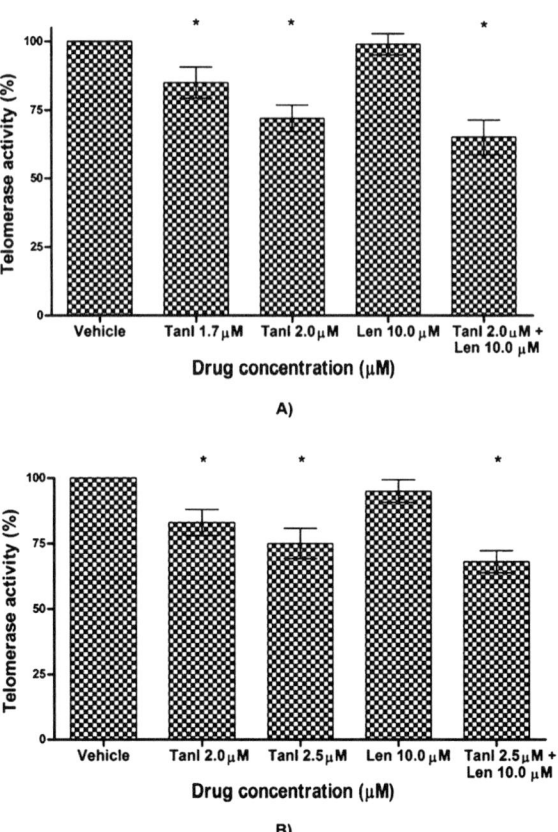

Fig. 14.6 Determination of relative telomerase activity after treatment of myeloma cells (RPMI 8226 and U266) with Tanshinone I for 24 hr. (Adapted with permission from Ref. (Kumar et al. 2018a))

14.7.5 Currently Used Inhibitors of Telomerase

Recently, a large number of chemically generated drugs targeting different sites of telomerase have been developed to fight against cancer cells. Among all, GRN163L, also called as Imetelstat, is by far the most widely used and successful chemotherapeutic drug that targets telomerase. It is 13-mer antisense oligonucleotides, which form a complementary pairing with the RNA template of the telomerase and inhibits its action (Chiappori et al. 2015). Hochreiter et al. observed that GRN163L dose dependently decreased telomerase activity and also reduced tumorigenicity by causing reduction in cell growth, metastasis, and invasiveness of breast cancer cells (Hochreiter et al. 2006). Similarly, in the breast and pancreatic cancer cells, GRN163L treatment resulted in the significant induction of apoptosis and impaired cellular growth. Simultaneously, the inhibitory effect was much more pronounced with the cancer cells with critically shortened telomeres (Burchett et al. 2014). Due to nonspecific toxicity and longer duration of treatment, these drugs showed limited clinical efficacy despite excellent inhibitory potential.

Another strategy aimed to inhibit the telomerase involved increased stabilization and generation of endogenous higher-order telomeric structures such as G-quadruplex. These naturally occurring secondary structures are rich in guanine, and they form planar structures which impede the movement of replication machinery through replication fork (Maestroni et al. 2017). Normally, these structures are resolved by telomerase, but utilization of G-quadruplex stabilizing ligands will prevent the access of telomerase and thus holds great promises for various malignant and progressive cancers. These ligands including telomestatin, BRACO-19, and RHPS4 significantly increased the G-quadruplex stability and ultimately enhanced the DDR pathways in cancer cells (Burger et al. 2005; Cookson et al. 2005). There are few indirect treatments available for targeting telomere or telomerase which includes tankyrase inhibitors and shelterin components inhibitors which have also shown promising effects in limiting the length of telomere and activating various DNA repair pathways.

14.8 Conclusion

At chromosomal ends, telomeres in association with telomeric proteins are organized into highly specialized nucleoprotein structures. These structures play significant role in maintaining telomere homeostasis. Normal somatic cells display limited replicative capacity due to end replication problem. However, for cancer cells to undergo continued proliferation, the maintenance of telomere length is critical. This cellular self-renewal capacity of cancer cells is regulated by reactivation of telomerase. Studies showed in this chapter supported that different mechanisms are involved in regulation of telomerase expression and reactivation. Literature supported that different genetic (promoter mutation) and epigenetic mechanisms (promoter methylation and miRNAs) are involved in reactivation of telomerase and further projected them as suitable diagnostic, prognostic, and therapeutic markers. Future research is focusing on cues and signals targeting these epigenetic changes and how these methylation patterns can be used as therapeutic targets. Further, due to selective expression of telomerase in cancer cells plethora of compounds targeting telomerase have been explored. Current literature provided ample evidences that there is a plethora of natural compounds that regulate telomerase levels at multiple genetic and epigenetic levels and also inhibits cellular proliferation and induces apoptosis. Few natural compounds such as resveratrol and curcumin and synthetic compounds such as imetelstat and G-quadruplex stabilizers have shown promising anticancer potential against multitude of cancers. This chapter concludes the telomerase biomarker and therapeutic implications in cancers and projected it as a suitable theranostic marker in future.

References

Abreu E, Aritonovska E, Reichenbach P, Cristofari G, Culp B, Terns RM, Lingner J, Terns MP (2010) TIN2-tethered TPP1 recruits human telomerase to telomeres in vivo. Mol Cell Biol 30(12):2971–2982. https://doi.org/10.1128/MCB.00240-10

Albertson DG (2006) Gene amplification in cancer. Trends Genet 22(8):447–455. https://doi.org/10.1016/j.tig.2006.06.007

Arnoult N, Karlseder J (2015) Complex interactions between the DNA-damage response and mammalian telomeres. Nat Struct Mol Biol 22(11):859–866. https://doi.org/10.1038/nsmb.3092

Avci CB, Yilmaz S, Dogan ZO, Saydam G, Dodurga Y, Ekiz HA, Kartal M, Sahin F, Baran Y, Gunduz C (2011) Quercetin-induced apoptosis involves increased hTERT enzyme activity of leukemic cells. Hematology 16(5):303–307. https://doi.org/10.1179/102453311X13085644680104

Bai L, Wang H, Wang A-H, Zhang L-Y, Bai J (2017) MicroRNA-532 and microRNA-3064 inhibit cell proliferation and invasion by acting as direct regulators of human telomerase reverse transcriptase in ovarian cancer. PLoS One 12(3):e0173912. https://doi.org/10.1371/journal.pone.0173912

Barczak W, Suchorska WM, Sobecka A, Bednarowicz K, Machczynski P, Golusinski P, Rubis B, Masternak MM, Golusinski W (2017) HTERT C250T promoter mutation and telomere length as a molecular markers of cancer progression in patients with head and neck cancer. Mol Med Rep 16(1):441–446. https://doi.org/10.3892/mmr.2017.6590

Bartlett TE, Zaikin A, Olhede SC, West J, Teschendorff AE, Widschwendter M (2013) Corruption of the intra-gene DNA methylation architecture is a hallmark of cancer. PLoS One 8(7):e68285. https://doi.org/10.1371/journal.pone.0068285

Baylin SB, Jones PA (2011) A decade of exploring the cancer epigenome—biological and translational implications. Nat Rev Cancer 11(10):726–734. https://doi.org/10.1038/nrc3130

Behjati M, Hashemi M, Kazemi M, Salehi M, Javanmard S (2017) Evaluation of energy balance on Human Telomerase Reverse Transcriptase (hTERT) alternative splicing by semi-quantitative RT-PCR in human umbilical vein endothelial cells. Adv Biomed Res 6(1):43. https://doi.org/10.4103/2277-9175.204591

Blackburn EH, Gall JG (1978) A tandemly repeated sequence at the termini of the extrachromosomal ribosomal RNA genes in Tetrahymena. J Mol Biol 120(1):33–53. https://doi.org/10.1016/0022-2836(78)90294-2

Brien TP, Kallakury BV, Lowry CV, Ambros RA, Muraca PJ, Malfetano JH, Ross JS (1997) Telomerase activity in benign endometrium and endometrial carcinoma. Cancer Res 57(13):2760–2764

Broccoli D, Young JW, de Lange T (1995) Telomerase activity in normal and malignant hematopoietic cells. Proc Natl Acad Sci 92(20):9082–9086. https://doi.org/10.1073/pnas.92.20.9082

Burchett KM, Yan Y, Ouellette MM (2014) Telomerase Inhibitor Imetelstat (GRN163L) limits the lifespan of human pancreatic cancer cells. PLoS One 9(1):e85155. https://doi.org/10.1371/journal.pone.0085155

Burger AM, Dai F, Schultes CM, Reszka AP, Moore MJ, Double JA, Neidle S (2005) The G-quadruplex-interactive molecule BRACO-19 inhibits tumor growth, consistent with telomere targeting and interference with telomerase function. Cancer Res 65(4):1489–1496. https://doi.org/10.1158/0008-5472.CAN-04-2910

Butler KS, Hines WC, Heaphy CM, Griffith JK (2012) Coordinate regulation between expression levels of telomere-binding proteins and telomere length in breast carcinomas. Cancer Med 1(2):165–175. https://doi.org/10.1002/cam4.14

Castelo-Branco P, Choufani S, Mack S, Gallagher D, Zhang C, Lipman T, Zhukova N, Walker EJ, Martin D, Merino D, Wasserman JD, Elizabeth C, Alon N, Zhang L, Hovestadt V, Kool M, Jones DT, Zadeh G, Croul S et al (2013) Methylation of the TERT promoter and risk stratification of childhood brain tumours: an integrative genomic and molecular study. Lancet Oncol 14(6):534–542. https://doi.org/10.1016/S1470-2045(13)70110-4

Castro-Vega LJ, Jouravleva K, Ortiz-Montero P, Liu W-Y, Galeano JL, Romero M, Popova T, Bacchetti S, Vernot JP, Londoño-Vallejo A (2015) The senescent microenvironment promotes the emergence of heterogeneous cancer stem-like cells. Carcinogenesis 36(10):1180–1192. https://doi.org/10.1093/carcin/bgv101

Cevatemre B, Erkısa M, Aztopal N, Karakas D, Alper P, Tsimplouli C, Sereti E, Dimas K, Armutak EII, Gurevin EG, Uvez A, Mori M, Berardozzi S, Ingallina C, D'Acquarica I, Botta B, Ozpolat B, Ulukaya E (2018) A promising natural product, pristimerin, results in cytotoxicity against breast cancer stem cells in vitro and xenografts in vivo through apoptosis and an incomplete autophagy in breast cancer. Pharmacol Res 129:500–514. https://doi.org/10.1016/j.phrs.2017.11.027

Chen R-J, Wu P-H, Ho C-T, Way T-D, Pan M-H, Chen H-M, Ho Y-S, Wang Y-J (2017) P53-dependent downregulation of hTERT protein expression and telomerase activity induces senescence in lung cancer cells as a result of pterostilbene treatment. Cell Death Dis 8(8):e2985–e2985. https://doi.org/10.1038/cddis.2017.333

Chiappori AA, Kolevska T, Spigel DR, Hager S, Rarick M, Gadgeel S, Blais N, Von Pawel J, Hart L, Reck M, Bassett E, Burington B, Schiller JH (2015) A randomized phase II study of the telomerase inhibitor imetelstat as maintenance therapy for advanced non-small-cell lung cancer. Ann Oncol 26(2):354–362. https://doi.org/10.1093/annonc/mdu550

Cho WC (2007) OncomiRs: the discovery and progress of microRNAs in cancers. Mol Cancer 6(1):60. https://doi.org/10.1186/1476-4598-6-60

Cohen SB, Graham ME, Lovrecz GO, Bache N, Robinson PJ, Reddel RR (2007) Protein composition of catalytically active human telomerase from immortal cells. Science 315(5820):1850–1853. https://doi.org/10.1126/science.1138596

Cong Y (1999) The human telomerase catalytic subunit hTERT: organization of the gene and characterization of the promoter. Hum Mol Genet 8(1):137–142. https://doi.org/10.1093/hmg/8.1.137

Cong Y-S, Wright WE, Shay JW (2002) Human telomerase and its regulation. Microbiol Mol Biol Rev 66(3):407–425. https://doi.org/10.1128/MMBR.66.3.407-425.2002

Cookson JC, Dai F, Smith V, Heald RA, Laughton CA, Stevens MFG, Burger AM (2005) Pharmacodynamics of the G-quadruplex-stabilizing telomerase inhibitor 3,11-Difluoro-6,8,13-trimethyl-8 H -quino[4,3,2- kl]acridinium methosulfate (RHPS4) in vitro: activity in human tumor cells correlates with telomere length and can be enhanced, or antagonized, with cytotoxic agents. Mol Pharmacol 68(6):1551–1558. https://doi.org/10.1124/mol.105.013300

Cui S-X, Qu X-J, Xie Y-Y, Zhou L, Nakata M, Makuuchi M, Tang W (2006) Curcumin inhibits telomerase activity in human cancer cell lines. Int J Mol Med 18(2):227–231

de Lange T (2005) Shelterin: The protein complex that shapes and safeguards human telomeres. Genes Dev 19(18):2100–2110. https://doi.org/10.1101/gad.1346005

de Lange T (2009) How telomeres solve the end-protection problem. Science 326(5955):948–952. https://doi.org/10.1126/science.1170633

de Lange, T., 2010. How Shelterin solves the telomere end-protection problem. Cold Spring Harb Symp Quant Biol, 75(0), 167–177. doi: https://doi.org/10.1101/sqb.2010.75.017

de Lange T (2015) A loopy view of telomere evolution. Front Genet 6. https://doi.org/10.3389/fgene.2015.00321

de Lange T, Shiue L, Myers RM, Cox DR, Naylor SL, Killery AM, Varmus HE (1990) Structure and variability of human chromosome ends. Mol Cell Biol 10(2):518–527. https://doi.org/10.1128/mcb.10.2.518

Deeb D, Gao X, Liu Y, Varma N, Arbab A, Gautam S (2013) Inhibition of telomerase activity by oleanane triterpenoid CDDO-Me in pancreatic cancer cells is ROS-dependent. Molecules 18(3):3250–3265. https://doi.org/10.3390/molecules18033250

Deeb D, Gao X, Liu Y, Pindolia K, Gautam SC (2015) Inhibition of hTERT/telomerase contributes to the antitumor activity of pristimerin in pancreatic ductal adenocarcinoma cells. Oncol Rep 34(1):518–524. https://doi.org/10.3892/or.2015.3989

Denchi EL, de Lange T (2007) Protection of telomeres through independent control of ATM and ATR by TRF2 and POT1. Nature 448(7157):1068–1071. https://doi.org/10.1038/nature06065

Deng J, Zhou D, Zhang J, Chen Y, Wang C, Liu Y, Zhao K (2015) Aberrant methylation of the TERT promoter in esophageal squamous cell carcinoma. Cancer Genet 208(12):602–609. https://doi.org/10.1016/j.cancergen.2015.10.004

Doksani Y, Wu JY, de Lange T, Zhuang X (2013) Super-resolution fluorescence imaging of telomeres reveals TRF2-dependent T-loop formation. Cell 155(2):345–356. https://doi.org/10.1016/j.cell.2013.09.048

Eckburg A, Dein J, Berei J, Schrank Z, Puri N (2020) Oligonucleotides and microRNAs targeting telomerase subunits in cancer therapy. Cancers 12(9):2337. https://doi.org/10.3390/cancers12092337

Feng J, Funk W, Wang S, Weinrich S, Avilion A, Chiu C, Adams R, Chang E, Allsopp R, Yu J et al (1995) The RNA component of human telomerase. Science 269(5228):1236–1241. https://doi.org/10.1126/science.7544491

Fernández-Marcelo T, Gómez A, Pascua I, de Juan C, Head J, Hernando F, Jarabo J-R, Calatayud J, Torres-García A-J, Iniesta P (2015) Telomere length and telomerase activity in non-small cell lung cancer prognosis: clinical usefulness of a specific telomere status. J Exp Clin Cancer Res 34(1):78. https://doi.org/10.1186/s13046-015-0195-9

Franceschin M, Rossetti L, D'Ambrosio A, Schirripa S, Bianco A, Ortaggi G, Savino M, Schultes C, Neidle S (2006) Natural and synthetic G-quadruplex interactive berberine derivatives. Bioorg Med Chem Lett 16(6):1707–1711. https://doi.org/10.1016/j.bmcl.2005.12.001

Frescas, D., and de Lange, T., 2014a. Binding of TPP1 Protein to TIN2 protein is required for POT1a,b protein-mediated telomere protection. J Biol Chem, 289(35), 24180–24187. doi: https://doi.org/10.1074/jbc.M114.592592

Frescas D, de Lange T (2014b) TRF2-Tethered TIN2 can mediate telomere protection by TPP1/POT1. Mol Cell Biol 34(7):1349–1362. https://doi.org/10.1128/MCB.01052-13

Fuggetta MP, Lanzilli G, Tricarico M, Cottarelli A, Falchetti R, Ravagnan G, Bonmassar E (2006) Effect of resveratrol on proliferation and telomerase activity of human colon cancer cells in vitro. J Exp Clin Cancer Res: CR 25(2):189–193

Fürtjes G, Köchling M, Peetz-Dienhart S, Wagner A, Heß K, Hasselblatt M, Senner V, Stummer W, Paulus W, Brokinkel B (2016) HTERT promoter methylation in meningiomas and central nervous hemangiopericytomas. J Neuro-Oncol 130(1):79–87. https://doi.org/10.1007/s11060-016-2226-6

Gao, K., Li, G., Qu, Y., Wang, M., Cui, B., Ji, M., Shi, B., and Hou, P., 2016. TERT promoter mutations and long telomere length predict poor survival and radiotherapy resistance in gliomas. Oncotarget 7(8), 8712–8725. doi: https://doi.org/10.18632/oncotarget.6007

Gerhardt D, Bertola G, Dietrich F, Figueiró F, Zanotto-Filho A, Moreira Fonseca JC, Morrone FB, Barrios CH, Battastini AMO, Salbego CG (2014) Boldine induces cell cycle arrest and apoptosis in T24 human bladder cancer cell line via regulation of ERK, AKT, and GSK-3β. Urol Oncol: Semin Original Investig 32(1):36.e1–36.e9. https://doi.org/10.1016/j.urolonc.2013.02.012

Glybochko PV, Zezerov EG, Glukhov AI, Alyaev YG, Severin SE, Polyakovsky KA, Varshavsky VA, Severin ES, Vinarov AZ (2014) Telomerase as a tumor marker in diagnosis of prostatic intraepithelial neoplasia and prostate cancer: telomerase-A tumor marker. Prostate 74(10):1043–1051. https://doi.org/10.1002/pros.22823

Greider CW (1991) Telomerase is processive. Mol Cell Biol 11(9):4572–4580. https://doi.org/10.1128/MCB.11.9.4572

Greider CW, Blackburn EH (1985) Identification of a specific telomere terminal transferase activity in tetrahymena extracts. Cell 43(2):405–413. https://doi.org/10.1016/0092-8674(85)90170-9

Greider CW, Blackburn EH (1989) A telomeric sequence in the RNA of Tetrahymena telomerase required for telomere repeat synthesis. Nature 337(6205):331–337. https://doi.org/10.1038/337331a0

Griffiths K, Aggarwal B, Singh R, Buttar H, Wilson D, De Meester F (2016) Food antioxidants and their anti-inflammatory properties: a potential role in cardiovascular diseases and cancer prevention. Diseases 4(4):28. https://doi.org/10.3390/diseases4030028

Guilleret I, Yan P, Grange F, Braunschweig R, Bosman FT, Benhattar J (2002) Hypermethylation of the human telomerase catalytic subunit (hTERT) gene correlates with telomerase activity. Int J Cancer 101(4):335–341. https://doi.org/10.1002/ijc.10593

Gunisova S, Elboher E, Nosek J, Gorkovoy V, Brown Y, Lucier J-F, Laterreur N, Wellinger RJ, Tzfati Y, Tomaska L (2009) Identification and comparative analysis of telomerase RNAs from Candida species reveal conservation of functional elements. RNA 15(4):546–559. https://doi.org/10.1261/rna.1194009

Gurung RL, Lim SN, Low GKM, Hande MP (2015) MST-312 Alters telomere dynamics, gene expression profiles and growth in human breast cancer cells. J Nutrigenet Nutrigenomics 7(4–6):283–298. https://doi.org/10.1159/000381346

Guterres AN, Villanueva J (2020) Targeting telomerase for cancer therapy. Oncogene 39(36):5811–5824. https://doi.org/10.1038/s41388-020-01405-w

Harrington L, Zhou W, McPhail T, Oulton R, Yeung DSK, Mar V, Bass MB, Robinson MO (1997) Human telomerase contains evolutionarily conserved catalytic and structural subunits. Genes Dev 11(23):3109–3115. https://doi.org/10.1101/gad.11.23.3109

Hayashi MT, Cesare AJ, Rivera T, Karlseder J (2015) Cell death during crisis is mediated by mitotic telomere deprotection. Nature 522(7557):492–496. https://doi.org/10.1038/nature14513

Hochreiter AE, Xiao H, Goldblatt EM, Gryaznov SM, Miller KD, Badve S, Sledge GW, Herbert B-S (2006) Telomerase template antagonist GRN163L disrupts telomere maintenance, tumor growth, and metastasis of breast cancer. Clin Cancer Res 12(10):3184–3192. https://doi.org/10.1158/1078-0432.CCR-05-2760

Hoffman H, Rice C, Skordalakes E (2017) Structural analysis reveals the deleterious effects of telomerase mutations in bone marrow failure syndromes. J Biol Chem 292(11):4593–4601. https://doi.org/10.1074/jbc.M116.771204

Horn S, Figl A, Rachakonda PS, Fischer C, Sucker A, Gast A, Kadel S, Moll I, Nagore E, Hemminki K, Schadendorf D, Kumar R (2013) TERT Promoter mutations in familial and sporadic melanoma. Science 339(6122):959–961. https://doi.org/10.1126/science.1230062

Hrdličková R, Nehyba J, Bargmann W, Bose HR (2014) Multiple tumor suppressor microRNAs regulate telomerase and TCF7, an important transcriptional regulator of the Wnt pathway. PLoS One 9(2):e86990. https://doi.org/10.1371/journal.pone.0086990

Hu H, Zhang Y, Zou M, Yang S, Liang X-Q (2010) Expression of TRF1, TRF2, TIN2, TERT, KU70, and BRCA1 proteins is associated with telomere shortening and may contribute to multistage carcinogenesis of gastric cancer. J Cancer Res Clin Oncol 136(9):1407–1414. https://doi.org/10.1007/s00432-010-0795-x

Huang FW, Hodis E, Xu MJ, Kryukov GV, Chin L, Garraway LA (2013) Highly recurrent TERT promoter mutations in human melanoma. Science 339(6122):957–959. https://doi.org/10.1126/science.1229259

Janoušková E, Nečasová I, Pavloušková J, Zimmermann M, Hluchý M, Marini V, Nováková M, Hofr C (2015) Human Rap1 modulates TRF2 attraction to telomeric DNA. Nucleic Acids Res 43(5):2691–2700. https://doi.org/10.1093/nar/gkv097

Ji X, Sun H, Zhou H, Xiang J, Tang Y, Zhao C (2012) The interaction of telomeric DNA and C-myc22 G-Quadruplex with 11 natural alkaloids. Nucleic Acid Ther 22(2):127–136. https://doi.org/10.1089/nat.2012.0342

Jin A, Xu J, Wang Y (2018) The role of TERT promoter mutations in postoperative and preoperative diagnosis and prognosis in thyroid cancer. Medicine 97(29):e11548. https://doi.org/10.1097/MD.0000000000011548

Kawashima M, Kojima M, Ueda Y, Kurihara S, Hiyama E (2016) Telomere biology including TERT rearrangements in neuroblastoma: a useful indicator for surgical treatments. J Pediatr Surg 51(12):2080–2085. https://doi.org/10.1016/j.jpedsurg.2016.09.042

Kazemi Noureini S, Wink M (2015) Dose-dependent cytotoxic effects of boldine in HepG-2 cells—telomerase inhibition and apoptosis induction. Molecules 20(3):3730–3743. https://doi.org/10.3390/molecules20033730

Killela PJ, Reitman ZJ, Jiao Y, Bettegowda C, Agrawal N, Diaz LA, Friedman AH, Friedman H, Gallia GL, Giovanella BC, Grollman AP, He T-C, He Y, Hruban RH, Jallo GI, Mandahl N, Meeker AK, Mertens F, Netto GJ et al (2013) TERT promoter mutations occur frequently in gliomas and a subset of tumors derived from cells with low rates of self-renewal. Proc Natl Acad Sci 110(15):6021–6026. https://doi.org/10.1073/pnas.1303607110

Kim M-Y, Gleason-Guzman M, Izbicka E, Nishioka D, Hurley LH (2003) The different biological effects of telomestatin and TMPyP4 can be attributed to their selectivity for interaction with intramolecular or intermolecular G-quadruplex structures. Cancer Res 63(12):3247–3256

Kinde, I., Munari, E., Faraj, S. F., Hruban, R. H., Schoenberg, M., Bivalacqua, T., Allaf, M., Springer, S., Wang, Y., Diaz, L. A., Kinzler, K. W., Vogelstein, B., Papadopoulos, N., Netto, G. J., 2013. TERT Promoter mutations occur early in urothelial neoplasia and are biomarkers of early disease and disease recurrence in Urine. Cancer Res, 73(24), 7162–7167. doi: https://doi.org/10.1158/0008-5472.CAN-13-2498

Kipling D, Cooke HJ (1990) Hypervariable ultra-long telomeres in mice. Nature 347(6291): 400–402. https://doi.org/10.1038/347400a0

Klobutcher LA, Swanton MT, Donini P, Prescott DM (1981) All gene-sized DNA molecules in four species of hypotrichs have the same terminal sequence and an unusual 3′ terminus. Proc Natl Acad Sci 78(5):3015–3019. https://doi.org/10.1073/pnas.78.5.3015

Kulić A, Plavetić ND, Gamulin S, Jakić-Razumović J, Vrbanec D, Sirotković-Skerlev M (2016) Telomerase activity in breast cancer patients: association with poor prognosis and more aggressive phenotype. Med Oncol 33(3):23. https://doi.org/10.1007/s12032-016-0736-x

Kumar R, Gupta N, Himani, and Sharma, A. (2018a) Novel combination of tanshinone I and lenalidomide induces chemo-sensitivity in myeloma cells by modulating telomerase activity and expression of shelterin complex and its associated molecules. Mol Biol Rep 45(6):2429–2439. https://doi.org/10.1007/s11033-018-4409-z

Kumar R, Khan R, Gupta N, Seth T, Sharma A, Kalaivani M, Sharma A (2018b) Identifying the biomarker potential of telomerase activity and shelterin complex molecule, telomeric repeat binding factor 2 (TERF2), in multiple myeloma. Leuk Lymphoma 59(7):1677–1689. https://doi.org/10.1080/10428194.2017.1387915

Kyo S, Takakura M, Fujiwara T, Inoue M (2008) Understanding and exploiting *hTERT* promoter regulation for diagnosis and treatment of human cancers. Cancer Sci 99(8):1528–1538. https://doi.org/10.1111/j.1349-7006.2008.00878.x

Leão R, Apolónio JD, Lee D, Figueiredo A, Tabori U, Castelo-Branco P (2018) Mechanisms of human telomerase reverse transcriptase (hTERT) regulation: clinical impacts in cancer. J Biomed Sci 25(1):22. https://doi.org/10.1186/s12929-018-0422-8

Lee JH, Chung IK (2010) Curcumin inhibits nuclear localization of telomerase by dissociating the Hsp90 co-chaperone p23 from hTERT. Cancer Lett 290(1):76–86. https://doi.org/10.1016/j.canlet.2009.08.026

Li J, Lei H, Xu Y, Tao Z (2015) MiR-512-5p Suppresses tumor growth by targeting hTERT in telomerase positive head and neck squamous cell carcinoma in vitro and in vivo. PLoS One 10(8):e0135265. https://doi.org/10.1371/journal.pone.0135265

Lillard-Wetherell K, Machwe A, Langland GT, Combs KA, Behbehani GK, Schonberg SA, German J, Turchi JJ, Orren DK, Groden J (2004) Association and regulation of the BLM helicase by the telomere proteins TRF1 and TRF2. Hum Mol Genet 13(17):1919–1932. https://doi.org/10.1093/hmg/ddh193

Liu L, Lai S, Andrews LG, Tollefsbol TO (2004) Genetic and epigenetic modulation of telomerase activity in development and disease. Gene 340(1):1–10. https://doi.org/10.1016/j.gene.2004.06.011

Liu YB, Gao X, Deeb D, Pindolia K, Gautam SC (2015a) Role of telomerase in anticancer activity of pristimerin in prostate cancer cells. J Exp Ther Oncol 11(1):41–49

Liu K, Li L, Rusidanmu A, Wang Y, Lv X (2015b) Down-regulation of MiR-1294 is related to dismal prognosis of patients with esophageal squamous cell carcinoma through elevating C-MYC expression. Cell Physiol Biochem 36(1):100–110. https://doi.org/10.1159/000374056

Liu L, Zuo J, Wang G (2017) Epigallocatechin-3-gallate suppresses cell proliferation and promotes apoptosis in Ec9706 and Eca109 esophageal carcinoma cells. Oncol Lett 14(4):4391–4395. https://doi.org/10.3892/ol.2017.6712

Lou M, Zhang L, Ji P, Feng F, Liu J, Yang C, Li B, Wang L (2016) Quercetin nanoparticles induced autophagy and apoptosis through AKT/ERK/Caspase-3 signaling pathway in human neuroglioma cells: in vitro and in vivo. Biomed Pharmacother 84:1–9. https://doi.org/10.1016/j.biopha.2016.08.055

Lue NF, Lin Y-C, Mian IS (2003) A conserved telomerase motif within the catalytic domain of telomerase reverse transcriptase is specifically required for repeat addition processivity. Mol Cell Biol 23(23):8440–8449. https://doi.org/10.1128/MCB.23.23.8440-8449.2003

Maestroni L, Matmati S, Coulon S (2017) Solving the telomere replication problem. Genes 8(2):55. https://doi.org/10.3390/genes8020055

McClintock B (1941) The stability of broken ends of chromosomes in Zea Mays. Genetics 26(2):234–282

Mitomo S, Maesawa C, Ogasawara S, Iwaya T, Shibazaki M, Yashima-Abo A, Kotani K, Oikawa H, Sakurai E, Izutsu N, Kato K, Komatsu H, Ikeda K, Wakabayashi G, Masuda T (2008) Downregulation of miR-138 is associated with overexpression of human telomerase reverse transcriptase protein in human anaplastic thyroid carcinoma cell lines. Cancer Sci 99(2):280–286. https://doi.org/10.1111/j.1349-7006.2007.00666.x

Muller HJ (1938) The remaking of chromosomes. Collect Nat 13

Naasani I, Seimiya H, Yamori T, Tsuruo T (1999) FJ5002: a potent telomerase inhibitor identified by exploiting the disease-oriented screening program with COMPARE analysis. Cancer Res 59(16):4004–4011

Naderlinger E, Holzmann K (2017) Epigenetic regulation of telomere maintenance for therapeutic interventions in gliomas. Genes 8(5):145. https://doi.org/10.3390/genes8050145

Nguyen THD, Collins K, Nogales E (2019) Telomerase structures and regulation: shedding light on the chromosome end. Curr Opin Struct Biol 55:185–193. https://doi.org/10.1016/j.sbi.2019.04.009

O'Connor CM, Lai CK, Collins K (2005) Two purified domains of telomerase reverse transcriptase reconstitute sequence-specific interactions with RNA. J Biol Chem 280(17):17533–17539. https://doi.org/10.1074/jbc.M501211200

Olovnikov AM (1973) A theory of marginotomy. J Theor Biol 41(1):181–190. https://doi.org/10.1016/0022-5193(73)90198-7

Opresko PL, von Kobbe C, Laine J-P, Harrigan J, Hickson ID, Bohr VA (2002) Telomere-binding protein TRF2 binds to and stimulates the Werner and bloom syndrome helicases. J Biol Chem 277(43):41110–41119. https://doi.org/10.1074/jbc.M205396200

Pan H, Jansson KH, Beshiri ML, Yin J, Fang L, Agarwal S, Nguyen H, Corey E, Zhang Y, Liu J, Fan H, Lin H, Kelly K (2017) Gambogic acid inhibits thioredoxin activity and induces ROS-mediated cell death in castration-resistant prostate cancer. Oncotarget 8(44):77181–77194. https://doi.org/10.18632/oncotarget.20424

Peifer M, Hertwig F, Roels F, Dreidax D, Gartlgruber M, Menon R, Krämer A, Roncaioli JL, Sand F, Heuckmann JM, Ikram F, Schmidt R, Ackermann S, Engesser A, Kahlert Y, Vogel W, Altmüller J, Nürnberg P, Thierry-Mieg J et al (2015) Telomerase activation by genomic rearrangements in high-risk neuroblastoma. Nature 526(7575):700–704. https://doi.org/10.1038/nature14980

Piscuoglio S, Ng CK, Murray M, Burke KA, Edelweiss M, Geyer FC, Macedo GS, Inagaki A, Papanastasiou AD, Martelotto LG, Marchio C, Lim RS, Ioris RA, Nahar PK, Bruijn ID, Smyth L, Akram M, Ross D, Petrini JH et al (2016) Massively parallel sequencing of phyllodes tumours of the breast reveals actionable mutations, and *TERT* promoter hotspot mutations and *TERT* gene amplification as likely drivers of progression: *TERT* alterations in Phyllodes tumors. J Pathol 238(4):508–518. https://doi.org/10.1002/path.4672

Ramachandran C, Fonseca HB, Jhabvala P, Escalon EA, Melnick SJ (2002) Curcumin inhibits telomerase activity through human telomerase reverse transcriptase in MCF-7 breast cancer cell line. Cancer Lett 184(1):1–6. https://doi.org/10.1016/S0304-3835(02)00192-1

Ramirez RD, Wright WE, Shay JW, Taylor RS (1997) Telomerase activity concentrates in the mitotically active segments of human hair follicles. J Investig Dermatol 108(1):113–117. https://doi.org/10.1111/1523-1747.ep12285654

Ren K-W, Li Y-H, Wu G, Ren J-Z, Lu H-B, Li Z-M, Han X-W (2017) Quercetin nanoparticles display antitumor activity via proliferation inhibition and apoptosis induction in liver cancer cells. Int J Oncol 50(4):1299–1311. https://doi.org/10.3892/ijo.2017.3886

Renaud S (2005) CTCF binds the proximal exonic region of hTERT and inhibits its transcription. Nucleic Acids Res 33(21):6850–6860. https://doi.org/10.1093/nar/gki989

Roake CM, Artandi SE (2020) Regulation of human telomerase in homeostasis and disease. Nat Rev Mol Cell Biol 21(7):384–397. https://doi.org/10.1038/s41580-020-0234-z

Robart AR, Collins K (2011) Human telomerase domain interactions capture DNA for TEN domain-dependent processive elongation. Mol Cell 42(3):308–318. https://doi.org/10.1016/j.molcel.2011.03.012

Rouda S, Skordalakes E (2007) Structure of the RNA-binding domain of telomerase: implications for RNA recognition and binding. Structure 15(11):1403–1412. https://doi.org/10.1016/j.str.2007.09.007

Rupaimoole R, Slack FJ (2017) MicroRNA therapeutics: towards a new era for the management of cancer and other diseases. Nat Rev Drug Discov 16(3):203–222. https://doi.org/10.1038/nrd.2016.246

Saito T, Matsuda Y, Suzuki T, Hayashi A, Yuan X, Saito M, Nakayama J, Hori T, Ishikawa F (1997) comparative gene mapping of the human and mouse TEP1 genes, which encode one protein component of telomerases. Genomics 46(1):46–50. https://doi.org/10.1006/geno.1997.5005

Sharma A, Rajappa M, Saxena A, Sharma M (2007) Telomerase activity as a tumor marker in Indian women with cervical intraepithelial neoplasia and cervical cancer. Mol Diag Ther 11(3):193–201. https://doi.org/10.1007/BF03256241

Shay JW, Wright WE (2011) Role of telomeres and telomerase in cancer. Semin Cancer Biol 21(6):349–353. https://doi.org/10.1016/j.semcancer.2011.10.001

Shin K-H, Kang MK, Dicterow E, Park N-H (2003) Hypermethylation of the hTERT promoter inhibits the expression of telomerase activity in normal oral fibroblasts and senescent normal oral keratinocytes. Br J Cancer 89(8):1473–1478. https://doi.org/10.1038/sj.bjc.6601291

Shin-ya K, Wierzba K, Matsuo K, Ohtani T, Yamada Y, Furihata K, Hayakawa Y, Seto H (2001) Telomestatin, a Novel telomerase inhibitor from *Streptomyces anulatus*. J Am Chem Soc 123(6):1262–1263. https://doi.org/10.1021/ja005780q

Smith S (1998) Tankyrase, a Poly(ADP-Ribose) Polymerase at human telomeres. Science 282(5393):1484–1487. https://doi.org/10.1126/science.282.5393.1484

Smith EM, Pendlebury DF, Nandakumar J (2020) Structural biology of telomeres and telomerase. Cell Mol Life Sci 77(1):61–79. https://doi.org/10.1007/s00018-019-03369-x

Spiegl-Kreinecker, S., Lötsch, D., Neumayer, K., Kastler, L., Gojo, J., Pirker, C., Pichler, J., Weis, S., Kumar, R., Webersinke, G., Gruber, A., and Berger, W., 2018. *TERT* promoter mutations are associated with poor prognosis and cell immortalization in meningioma. Neuro-Oncology, 20(12), 1584–1593. doi: https://doi.org/10.1093/neuonc/noy104

Svahn F, Paulsson J, Stenman A, Fotouhi O, Mu N, Murtha T, Korah R, Carling T, Bäckdahl M, Wang N, Juhlin C, Larsson C (2018) TERT promoter hypermethylation is associated with poor prognosis in adrenocortical carcinoma. Int J Mol Med. https://doi.org/10.3892/ijmm.2018.3735

Szostak JW, Blackburn EH (1982) Cloning yeast telomeres on linear plasmid vectors. Cell 29(1): 245–255. https://doi.org/10.1016/0092-8674(82)90109-X

Tahara H, Kuniyasu H, Yokozaki H, Yasui W, Shay JW, Ide T, Tahara E (1995) Telomerase activity in preneoplastic and neoplastic gastric and colorectal lesions. Clin Cancer Res 1(11): 1245–1251

Tiedemann RE, Schmidt J, Keats JJ, Shi C-X, Zhu YX, Palmer SE, Mao X, Schimmer AD, Stewart AK (2009) Identification of a potent natural triterpenoid inhibitor of proteosome chymotrypsin-like activity and NF-κB with antimyeloma activity in vitro and in vivo. Blood 113(17): 4027–4037. https://doi.org/10.1182/blood-2008-09-179796

Tippani R, Prakhya L, Porika M, Sirisha K, Abbagani S, Thammidala C (2014) Pterostilbene as a potential novel telomerase inhibitor: molecular docking studies and its in vitro evaluation. Curr Pharm Biotechnol 14(12):1027–1035. https://doi.org/10.2174/1389201015666140113112820

Tsujioka T, Yokoi A, Itano Y, Takahashi K, Ouchida M, Okamoto S, Kondo T, Suemori S, Tohyama Y, Tohyama K (2015) Five-aza-2′-deoxycytidine-induced hypomethylation of cholesterol 25-hydroxylase gene is responsible for cell death of myelodysplasia/leukemia cells. Sci Rep 5(1):16709. https://doi.org/10.1038/srep16709

Turgut Cosan D, Soyocak A, Basaran A, Degirmenci İ, Gunes HV, Mutlu Sahin F (2011) Effects of various agents on DNA fragmentation and telomerase enzyme activities in adenocarcinoma cell lines. Mol Biol Rep 38(4):2463–2469. https://doi.org/10.1007/s11033-010-0382-x

Umbricht CB, Saji M, Westra WH, Udelsman R, Zeiger MA, Sukumar S (1997) Telomerase activity: a marker to distinguish follicular thyroid adenoma from carcinoma. Cancer Res 57(11):2144–2147

Venteicher AS, Meng Z, Mason PJ, Veenstra TD, Artandi SE (2008) Identification of ATPases pontin and reptin as telomerase components essential for holoenzyme assembly. Cell 132(6): 945. https://doi.org/10.1016/j.cell.2008.01.019

Vinagre J, Almeida A, Pópulo H, Batista R, Lyra J, Pinto V, Coelho R, Celestino R, Prazeres H, Lima L, Melo M, da Rocha AG, Preto A, Castro P, Castro L, Pardal F, Lopes JM, Santos LL, Reis RM (2013) Frequency of TERT promoter mutations in human cancers. Nat Commun 4(1): 2185. https://doi.org/10.1038/ncomms3185

Vincent K, Pichler M, Lee G-W, Ling H (2014) MicroRNAs, genomic instability and cancer. Int J Mol Sci 15(8):14475–14491. https://doi.org/10.3390/ijms150814475

Wang J, Zhang P, Tu Z (2003) Effects of quercetin on proliferation of lung cancer cell line A549 by down-regulating hTERT gene expression. J Third Mil Med Univ 24

Wang, X. Y., Fan, Y., Zhang, Y. L., Zhong, X. M. (2010) *Effect of resveratrol on promoter and human telomerase reverse transcriptase(hTERT) expression of human colorectal carcinoma cell—《Journal of Jiangsu University(Medicine Edition)》2010年01期*. https://en.cnki.com.cn/Article_en/CJFDTotal-ZJYZ201001015.htm

Wang, N., Liu, T., Sofiadis, A., Juhlin, C. C., Zedenius, J., Höög, A., Larsson, C., and Xu, D., 2014. *TERT* promoter mutation as an early genetic event activating telomerase in follicular thyroid adenoma (FTA) and atypical FTA: *TERT* Promoter Mutation in FTA/AFTA. Cancer, 120(19), 2965–2979. doi: https://doi.org/10.1002/cncr.28800

Wang N, Kjellin H, Sofiadis A, Fotouhi O, Juhlin CC, Bäckdahl M, Zedenius J, Xu D, Lehtiö J, Larsson C (2016) Genetic and epigenetic background and protein expression profiles in relation to telomerase activation in medullary thyroid carcinoma. Oncotarget 7(16):21332–21346. https://doi.org/10.18632/oncotarget.7237

Watson JD (1972) Origin of concatemeric T7DNA. Nat New Biol 239(94):197–201. https://doi.org/10.1038/newbio239197a0

Weber JA, Baxter DH, Zhang S, Huang DY, How Huang K, Jen Lee M, Galas DJ, Wang K (2010) The MicroRNA spectrum in 12 body fluids. Clin Chem 56(11):1733–1741. https://doi.org/10.1373/clinchem.2010.147405

Wei, J. W., Fan, Y., Y. L. Zhang, Wu, Y., Wang, X. Y., & Chen, P. (2007) *Effects of Quercetin on telomerase activity and apoptosis in gastric cancer cells.*《Shandong Medical Journal》2007年 35期. https://en.cnki.com.cn/Article_en/CJFDTotal-SDYY200735008.htm

Wright WE, Pereira-Smith OM, Shay JW (1989) Reversible cellular senescence: implications for immortalization of normal human diploid fibroblasts. Mol Cell Biol 9(7):3088–3092. https://doi.org/10.1128/MCB.9.7.3088

Wu K-J, Grandori C, Amacker M, Simon-Vermot N, Polack A, Lingner J, Dalla-Favera R (1999a) Direct activation of TERT transcription by c-MYC. Nat Genet 21(2):220–224. https://doi.org/10.1038/6010

Wu HL, Hsu CY, Liu WH, Yung BY (1999b) Berberine-induced apoptosis of human leukemia HL-60 cells is associated with down-regulation of nucleophosmin/B23 and telomerase activity. Int J Cancer 81(6):923–929. https://doi.org/10.1002/(sici)1097-0215(19990611)81:6<923::aid-ijc14>3.0.co;2-d

Wu Y, Tang Y, Jiang ZQ (2000) Diagnosis of human bladder cancer by detecting the telomerase activity in exfoliated urothelial cells. Hunan Yi Ke Da Xue Xue Bao = Hunan Yike Daxue Xuebao = Bulletin of Hunan Medical University 25(6):599–600

Wu S, Huang P, Li C, Huang Y, Li X, Wang Y, Chen C, Lv Z, Tang A, Sun X, Lu J, Li W, Zhou J, Gui Y, Zhou F, Wang D, Cai Z (2014) Telomerase reverse transcriptase gene promoter mutations help discern the origin of urogenital tumors: a genomic and molecular study. Eur Urol 65(2):274–277. https://doi.org/10.1016/j.eururo.2013.10.038

Xie M, Mosig A, Qi X, Li Y, Stadler PF, Chen JJ-L (2008) Structure and function of the smallest vertebrate telomerase RNA from teleost fish. J Biol Chem 283(4):2049–2059. https://doi.org/10.1074/jbc.M708032200

Xu X, Chen W, Miao R, Zhou Y, Wang Z, Zhang L, Wan Y, Dong Y, Qu K, Liu C (2015) MiR-34a induces cellular senescence via modulation of telomerase activity in human hepatocellular carcinoma by targeting FoxM1/c-Myc pathway. Oncotarget 6(6):3988–4004. https://doi.org/10.18632/oncotarget.2905

Ye JZ-S (2004) POT1-interacting protein PIP1: a telomere length regulator that recruits POT1 to the TIN2/TRF1 complex. Genes Dev 18(14):1649–1654. https://doi.org/10.1101/gad.1215404

Yu J, Guo Q-L, You Q-D, Zhao L, Gu H-Y, Yang Y, Zhang H-W, Tan Z, Wang X (2006) Gambogic acid-induced G2/M phase cell-cycle arrest via disturbing CDK7-mediated phosphorylation of CDC2/p34 in human gastric carcinoma BGC-823 cells. Carcinogenesis 28(3):632–638. https://doi.org/10.1093/carcin/bgl168

Yuan P, Cao J, Abuduwufuer A, Wang L-M, Yuan X-S, Lv W, Hu J (2016) Clinical characteristics and prognostic significance of TERT promoter mutations in lung cancer: a cohort study and a meta-analysis. PLoS One 11(1):e0146803. https://doi.org/10.1371/journal.pone.0146803

Zamin LL, Filippi-Chiela EC, Dillenburg-Pilla P, Horn F, Salbego C, Lenz G (2009) Resveratrol and quercetin cooperate to induce senescence-like growth arrest in C6 rat glioma cells. Cancer Sci 100(9):1655–1662. https://doi.org/10.1111/j.1349-7006.2009.01215.x

Zhang A, Zheng C, Lindvall C, Hou M, Ekedahl J, Lewensohn R, Yan Z, Yang X, Henriksson M, Blennow E, Nordenskjöld M, Zetterberg A, Björkholm M, Gruber A, Xu D (2000) Frequent amplification of the telomerase reverse transcriptase gene in human tumors. Cancer Res 60(22):6230–6235

Zhang W, Yang P, Gao F, Yang J, Yao K (2014) Effects of epigallocatechin gallate on the proliferation and apoptosis of the nasopharyngeal carcinoma cell line CNE2. Exp Ther Med 8(6):1783–1788. https://doi.org/10.3892/etm.2014.2020

Zhang H, Weng X, Ye J, He L, Zhou D, Liu Y (2015) Promoter hypermethylation of TERT is associated with hepatocellular carcinoma in the Han Chinese population. Clin Res Hepatol Gastroenterol 39(5):600–609. https://doi.org/10.1016/j.clinre.2015.01.002

Zhao X, Tian X, Kajigaya S, Cantilena CR, Strickland S, Savani BN, Mohan S, Feng X, Keyvanfar K, Dunavin N, Townsley DM, Dumitriu B, Battiwalla M, Rezvani K, Young NS, Barrett AJ, Ito S (2016) Epigenetic landscape of the *TERT* promoter: a potential biomarker for high risk AML/MDS. Br J Haematol 175(3):427–439. https://doi.org/10.1111/bjh.14244

Zhao T, Wang H-J, Zhao W-W, Sun Y-L, Hu L-K (2017) Gambogic acid improves non-small cell lung cancer progression by inhibition of mTOR signaling pathway. Kaohsiung J Med Sci 33(11):543–549. https://doi.org/10.1016/j.kjms.2017.06.013

Zheng D-S, Chen L-S (2017) Triterpenoids from Ganoderma lucidum inhibit the activation of EBV antigens as telomerase inhibitors. Exp Ther Med 14(4):3273–3278. https://doi.org/10.3892/etm.2017.4883

Zhou XZ, Lu KP (2001) The Pin2/TRF1-interacting protein PinX1 Is a potent telomerase inhibitor. Cell 107(3):347–359. https://doi.org/10.1016/S0092-8674(01)00538-4

Zhou J, Dai W, Song J (2016) MiR-1182 inhibits growth and mediates the chemosensitivity of bladder cancer by targeting hTERT. Biochem Biophys Res Commun 470(2):445–452. https://doi.org/10.1016/j.bbrc.2016.01.014

Zimmermann M, Kibe T, Kabir S, de Lange T (2014) TRF1 negotiates TTAGGG repeat-associated replication problems by recruiting the BLM helicase and the TPP1/POT1 repressor of ATR signaling. Genes Dev 28(22):2477–2491. https://doi.org/10.1101/gad.251611.114

Zinn RL, Pruitt K, Eguchi S, Baylin SB, Herman JG (2007) HTERT Is Expressed in Cancer Cell Lines Despite Promoter DNA methylation by preservation of unmethylated DNA and active chromatin around the transcription start site. Cancer Res 67(1):194–201. https://doi.org/10.1158/0008-5472.CAN-06-3396

Pain Management in Oncology

15

Subrata Goswami, Debolina Ghosh, Gargi Nandi, Sayanee Mukherjee, and Biplab Sarkar

Abstract

Cancer pain is one of the most dreadful types of pain experienced by the patients and difficult one to treat by the pain physicians; 60–80% cancer pain is inadequately treated, 40% dies with severe pain. But multimodal, multidisciplinary management can cure more than 90% of cancer pain. Barriers of pain management are lack of awareness of health workers, medical personals, policy makers, and public, lack of infrastructure as compared to the growing number of cancer patients, fear of drug abuse/dependency of the patients, and nonavailability of opioid due to legal restrictions. Various validated scales are used for qualitative and quantitative assessment of cancer pain. History, clinical examinations, and relevant imaging are used for diagnosis. Treatment is done following WHO Analgesic ladder. In severe cancer pain, interventional procedures are considered to relieve pain and to improve the quality of life. Advanced interventional procedures (e.g., spinal cord stimulators, vertebroplasty, and balloon kyphoplasty) and neurosurgical palliative techniques (e.g., commissural myelotomy and percutaneous cordotomy) are considered in advanced cancer with excruciating pain. Cognitive-behavioral therapy, hypnosis, relaxation therapy, physical therapy, and psychological counseling are the adjuvant therapy techniques. Palliative care begins at any point along the cancer care continuum, extended even after death of the patient. Hospice-based, home-based cancer pain management and respite care are given in final days of life.

Keywords

Pain management · Cancer pain · Intervention

S. Goswami (✉) · D. Ghosh · G. Nandi · S. Mukherjee · B. Sarkar
ESI Institute of Pain Management, ESI Hospital Sealdah Premises, Kolkata, West Bengal, India

© The Author(s), under exclusive license to Springer Nature Singapore Pte Ltd. 2022
S. K. Basu et al. (eds.), *Cancer Diagnostics and Therapeutics*,
https://doi.org/10.1007/978-981-16-4752-9_15

15.1 Introduction and Epidemiology

Cancer has huge and increasing global impact on public health and is considered to be the second most common cause of death; 1 in 6 deaths are due to cancer; 9.6 million deaths in 2018 are due to cancer. According to the statistics given by WHO (World Health Organization), the incidence of cancer is set to increase by 50%, so that there will be 15 million newly diagnosed cases of malignancy in 2020 (Stannard and Booth 2004). The prevalence of cancer is higher not only in the richer nations, but also in the developing world like India. In the present scenario, in India about 1 million patients each year are diagnosed to have cancer, which is almost equal to 1.04 million of new cases in USA (Silverberg et al. 1990).

Cancer has a huge economic impact. In 2010, approximately US$ 1.16 trillion was the annual expenditure for cancer treatment in USA (Wu et al. 2020). Cancer patients very commonly present with pain, particularly in the advanced stage the prevalence of which is more than 70% (Portenoy 2011). These lead to poor physical and emotional well-being. With life-prolonging or curative treatment, patients experience persistent pain due to treatment or because of the disease itself or a combination of both (Glare et al. 2014). Approximately 5–10% of cancer survivors have severe chronic pain that significantly interferes with the functioning of the individual (Brown et al. 2014).

Though there are guidelines and recommendations of World Health Organization (WHO), Agency for Health Care Policy and Research (AHCPR), and Expert Working Group of the European Association for the Palliative Care, and even if effective treatments are available for 70–90% of cases (Cleeland et al. 1994), about 40% cases are still undertreated (Cohen et al. 2003). Inappropriate and inadequate use of opioids due to unavailability or difficulty in availability of drugs to the patient, family, healthcare provider, institution, and society (Maltoni 2008). In 2014 (Greco et al. 2014), a systemic review was published using Pain Management Index (PMI) (Cleeland et al. 1994), showed that approximately one-third of patients do not receive appropriate analgesia proportional to their pain intensity (PI).

IASP (International society for the study of pain) declared October 2008–October 2009 as a *global year against cancer pain*. They observed that there are number of barriers related to the healthcare professionals, the patients, and the healthcare system, for which only 50% patients achieved adequate pain control. IASP defines pain as an unpleasant sensory and emotional experience associated with actual or potential tissue damage. Pain is a subjective perception. It is said to be the *fifth vital sign* and *biopsychosocial phenomenon* that includes sensory, emotional, cognitive, developmental, behavioral, spiritual, and cultural components. So, the treatment should be directed toward relief of *total pain* (Brant 2017).

In 1960, Dame Cicely Saunders conceptualized the cancer pain as *total pain.* This is a holistic approach to pain considering the interplay of social, psychological, spiritual, and cultural well-being along with physical pain (Fig. 15.1). Emotional distress, depression, anxiety, and hopelessness are all forms of psychological pain that can co-occur with physical pain. The experience of *total pain* is also

15 Pain Management in Oncology

Fig. 15.1 Concept of total pain

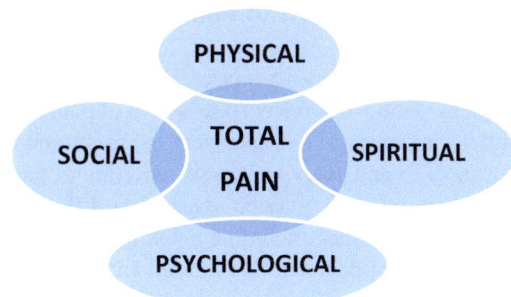

Table 15.1 Eastern Cooperative Oncology Group Scale

Score	Performance status
0	Fully active; able to carry on all predisease performance without restriction
1	Restricted in physically strenuous activity but ambulatory and able to carry out light or sedentary
2	Ambulatory and capable of self-care but unable to work; up and about more than 50% of walking hours
3	Capable of only limited self-care; confined to bed or chair more than 50% walking hours
4	Completely disabled; cannot carry out any self-care; totally confined to bed or chair
5	Dead

individualized and specific to one's situation. Culture and ethnicity have definite influence on individual pain perception, expression, and coping abilities.

15.2 Assessment of Performance

ECOG (Eastern Cooperative Oncology Group) and Karnofsky scales are two methods of assessment of the performance status. These are the global indicators of overall functional status of a patient and are also consistent with the patients' overall pain relief. Pain management improves nutrition, rest, mood, and outcome of anticancer treatment.

15.2.1 Assessment of Performance Status

ECOG scale of performance status is a standard way of measuring the impact of a disease on patient's daily life and abilities, used worldwide. With this tool, physicians also measure the improvement of the patients' well-being after treatment. Table 15.1 describes the performance status of each score defined in the ECOG scale.

15.2.2 Assessment of Pain Intensity

Pain can be nociceptive (visceral/somatic) or neuropathic; intensity of pain can be mild, moderate, or severe. Assessment of pain helps to improve the choice of therapy and its outcome. So it is necessary to assess pain intensity regularly and consistently, during each and every visit. Self-reporting of pain is the standard way of evaluating pain. Pain intensity may be assessed at different point of time. During assessment, psychosocial issues should be taken into account (Turk and Okifuji 2002).

Intensity of pain can be assessed and monitored by several instruments:

(a) Visual analog scale (VAS): (0–100 mm: 0 mm = no pain, 100 mm = worst pain imaginable) (Fig. 15.2)
(b) Numerical pain rating scale (NRS): (0–10: 0 = no pain, 10 = worst pain imaginable) (Fig. 15.2)
(c) Verbal rating scale (VRS): (none, mild, moderate, severe, excruciating) (Fig. 15.2)
(d) Facial scales (used for pediatric patients or when literacy rate low) (Fig. 15.3)

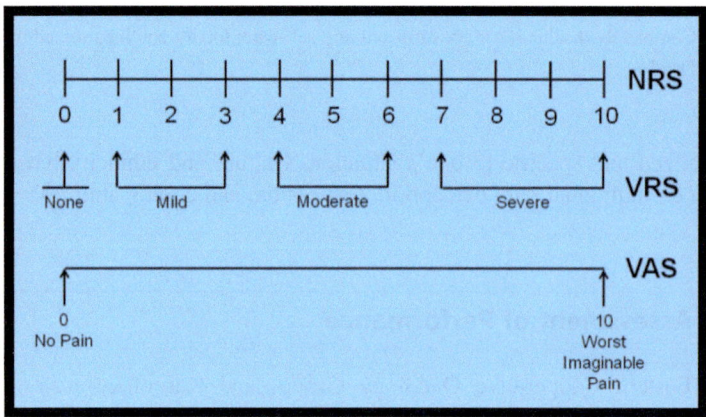

Fig. 15.2 Pain assessment scales: NRS, VRS, VAS

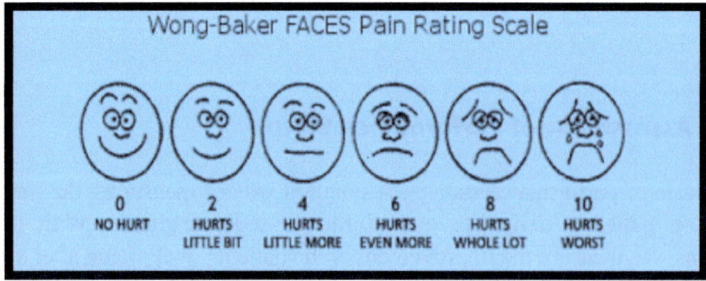

Fig. 15.3 Wong–Baker faces pain rating scale

(e) Multidimensional pain assessment tools—brief pain inventory (Cleeland and Ryan 1994)
(f) McGill pain questionnaire (Melzack 1987) (Fig. 15.4) and memorial pain assessment cards (Fishman et al. 1987).

According to Edmonton Classification System for Cancer Pain (ECS-CP) (Fainsinger et al. 2005, 2009), the presence of incident pain, neuropathic pain, addiction, psychological distress, and cognitive impairment makes the control of pain more difficult. This ECS-CP has been validated in various cancer pain settings (Fainsinger and Nekolaichuk 2008).

Cancer Pain Prognostic Scale (CPPS) predict the likelihood of pain relief in moderate to severe cancer pain. Score ranges from 0 to 17 (Hwang et al. 2003). Higher the score, higher is the possibility of pain relief. This scale includes worst pain severity (brief pain inventory), emotional well-being, and pain characteristics.

Among elderly patients, pain assessment can be done by direct observation of patient behavior, report from family/caregiver, and evaluation of response to pain relief interventions (Herr et al. 2006; Warden et al. 2003; Fuchs-Lacelle and Hadjistavropoulos 2004; Regnard et al. 2007). Though these tools are often used in cognitive impairment and dementia patients, still validity and reliability of these tools are questionable.

In observational scale (Van Herk et al. 2007) (yet not validated), facial expression, body movements, verbalization or vocalizations, changes in interpersonal interactions, changes in routine activity, etc., are observed. For Faces Pain Scale (Bieri et al. 1990), Colored Analog Scale (McGrath et al. 1996), and numeric rating scale, vertical instead of horizontal orientation of scales may be preferable (Kremer et al. 1981). As it is seen culture plays a role in pain experience, it is seen some Asian cultured patients tend not to report pain (Duke and Petersen 2015).

15.3 Evaluation of the Patient with Cancer Pain

Evaluation of cancer pain starts with history taking of the underlying risk factors, malignancy itself, elaborate history of associated pain, and other general symptoms history.

15.3.1 History of the Malignancy

Under this, possible causes of risk like tobacco habit, alcohol, unhealthy dietary habit, and sedentary lifestyle, chronic infections with Helicobacter pylori, Human papillomavirus (HPV), Hepatitis B virus, Hepatitis C virus, and Epstein–Barr virus (Plummer et al. 2016) are to be noted. The concluded diagnosis, stage of the disease, history of treatment like chemotherapy, radiotherapy, surgery and the consequences, and the prognosis are recorded.

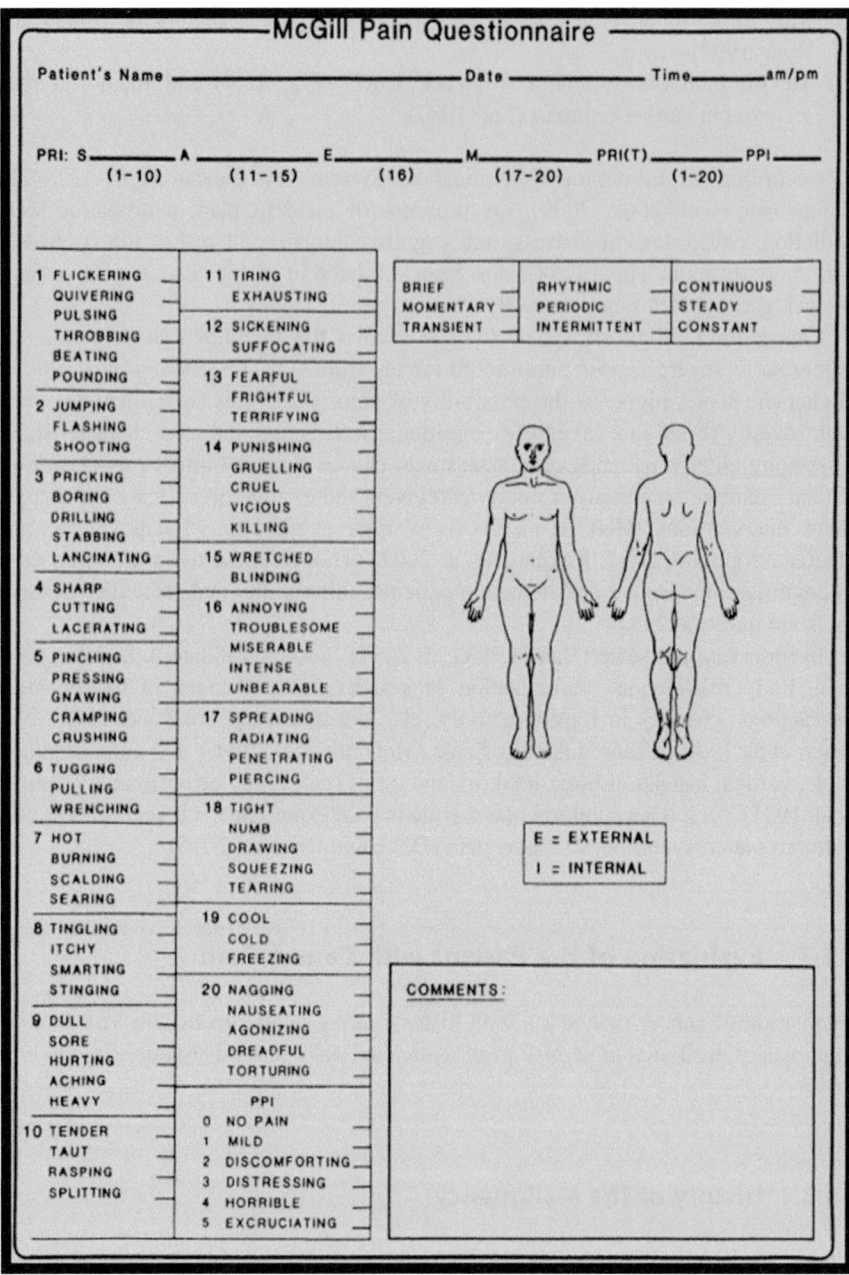

Fig. 15.4 McGill Pain Questionnaire

15.3.2 Pain History

Elaborate history of pain includes its site of onset and radiation, mode of onset (acute, subacute, chronic), pattern of pain (constant, intermittent, breakthrough or incident), and aggravating and relieving factors. Character of pain whether nociceptive (somatic—aching, stabbing, throbbing/visceral—cramping, aching, sharp, throbbing) or neuropathic (burning, tingling, numbing, laminating or electric shock like shooting pain) or referred (e.g., back from pancreas, shoulder from liver) or idiopathic are to be noted. Psychological and personal history includes marital, occupational status, past history of drug/alcohol abuse, anxiety, depression, and irritability.

15.3.3 General History

History of anorexia, weight loss, fatigue, weakness, insomnia, dysphagia, nausea, vomiting, constipation, and diarrhea is to be taken. Neurological (sedation, headache, confusion, hallucination), respiratory (dyspnea, cough, shortness of breath), and genitourinary (incontinence, retention, hematuria) symptoms are to be ruled out.

15.3.4 Cancer Pain Syndromes

Cancer pain syndromes are defined by the association of particular pain characteristics and physical signs with specific consequences of the underlying disease or its treatment. Pain syndromes associated with cancer can be either acute or chronic. It may be related to diagnostic and therapeutic interventions and negative consequences of surgery, chemotherapy, and radiotherapy or due to tumor infiltration to viscera, bone, and neural structures.

15.3.4.1 Bone Pain Syndrome
Bone metastasis is common from breast, prostate, thyroid, kidney, and lung tumors (Coleman et al. 2006; Coleman 1997). So, majority of patients complain of moderate to severe bone pain with progression of the disease. Multiple myeloma, leukemia, and primary bone tumor are also common causes of bone pain. As tumor cells invade the skeletal tissue, constant, dull aching bone pain starts, which may gradually increase with time (Mercadante 1997). However, it may be breakthrough pain. The other possible symptoms are fever, fatigue, swelling, inflammation, lump, and fractures after minor injury or fall, unexplained weight loss.

Mechanism of Bone Pain
The tumor cells proliferate rapidly in the bone marrow and replace the original hemopoietic cells (Honore et al. 2000; Schwei et al. 1999). Gradually, the entire bone marrow gets filled with tumor cells and inflammatory/immune cells (Mantyh 2006; Joyce and Pollard 2009). Prostaglandins, bradykinin, tumor necrosis factor-

alpha, endothelin, interleukins-1 and 6, epidermal growth factor, transforming growth factor-alpha, platelet-derived growth factor, and nerve growth factor (NGF) (Joyce and Pollard 2009; Von Moos et al. 2008) are liberated, which stimulate primary afferent neurons causing bone pain.

Osteoclast has important role in cancer-induced bone loss and bone pain (Von Moos et al. 2008). Both osteolytic (bone destroying) and osteoblastic (bone forming) cancers are characterized by osteoclast proliferation and hypertrophy (Honore et al. 2000; Drake et al. 2008; Halvorson et al. 2006), leading to remodeling of bone. This whole process leads to massive production of extracellular protons (Teitelbaum 2007), which are known to be potent activators of nociceptors (Julius and Basbaum 2001).

Acidic microenvironment produced by osteoclasts activates acid-sensitive nociceptors that innervate the marrow, mineralized bone, and periosteum (Ghilardi et al. 2005). Two acid-sensing ion channels expressed by nociceptors are transient receptor potential vanilloid1 (TRPV1) and acid-sensing ion channel-3 (ASIC-3) (Julius and Basbaum 2001). While invading the tumor stroma, inflammatory and immune cells produce local acidosis (Mantyh 2006; Joyce and Pollard 2009; Von Moos et al. 2008; Halvorson et al. 2006; Teitelbaum 2007; Ghilardi et al. 2005). Stretching of periosteum, tumor growth into neighboring soft tissue (e.g., muscles), infiltration to nerves, weakening of bone, collapse or fracture all lead to activate nociceptive or neuropathic pain.

Investigation

Basic investigation is X-ray (50% decalcification necessary), followed by CT scan, MRI, isotope scan, CSF study (protein, glucose, malignant cells)—as and when required.

Management

Treatment aims to relieve pain, prevent or treat any fractures, and prevent or delay further bone complications. Pain management is done by following WHO analgesic ladder. Other specific treatment options are **surgery (if operable), chemotherapy**, **radiotherapy** (localized metastatic bone pain), hormone therapy (e.g., anastrozole for breast tumor), and steroid (to combat acute edema and nerve involvement). One of the most effective and approved drugs are bisphosphonates (clodronate, pamidronate, zoledronate, ibandronate) for reducing tumor-induced bone pain and destruction. Bisphosphonate reduce the function of osteoclasts and induce its apoptosis (Drake et al. 2008) by binding to RANKL (receptor activator for nuclear factor κB ligand) (Lipton et al. 2006).

Preservation of the mechanical strength of bone should reduce movement-induced incident pain, as this pain is probably driven by activation of normally silent mechanosensitive nociceptors that innervate the bone. Denosumab (anti-RANKL) (Body et al. 2006) reduces cancer-induced bone loss in breast and prostate cancers (Steger and Bartsch 2011), multiple myeloma, and multiple other solid tumors. Recent pharmacological studies showed that selective TRPV1 antagonists significantly decreased ongoing pain. Tanezumab (anti-NGF) and pregabalin are under clinical trials. **Mifamurtide** is a drug, often used in osteosarcoma.

15.3.4.2 Back Pain Syndrome

Back pain is so common in general population that serious causes of back pain in cancer often goes undetected until it is advanced. Since 90% patients with spinal tumors present with back pain (Gilbert et al. 1978), neoplasia must be an important differential diagnosis of any persistent, unremitting back pain. But if cancer patients present with painless neurological deficit, prompt action must be taken.

Spine is one of the most common sites of metastasis, especially in lung, breast, prostate, and thyroid cancer. Pain is generally dull aching, persistent, progressive in character, exacerbated by sneezing, coughing, lying down, and worsen at night so much that it can make the patient awake from sleep. This pain is not relieved by taking rest. Back pain may present focally at the level of lesion, with or without radiculopathy.

Spinal tumor can give rise to neurological signs or symptoms due to spinal cord or nerve roots compression (Deyo and Diehl 1988). Loss of bowel, bladder control, or any significant change in bowel–bladder habit is always to be treated with urgency.

So, thorough neurological examination is mandatory, which may often reveal altered touch, temperature, and joint position sense. Lhermitte's sign (sudden electric shocks along the length of spine during cervical flexion) may be positive. Pathologic fractures unfortunately may be the first and late presentation in vertebral metastasis. Fatigue, weight loss, cachexia, anemia, and gait abnormalities are constellation of few systemic symptoms.

15.3.4.3 Degenerative Disk Disease and Neoplastic Disease

Degenerative disk disease pain often starts after spine surgery or any other injury, aggravated with activity and relieved by taking rest, physical therapy, and nonsteroidal anti-inflammatory drugs. Imaging studies are typically normal. On the contrary, neoplastic disease generally occurs for individuals with age more than 50 years, where pain is not relieved by taking rest or in recumbent posture. Even pain may worsen at night. The patient may be anemic with elevated erythrocyte sedimentation rate (ESR) and having C-reactive protein level (CRP). Conservative therapy does not improve pain condition. The patient has some typical history of cancer.

Causes of Neoplastic Disease

Spinal cord compression is mostly due to retropulsion of the vertebral body due to metastasis (Fig. 15.5). In *odontoid process syndrome* neck pain generally radiates to posterior vertex, where MRI is the most useful tool for diagnosis. *C7-T1 syndrome* presents with dull aching constant pain which radiates to shoulders, interscapular region, upper arm, elbow, and ulnar distribution. Horner's syndrome may be present with C7-T1 syndrome. In *T12-L1 syndrome,* patient often complains of mid-back pain, exacerbated by lying down or sitting position, which may be referred to sacroiliac or iliac crest area. *Sacral syndrome* is mostly seen with gynecological,

Fig. 15.5 Cord compression (taken from ESI Institute of Pain Management file images with permission)

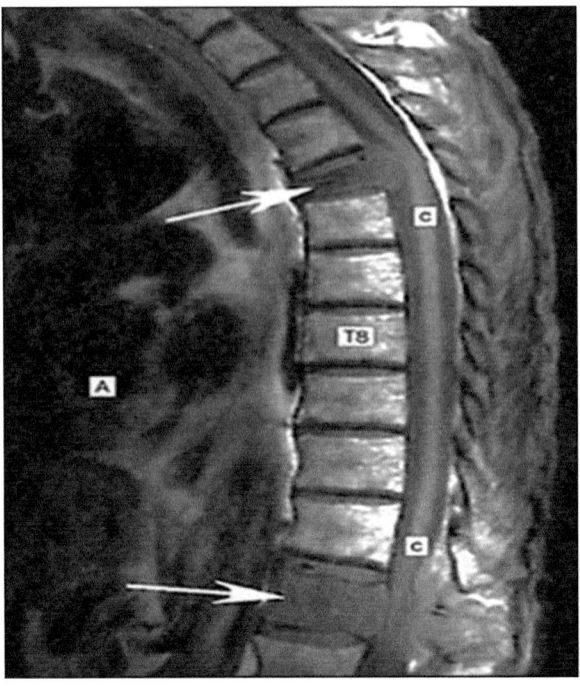

genitourinary, colon, or breast cancer. In cauda equina syndrome, there is loss of bowel and bladder control and patients present with incontinence associated with low back pain, radiating to both the legs, paresthesia, and numbness in the perianal region. Low back pain may be the side effect of chemotherapy or other cancer therapies, for example, Herceptin therapy to treat breast and gastric cancer.

Investigations

Complete blood cell count (CBC), erythrocyte sedimentation rate (ESR), C-reactive protein (CRP) level (in cancer usually ESR, CRP raised) prostate-specific antigen level (PSA), urinalysis, and fecal occult blood testing are the usual relevant tests. Anemia, hypercalcemia, and elevated levels of alkaline phosphatase should be given special attention.

If laboratory results are abnormal, chest radiography (CXR), computed tomography (CT), and mammography should be done for the women. Plain radiograph of spine is the first imaging study to obtain to rule out malignancy of spine. Compression fractures, focal loss of bone mineralization, and soft tissue calcifications suggest tumor. In multiple myeloma or plasmacytoma, there are abnormal results on serum and urine protein electrophoresis. Magnetic resonance imaging (MRI) is the best imaging tool to evaluate spinal tumors (Citrin et al. 1977) showing bone marrow infiltration and soft issue masses in and around the spinal column. Myelography can be done when indicated.

Management

Pain is controlled following WHO analgesic ladder. Dexamethasone often provides analgesia in acute cases and reduces edema. Antineuropathic and antidepressant medications should be given. The use of braces and physical therapy are often helpful. Radiotherapy is the most effective way of treatment for cancer-related bone pain, with which up to 80% of patients have significant, long-lasting, or total reduction in pain (Chow et al. 2007). In recurrence of pain, radiotherapy can be repeated as required. Stereotactic radiotherapy is a specialized form of highly conformal ablative radiotherapy, which may be recommended in highly selected cases.

Palliative radiotherapy is well tolerated, having self-limiting side effects. Patients may experience a transient flare of pain, mild nausea, fatigue, and skin erythema which can be actively managed by radiation oncologist. Referral for neurosurgical opinion is to be taken, particularly when spinal stability is a concern and red flag signs are present.

Pain Due to Direct Tumor Involvement of Nerve

When tumor invades the nervous tissue, it causes inflammation or compression of the nerves; pain is produced which may be associated with other symptoms like paresthesia, numbness, and bowel–bladder disturbances. As tumor compresses or infiltrates the tissue or nervous system, it liberates chemicals which make normally nonpainful stimuli painful. Brain does not contain any pain sensors, but as the tumor compresses or infiltrates the brain tissue, compression on blood vessels and meninges, edema of brain tissue compressing on pain-sensitive structures causes pain.

Epidural Spinal Cord and Cauda Equine Compression

In oncologic emergency, MRI is the investigation of choice. There is motor weakness followed by paraplegia, and finally loss of bladder-bowel function happens.

In *tumor-related mononeuropathy,* there is constant burning pain, hypoesthesia, and dysesthesia (e.g., intercostal nerve in rib metastasis) along the nerve distribution. Constant and paroxysmal dysesthesia (small cell CA of lung) is seen in *paraneoplastic painful neuropathy.*

In *cervical plexopathy,* C1–4 infiltration (e.g., metastatic lymph node) is common, along with lancinating, aching, or burning pain felt in the preauricular and postauricular area, anterior part of neck, lateral side of face, neck, and shoulder.

In *brachial plexopathy lower plexus,* C8-T1 involvement is more common, associated with carcinoma lung (Pancoast), breast, and lymphoma. CT and MRI are confirmatory diagnostic tests.

In *lumbosacral plexopathy,* involvement is mainly of L1-4 and L4/5-S1,2,3 nerves, which is commonly seen in gynecological, genitourinary, colorectal cancer, sarcoma, and lymphoma. Clinical presentation is mainly pain in back, lower abdomen, flank, iliac crest, and anterolateral portion of the thigh. There is painful flexion of ipsilateral hip, which is known as painful psoas syndrome.

Sacral plexopathy is common in colorectal, genitourinary, and gynecological cancer, where sensory loss starts in perianal area. Numbness felt over dorsal medial foot and sole with weakness of ankle. Impotence and bladder–bowel dysfunction may be possible. CT and MRI are necessary to confirm the diagnosis.

In *leptomeningeal metastasis,* mainly there is nerve root involvement and diffuse multifocal involvement of subarachnoid space happens by metastatic tumor in the diseases like non-Hodgkin's lymphoma (NHL), acute lymphocytic leukemia (ALL), breast cancer, and lung cancer. Constant headache is a common presentation, meningism may present, and low back pain (LBP) and buttock pain may occur. Nausea, vomiting, and paroxysmal headache are often seen. CSF examination and gadolinium-enhanced MRI are the confirmatory tests for diagnosis.

15.3.4.4 Base of Skull Disease Syndrome

Few syndromes are categorized under the base of the skull disease syndrome, which mainly originate from the structures related to base of the skull. *Orbital syndrome* is one of the important among those in which patients present with retro-orbital pain or supraorbital pain, blurred vision, and diplopia. On examination, proptosis, external ophthalmoparesis, and decreased sensation in V2 area are generally noted. Secondly, *parasellar/cavernous sinus syndrome* patients present with unilateral headache, ocular paresis without proptosis, visual field defects, and papilledema. Thirdly, *middle fossa/Gasserian ganglion syndrome* patients have numbness, paresthesia, and pain at V2/V3 distribution. In *jugular foramen syndrome,* hoarseness, dysphagias, dysarthria, occipital pain behind ear, and glossopharyngeal neuralgia are common presentations. Horner's syndrome may be present in those patients. *Clivus syndrome* is characterized by vertex headache which is aggravated by neck flexion, lower cranial nerve dysfunction of VI-XII, facial weakness, hoarseness, dysphagia, dysarthria, and trapezius weakness. Severe bifrontal headache is common in *sphenoid sinus syndrome* which radiates to both temples along with nasal stiffness, fullness in head, diplopia unilateral, or bilateral VI nerve palsy. XII nerve involvement is possible in *occipital condyle syndrome* with severe localized continuous unilateral occipital pain exacerbated by neck flexion.

15.3.4.5 Postsurgical Syndromes

Postthoracotomy pain is one of the most common and nagging complication of thoracic surgery. It mainly happens due to injury to intercostal nerve by incision or retraction, liberal use of electrocautery, and mal-handling of tissue. Patients have burning and intermittent stabbing pain with dysesthesia and allodynia.

Postmastectomy pain happens due to interruption of T1–T3 nerves usually by axillary dissection. A tight, constricting, burning pain is experienced mostly at posteromedial aspect of arm, axilla, and anterior chest wall. Postradical neck dissection pain is felt due to damage of cervical plexus. Pain is tingling, burning type with sensory loss to certain area. Phantom limb syndrome is the pain seemed to arise at the stump or perceived from the amputated limb or breast.

15.3.4.6 Postradiotherapy Pain Syndromes

Radiation fibrosis of brachial plexus may occur 6 months to 20 years after radiotherapy, which may present with paresthesia and weakness in C5/6 area. Radiation myelopathy is a burning dysaesthetic pain because of irradiation process months to many years after radiotherapy. Months to years after radiotherapy, patients may have mucositis, enteritis, proctitis, and osteoradionecrosis due to ischemic tissue damage.

15.3.4.7 Postchemotherapy Pain Syndrome

Various chemotherapeutic agents are used in different types of malignancy. Though these cytotoxic drugs have therapeutic effects on tumor cells, these cause multiple side effects in patient's body. Aggressive chemotherapy with methotrexate, doxorubicin, bleomycin, and etoposide often leads to oral mucositis, which usually peaks after 2–3 weeks. Symmetrical peripheral polyneuropathy is common with the use of vincristine, vinblastine, paclitaxel, and bortezomib. Cisplatin mostly causes painful dysesthesias at feet or hands.

Steroid pseudorheumatism occurs after rapid steroid dose reduction or withdrawal. It presents with muscle and joint tenderness without signs of inflammation, fatigue, and neuropsychiatric disorders. Aseptic necrosis of femoral head and osteoporosis are the usual complication of steroid therapy. Postherpetic neuralgia is not uncommon after chemotherapy, as immunosuppression is inevitable in malignancy. Many elderly patients after chemotherapy present with herpes zoster, which if not treated early with antiviral may lead to severe post herpetic neuralgia (PHN).

15.3.4.8 Neuropathic Pain

Neuropathic pain in cancer patients is very common. It may be due to preexisting condition, associated comorbidities, due to the disease itself or related to the varied effect of cancer treatment. For example, 90% of patients receiving neurotoxic chemotherapy or radiation therapy have peripheral neuropathy. There may be increased pain sensation in non-noxious stimulus (allodynia) or less noxious stimulus (hyperalgesia). Cancer pains are often mixed in nature. It is often associated with somatic pain, mostly in elderly populations with other comorbidities, e.g., diabetes mellitus. Patients should be carefully examined. Morphine is effective in both somatic and neuropathic pain, but neuropathic medicines like anticonvulsants (gabapentin, pregabalin) and tricyclic antidepressants (amitriptyline , nortriptyline) are added as adjuvants for better result. Also sometime topical treatment (capsaicin, lidocaine) helps.

15.4 Management of Pain

Cancer pain can be managed by antineoplastic drugs, analgesic medications, opioids, radiation therapy, percutaneous and surgical interventions, psychological support, and spiritual motivation. The mode of treatment is always interdisciplinary, multidisciplinary, and multimodal.

15.4.1 Antineoplastic Treatment

Treatment of the disease in most cases will eliminate the pain. Treatment modalities are surgical resection, radiation therapy, whole body radiation, chemotherapy, and hormonal treatment.

15.4.2 Pharmacologic Management

Along with the treatment of cancer itself by antineoplastic, cancer pain is treated following WHO analgesic ladder. It has been seen that 70–90% of the patients become pain free through this guideline-based oral therapy. The four basic principles of WHO guidelines are as follows:

(i) By the mouth: Oral route is always preferred, it is simple and convenient, and it can be taken at home.
(ii) By the clock: For continuous pain relief, fixed dose in fixed interval is preferred to prevent recurrence of pain rather than *as and when needed* dosing.
(iii) By the ladder: Three step analgesic ladder formulated by WHO in 1986 is generally followed (Fig. 15.6)
(iv) By the individual: Prescribe according to the comfort of patient and medical problems. Drugs and the dosage should always be individualized (Fig. 15.7)

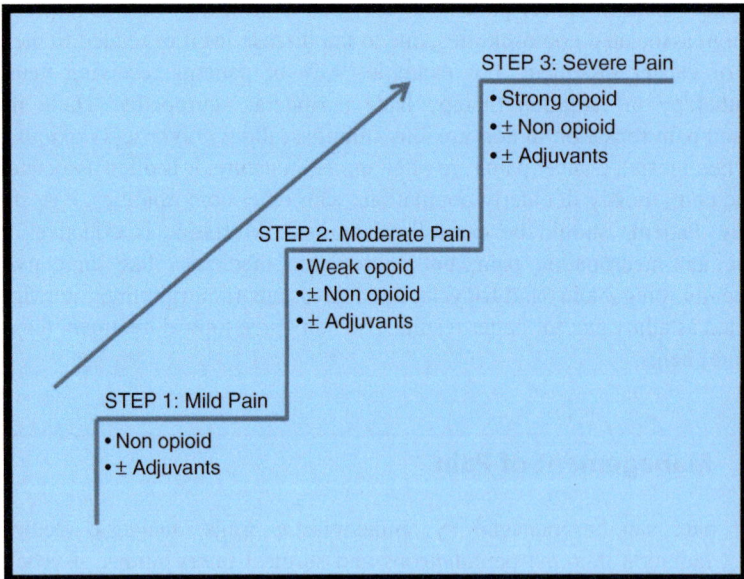

Fig. 15.6 WHO Analgesic ladder

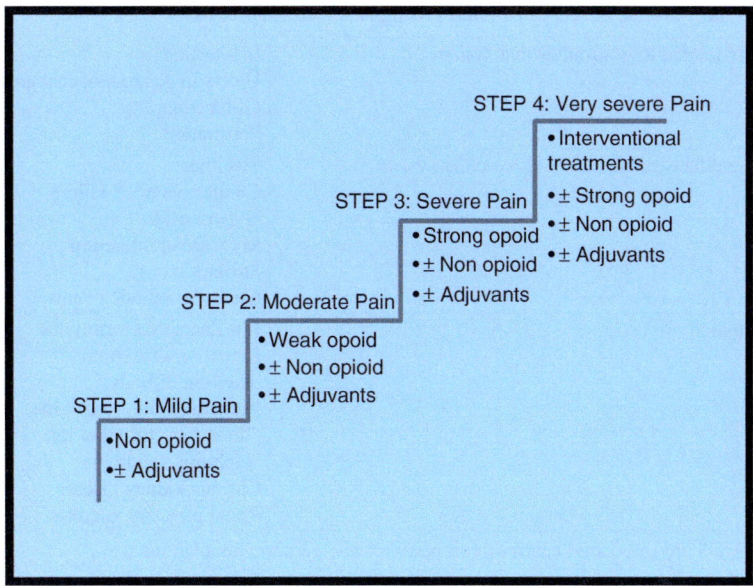

Fig. 15.7 Modified WHO Analgesic ladder

For quantification of pain, numeric rating scale (NRS) is a validated scale, based on which this WHO analgesic ladder was made. According to the guideline, "mild pain" (NRS: 1–4) can be treated by paracetamol and/or NSAIDs and/or adjuvants, and co-analgesics such as antidepressant and antiepileptic.

Moderate pain (NRS: 5–7) can be managed by paracetamol and/or NSAIDs with weak opioid such as codeine, dihydrocodeine, tramadol or transdermal buprenorphine, and fentanyl. Adjuvants and co-analgesics (e.g., antidepressant, antiepileptic) may be added.

For severe pain (NRS: 8–10), morphine, hydromorphone, and oxycodone are to be prescribed.

Fentanyl, buprenorphine transdermal, methadone, and parental morphine can be tried to those patients who are unable to swallow. Adjuvants, co-analgesics (antidepressant, antiepileptic), paracetamol and/or NSAIDs, bisphosphonates, steroids, and NMDA receptor antagonists may be considered.

If the pain is still resistant, then interventional pain procedures are to be considered.

15.4.2.1 Drugs

Acetaminophen/paracetamol: Most of the time, it is prescribed as weak analgesics for acute as well as chronic pain. It has both antipyretic and analgesic functions without any peripheral anti-inflammatory and antiplatelet actions. This noble drug has limited side effects on the GI tract as compared to NSAIDs (Lucas et al. 2005).

Table 15.2 Adverse effects of NSAIDs (Wongrakpanich et al. 2018)

Toxicity related to gastrointestinal system	Indigestion Ulcers in gastroduodenal area GI bleeding Perforation
Adverse effects related to cardiovascular system	Swelling Congestive heart failure Hypertension Myocardial infarction Stroke Other thrombotic events
Renal toxicity	Imbalance of electrolytes Retention of sodium Swelling of body Acute interstitial nephritis Glomerular filtration rate reduction Nephrotic syndrome Chronic kidney disease Renal papillary necrosis

***NSAIDS are to be used cautiously in pediatric and geriatric populations

Recent researches have proposed that central analgesic property may be due to down regulation of β-endorphin, which is a subtype of endogenous opioid peptide.

Metabolism of 90% drugs occurs in liver by the process of glucuronidation as well as sulfate conjugation to form nontoxic metabolites, while rest 10% drugs are metabolized by oxidative metabolism leading to formation of a hepatotoxic and nephrotoxic compound; 85% of oral dosing is predominantly excreted via urine within one day of oral dosing. This drug need to be used cautiously in hepatic and renal derangement, by reducing the dose or increasing the dose interval, and chronic use must be avoided.

Nonsteroidal anti-inflammatory drugs (NSAIDs): They act by prostaglandin inhibition and leukotriene production. Most commonly used NSAIDS are diclofenac and ibuprofen. Major advantages of this group of drug are lack of its respiratory depression, abuse potential, and good analgesic property. But while using the following adverse effects are to be kept in mind. Table 15.2 provides a summarized view of adverse effects of NSAIDs.

Caution: Patients suffering from liver and kidney disease have to be cautious about the use of NSAIDs. Higher incidence of hepatotoxicity of diclofenac than other NSAIDs is noticed. Conjugate accumulates in end-stage renal disease (ESRD).

Side effects: Some of the NSAIDs cause are GI side effects, e.g. nausea and dyspepsia. Nephrotoxicity is mainly observed in patients with reduced intravascular volume and cardiac output, especially in geriatric patients. It has increased cardiovascular risk (Catella-Lawson et al. 2001).

Contraindications: Peptic ulcer disease; GI bleeding; coagulopathy; hepatic, and renal insufficiency. Nowadays, many pharmacological preparations of acetaminophen and diclofenac or ibuprofen combinations are available. If both the drugs are separately used together and combination drugs are used, it is seen much lesser individual drugs are required for effective pain relief. Moreover, there is comparatively lesser chances of NSAIDS-related adverse effect and acetaminophen related hepatic toxicity.

Aspirin: It effectively relieves acute pain of mild to moderate variety. It should be given 4 h. Aspirin can cause gastrointestinal hemorrhage and ulcer (Derry and Moore 2012), bronchospasm (Carvajal et al. 1996), and platelet inhibition. Co-administration of aspirin and NSAIDS increase the chance of GI bleeding.

Cox-2 inhibitors: Cancer prevention property of rofecoxib and celecoxib in breast and NSAIDs in colorectal cancer are under trial (Hartmann et al. 2005).

Opioids: Natural opioids examples are endorphin, enkephalin, and dynorphin. Opioid receptor are mu (mechanical, chemical, thermal painful stimuli, and also respiratory depression and constipation), kappa (thermal), delta (mechanical and inflammatory)—seen predominantly in CNS, particularly in dorsal horn of spinal cord, substantia gelatinosa, dorsal root ganglion (DRG), and peripheral nerves.

Morphine: It is a prototypical opioid with strong agonist to mu type of opioid receptor which is situated in both central nervous system (CNS) and peripheral nervous system. It is hydrophilic, phenanthrene derivative. It is the drug of choice for the management for moderate to severe type of pain. As it has hydrophilic property, it crosses the blood–brain barrier slowly. For that reason, onset of action is delayed. Oral bioavailability is 20–30%.

Routes of Administration

Morphine can be administered by several routes like intravenous, subcutaneous, intrathecal (spinal or epidural), and oral. Intravenous or intrathecal route is preferred only in hospitalized patients with severe pain or for those unable to take orally. Oral route is preferred for all other patients, as it is safe, convenient, and preferred by patients. Onset of action is 15–30 min, and duration of action is 4–5 h. Peak plasma concentration is achieved within 15–60 min in case of oral administration and 10–20 min in case of IM/SC.

Chronic intake of morphine ultimately leads to higher circulating concentration of morphine-3 glucuronide and morphine 6 glucuronide metabolites than the morphine (Gadisseur et al. 2003; Wilcox et al. 2005). It should be prescribed with caution in patients with decreased liver function, such as the patients suffering with cirrhosis (Wilcox et al. 2005). Patients with renal impairment should also be monitored properly because of the fear of accumulation of morphine's toxic metabolites.

Normal-release (NR) morphine is dosed 4–6 h; sustained release (SR) morphine is dosed 12 h. NR and SR morphine has the same efficacy and side effect. It may be given in different routes such as oral, transdermal, rectal, subcutaneous, intrathecal, and epidural. It may cause hepatic and renal dysfunction, where dose modification is needed in view of cumulative effect of active metabolites. There are no ceiling effect and no maximum dose of morphine. However, sometimes side effects prevent to achieve adequate analgesia. Dose is highly individual patient-dependent.

Side effects and complications of morphine:

(i) Constipation (MC)—Prescribe laxatives, that is, bisacodyl, unless the patient is having diarrhea. cremaffin, dulcolax tablets/enema/suppository.
(ii) Nausea, vomiting—antiemetic: tab metoclopramide/haloperidol/ondansetron pruritus—antihistaminic: chlorpheniramine/hydroxyzine/cetirizine
(iii) CNS side effects—euphoria/dysphoria.
(iv) Respiratory depression:
   (a) Mild: reduce the dose.
   (b) Moderate to severe: naloxone to be used in incremental dose, as required.
(v) Drowsiness may be noticed in the first few days.
(vi) Tolerance to morphine may occur.
(vii) Physical dependence to morphine in cancer patients is rare but dreadful complication.
(viii) Addiction (psychological dependence)—uncommon in cancer patients.

Morphine dose as rescue analgesic: 5–15% of the total daily dose is recommended to be safe.

Caution about the side effects should be taken if the patient is taken high daily doses of opioid (>200 mg oral morphine or equivalent) or patients is in opioid rotation. Hepatic and renal derangements are not a contraindication to use opioids. But caution is required in these cases because of the fear of accumulation of drugs or active metabolites. Dose interval of morphine to be increased, so that the morphine is to be given at 6, 8, or even 12 hrs interval. In the case of renal impairment, it is better to avoid.

Different formulations of morphine are mainly of two types. Immediate release formulations are morphine elixir, immediate release tablets, and suppositories. These are prescribed 4 h. Another type available is controlled release formulation. It is administered 12 h or 24 h.

These are effective in wide dose range when administered orally. Normally, it is started in lower dose and increased gradually over time as required. But it is to be kept in mind that the dose is appropriate for a patient with best possible analgesia with least possible/manageable side effects.

Opioid Medicines

The easiest method of dose titration is with immediate release morphine given 4 h, and the same dose is administered to alleviate breakthrough pain as per European Association for Palliative Care (EAPC) guideline. The EAPC supports the use of a double dose of immediate release morphine at bedtime. At the time of dose adjustment of morphine, increase the dose by 25%, 75%, or 100%. Instructions must be given clearly that any extra dose of morphine given for breakthrough pain does not imply that the next regular dose should be omitted. The breakthrough dose must be considered whenever the regular dose of morphine is increased.

Morphine in Children

It is safe to use opioids in children, and pain physician should not withhold analgesia from children who are suffering from pain. Starting doses are to be calculated according to the weight of the child. Management of pain must be individualized according to the child's need for pain relief by titration of the dose. Extra cautions should be considered in case of neonates, because they are more prone to the central nervous system depression due to the presence of a poorly developed blood brain barrier and accumulation of metabolites. So it is better to avoid.

Signs of Morphine Toxicity
This includes drowsiness, confusion, respiratory depression, hallucinations, pinpoint pupils, and myoclonus, but these are seldom seen with oral morphine dosing.

Management of Morphine Toxicity
Reduce the dose of morphine by 50%, and consider parenteral fluids to increase excretion of it. In severe cases, stop the morphine and administered naloxone, an opioid antagonist. Start with (0.4–2) mg which can be titrated up to (6–10) mg over a short interval. Haloperidol can be given at night to decrease hallucinations or confusion caused by the morphine.

Codeine: It is a weak μ opioid agonist. Analgesic effect of codeine depends on conversion to morphine and genetic variations in the enzymes.

Oxycodone: It is a semisynthetic opioid, two times potent than morphine. It is a prodrug, metabolized in liver by CYP2D6 isoenzyme to oxymorphone, which is a μ receptor agonist.

Hydromorphone: It is hydrogenated ketone analog of morphine hydrophilic in nature. It is five times more potent than oral morphine. It is also effective if given subcutaneously at a particular rate.

Methadone: It is synthetic opioid with complex pharmacology; μ and delta opioid agonist, NMDA antagonist, and serotonin reuptake inhibitor. It is not metabolized to neurotoxic or active metabolites, rather metabolized in the liver in to inactive metabolites. But, there is increased incidence of methadone-related deaths (overdose, accumulation, cardiac arrhythmias) making it potential for adverse outcomes and controversial. For using this drug, specialized palliative care expertise is needed.

Meperidine: It is weaker μ opioid agonist than morphine. Potentially neurotoxic metabolite and nor-meperidine are produced, rendering it unpopular. If this drug is administered with MAOI, SSRI, tramadol, methadone – neurotoxicity increased. This is one of the most widely used opioid analgesic agents used in emergency department for acute pain management. Due to short duration of action, it is better to be avoided in long-term use, especially in renal insufficiency and elderly.

Fentanyl: It is synthetic opioid, highly lipophilic agent having high affinity for μ opioid receptor. It is 75–125 times more potent and faster onset of action than morphine. Transmucosal fentanyl avoids first-pass metabolism. Onset of action is within 5–10 min.

Transdermal Fentanyl
It is long-acting mainly used for chronic pain and cancer pain, typically placed on a clean, dry, undamaged, and nonhairy flat part of the body. Site of application may be varied with each change of the patch. Patches available are 12/25/50/75/100 mcg/h;

determined doses are delivered in 72 hrs. After the patch is removed after 72 h, blood concentration falls slowly. After 17 h of removal, it becomes 50%. Patients should be instructed to avoid submerging it in hot water or placing a heating pad over the area, because this hampers absorption.

It is always challenging to titrate the dose due to the variation in each patient's skin perspiration, skin temperature, fat layer, and muscle bulk (Wilcox et al. 2005). Most common side effects are local skin erythema or irritation. It has less GI issues like constipation than oral opioids because it avoids direct exposure to the GI tract. More practical approach is to use morphine formulation to calculate the required dose of opioid and then convert it to fentanyl patches. The patient must have morphine tablets available for breakthrough pain.

Oral Transmucosal Fentanyl Citrate (OTFC)

This type is available in lozenges form in 200, 400, 600, 800, 1200, 1600 mcg strengths. It avoids GIT and hepatic first-pass metabolism. It has rapid onset of action (10–15 min, crosses BBB rapidly). But it should not be used for ≥ 4 times in 24 h. It is very effective for breakthrough pain. For one episode of breakthrough pain, keep OTFC 15 min in mouth and then remove. It can be inserted once again after 15 min, but not more than that. If the pain is not relieved, use next higher dose lozenge for next break through episode. Intravenous fentanyl is used for patient-controlled analgesia (PCA) to alleviate breakthrough pain.

Buprenorphine: It is a semisynthetic analog of thebaine (Heel et al. 1979) nonselective, mixed agonist–antagonist opioid receptor modulator (Ehrich et al. 2015). It acts differently on different receptors. It has μ-opioid receptor (MOR) partial agonistic, κ-opioid receptor (KOR) antagonistic, and δ-opioid receptor (DOR) antagonistic property. Very weak partial agonistic action is present on nociceptin receptor (NOP, ORL-1) (Lutfy and Cowan 2005; Kress 2009; Robinson 2002; Bidlack 2014; Ehrich et al. 2015). Lack of respiratory depression is seen with buprenorphine in overdose. It also blocks voltage-gated sodium channels via the local anesthetic binding site. This is the cause for its potent local anesthetic properties (Leffler et al. 2012).

In 1978, it was first launched in the UK as an injection to treat severe pain. In 1982, sublingual formulation was released. It can be used both in acute and chronic pain (Heel et al. 1979). It can be administered under the tongue, intramuscularly, transdermal, or as an implant (Heel et al. 1979). Maximum pain relief is generally achieved within an hour with effects last up to 24 h. A transdermal patch is indicated for the treatment of chronic pain. It is metabolized via liver into nor-buprenorphine and then eliminated through bile. The elimination half-life is varied from 20 to 73 h. As it is eliminated through liver, there is no risk of accumulation of drugs in patients with renal impairment (Moody et al. 2009). It has slow onset of action and a long half-life of 24–60 h.

Common side effects are nausea, vomiting, dizziness, drowsiness, headache, memory loss, itchiness, cognitive and neural inhibition, perspiration, dry mouth, meiosis, orthostatic hypotension, decreased libido, male ejaculatory difficulty, and urinary retention. Constipation and CNS effects are noticed less frequently than morphine. Others side effects are respiratory depression, sleepiness, low blood pressure, allergic reactions, adrenal insufficiency, QT prolongation, and opioid addiction. Risk of psychological or physical dependence is always there on chronic use. So the patients must be supervised during its use, and medically supervised withdrawal is a must. The reversal agents for opioids, such as naloxone, act partially, and additional efforts must be given to support breathing.

15.4.2.2 Weak Opioids

Tramadol: It is a weak synthetic opioid (acting on μreceptor), mainly used for its centrally acting analgesic properties, weak synthetic opioid mostly on mu receptors, with mild serotonin (at periaqueductal gray, nucleus raphe magnus inhibits 5HT reuptake) and norepinephrine (alpha adrenergic in locus coeruleus of pons) reuptake inhibiting effects. This drug, though is an opioid, seems to have an excellent safety profile and low abuse potential. Serotonin toxicity may be seen if concomitantly used with alcohol (Clarot et al. 2003; Mannocchi et al. 2013; Michaud et al. 1999; Tjäderborn et al. 2007), TCA, and SSRI. Intravenous naloxone has been used to reverse the effects of tramadol overdose (Marquardt et al. 2005; Sachdeva and Jolly 1997).

Normal-release formulations may be administered 4–6 h, and the extended-release preparations should be prescribed every 12–24 h. To minimize adverse effects and to improve tolerability, 10–16 days dose titration schedule of tramadol (IR) is standard recommendation. ER preparations are always better tolerated and are given once or twice daily. It is also available in combination preparation with acetaminophen (paracetamol). When the combination of tramadol and acetaminophen are used, both the drugs are needed in comparatively lesser amount for adequate pain relief, which helps to curtail down the known side effects of individual drug.

Dihydrocodeine: It is semisynthetic opioid analgesic administered for moderate to severe pain. It has similar chemical structure to codeine but is twice as strong as codeine. It can either be used alone or in combination with paracetamol or aspirin. It was developed in Germany in 1908 and first marketed in 1911 (*Encyclopedia of Psychopharmacology | Ian P. Stolerman | Springer*, December 15, 2020), approved for medical use in 1948.

Commonly this drug is available as tablets, capsules, solutions, elixirs, sublingual drops, and other oral forms, injectable solution for deep subcutaneous and intramuscular administration (in some countries). Controlled-release preparations are also available—60, 90, and 120 mg modified release tablets. As with other opioids, tolerance and physical and psychological dependence develop with chronic use. Itching, flushing, and constipation are the common side effects.

Dextropropoxyphene: It is a mild synthetic opioid (32.5, 65 mg tablets), used to treat mild pain. It is often used in combination with paracetamol (Li Wan Po and Zhang 1997; Collins et al. 1998). India is not the only country to withdraw this drug, but various countries like European Union, Australia, Canada, UK, New Zealand, and USA had withdrawn it from market. Later, it was withdrawn from market across the European Union in 2009. Subsequently, US Food and Drug Administration

(FDA), New Zealand, and Canada in the year 2010 banned its use. Australia issued a notification against its use in 2012 (Balhara 2014). The reasons cited for the withdrawal from these countries include its overdose related deaths and its impact on cardiovascular electrophysiology even within therapeutic window. Additionally, concerns have been expressed about its utility as an analgesic as well as high suicide rate (Simkin et al. 2005; Jang et al. 2011).

15.4.2.3 Adjuvants
Adjuvants are very much necessary part in cancer pain management.

Antiepileptic drugs, such as carbamazepine (Na channel blocker), oxcarbamazepine (Na-channel blocker), gabapentin (voltage-gated Ca-channel blocker), pregabalin (voltage-gated Ca-channel blocker), are used.

Antidepressants drugs such as amitriptyline (TCA), nortriptyline (TCA), fluoxetine (SSRI), and duloxetine (SNRI) are frequently prescribed.

Neuroleptics such as haloperidol and chlorpromazine can be given.

NMDA antagonist, ketamine, in subanesthetic dose is helpful.

Corticosteroids (methyl prednisolone, dexamethasone, etc.) are used in tumor-induced bone pain, hepatic capsular stretching, and neural compression, which are generally started in higher dose and then gradually tapered to lowest effective dose.

Benzodiazepines and sedatives are also helpful. A good sleep can reduce pain and also painful muscle spasms. For example, tension headache is also relieved. These are also helpful in terminal period and resistant types of pain. Clonazepam at night may be given and to be started at lower dose. Propofol and midazolam are also helpful as sedative.

Desensitization: Few cancer pain patients get relief with intravenous lidocaine given as slow infusion, especially those with wide spread pain and neuropathic pain. But it should be given with caution and only for in-patients with all monitoring facilities.

15.5 Interventional Pain Management

There are various routes other than oral for drug administration for pain relief like intramuscular, intravenous, subcutaneous, or neuraxial block.

Intramuscular: Commonest intervention is intramuscular injection. Drug is injected inside the muscle from where it is absorbed in the systemic circulation.

Intravenous: Drug is injected directly into the vein. It may be of two types. (1) Bolus, for immediate relief. (2) Continuous: It is also called patient-controlled analgesia (PCA). Initially, total daily requirement of opioids is calculated, and it is delivered intravenously through IV cannula and continuous syringe drivers or infusion pumps. Patient can control the dose of the drug up to certain safety limit. Nowadays, even family controlled analgesia is helpful at domestic environment for terminally ill patients.

Subcutaneous: It is continuous opioid infusion subcutaneously. It is very easy to administer. Even patient relative also can administer.

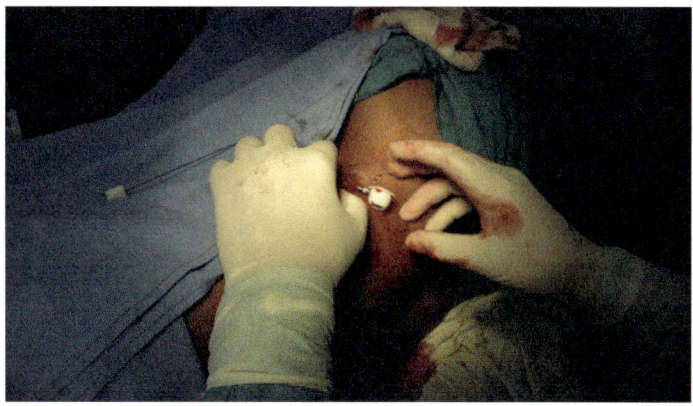

Fig. 15.8 Implantation of epidural port for continuous epidural infusion (taken from ESI Institute of Pain Management, Sealdah file images with permission)

Neuraxial: Drug is given directly at the spinal cord or nerves which carry the pain sensation to the brain.

(a) **Epidural**: Opioid or steroid is given in the epidural space as infusion or bolus (as required). Epidural catheter with a port can be implanted in patients who require long-term pain relief or repeated epidural injections (Fig. 15.8).
(b) **Intrathecal:** Morphine is given into theca in cerebrospinal fluid. Bolus is generally given for diagnostic purpose, before implantation of long-term intrathecal morphine pump. Intrathecal pump is used to deliver minute quantities morphine continuously at the level of receptors, e.g., substantia gelatinosa. The mechanical pump uses pacemaker technology to generate programmability. The pump (Fig. 15.9) is to be implanted in a subcutaneous pocket at the abdominal wall. The pump has to be refilled with drugs when its reservoir of 18 mL gets empty. Bupivacaine, clonidine, dexmedetomidine, ketamine, midazolam, etc., also can be added as adjuncts if indicated.
(c) **Nerve or ganglion blocks**: Peripheral nerve or ganglion block is done when a specific nerve pathway of a particular branch is responsible for carrying pain.
For oral and head neck region cancer: Many interventional procedures can be done which can reduce the pain drastically and relieve the patients from severe excruciating pain. Some of the common interventions are greater and lesser occipital block, trigeminal ganglion and/or the divisions neurolysis or nerve ablation, sphenopalatine block (Fig. 15.10a, stellate ganglion block (Fig. 15.10b, c)), glossopharyngeal nerve block, and cervical plexus block.

For pain arising from tumors involving hands, brachial plexus, axillary, brachial or branch block, and stellate ganglion block can be done.

For chest wall tumors, intercostal nerve block or neurolysis, T2 and T3 blocks (Fig 15.10b), can give good relief to pain.

Fig. 15.9 Intrathecal morphine pump (taken from ESI Institute of Pain Management, Sealdah file images with permission)

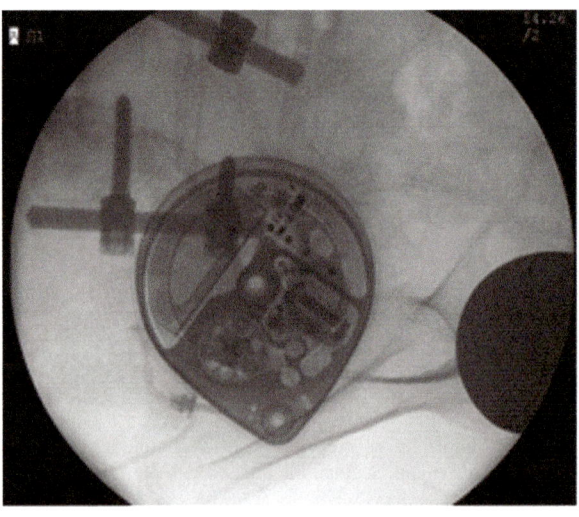

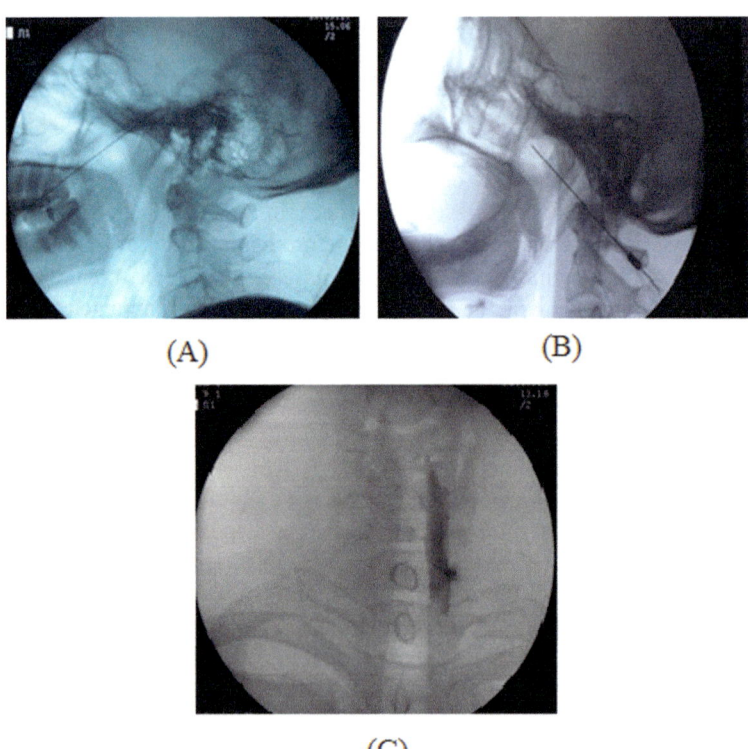

Fig. 15.10 (**a**) Trigeminal ganglion block; (**b**) sphenopalatine ganglion block; (**c**) a stellate ganglion block (taken from ESI Institute of Pain Management file images with permission)

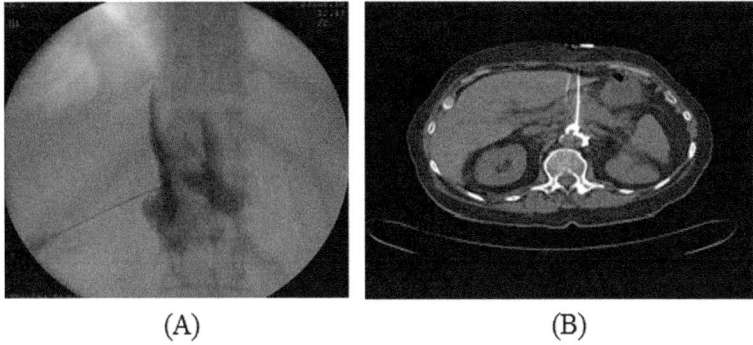

Fig. 15.11 (a) Celiac plexus block; (b) celiac plexus block, anterior approach

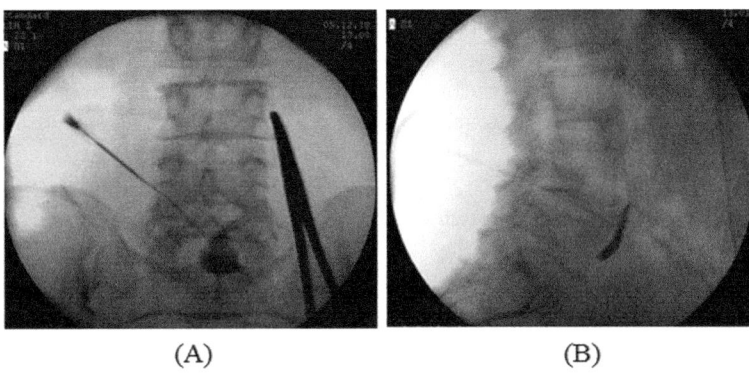

Fig. 15.12 (a) Superior hypogastric plexus block (AP view), (b) superior hypogastric plexus block (lateral view)

For pain from Upper Abdominal cancers like stomach CA, Pancreatic Cancer, Celiac plexus block (Fig 15.11a, b) and/or with Splanchnic nerve RF is usually preferred.

In case of pain from tumors involving lower abdomen, superior hypogastric plexus block (Fig 15.12a, b) and ganglion impar block (Fig 15.13a) can be tried.

For pain from tumors involving lower extremity, lumbar plexus block, lumbar sympathetic block (Fig 15.13c), ilioinguinal and iliohypogastric, obturator, femoral, popliteal, and branches can be tried.

Local anesthetics (LA): LAs are used during interventional pain procedures to anesthetize the nerves supplied to the painful regions. Commonly used LAs are lignocaine (1–2%), bupivacaine (0.25%), and ropivacaine (0.125–25%).

(i) **Lidocaine** (Na channel, NMDA blocker): It is the first amino-amide used clinically. It is also used for local infiltration, nerve block, epidural, and intrathecal procedures. Intravenous (5 mg/kg for 30–60 min and then

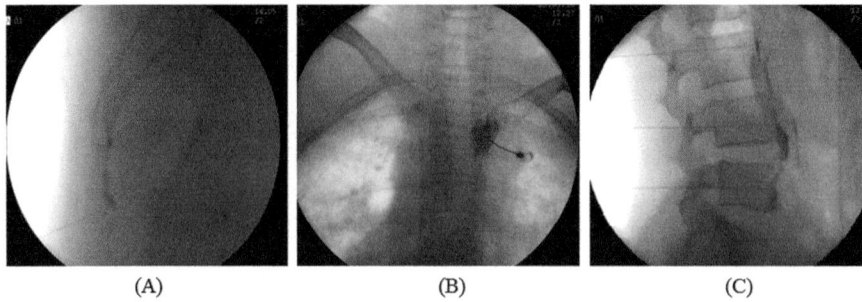

Fig. 15.13 (a) Ganglion impar block (lateral view); (**b**) T2 and T3 sympathetic blocks; (C) lumbar sympathetic block (Fig. 15.11 through Fig. 15.13 are taken from ESI institute of pain management file images with permission)

1–2 mg/kg/h for 4–8 h) lignocaine is used for desensitization in any chronic pain.
(ii) **Bupivacaine:** It is longer acting LA. It is used commonly but to be used with caution due to its cardiotoxicity and neurotoxicity.
(iii) **Ropivacaine:** It is 30–40% less cardio toxic than bupivacaine.

Adjuvants: Adjuvants are often used with local anesthetics for prolonging the effect of blocks and increase the intensity of nerve block. $Alpha_2$ adrenergic agonists—clonidine and dexmedetomidine—are commonly used as adjuvant. They inhibit release of presynaptic excitatory neurotransmitters, e.g., substance P and glutamate. They also can block A-alpha an C-fiber conduction like local anesthetics.

Neurolytics: Neurolytic therapy is the last resort of treatment, where appropriate cancer patients get significant benefit from pain. Peripheral, visceral, or neuraxial neurolysis can be done. Commonly used neuraxial agents are as follows:

(i) Alcohol: Ethyl alcohol produces Wallerian degeneration of nerve fibers; 50–70% concentration is used for nerve block. Sympathetic nerves regenerate within 3–5 months.
(ii) Phenol: 6–10% is used; initial local anesthetic effect makes injection painless. Neurolysis is evident within 3–7 days.
(iii) Glycerol: Trihydric alcohol absorbs water from atmosphere, is a mild neurolytic agent, and is primarily used for trigeminal ganglion block with 100% glycerol.

15.5.1 Radio-Frequency Ablation

Radio-frequency waves are delivered by a generator, percutaneously in continuous or pulse mode close to the targeted nerve for prolonged pain relief. The radio-frequency waves produce an electromagnetic field which causes ionic movement s in the nerve and cause ablation of the nerve. The nerve takes three months to three years to regenerate. So radio frequency can provide long-term pain relief. Radio frequency of the nerves are done in trigeminal neuralgia, cluster headache, sphenopalatine ganglion neuralgia, cancer pain, intractable ocular pain, occipital neuralgia, cervicogenic headache, facetogenic pain, chronic regional pain syndrome, abdominal pain, discogenic pain, perineal pain, peripheral vascular disease, chronic neuralgic pain, etc.

15.5.2 Intrathecal Saddle Neurolysis

Neurolytics can be placed in theca for terminal stage of oncologic pain. As various side effects are there, it should be used in terminal stages only with precautions. Dogliotti in 1931 first used intrathecal neurolytic block. Recently, 100% alcohol and 10–12% phenol are used most commonly for intrathecal neurolysis (Narouze and Kapural 2007).

15.5.2.1 Advanced Interventional Pain Management Procedures

Many advanced technologies and devices have been developed in recent times which are designed to combat the intractable pain. Some of them are described below.

Spinal cord stimulation: The stimulator leads are placed in the epidural space either percutaneously or by laminectomy surgery. Spinal cord is stimulated with the help of those leads by a device called stimulator (Figs. 15.14a and 15.15b). Stimulator can be implanted in a subcutaneous pocket after a period of satisfactory pain relief by external stimulation generator. This device stimulates the spinal cord by

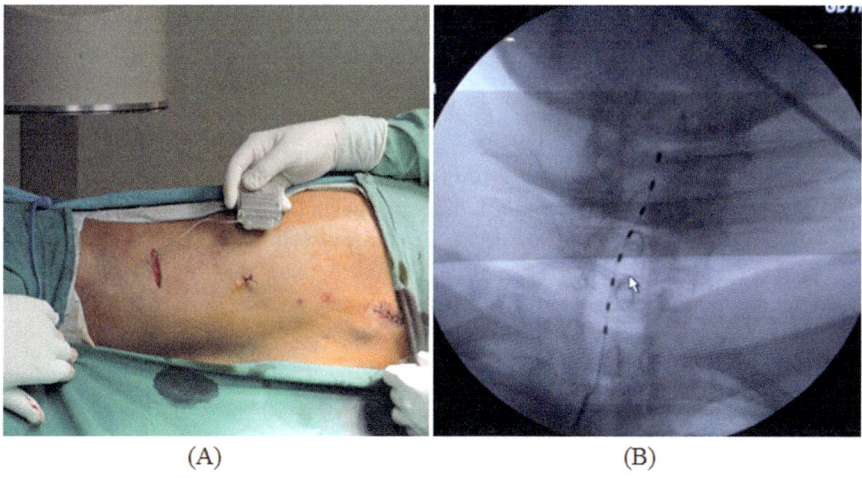

Fig. 15.14 (**a**) spinal cord stimulator; (**b**) spinal cord stimulator leads (from the personal archive of the first author with permission of the patient)

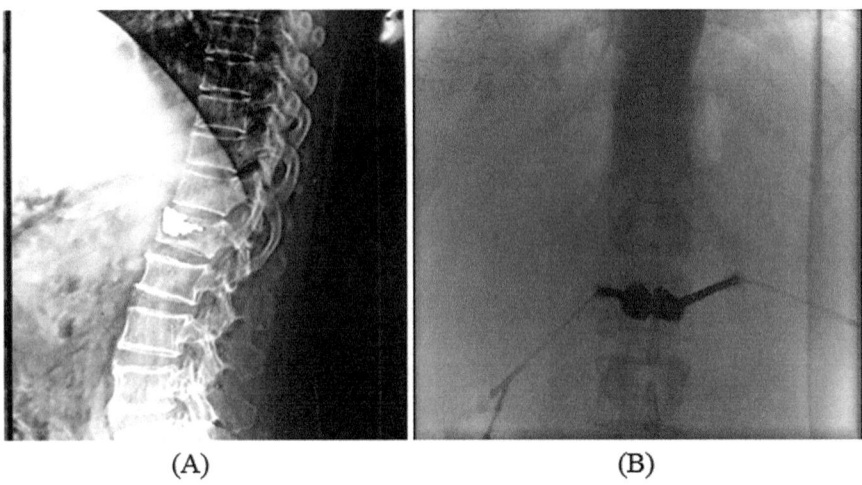

Fig. 15.15 (**a**) Postvertebroplasty X-ray; (**b**) balloon kyphoplasty (taken from ESI Institute of Pain Management file images with permission)

delivering pulse frequencies of 40–60 Hz in such a way that a soothing buzzing sensation blocks the pain sensation to be carried through the spinal cord. FDA has approved this spinal cord stimulator for treatment of chronic intractable low back pain, leg pain, and pain from failed back surgery syndrome. In Europe, its use has been additionally approved for refractory angina pectoris and peripheral limb ischemia (Slatkin and Rhiner 2003).Though in cancer pain its use is limited due to high cost, it may be used for refractory neuropathic states associated with cancer and it can reduce the dose of opioid significantly.

Recently, many modes of spinal cord stimulators have been developed.

(i) DRG SCS: It is highly targeted form of neuromodulation therapy by applying an innovative lead configuration and delivery system around the DRG (Shanthanna et al. 2014; Deer et al. 2014; Liem et al. 2015).
(ii) High-frequency 10 SCS (HF10) (Tiede et al. 2013; Van Buyten et al. 2013): It applies a unique waveform at 10,000 Hz at a subthreshold level and provides pain relief but does not produce any paresthesia (Kapural et al. 2015).
(iii) Burst SCS: Burst frequency is 40 Hz, and pulse frequency is 500 Hz. It is a novel mode of stimulation because by delivering subthreshold stimulation pain relief is provided, with either reduced or no paresthesia (De Ridder et al. 2010, 2013; De Vos et al. 2014).

15.5.2.2 Peripheral Nerve Stimulation

After accurate diagnosis and medical management, if pain persists, then peripheral nerve stimulation is one of the proven effective ways to relieve pain. Transcutaneous electrical nerve stimulation (TENS) is delivered by electrode placement on the affected area and delivery of measured amount of current to the body. Implantable peripheral nerve stimulator is placed in close proximity of peripheral nerves, and either an external or implanted battery is there to stimulate the nerve. In acute musculoskeletal injury and acute postoperative pain, it can be used with good efficacy (Long 1983). It is also effective in chronic trigeminal neuropathic pain (Feletti et al. 2013 ; Klein et al. 2016), episodic cluster headache (supraorbital nerve stimulation) (Asensio-Samper et al. 2008; Narouze and Kapural 2007), chronic migraine/headache disorders (occipital nerve stimulation) (Matharu et al. 2004), fibromyalgia (C2 area stimulation) (Thimineur and De Ridder 2007), postherpetic neuralgia (Dunteman 2002; Lerman et al. 2015; Upadhyay et al. 2010), complex regional pain syndrome type I (Hassenbusch et al. 1996) and type II (Jeon et al. 2009), isolated peripheral neuropathy (Eisenberg et al. 2004), ilioinguinal, iliohypogastric, and lateral femoral cutaneous neuralgia (Stinson et al. 2001), back pain (Kapural et al. 2018), foot pain (tibial nerve stimulation) (Chan et al. 2010), and coccydynia (Slavin 2007; Granville et al. 2017).

In patients with coagulopathy, psychiatric illness, infection in the surgical site, complete sensory loss, a failed diagnostic trial, and who require periodic MRIs, peripheral nerve stimulation should not be used.

15.5.2.3 Vertebroplasty and Balloon Kyphoplasty (McCall et al. 2008)
Vertebral compression fractures commonly occur after trauma or in severe osteopenic conditions also in vertebral metastasis. Sometimes, it is very painful and pain remains even after adequate conservative therapy. Vertebroplasty and balloon kyphoplasty can reduce the pain. Bone cement (PMMA: polymethylmethacrylate) is injected percutaneously into the cancellous bone of the compressed vertebral body for vertebroplasty, whereas in kyphoplasty first a cavity is created inside the collapsed vertebrae with an inflatable balloon and then bone cement is filled inside that vertebrae (Fig. 15.15a, b) Pain is reduced by the bone cement or due to correction in alignment.

Paralysis or radiculopathy may occur as a complication of these procedures due to extravasation of the cement into the epidural space and compression of neural elements. Pulmonary embolism is another severe complication and it occurs when cement flows into venous channels (Lee et al. 2002; Ratliff et al. 2001; Chen et al. 2002; Jang et al. 2002; Tozzi et al. 2002). The chance of extravasation of cement is less in kyphoplasty than vertebroplasty. Other complications of these procedures are bleeding, infection, rib fracture, or pneumothorax in thoracic cases. Pedicle fractures (Nussbaum et al. 2004) and osteomyelitis are rare complication (Walker et al. 2004).

15.5.2.4 Neurosurgical Palliative Techniques
Recently, some neurosurgical palliative interventions are being done to control the unbearable pain of the cancer patients.

15.5.2.5 Pituitary Ablation
In the case of severe pain from **endocrinological** tumors, mostly bone metastasis and deafferentation pain stereotactic trans-sphenoidal microsurgical hypophysectomy are indicated. It may be done by radio frequency, by chemical agents, or by ionizing

radiation. In severe widespread cancer pain when other methods of pain management fail, ablation of the pituitary gland may be tried. Pain relief can occur within hours though the mechanism is not clear. Injection of alcohol in pituitary is a relatively simple procedure. The transnasal trans-sphenoidal route was described by Bonica in 1968. It requires a light general anesthesia and is performed under x-ray image intensifier; 2ml of alcohol is injected. A cisternal puncture is then performed to inject 40 ml hydrocortisone to prevent damage to the optic nerve. Pan-hypopituitarism, diabetes insipidus, and visual dysfunction are common side effects with this procedure.

This historical evidence prompted pain physicians to perform gamma knife surgery (GKS) using a high irradiation dose to the pituitary stalk/gland (Hayashi et al. 2003). This GKS is a non-invasive method of partial pituitary ablation without significant side effects.

Stereotactic trans-sphenoidal microsurgical hypophysectomy by radio-frequency ablation has advantage of fast resolution of intense pain and low rate of complications with diminishing hospitalization days (Peixoto dos Santos et al. 2016).

15.5.2.6 Commissural Myelotomy

Through the center of the spinal cord runs the polysynaptic pain pathway, as well as the pain conducting fibers. Commissural myelotomy disrupts this pathway. Indications are bilateral and midline pelvic or perineal pain. Complications are sphincter or motor dysfunction. In open myelotomy, multilevel laminectomy is done to expose of the appropriate lumbar or sacral segments of the spinal cord. From the study, it has been seen that open midline commissural myelotomy can provide effective pain relief with acceptable postoperative morbidity (Viswanathan et al. 2010).

15.5.2.7 Percutaneous Cordotomy

For unilateral intractable pain below C5, of the lateral chest/abdomen, lumbosacral plexus, pelvis, or lower extremities, cordotomy is the best way to reduce pain. Cordotomy is done at high cervical level. With the use of stereotactic percutaneous technique, this procedure has become simplified. Commonest complication is respiratory, which can be well reduced by careful selection and evaluation of patients.

15.5.3 Biopsychosocial Model of Pain (Dueñas et al. 2016)

Biopsychosocial model of pain accounts for multidimensional nature of pain. It describes complex interactions between biological factors (e.g., genetical, hormonal, endogenous opioids, injury, trauma, infection, illness, and nerve damage), psychological factors (e.g., anxiety, depression, fear, grief, and coping skills), and social factors (e.g., gender, ethnicity, healthcare provider bias, work related issues, family circumstances, and economic conditions) that may contribute to pain. The biopsychosocial model of pain is shown in Fig. 15.16.

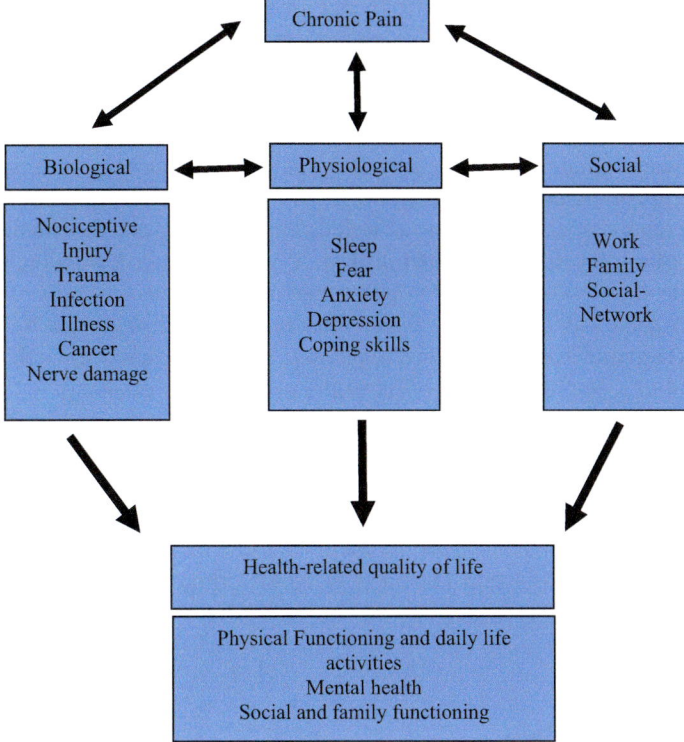

Fig. 15.16 Biopsychosocial model of pain and consequences on quality of life

Thus, biopsychosocial phenomenon includes sensory, emotional, cognitive, developmental, behavioral, spiritual, and cultural components of pain. Ultimately, all these lead to compromise in quality of life, mental health got affected, and physical–social functioning got jeopardized.

This model of pain leads to the concept of *total pain*, which needs to be treated from multidisciplinary, multifactorial aspect.

15.5.3.1 Behavioral Pain Management Techniques

Positive thinking and relaxed state of mind minimize pain sensation, whereas negativity makes pain experience worse. So, cognitive modifications and relaxation techniques are taught to the patients. In relaxation techniques, health psychologist teaches the patients to relax physically and mentally. Pain patients are taught to accept pain as the part of their life and work to reduce the severity and to lead a constructive life, to improve the quality of life.

(i) **Hypnosis:** Recent reviews reveal that hypnosis produce significant reduction in pain in chronic pain conditions and is relatively better alternative than nonhypnotic techniques, i.e., physical therapy and educating the patients.

The method was seen to be effective in management of pain in metastatic breast cancer and adult cancer patients undergoing bone marrow transplant therapy (Spiegel and Bloom 1983).

(ii) **Relaxation:** Progressive muscle relaxation technique helps to break pain–muscle–tension–anxiety cycle. The calming effect facilitates pain relief. This self-care technique is very easy to learn and can be taught by nurses.

(iii) **Cognitive–behavioral therapy:** It helps to reduce pain, depression, anxiety, and other associated distress, as well as improve sleep and functioning.

(iv) **Treating psychiatric comorbidities:** Any of anxiety, depression, posttraumatic stress disorder [PTSD], and somatoform disorders is most of the time associated with cancer pain, which must be taken care of along with the pain.

(v) **Therapeutic exercise:** It helps to increase aerobic capacity, strength, balance, flexibility, improvement of posture, and generalized well-being.

(vi) **Physical therapy (PT):** It helps to increase range of movement, functioning, and strength of the muscles.

(vii) **Delivery of care:** It is given by in-patient or out-patient based, which requires a team work of doctors, nurses (core clinical team), physiotherapist, occupational therapist and other specific therapist, social worker, chaplain/priest/guru, and volunteers (trained). Patient is given the prime importance, followed by family.

15.5.4 Palliative Care

Palliative is derived from the word *pallium*, which means a cloak. According to the definition of WHO, palliative care is an approach that improves the quality of life of patients and their families facing the problem associated with life-threatening illness, through the prevention and relief of suffering by means of early identification and impeccable assessment and treatment of pain and other problems: physical, psychosocial, and spiritual (Fig. 15.17). WHO also addressed children as a special group.

Palliative care works on the concept of *total care*, with an objective to provide relief from pain and other distressing symptoms, and does not intend to hasten or postpone death. Affirmation of life is important, whereas dying here is taken as a normal process. It integrates the psychological and spiritual aspects of patient care and helps the patients to live as actively as possible until death. Death is inevitable, but not the sufferings. Terminally ill cancer patients are more afraid of pain than death. So, in palliative care pain is taken care of, with providing best possible quality of life. This leads to death with dignity. Bereavement care is given to the family even after the death of the patient. Emotional distress—anxiety, depression, worry, fear, sadness, and hopelessness—of the patient is given prime importance in pain management and palliative care. It is a multidisciplinary approach, involving doctors and nurses (core clinical team), physiotherapist, occupational therapist, and other specific therapist, social worker, chaplain/priest/guru, and volunteers (trained).

Ethics followed in palliative care are

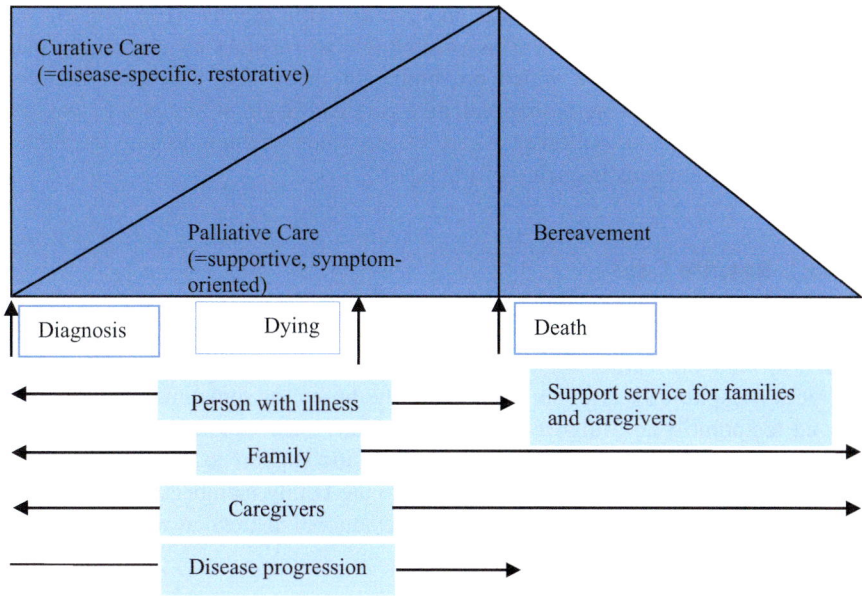

Fig. 15.17 WHO model of palliative care

(i) Beneficence—treatment to be given in the best interest of the patient.
(ii) Nonmaleficence—weigh benefit against adverse effects for every intervention.
(iii) Autonomy—acknowledgement of patients' right to take decision regarding treatment.
(iv) Justice—equitable allocation of health resources according to need, irrespective of class, creed, race, and color.

In some recent studies, it has been shown that, if palliative care can be incorporated soon after the diagnosis of advanced cancer, quality of life and mood can be improved along with prolong survival (Temel et al. 2010). Palliative care in advanced cancer is also a recommendation of "The American Society of Clinical Oncology" (Ferrell et al. 2017).

15.6 Hospice-Based and Home-Based Pain Management

Hospice care is an approach of care, which provides comprehensive care to the terminally ill patients as well as supports the emotional and spiritual needs of the family members. Many times, it refers to a institute which gives special care to these patients by medical professionals as well as health workers. The term "hospice" originates from a Latin word named "*hospes*" which mean "host." Hospice actually hosts a terminally ill person in their final days. They provide medical, psychological, and spiritual support to the patients. The goal of the care is to help people to die peacefully, comfortably, and more importantly with dignity. The aim of the caregivers is to make the patient as comfortable as possible by controlling pain and other distresses, those which are bothering them. The difference between hospice and palliative care is that palliative care can begin at any point along the cancer care continuum, but hospice care begins when attempts to cure the illness stops and the sole focus is quality of life (QOL).

15.6.1 Respite Care

It is a short-term, temporary break for the primary care giver and other family members who are caring for terminally ill patients, and it is often offered by hospices. Respite care is often provided in the home, or sometimes they may be moved temporarily to a higher care facility. Many times trained volunteers offer respite care for family members as well as provide support to the patient care. It provides the much needed, temporary break to the family members, without which families not only suffer from economical and emotional distress, as well as may face serious health hazards.

15.7 Conclusion

With the development and continuous progress in medical science, new diagnostic and therapeutic modalities are coming in the way. All of these help to diagnose the type, variety, and stage of different cancers. Well known and newer drugs try to combat with the spread of cancer. Surgical interventions are done with an intention to eliminate cancer from the body. Whole lot of these endeavors is one side of the spectrum.

On the other end is the patient, suffering from terminal stage of cancer with severe pain, anorexia, nausea, vomiting, respiratory distress, emotional breakdown, and other constellation of symptoms. There remains a physical, social, psychological, spiritual, and financial crisis to the patient as well as his/her family members with severe compromise in the quality of life, loss of relationships, disfigurement, and isolation.

Of all these sufferings pain is the most common (67%) and dreadful one, for which patient often becomes more prone to suicide. Many of the times, they are not bothered about the cancer, but their life become miserable because of pain. Albert Schweitzer, a famous philosopher, physician, noble laureate (1952) said, "We all must die. But if I can save him from days of torture, that is what I feel as my great and ever new privilege. Pain is a more terrible lord of mankind than even death himself."

In the modern era, where cancer has become an epidemic, oncologists and oncosurgeons are struggling to cure this evil; we the pain physicians have the greatest responsibilities to save those moribund faces from their suffering from pain. Best effort to be put is to candle the awareness in general people, to shake the government with the opioid supply policies, and to generate more and more pain physicians to maintain considerable doctor–patient ratio to combat the immense load of cancer pain.

As in democracy decision of the people is final, similarly the baton of pain free society lies not only within the physicians but among those common people, their awareness, will, and urge to get victory over pain.

References

Asensio-Samper JM et al (2008) Peripheral neurostimulation in supraorbital neuralgia refractory to conventional therapy. Pain Practice 8(2):120–124. https://doi.org/10.1111/j.1533-2500.2007.00165.x

Balhara YPS (2014) Dextropropoxyphene ban in India: is there a case for reconsideration? J Pharmacol Pharmacother 5(1):8–11

Bidlack JM (2014) Mixed Kappa/Mu partial opioid agonists as potential treatments for cocaine dependence. In: Advances in pharmacology. Academic Press Inc., pp 387–418. https://doi.org/10.1016/B978-0-12-420118-7.00010-X

Bieri D et al (1990) The faces pain scale for the self-assessment of the severity of pain experienced by children: development, initial validation, and preliminary investigation for ratio scale properties. Pain 41(2):139–150. https://doi.org/10.1016/0304-3959(90)90018-9

Body JJ et al (2006) A study of the biological receptor activator of nuclear factor-κ ligand inhibitor, denosumab, in patients with multiple myeloma or bone metastases from breast cancer. Clin Cancer Res 12(4):1221–1228. https://doi.org/10.1158/1078-0432.CCR-05-1933

Brant JM (2017) Palliative medicine and hospice care holistic total pain management in palliative care: cultural and global considerations article history'. Palliative Med Hosp Care Open J 1:S32–S38. https://doi.org/10.17140/PMHCOJ-SE-1-108

Brown MRD, Ramirez JD, Farquhar-Smith P (2014) Pain in cancer survivors. Br J Pain 8(4):139–153. https://doi.org/10.1177/2049463714542605

Carvajal A et al (1996) Aspirin or acetaminophen? A comparison from data collected by the Spanish Drug Monitoring System. J Clin Epidemiol 49(2):255–261. https://doi.org/10.1016/0895-4356(95)00539-0

Catella-Lawson F et al (2001) Cyclooxygenase inhibitors and the antiplatelet effects of aspirin. N Engl J Med 345(25):1809–1817. https://doi.org/10.1056/nejmoa003199

Chan I et al (2010) Ultrasound-guided, percutaneous peripheral nerve stimulation: technical note. Neurosurgery 67(1). https://doi.org/10.1227/01.NEU.0000383137.33503.EB

Chen H-L et al (2002) A lethal pulmonary embolism during percutaneous vertebroplasty. Anesth Analg 95(4):1060–1062. https://doi.org/10.1097/00000539-200210000-00049

Chow E et al (2007) Palliative radiotherapy trials for bone metastases: a systematic review. J Clin Oncol:1423–1436. https://doi.org/10.1200/JCO.2006.09.5281

Citrin DL, Bessent RG, Greig WR (1977) A comparison of the sensitivity and accuracy of the 99Tcm-phosphate bone scan and skeletal radiograph in the diagnosis of bone metastases. Clin Radiol 28(1):107–117. https://doi.org/10.1016/S0009-9260(77)80137-2

Clarot F et al (2003) Fatal overdoses of tramadol: Is benzodiazepine a risk factor of lethality. Forensic Sci Int 134(1):57–61. https://doi.org/10.1016/S0379-0738(03)00100-2

Cleeland CS, Ryan KM (1994) Pain assessment: global use of the brief pain inventory. Annals, Academy of Medicine, Singapore 23(2):129–138

Cleeland CS et al (1994) Pain and its treatment in outpatients with metastatic cancer. N Engl J Med 330(9):592–596. https://doi.org/10.1056/nejm199403033300902

Cohen MZ et al (2003) Cancer pain management and the JCAHO's pain standards: an institutional challenge. J Pain Symp Manage 25(6):519–527. https://doi.org/10.1016/S0885-3924(03)00068-X

Coleman RE (1997) Skeletal complications of malignancy. In: Cancer. John Wiley and Sons Inc., pp 1588–1594. https://doi.org/10.1002/(sici)1097-0142(19971015)80:8+<1588::aid-cncr9>3.0.co;2-g

Coleman RE et al (2006) Clinical features of metastatic bone disease and risk of skeletal morbidity. Clin Cancer Res. https://doi.org/10.1158/1078-0432.CCR-06-0931

Collins SL et al (1998) Single-dose dextropropoxyphene in post-operative pain: a quantitative systematic review. Eur J Clin Pharmacol 54(2):107–112. https://doi.org/10.1007/s002280050430

De Ridder D et al (2010) Burst spinal cord stimulation: toward paresthesia-free pain suppression. Neurosurgery 66(5):986–990. https://doi.org/10.1227/01.NEU.0000368153.44883.B3

De Ridder D et al (2013) Burst spinal cord stimulation for limb and back pain. World Neurosurg. https://doi.org/10.1016/j.wneu.2013.01.040. Elsevier Inc.

De Vos CC et al (2014) Burst spinal cord stimulation evaluated in patients with failed back surgery syndrome and painful diabetic neuropathy. Neuromodulation 17(2):152–159. https://doi.org/10.1111/ner.12116

Deer TR et al (2014) The appropriate use of neurostimulation of the spinal cord and peripheral nervous system for the treatment of chronic pain and ischemic diseases: the neuromodulation appropriateness consensus committee. Neuromodulation 17(6):515–550. Blackwell Publishing Inc. https://doi.org/10.1111/ner.12208

Derry S, Moore RA (2012) Single dose oral aspirin for acute postoperative pain in adults. Cochrane Database Syst Rev (4) John Wiley and Sons Ltd. https://doi.org/10.1002/14651858.CD002067.pub2

Deyo RA, Diehl AK (1988) Cancer as a cause of back pain—frequency, clinical presentation, and diagnostic strategies. J Gen Internal Med 3(3):230–238. https://doi.org/10.1007/BF02596337

Drake MT, Clarke BL, Khosla S (2008) Bisphosphonates: mechanism of action and role in clinical practice. In: Mayo clinic proceedings. Elsevier Ltd., pp 1032–1045. https://doi.org/10.4065/83.9.1032

Dueñas, M. et al. 2016. A review of chronic pain impact on patients, their social environment and the health care system. J Pain Res, 457–467. Dove Medical Press Ltd. doi: https://doi.org/10.2147/JPR.S105892.

Duke G, Petersen S (2015) Perspectives of Asians living in Texas on pain management in the last days of life. Int J Palliative Nurs 21(1):24–34. https://doi.org/10.12968/ijpn.2015.21.1.24

Dunteman E (2002) Peripheral nerve stimulation for unremitting ophthalmic postherpetic neuralgia. *Neuromodulation* 5(1):32–37. https://doi.org/10.1046/j.1525-1403.2002._2006.x

Ehrich E et al (2015) Evaluation of opioid modulation in major depressive disorder. Neuropsychopharmacology 40(6):1448–1455. https://doi.org/10.1038/npp.2014.330

Eisenberg E, Waisbrod H, Gerbershagen HU (2004) Long-term peripheral nerve stimulation for painful nerve injuries. Clin J Pain 20(3):143–146. https://doi.org/10.1097/00002508-200405000-00003

Encyclopedia of Psychopharmacology | Ian P. Stolerman | Springer (2020). Available at: https://www.springer.com/gp/book/9783540687061 (Accessed: 12 April 2021).

Fainsinger RL, Nekolaichuk CL (2008) A "TNM" classification system for cancer pain: The Edmonton Classification System for Cancer Pain (ECS-CP). Support Care Cancer:547–555. https://doi.org/10.1007/s00520-008-0423-3

Fainsinger RL et al (2005) A multicenter study of the revised Edmonton Staging System for classifying cancer pain in advanced cancer patients. J Pain Sympt Manage 29(3):224–237. https://doi.org/10.1016/j.jpainsymman.2004.05.008

Fainsinger RL et al (2009) Is pain intensity a predictor of the complexity of cancer pain management. J Clin Oncol 27(4):585–590. https://doi.org/10.1200/JCO.2008.17.1660

Feletti A et al (2013) Peripheral trigeminal nerve field stimulation: report of 6 cases. Neurosurg Focus 35(3). https://doi.org/10.3171/2013.7.FOCUS13228

Ferrell BR et al (2017) Integration of palliative care into standard oncology care: American society of clinical oncology clinical practice guideline update. J Clin Oncol 35(1):96–112. https://doi.org/10.1200/JCO.2016.70.1474

Fishman B et al (1987) The memorial pain assessment card. A valid instrument for the evaluation of cancer pain. Cancer 60(5):1151–1158. https://doi.org/10.1002/1097-0142(19870901)60:5<1151::AID-CNCR2820600538>3.0.CO;2-G

Fuchs-Lacelle S, Hadjistavropoulos T (2004) 'Development and preliminary validation of the Pain Assessment Checklist for Seniors with Limited Ability to Communicate (PACSLAC). Pain Manage Nurs 5(1):37–49. https://doi.org/10.1016/j.pmn.2003.10.001

Gadisseur APA, Van Der Meer FJM, Rosendaal FR (2003) Sustained intake of paracetamol (acetaminophen) during oral anticoagulant therapy with coumarins does not cause clinically important INR changes: a randomized double-blind clinical trial. J Thromb Haemostasis 1(4):714–717. https://doi.org/10.1046/j.1538-7836.2003.00135.x

Ghilardi JR et al (2005) Selective blockade of the capsaicin receptor TRPV1 attenuates bone cancer pain. J Neurosci 25(12):3126–3131. https://doi.org/10.1523/JNEUROSCI.3815-04.2005

Gilbert RW, Kim J-H, Posner JB (1978) Epidural spinal cord compression from metastatic tumor: diagnosis and treatment. Ann Neurol 3(1):40–51. https://doi.org/10.1002/ana.410030107

Glare PA et al (2014) Pain in cancer survivors. J Clin Oncol:1739–1747. https://doi.org/10.1200/JCO.2013.52.4629

Granville M, Brennan P, Jacobson RE (2017) Bilateral peripheral nerve field stimulation for intractable coccygeal pain: a case study using dual lead intercommunication. Cureus 9(11). https://doi.org/10.7759/cureus.1832

Greco MT et al (2014) Quality of cancer pain management: an update of a systematic review of undertreatment of patients with cancer. J Clin Oncol:4149–4154. https://doi.org/10.1200/JCO.2014.56.0383

Halvorson KG et al (2006) Similarities and differences in tumor growth, skeletal remodeling and pain in an osteolytic and osteoblastic model of bone cancer. Clin J Pain:587–600. https://doi.org/10.1097/01.ajp.0000210902.67849.e6

Hartmann LC et al (2005) Benign breast disease and the risk of breast cancer. N Engl J Med 353(3):229–237. https://doi.org/10.1056/nejmoa044383

Hassenbusch SJ et al (1996) Long-term results of peripheral nerve stimulation for reflex sympathetic dystrophy. J Neurosurg 84(3):415–423. https://doi.org/10.3171/jns.1996.84.3.0415

Hayashi M et al (2003) Role of pituitary radiosurgery for the management of intractable pain and potential future applications. Stereotactic Funct Neurosurg 81(1–4):75–83. https://doi.org/10.1159/000075108

Heel RC, Brogden RN, Speight TM, Avery GS (1979) Buprenorphine: a review of its pharmacological properties and therapeutic efficacy. Drugs 17(2):81–110

Herr K et al (2006) Pain assessment in the nonverbal patient: position statement with clinical practice recommendations. Pain Manage Nurs 7(2):44–52. https://doi.org/10.1016/j.pmn.2006.02.003

Honore P et al (2000) Osteoprotegerin blocks bone cancer-induced skeletal destruction, skeletal pain and pain-related neurochemical reorganization of the spinal cord. Nat Med 6(5):521–528. https://doi.org/10.1038/74999

Hwang SS et al (2003) Development of a cancer pain prognostic scale. J Pain Symp Manage 24(4):366–378. https://doi.org/10.1016/S0885-3924(02)00488-8

Jang JS, Lee SH, Jung SK (2002) Pulmonary embolism of polymethylmethacrylate after percutaneous vertebroplasty. Spine 27(19):E416–E418. https://doi.org/10.1097/00007632-200210010-00021

Jang DH et al (2011) Fatal outcome of a propoxyphene/acetaminophen (Darvocet) overdose: should it still be used in the United States? Ann Emerg Med:421–422. https://doi.org/10.1016/j.annemergmed.2010.11.007

Jeon IC, Kim MS, Kim SH (2009) Median nerve stimulation in a patient with complex regional pain syndrome type II. J Korean Neurosurg Soci 46(3):273–276. https://doi.org/10.3340/jkns.2009.46.3.273

Joyce JA, Pollard JW (2009) Microenvironmental regulation of metastasis. Nat Rev Cancer:239–252. https://doi.org/10.1038/nrc2618

Julius D, Basbaum AI (2001) Molecular mechanisms of nociception. Nature:203–210. https://doi.org/10.1038/35093019

Kapural L et al (2015) Novel 10-kHz high-frequency therapy (HF10 therapy) is superior to traditional low-frequency spinal cord stimulation for the treatment of chronic back and leg pain. Anesthesiology 123(4):851–860. https://doi.org/10.1097/ALN.0000000000000774

Kapural L et al (2018) Percutaneous peripheral nerve stimulation for the treatment of chronic low back pain: two clinical case reports of sustained pain relief. Pain Practice 18(1):94–103. https://doi.org/10.1111/papr.12571

Klein J et al (2016) Peripheral nerve field stimulation for trigeminal neuralgia, trigeminal neuropathic pain, and persistent idiopathic facial pain. Cephalalgia 36(5):445–453. https://doi.org/10.1177/0333102415597526

Kremer E, Hampton Atkinson J, Ignelzi RJ (1981) Measurement of pain: patient preference does not confound pain measurement. Pain 10(2):241–248. https://doi.org/10.1016/0304-3959(81)90199-8

Kress HG (2009) Clinical update on the pharmacology, efficacy and safety of transdermal buprenorphine. Eur J Pain:219–230. https://doi.org/10.1016/j.ejpain.2008.04.011

Lee B-J, Lee S-R, Yoo T-Y (2002) Paraplegia as a complication of percutaneous vertebroplasty with polymethylmethacrylate. Spine 27(19):E419–E422. https://doi.org/10.1097/00007632-200210010-00022

Leffler A et al (2012) Local anesthetic-like inhibition of voltage-gated Na+ channels by the partial μ-opioid receptor agonist buprenorphine. Anesthesiology 116(6):1335–1346. https://doi.org/10.1097/ALN.0b013e3182557917

Lerman IR et al (2015) Novel high-frequency peripheral nerve stimulator treatment of refractory postherpetic neuralgia: a brief technical note. Neuromodulation 18(6):487–493. https://doi.org/10.1111/ner.12281

Li Wan Po A, Zhang WY (1997) Systematic overview of co-proxamol to assess analgesic effects of addition of dextropropoxyphene to paracetamol. Br Med J 315(7122):1565–1571. https://doi.org/10.1136/bmj.315.7122.1565

Liem L et al (2015) One-year outcomes of spinal cord stimulation of the dorsal root ganglion in the treatment of chronic neuropathic pain. Neuromodulation 18(1):41–49. https://doi.org/10.1111/ner.12228

Lipton A et al (2006) Future treatment of bone metastases. Clin Cancer Res. https://doi.org/10.1158/1078-0432.CCR-06-1157

Long DM (1983) Stimulation of the peripheral nervous system for pain control. Clin Neurosurg 31:323–343. https://doi.org/10.1093/neurosurgery/31.cn_suppl_1.323

Lucas R et al (2005) Cellular mechanisms of acetaminophen: role of cyclo-oxygenase. FASEB J 19(6):1–15. https://doi.org/10.1096/fj.04-2437fje

Lutfy K, Cowan A (2005) Buprenorphine: a unique drug with complex pharmacology. Curr Neuropharmacol 2(4):395–402. https://doi.org/10.2174/1570159043359477

Maltoni M (2008) Opioids, pain, and fear. Ann Oncol 19(1):5–7. https://doi.org/10.1093/annonc/mdm555. Oxford University Press

Mannocchi G et al (2013) Fatal self administration of tramadol and propofol: a case report. J Forensic Legal Med 20(6):715–719. https://doi.org/10.1016/j.jflm.2013.04.003

Mantyh PW (2006) Cancer pain and its impact on diagnosis, survival and quality of life. Nat Rev Neurosci:797–809. https://doi.org/10.1038/nrn1914

Marquardt KA, Alsop JA, Albertson TE (2005) Tramadol exposures reported to statewide poison control system. Ann Pharmacother:1039–1044. https://doi.org/10.1345/aph.1E577

Matharu MS et al (2004) Central neuromodulation in chronic migraine patients with suboccipital stimulators: a PET study. Brain 127(1):220–230. https://doi.org/10.1093/brain/awh022

McCall T, Cole C, Dailey A (2008) Vertebroplasty and kyphoplasty: a comparative review of efficacy and adverse events. Curr Rev Musculoskeletal Med 1(1):17–23. https://doi.org/10.1007/s12178-007-9013-0

McGrath PA et al (1996) A new analogue scale for assessing children's pain: an initial validation study. Pain 64(3):435–443. https://doi.org/10.1016/0304-3959(95)00171-9

Melzack R (1987) The short-form McGill pain questionnaire. Pain 30(2):191–197. https://doi.org/10.1016/0304-3959(87)91074-8

Mercadante S (1997) Malignant bone pain: pathophysiology and treatment. Pain:1–18. https://doi.org/10.1016/S0304-3959(96)03267-8

Michaud K et al (1999) Fatal overdose of tramadol and alprazolam. Forensic Sci Int 105(3):185–189. https://doi.org/10.1016/S0379-0738(99)00118-8

Moody DE et al (2009) Effect of rifampin and nelfinavir on the metabolism of methadone and buprenorphine in primary cultures of human hepatocytes. Drug Metab Dispos 37(12):2323–2329. https://doi.org/10.1124/dmd.109.028605

Narouze SN, Kapural L (2007) Supraorbital nerve electric stimulation for the treatment of intractable chronic cluster headache: a case report. Headache 47(7):1100–1102. https://doi.org/10.1111/j.1526-4610.2007.00869.x

Nussbaum DA, Gailloud P, Murphy K (2004) A review of complications associated with vertebroplasty and kyphoplasty as reported to the food and drug administration medical device related web site. J Vascular Intervent Radiology 15(11):1185–1192. https://doi.org/10.1097/01.RVI.0000144757.14780.E0

Peixoto dos Santos JGR, Duarte KP, Teixeira MJ (2016) Stereotactic transsphenoidal hypophysectomy by radiofrequency for chronic pain from hormone-independent metastatic tumors: A new perspective'. Journal of Pain & Relief 5(5):1–4. https://doi.org/10.4172/2167-0846.1000262

Plummer M et al (2016) Global burden of cancers attributable to infections in 2012: a synthetic analysis. The Lancet Global Health 4(9):e609–e616. https://doi.org/10.1016/S2214-109X(16)30143-7

Portenoy RK (2011) Treatment of cancer pain. Lancet:2236–2247. https://doi.org/10.1016/S0140-6736(11)60236-5

Ratliff J, Nguyen T, Heiss J (2001) Root and spinal cord compression from methylmethacrylate vertebroplasty. Spine 26(13). https://doi.org/10.1097/00007632-200107010-00021

Regnard C et al (2007) Understanding distress in people with severe communication difficulties: Developing and assessing the Disability Distress Assessment Tool (DisDAT). J Intellect Disability Res 51(4):277–292. https://doi.org/10.1111/j.1365-2788.2006.00875.x

Robinson SE (2002) Buprenorphine: An analgesic with an expanding role in the treatment of opioid addiction. CNS Drug Rev:377–390. https://doi.org/10.1111/j.1527-3458.2002.tb00235.x. Neva Press Inc.

Sachdeva DK, Jolly BT (1997) Tramadol overdose requiring prolonged opioid antagonism [9]. Am J Emerg Med:217–218. https://doi.org/10.1016/S0735-6757(97)90116-9. W.B. Saunders

Schwei MJ et al (1999) Neurochemical and cellular reorganization of the spinal cord in a murine model of bone cancer pain. J Neurosci 19(24):10886–10897. https://doi.org/10.1523/jneurosci.19-24-10886.1999

Shanthanna H et al (2014) Pulsed radiofrequency treatment of the lumbar dorsal root ganglion in patients with chronic lumbar radicular pain: a randomized, placebo-controlled pilot study. J Pain Res 7:47–55. https://doi.org/10.2147/JPR.S55749

Silverberg E, Boring CC, Squires TS (1990) Cancer statistics, 1990. CA: Cancer J Clin 40(1):9–26. https://doi.org/10.3322/canjclin.40.1.9

Simkin S et al (2005) Co-proxamol and suicide: Preventing the continuing toll of overdose deaths. QJM: Monthly J Assoc Phys:159–170. https://doi.org/10.1093/qjmed/hci026

Slatkin NE, Rhiner M (2003) Phenol saddle blocks for intractable pain at end of life: report of four cases and literature review. Am J Hospice Palliative Med:62–66. https://doi.org/10.1177/104990910302000114. Weston Medical Publishing

Slavin KV (2007) Peripheral neurostimulation in fibromyalgia: a new frontier?! Pain Med:621–622. https://doi.org/10.1111/j.1526-4637.2007.00328.x

Spiegel D, Bloom JR (1983) Group therapy and hypnosis reduce metastatic breast carcinoma pain. Psychosomatic Med 45(4):333–339. https://doi.org/10.1097/00006842-198308000-00007

Stannard C, Booth S (2004) Pain, 2nd edn. Churchill Livingstone

Steger GG, Bartsch R (2011) Denosumab for the treatment of bone metastases in breast cancer: Evidence and opinion. Ther Adv Med Oncol:233–243. https://doi.org/10.1177/1758834011412656

Stinson LW et al (2001) Peripheral subcutaneous electrostimulation for control of intractable postoperative inguinal pain: A case report series. Neuromodulation 4(3):99–104. https://doi.org/10.1046/j.1525-1403.2001.00099.x

Teitelbaum SL (2007) Osteoclasts: what do they do and how do they do it? Am J Pathol. American Society for Investigative Pathology Inc.:427–435. https://doi.org/10.2353/ajpath.2007.060834

Temel JS et al (2010) Early Palliative Care for Patients with Metastatic Non–Small-Cell Lung Cancer. N Engl JMed 363(8):733–742. https://doi.org/10.1056/nejmoa1000678

Thimineur M, De Ridder D (2007) C2 area neurostimulation: A surgical treatment for fibromyalgia. Pain Med 8(8):639–646. https://doi.org/10.1111/j.1526-4637.2007.00365.x

Tiede J et al (2013) Novel spinal cord stimulation parameters in patients with predominant back pain. Neuromodulation 16(4):370–375. https://doi.org/10.1111/ner.12032

Tjäderborn M et al (2007) Fatal unintentional intoxications with tramadol during 1995–2005. Forensic Sci Int 173(2–3):107–111. https://doi.org/10.1016/j.forsciint.2007.02.007

Tozzi P et al (2002) Management of pulmonary embolism during acrylic vertebroplasty. Ann Thorac Surg 74(5):1706–1708. https://doi.org/10.1016/S0003-4975(02)03962-0

Turk DC, Okifuji A (2002) Psychological factors in chronic pain: evolution and revolution. J Consult Clin Psychol 70(3):678–690. https://doi.org/10.1037//0022-006x.70.3.678

Upadhyay SP et al (2010) Successful treatment of an intractable postherpetic neuralgia (PHN) using peripheral nerve field stimulation (PNFS). Am J Hospice Palliative Med 27(1):59–62. https://doi.org/10.1177/1049909109342089

Van Buyten JP et al (2013) High-frequency spinal cord stimulation for the treatment of chronic back pain patients: results of a prospective multicenter European clinical study. *Neuromodulation*. Blackwell Publishing Inc., pp. 59–66. https://doi.org/10.1111/ner.12006

Van Herk R et al (2007) Observation scales for pain assessment in older adults with cognitive impairments or communication difficulties. Nurs Res 56(1):34–43. https://doi.org/10.1097/00006199-200701000-00005

Viswanathan A et al (2010) Commissural myelotomy in the treatment of intractable visceral pain: technique and outcomes. Stereotactic Funct Neurosurg 88(6):374–382. https://doi.org/10.1159/000319041

Von Moos R et al (2008) Metastatic bone pain: Treatment options with an emphasis on bisphosphonates. Support Care Cancer:1105–1115. https://doi.org/10.1007/s00520-008-0487-0

Walker DH, Mummaneni P, Rodts GE (2004) Infected vertebroplasty. Report of two cases and review of the literature. Neurosurg Focus. https://doi.org/10.3171/foc.2004.17.6.6

Warden V, Hurley AC, Volicer L (2003) Development and psychometric evaluation of the pain assessment in advanced dementia (PAINAD) scale. J A Med Direct Assoc 4(1):9–15. https://doi.org/10.1097/01.JAM.0000043422.31640.F7

Wilcox CM, Cryer B, Triadafilopoulos G (2005) Patterns of use and public perception of over-the-counter pain relievers: focus on nonsteroidal antiinflammatory drugs. J Rheumatol 32(11)

Wongrakpanich S et al (2018) A comprehensive review of non-steroidal anti-inflammatory drug use in the elderly. Aging Dis:143–150. https://doi.org/10.14336/AD.2017.0306. International Society on Aging and Disease

Wu T et al (2020) Healthcare utilization and direct medical cost in the years during and after cancer diagnosis in patients with type 2 diabetes mellitus. J Diabetes Invest 11(6):1661–1672. https://doi.org/10.1111/jdi.13308

Part IV
Emerging Trends in Cancer Research

New Approaches in Cancer Research: Stem Cell Research, Translational Research, Immunotherapy, and Others

Soumyadeep Mukherjee, Ashesh Baidya, and Subhasis Barik

Abstract

Despite relentless efforts from the scientific community, emergence of an optimal anticancer medicament is yet an elusive concept. In this chapter, in retrospection of the pernicious aftermath of conventional chemo- and radio-therapeutic cancer treatment techniques, the authors reflect upon new-age strategies for treating various types of cancer. Starting off with a reductionistic view, the chapter highlights the fundamentals of different up-to-date cancer treatment modalities. From stem cell-mediated site-directed drug delivery or cancer stem cell targeting to tailored modifications in the immuno-environment of the neoplastic mass, the focus glides over multiple corners of futuristic cancer-targeting protocols. The attention gradually shifts toward a more translational angle, taking patient-to-patient variability into account, and inching toward personalized therapy modules. In a nutshell, this chapter describes how the amalgamation of different perspectives of cancer research like immunotherapy, stem cell therapy, and personalized medicine creates a notion of switching from nonspecific, unifactorial targeting of cancer toward a specific, nontoxic, multimodal therapeutic regimen.

Keywords

Stem cell therapy · Cancer stem cells · Translational research · Cancer immunotherapy · Personalized therapy

S. Mukherjee · A. Baidya · S. Barik (✉)
Department of In Vitro Carcinogenesis and Cellular Chemotherapy, Chittaranjan National Cancer Institute, Kolkata, West Bengal, India

© The Author(s), under exclusive license to Springer Nature Singapore Pte Ltd. 2022
S. K. Basu et al. (eds.), *Cancer Diagnostics and Therapeutics*,
https://doi.org/10.1007/978-981-16-4752-9_16

Abbreviations

ADCC	Antibody-dependent cell cytotoxicity
AIDS	Acquired immunodeficiency syndrome
AML	Acute myeloid leukemia
ASC	Adult stem cells
BCG	Bacillus Calmette–Guérin
CAR	Chimeric antigen receptor
CINC-1	Cytokine-induced neutrophil chemoattractant 1
CSCs	Cancer stem cells
CTL	Cytotoxic T lymphocyte
CTLA-4	Cytotoxic T lymphocyte-associated protein 4
DC	Dendritic cells
DNA	Deoxyribonucleic acid
EGFR	Epidermal growth factor receptor
ESC	Embryonic stem cells
EUSTM	European Society for Translational Medicine
Fab	Fragment antigen-binding
Fc	Fragment crystallizable
FDA	Food and Drug Administration, United States
GBM	Glioblastoma multiforme
GM-CSF	Granulocyte macrophage–colony-stimulating factor
HSCs	Hematopoietic stem cells
IFN	Interferon
IL	Interleukin
iPSCs	Induced pluripotent stem cells
ITAMs	Immunoreceptor tyrosine-based activation motifs
MDSCs	Myeloid-derived suppressor cells
MHC	Major histocompatibility complex
MSCs	Mesenchymal stem cells
NIH USA	National Institute of Health, United States of America
NK	Natural killer
NOD	Nonobese diabetic
NSC	Neural stem cells
OV	Oncolytic viruses
PD-L1	Programmed death ligand 1
RNAs	Ribonucleic acids
SCID	Severe combined immunodeficiency
SDF-1	Stromal cell-derived factor 1
TAMs	Tumor-associated macrophages
TCR	T-cell receptor
TGF-β	Transforming growth factor β
TIMP-1	Tissue inhibitor matrix metalloproteinase 1
TLR	Toll-like receptor

TNF	Tumor necrosis factor
TRAIL	Tumor necrosis factor-related apoptosis-inducing ligand
WHO	World Health Organization

16.1 Introduction

Cancer, in brief, can be defined by the state of one or more cells whose genetic regulatory circuitry has been ruffled. These cells often have erratic patterns of proliferation and other biological processes associated with it. Due to such metabolic aggression coupled with the ability to secrete a variety of signaling molecules, which can evoke a local and systemic damage response, these cells can turn out to be potential threats to the whole organ and, often, the whole system it belongs to. Unfortunately, the protective measures taken against their iniquity are seldom foolproof (Król et al. 2010). Therapeutic interventions against cancer mostly rely on application of cytotoxic doses of radiation or chemical agents through different routes. Location-specific delivery and target-specific action of most chemotherapeutic compounds in the tumor microenvironment are hardly achievable because of multiple anatomical barriers and phenotypic plasticity of the cancer cells. Radiation therapy, on the other hand, suffers from the hazards of collateral tissue damage (De Ruysscher et al. 2019). Natural selection always favors overgrowth of drug-resistant (and often multiple drug-resistant) clones of cancer cells amidst an environment of drug-mediated cancer suppression (Luqmani 2005). Moreover, the most anticancer drugs have insufficient lethality against a rare group of cancer cells called cancer stem cells, which retain multipotency to an extent whereby they can regenerate any type of cells associated with a cancerous mass once needed (Yu et al. 2012; Zhao 2016). Therefore, it is necessary to target cancers using novel, modified approaches, which utilize alternative yet physiologically relevant pathways and hurt the core of the evasion mechanisms used by different types of cancers. This chapter reviews the important facets of many such cutting-edge strategies and chronicles them on a reductionist-to-holistic acclivity.

16.2 Stem Cell Research

In recent times, conventional chemotherapy and radiotherapy have often fallen short of desired success due to emergence of resistant populations of cancer cells (Luqmani 2005; Barker et al. 2015). Hence, it has become necessary to implement novel biochemical tools to keep up with these multifactorial resistance mechanisms. Stem cells are pluripotent cells, capable of self-renewal. In lucid terms, they are a group of cells, which can give rise to a clonal population of many different types of cells, and at the same time can maintain their own pool by means of creating new copies of themselves. This process of generating multiple kinds of cells occurs through progressive loss of potency at each stage of differentiation and culminates in restriction into a particular lineage (Becker et al. 1963). Stem cells usually reside in specified microenvironments, called "stem cell niché," where they receive their nutrients and growth and differentiation cues from dedicated feeder cells and different physiological sources, respectively. These versatile cells can be classified based on their origin. Keeping aside the embryonic stem cells (ESCs), the stem cells which remain inside a fully formed organism are called adult stem cells (ASCs). These cells have multiple potency and can differentiate into any specific lineage, like mesenchymal stem cells (MSCs; give rise to musculoskeletal connective tissues), hematopoietic stem cells (HSCs; give rise to blood cells), and neural stem cells (NSCs; give rise to neurons and glial lineage cells). (Stem cells: What they are and what they do 2019). Because of the wide range of fates these cells can adopt depending upon their ever-changing microenvironment, they have become a lucrative model system to work on various developmental and clinical aspects.

16.2.1 Stem Cell Therapy: A New Messiah against Cancer

One interesting fact about stem cells is that they can express and secrete various growth factors, cytokines, and chemokines, which are of great importance in battling otherwise indestructible cancer cells, and also for regenerative growth of damaged tissue. Bone marrow-derived mesenchymal stem cells secrete high levels of TIMP-1 and CINC-1, and directly promote apoptosis in K562 cells by Bax and caspase-3 upregulation (Fathi et al. 2019).

However, to exert their tumoricidal actions, stem cells must reach the specific locale where the cancerous tissue is present. This is achieved by the fascinating tumor-tropic property of various stem cells. This property was first identified in xenograft mouse models, where neural stem cells were found to be extremely efficient in migrating toward glioma tissue (Aboody et al. 2000). Again, stem cell-secreted chemokines and other factors play a key role in this homing process, whose expression is triggered by hypoxia (Bagó et al. 2016). Migration of different stem cells like HSCs (Khurana et al. 2013) and MSCs (Eseonu and De Bari 2015) has been shown to be dependent upon interaction between various cytokines and chemokines and their receptors. Engineering this ligand–receptor interaction by overexpression of chemokine receptor CXCR4 in MSCs resulted in improved migration toward glioma cells in vitro and in a xenografted mouse model exhibiting human glioma (Park et al. 2011). Biomaterial platforms capable of releasing migration factors like SDF-1 in a continuous, regulated fashion have also been proved to be a promising tool in promoting stem cell recruitment to tumors in vivo (Schantz et al. 2007).

Many stem cells also carry the ability of tinkering with the immune system (Bernardo and Fibbe 2013). Mesenchymal stem cells are capable of suppressing CD4$^+$ and CD8$^+$ T-cell proliferation (Le Blanc et al. 2004), which is partly due to the expression of cyclooxygenase (COX1/COX2) and consequent prostaglandin E2 (PGE2) production (Najar et al. 2010). Involvement of lymphocyte regulatory cytokines such as IL-10 (Yang et al. 2009) and growth factors such as hepatocyte growth factor and transforming growth factor-β (Di Nicola et al. 2002) is also crucial in this MSC-mediated T-cell suppression. TGF-β, on a precise note, skews the equilibrium in the T-cell repertoire toward regulatory T (T_{reg})-cell phenotype, in conjunction with PGE2 (English et al. 2009). Additionally, activation of dendritic cells (Nauta et al. 2006) and NK cells (Carosella et al. 2008) is also inhibited by MSCs. Together, these activities dampen the adaptive and innate immune responses such that stem cells are not attacked by these patrolling immune machinery during their tumor-tropic migration, and simultaneously aid the healing process in the inflammation-stricken cancer microenvironment.

16.2.2 Anticancer Applications of Stem Cells

Since stem cells, especially MSCs and NSCs, can efficiently traffic to the cancerous tissue in vivo without alarming the immune system, they can be modified by different strategies to enhance their tumoricidal activity. Some common modifications along with their underlying principles are discussed below.

16.2.2.1 Enzyme/pro-Drug Therapy

Stem cells are well known to localize in cancer tissues. When these cells are modified using recombinant technology to overexpress an enzyme that converts a nontoxic pro-compound into a toxic, tumoricidal compound, or to overexpress the pro-drug itself, which can be biologically converted into the toxic product by the action of an enzyme in that anatomical location, it becomes a tumor-specific way of stem cell therapy. Murine NSCs expressing cytosine deaminase, when administered along with its substrate 5-fluorocytosine (which is converted into 5-fluorouracil, that acts as a replication inhibitor and point mutagen), inhibited growth of glioblastoma cells (Aboody et al. 2000). Intraperitoneal ganciclovir monophosphate injection in rats after intratumoral injection with NSCs expressing herpes simplex virus–thymidine kinase (HSV-TK) also significantly improved probability of survival (Li et al. 2005).

16.2.2.2 Secretion of Antitumor Agents

Engineered stem cells can act as a source of multiple antitumor agents. The prime example of such therapy makes use of interferon-β (IFN-β), a type I interferon, a protein that conditions cells against viral infections. Engraftment of IFN-β expressing MSCs into primary breast tumor sites caused suppression of tumor growth in a STAT3-dependent mechanism (Ling et al. 2010). Secretion of IFN-β was found to be high only in the tumor microenvironment, not in the circulation, again highlighting the specificity in targeting by the stem cells.

16.2.2.3 Use of Viruses

Transduction of tumor-tropic viruses through stem cells is another effective therapeutic anticancer technique. Oncolytic viruses (OVs), like certain mutants of herpes simplex virus (Liu et al. 2003) and vaccinia virus (Dai et al. 2017), can replicate only in the tumor cells and act as a two-edged sword by activating the antiviral immune response in a way that becomes antitumor as well. These viruses exhibit variable anticancer efficacy, even in phase I phase III clinical trials (Twumasi-Boateng et al. 2018). OVs delivered by NSCs improved survival of mice intracranially implanted with GBM43 glioma cells, postradiotherapy, and temozolomide-based chemotherapy (Tobias et al. 2013). This finding was of particular interest since both radiotherapy and chemotherapy combined also did not affect viral replication inside the cells. MSCs, infected with attenuated measles virus and delivered systemically, were found to migrate toward orthotopically implanted tumors in liver and allow MV infection in cancerous hepatocytes through heterofusion, resulting in reduced hepatocarcinoma (Ong et al. 2013).

16.2.3 Stem Cells as Regenerative Medicine for Cancer

Embryonic stem cells (ESCs) derived from the embryonic inner cell mass are pluripotent in nature. Not only they can generate all different kinds of cells in a tissue, but they can also organize those cells in a tissue-like layout, and can subsequently scaffold the tissues in an organ-like arrangement. Despite immunosuppressive ventures exerted by tumor cells in their own microenvironment, which keep cytotoxic immune responses at a check, tumor tissues sustain significant injury due to radiotherapy (Prasanna et al. 2012) and chemotherapy (Vyas et al. 2014). Transplantation of pluripotent stem cells onto these sites is an ideal remedy for this, since the stem cell can then heal and regenerate the damaged parts. However, due to restrictions on direct use of ESCs for scientific research on ethical grounds, differentiated somatic cells genetically reprogrammed to regain their pluripotency, called induced pluripotent stem cells (iPSCs), can be potential alternatives (Takahashi and Yamanaka 2006). Patient-specific autologous iPSCs are an attractive prospect due to their lack of immunogenicity and, consequently, no need for immune suppression during transplantation (Guha et al. 2013). However, contradictory reports do exist (Zhao et al. 2011). Therefore, it is still an open area of exploration to find out definitive approaches in stem cell-mediated tissue regeneration postradiotherapy and chemotherapy.

16.2.4 Factors Governing Success of Stem Cell Therapy

Despite huge potential of stem cells in anticancer therapeutic purposes, certain factors regulate their clinical outcomes. For example, in a glioma model, NSCs supported more oncolytic viral replication than MSCs. Additionally, despite comparable migration capacities, therapeutic efficacy of NSCs against intracranial tumors was far better than MSCs (Miska and Lesniak 2015), highlighting the fact that *"stem cells of greater physiological relevance are better therapeutic dopes."* Again, *transplantation route* (Miska and Lesniak 2015) and *timing* (Ahmed et al. 2011) of NSCs affect their tumoricidal efficiency too. *Number of stem cells* transplanted also dictates success of therapy. Transplantation of less than optimum number of HSCs in patients with hematological malignancies leads to incomplete regeneration of hematopoietic machinery and an early relapse (Golubeva et al. 2014). Identification of these factors and their careful execution are an essential part of designing a stem cell-based therapy against all physiological kinds of cancer.

16.2.5 Cancer Stem Cell: A Hero-Turned-Villain?

In 1994, a study by John Edgar Dick's group first reported a unique cell from a tumor mass of acute myeloid leukemia (AML), which, upon transplantation into severe combined immune-deficient (SCID) mice, migrated to the bone marrow and repopulated the leukemic tissue with the help of suitable cytokines (Lapidot et al. 1994). They identified this cell as "leukemia-initiating cell" and fractionated it based on surface marker expression ($CD34^+CD38^-$). In later studies, they proved the presence of this kind of "initiating cells" in other types of leukemias and hypothesized a hierarchy within a tumor microenvironment where different cells have different priorities based on the surroundings of the tumor (Bonnet and Dick 1997). These initiating cells are called cancer stem cells (CSCs). Their presence has been proved not only in leukemia, but also in other types of cancers too (Al-Hajj et al. 2003; Li et al. 2007; O'Brien et al. 2007; Patel et al. 2012; Schatton et al. 2008; Singh et al. 2003; Zhang et al. 2008). However, exactly how they originate is still an unsolved mystery. Among the many standing hypothesis regarding this, one says these cancer stem cells are mutant forms of healthy stem cells (Wang et al. 2009), while another states that CSCs are dedifferentiated from cancer cells (Nouri et al. 2017). Since these cells are believed to be the primal source of all cancer cells, targeting them is always a better option than targeting the heterogeneous population of cells inside the cancer tissue. Moreover, after chemotherapy, CSCs can cause tumor relapse by promoting regrowth of the cancer cells and guiding their metastatic activities (Yu et al. 2012). Figure 16.1 shows a lineage tree for normal and cancer stem cells.

16.2.6 Therapeutic Approaches against Cancer Stem Cells

Despite significant progress in terms of identification of CSCs in many types of cancer, few therapies on them have been successfully applied. The reason for this is the extensive chemoresistance property of these cells. Many layers of cellular regulation act as key players in this process (Zhao 2016). The cells themselves slow down their division process, thereby making themselves immune against chemotherapeutics targeting rapidly dividing cells. There is a marked upregulation of anti-apoptotic proteins, enzymes involved in DNA repair, and transmembrane efflux pumps for actively exporting drugs from the cells. Furthermore, the cellular niche also protects the cells from exposure to huge concentration of anticancer drug

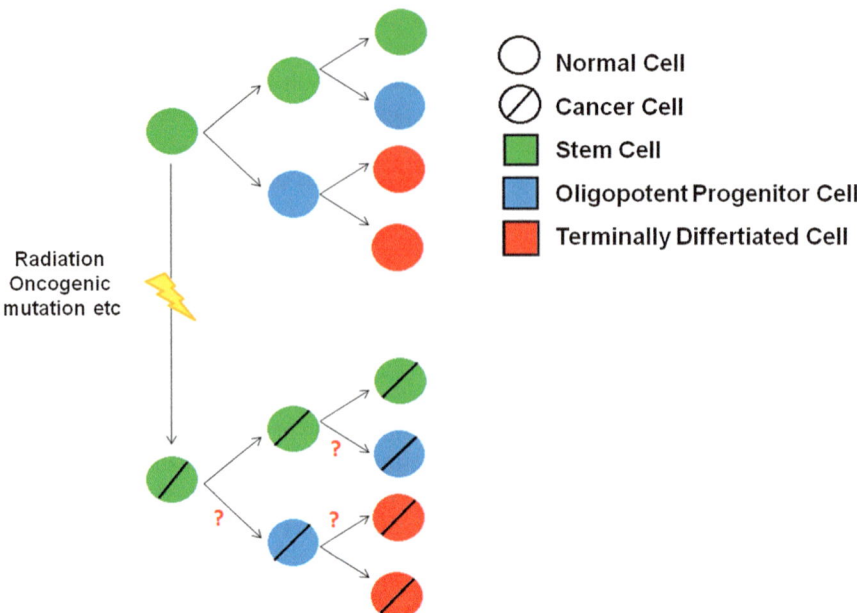

Fig. 16.1 Lineage tree for normal and cancer stem cells. Normal stem cells are present in every tissue system. During their division, one daughter cell retains its pluripotency, while the other undergoes a commitment to some extent, and becomes an oligopotent progenitor cell, which again differentiates into terminally differentiated cells. Oncogenic mutations or other factors convert stem cells to cancer stem cells, which further supply malignant cells in the cancer tissue. However, the symmetry and kinetics of these divisions and differentiations are not well-established. Many cancer therapies fail to prevent relapses due to their inability to eliminate these cancer stem cells

molecules, by means of extracellular matrix modifications. Hence, alternate strategies are necessary for efficient killing of these stem cells.

Although CSCs are an integral part of almost any cancer, their surface marker expression pattern often remarkably differs from that of other metastatic or nonmetastatic cancer cells. For example, interleukin-3 receptor-alpha (IL-3Rα) overexpression is associated with $CD34^+CD38^-$ leukemic stem cells from AML, but not observed in normal $CD34^+CD38^-$ bone marrow cells (Jordan et al. 2000). Monoclonal antibody-mediated targeting of IL-3Rα drastically reduced leukemic stem cell homing to bone marrow and relapse rates in AML-treated NOD/SCID mice (Jin et al. 2009). Another prospective way to target these cells is to fix on the biological pathways outside the range of conventional chemotherapeutic targets, which are essential for their survival. Salinomycin, a potent autophagy inhibitor, selectively kills breast cancer CSCs with a 100-time greater efficiency than paclitaxel, a conventional anticancer drug (Gupta et al. 2009), because those CSCs considerably rely on autophagy for their survival. However, generalized application of a single protocol against all CSCs may not be effective, since each of them has their own unique biological traits. For an instance, the same salinomycin treatment, as described above, loses its efficacy against some CSCs, as they can use acidic lysosomes instead of autophagosomes for recycling cellular debris (Jangamreddy et al. 2013).

16.2.7 Choosing the Right Path

All in all, research in the field of stem cell biology has unfurled many possibilities in the long-fought battle against cancer. However, the question remains, is it a complete cure? The huge variability and lack of reproducibility in the datasets obtained so far pose a serious problem in translational application of stem cells in cancer treatment. Rest aside the ethical controversies, the procedure and quantitative output of stem cell harvesting methods are another serious concern. Adult stem cells, which exhibit lower chances of histo-incompatibility to some extent, are very slow-growing in vitro and difficult to maintain. Using pluripotent stem cells like ESCs or iPSCs poses another problem: These cells often transform into malignancy and create teratomas. iPSCs, especially, carry this major drawback since oncogenic viral constructs are used to introduce embryonic genes into somatic cells to create them (Leventhal et al. 2012). Selective targeting of cancer stem cells also comes with its own sets of difficulties associated with it. Apart from their extreme chemo- and radioresistance, marker-based or metabolic targeting can also have gruesome side effects unless maneuvered carefully. The elemental reason for this is the plasticity of the CSCs, which ultimately leads to the intratumor heterogeneity (Eun et al. 2017). Hence, the success stories of stem cell therapies should always be taken with a pinch of salt and the newfound avenues must be trodden carefully, to make these therapies as much efficacious as possible. Devising the right strategies to clear the gray areas can unbar various other modes of stem cell applications to treat diverse types of cancers with no effective therapy yet and can also become an inseparable part of future ventures such as personalized medicine.

16.3 Translational Research

Opening a bottle of medicine and reaching for medication sound very simple to us. But it is not as simple as it seems. Hard work of basic scientists and clinical investigators has made it possible for us. But how do treatment like drug, device, and diagnostics from scientific expert come from the laboratory bench to clinic to improve patient health? The answer for this vital question is translational research. The term "translational research" is interchangeable with translational medicine.

16.3.1 Basic Research Versus Applied Research

Translational research is divided into two types: basic research and applied research. Basic research or fundamental research is driven by scientists' curiosity and interest, which direct the scientists toward the systematic study to understand the fundamental phenomena of the aspect (Schauz 2014). It is not performed by considering the applicative and practical ends. It only provides deep understanding and knowledge toward the nature and law of the investigated subject, for example, how the universe began, what are the building blocks of the elements, what are proton, neutron, and electron composed of, what is basic feature of life. Most of the scientists consider that basic and fundamental understanding of all branches of science is needed in order to progress in the pursuit of knowledge. In other words, basic research is the fundamental platform of applied research. Basic research should be done first in order to reach the fruit of applied research. On the other hand, applied research is designed to solve the practical problems of modern world. Rather than acquiring knowledge from the sea of information, the goal of applied science is to transform human life toward betterment. It deals with systematic inquiry, which follows practical application of science for human welfare. It investigates acquired knowledge from previous theories and comes up with new method and techniques for client-driven purposes like improving agricultural crop production, coming up with new treatment strategies to cure diseases, and improving energy efficiency. Some group of scientists believe that the time has come to shift from basic research toward applied research. This trend has arisen from global problem of population and their problems.

> *Basic research is performed without thought of practical ends. It results in general knowledge and an understanding of nature and its laws. This general knowledge provides the means of answering a large number of important practical problems, though it may not give a complete specific answer to any one of them. The function of applied research is to provide such complete answers.*—(1943, US Office of Scientific Development and Research)

16.3.2 Definition of Translational Research

The term translational research appeared early in 1993. In early 1990, this term was used as a synonym of cancer research. Today, the literature includes a collection of attempts in various fields to define the term (*Translational science spectrum* 2018).

Translational research is an effort by scientists to come up with new therapies, drugs, and diagnosis procedure for human well-being and commercialization, by keeping basic scientific research as their milestone. As coded by NIH USA, "Translational research is defined as the process of applied ideas, insights and discoveries generated through basic scientific inquiries to treatment or prevention of human disease." The branch of translational research is a new research discipline, which describes the process where the basic science discoveries, clinical insights, and health policies are taken from laboratory to clinic and brought to the real world. It applies the discoveries from basic science in order to enhance human health and their well-beings. Actually, translational research is branch of applied science, which investigates the problems raised by fundamental researcher in medical context and applies the finding to medical practices for meaningful health outcomes. In other word, this branch of science harvests not only the knowledge from current health challenges but also its implementation of practical application, which gives new promising treatment to human health. By keeping the milestone of fundamental idea, it aims to translate new drugs, devices, and treatment procedures to cure patients. Hence, it deals with the idea "bench to bedside" where laboratory experiments are followed by clinical trials so that it can be plausible to come up with a new and better procedure of treatment (Dilmore et al. 2013).

As it is a relatively new branch of biomedical science, it supports interdisciplinary science with rewarding outcomes in favor of public health. According to European Society for Translational Medicine (EUSTM), it is an interdisciplinary branch of medical science, which supports three basic areas those are bench side, bedside, and community. For its current necessity to modern world, it is becoming more effective and popular to many universities and research centers. Since 2009, specialized journals like American Journal of translational Research and Science Translational Medicine are being popular.

16.3.3 Translational Research in Cancer Diagnosis and Drug Development

This wing of research can be approached in various ways in a collaborative manner. As its general definition suggests that it procures with clinical implementations, it can be utilized from a multidimensional viewpoint. One area of this research can be applied in diagnosis of cancer. Through new diagnostic approaches, one can identify the specific target and biomarkers of cancer. By knowing this information, physicians can better understand about the target, as cancers in the same organs, or even tissues, may have different biomarkers and different mutations. For example, in some cases it might carry a p53 mutation, some may have *RAS* overexpression, or some may have tyrosine kinase overexpression. So, by the help of advanced diagnosis techniques one may get the exact biomarker of the patient's cancer, which may help further for better prognosis.

Translational research not only tells us about the biomarkers and targets of cancer, but through its high output process scientists are developing new promising drugs, effective for cancers of different biological origins sharing the same biomarkers. As the drugs are being designed according to their respective targets, they may have enhanced efficiency toward the targets. Nowadays, many new approaches like gene therapies and immune therapies are also in use to cure cancer. Hence, one might say that translational research is an essential tool in personalizing the battle against cancer (Nagahiro 2002).

16.4 Immunotherapy in Cancer

In 1957, a group of researchers from Washington University School of Medicine proposed that lymphocytes act as surveillance agents in the body by recognizing and fighting nascent transformed cells (Dunn et al. 2002). However, they also hypothesized that this process is like a cat-and-mouse relationship where malignant cells also continuously evolve to edit their antigenic repertoire, so that the immune system fails to identify them. This constant arms race between the cancer cells and the immune system eventually shapes the immunological landscape in cancer. Malignant cells express various antigens, which are unique to them. These antigens, called "tumor-specific antigens," are the prime targets of the immune system in their anticancer venture. However, many other antigens are expressed in larger amounts in cancer cells compared to their nonmalignant counterparts. These "tumor-associated antigens" also carry enough potential to activate the immune system (Storkus et al. 2003). Directed targeting of various immune compartments and activating them with a desired level of specificity against tumors is the basic goal of immunotherapy, which is the need of time due to the limitations of chemo- and radiotherapy protocols.

16.4.1 Immune Scenario in Cancer

Both branches of the vertebrate immune system, adaptive and innate, actively participate in the immune response against cancer. The $CD8^+$ T lymphocytes are the major combatants in the adaptive branch. These cells, in their naive state, express a transmembrane glycoprotein, called "T-cell receptor (TCR)" on their surface. These TCRs have extracellular domains, which can recognize antigenic peptides combined with a glycoprotein named "major histocompatibility complex (MHC)" class I, expressed on surfaces of altered self-cells, like transformed or pathogen-infected cells. Recognition and binding to this complex through TCR activate the T cells, whereby they differentiate into effector cytotoxic T lymphocytes (CTL) and memory $CD8^+$ cells (Moss et al. 1992). The effector cells exocytose two proteins: perforin and granzyme. Perforin inserts itself across the target cell membrane and acts as a passageway for granzyme entry into the target cell. Inside the cell, granzyme cleaves initiator caspase-8 and caspase-10, and executioner caspase-3 and caspase-7, thereby activating apoptosis in the cells. In addition, it also cleaves BID to promote BAX/BAK oligomerization on mitochondrial membrane and cytochrome c release, which is another essential facet of activation of apoptosis (Bots and Medema 2006). The memory cell, on the other hand, retains the characteristics to activate and clonally propagate itself upon receiving the same immunogenic signal, which becomes a necessary antitumor defense during relapse phases (Martin and Badovinac 2018). However, a humoral response is also generated by intratumor infiltration of dendritic cells (DCs). These cells internalize soluble neoantigens secreted by cancerous cells, process them, and present them to helper T cells via MHC class II. The resulting activation of T cells and subsequent cytokine release

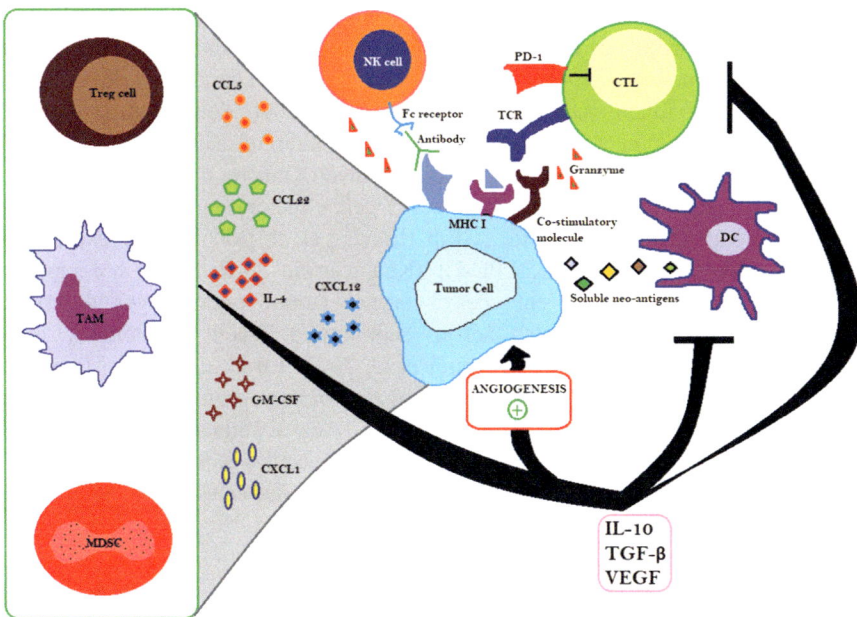

Fig. 16.2 Immune landscape of cancer. Upon detection of a tumor antigen, many immune cells, especially cytotoxic T lymphocytes (CTL), dendritic cells, and natural killer cells employ different cytotoxic mechanisms (including granzyme secretion) to eliminate the tumor cells. However, cytokines secreted by the tumor cells promote the development of various suppressor cells such as regulatory T (Treg) cells, tumor-associated macrophages (TAMs), myeloid-derived suppressor cells (MDSCs), which try to suppress this immune response and aggravate angiogenesis and metastatic spread of the tumor. Additionally, signaling through PD-1 and related proteins induce exhaustion of activated T cells

helps DC maturation, and enhancement of their antitumor activities (Hanke et al. 2013). Involvement of nitric oxide (Shimamura et al. 2002), tumor necrosis factor-related apoptosis-inducing ligand (TRAIL) (Fanger et al. 1999), etc., has also been reported in tumor-killing endeavors of these cells. In the innate branch, natural killer (NK) cells are most active in eliminating cancer cells. These cells identify somatic cells with less-than-optimum level of MHC I and kill them using cytotoxic mechanisms very similar to those employed by CTLs (O'Leary et al. 2006). Various pro-inflammatory cytokines such as IL-1 and TNF act as chemokines, directing neutrophil migration into the cancer tissue, albeit at a cost of consequent extracellular matrix degeneration and aggravated tumor invasion (Dinarello 2006). Figure 16.2 depicts the immune landscape of cancer.

However, malignant cells have also developed strategies to evade the crushing claws of the immune defense machinery. One of the major tactics here is suppression of T-cell response. Several oncoviruses like hepatitis B virus (Chen et al. 2006), Epstein–Barr virus (Croft et al. 2009), human papillomavirus (Ashrafi et al. 2005), and human cytomegalovirus (Yamashita et al. 1993) have been reported to reduce

MHC class I expression on host cell surface, an excellent strategy to prevent T cells from being alarmed. Cells from various cancer types like colon and bladder cancer express lower amounts of co-stimulatory molecules on their surfaces, without which T cells cannot be functionally activated (Tirapu et al. 2006; Pettit et al. 1999). The tumor microenvironment always promotes maturation of naive T cells toward regulatory T (T_{reg}) phenotype and boosts tolerance (Knutson et al. 2007). Tumor-associated macrophages (TAMs) suppress T-cell effector functions by secreting IL-10 (Bolpetti et al. 2010) and TGF-β (Biswas et al. 2006) or by expressing anergy-inducing ligands such as PD-L1 (Kuang et al. 2009) or B7 homologues (Kryczek et al. 2006), in different types of cancers. Tumor-resident myeloid-derived suppressor cells (MDSCs) also act in suppressing T-cell functionality in a peroxynitrite-dependent fashion (Nagaraj et al. 2007; Lu et al. 2011). Moreover, these immune cells assist tumor metastasis by dint of their pro-angiogenic cytokine secretome (Lewis et al. 1995; Binsfeld et al. 2016). Eventually, this multisided interaction between immune cells and malignant cells creates an equilibrium, where only tumor cells able to escape immune surveillance can sustain and proliferate (Dunn et al. 2004).

16.4.2 Immunotherapy against Cancer: Enforcing the Immune System in the Right Way

Since the immune system can detect and target neo-antigens and eliminate malignant cells using various modes of cell killing, if they are properly primed so that their specificity and efficiency against tumor antigens get enhanced, they can become better fighters against cancer and tide over the immunosuppressive mechanisms employed by the tumors and their microenvironment. With an eye to this, various strategies have been developed to instructively drive the immune system for a particular therapeutic purpose. All these strategies, collectively, can be termed immunotherapy. This kind of goal-oriented immune modulation can be done actively (i.e., by instructing the host's immune system by imparting a target specificity and memory to it) or passively (i.e., by introducing immune components from outside into the body with the desired specificity). Naturally, active therapies involve in situ activation of host's own immune cells, whereas passive therapies are based on specific antibodies or other cell surface immunogens.

16.4.3 Cell-Based Immunotherapy: Invigorating the Sentinels

The basic principle of cell-based immunotherapy is to fetch immune cells with tumoricidal activity from the host himself, culture them in vitro, if required, genetically alter them to express high-affinity receptors for tumor cells, and finally reintroduce them to the host body so that they can now fight cancer with greater efficiency (Armstrong et al. 2001). However, in many cases, in situ activation of the cells can also be carried out. The prime advantage of this kind of therapy is the subsequent memory response generation, which lowers the chances of tumor relapse. There are multiple therapeutic methods for this kind of therapy, as described below.

16.4.3.1 Whole Tumor Cell Vaccine

Whole tumor cells have been a choice for immunization against cells of its own kind. The cells are first radiation-treated to curb their malignant properties and then mixed with adjuvants to enhance their immunogenicity before administration as vaccine. In a randomized trial, postoperative active specific immunotherapy with an autologous tumor cell vaccine coupled with BCG adjuvant showed 44% risk reduction in recurrence of stage II and stage III colon cancer, compared to control population (Vermorken et al. 1999). Positive impact was also obtained in vaccination against melanoma (Berd et al. 1990), renal cell carcinoma (Repmann et al. 1997), etc.

16.4.3.2 Dendritic Cells

Boosting efficiency of antigen-presenting cells and consequently enhanced T-cell activation are the principal goal of any vaccine. Since DCs are primary antigen-presenting cells in the tumor microenvironment by virtue of their robust expression of MHC, co-stimulatory and adhesion molecules attempt to strengthen their functions is an ideal option. This has been supported by the advances in protocols for ex vivo generation of antigen-specific functional DCs from autologous hematopoietic stem cells and peripheral blood (Bernhard et al. 1995). In vivo activation of DCs can be carried out by introduction of whole tumor lysates (Hirayama and Nishimura 2016) or short peptides derived from tumor antigens, in combination with adjuvants and/or chemoattractants. A potent vaccine against aggressive glioblastoma has been developed based on this rationale, with tumor lysate, cytomegalovirus RNA, and tumor-associated peptides such as EGFRvIII as the primary immunogen (Dastmalchi et al. 2018). Genetic manipulation of tumor cells or tumor-homing cells like stem cells to make them robust producers of GM-CSF is another useful strategy, since GM-CSF is a strong activator of DCs (Van de Laar et al. 2012). Antibodies against DC surface receptors like multiple TLRs and CD40 have also been used to nonspecifically mature and activate them and strengthen their phagocytic capabilities (Palucka and Banchereau 2013). However, the only FDA-approved, clinically successful DC-based anticancer vaccine so far is Sipuleucel-T (trade name: Provenge) (Plosker 2011), targeted toward metastatic, asymptomatic, hormone-refractory prostate cancer. Working procedure of this vaccine relies on the extraction of patient's DCs by leukapheresis, incubation of those cells with a fusion protein PA2024, consisting of prostatic acid phosphatase (PAP), a cancer antigen and GM-CSF, and finally reinfusion of blood containing the active DCs into the patient to impart the DC-mediated immunity to the patient.

16.4.3.3 CAR-T Cells

The T-cell receptor (TCR) is the key factor maintaining the antigenic specificity of each and every T-cell clone. Therefore, T cells have been engineered to express chimeric TCRs capable of targeting a specific antigen and activating the T cell.

These T cells, called chimeric antigen receptor (CAR)-T cells, are an effective adaptive therapy against various types of cancer, because they can be modified in a way, which can boost their specificity against cancer antigens. To produce typical CAR-T cells, autologous T cells from patients are collected by leukapheresis, activated and clonally propagated in vitro by IL-2 and anti-CD3 antibody (Makita et al. 2017), transduced with a gene coding for the engineered CAR via gamma-retroviral or lentiviral vector (Jin et al. 2016). Before adoptive transfer of these T cells to the patients, they need to undergo specific chemotherapy regimen, which depletes the number of existing leukocytes in their circulation, which in effect promotes expansion of the newly introduced T cells inside the patients' bodies (Muranski et al. 2006). The CAR is a transmembrane protein. Its ectodomain is made up of variable regions from heavy- and light-chain polypeptides of a cancer antigen-specific monoclonal antibody, which act as the antigen-binding domain and binds only the cancer antigen it is directed against. The endodomain contains immunoreceptor tyrosine-based activation motifs (ITAM) from CD3ζ, which relay the activation signal to the nuclei of T cells, along with chimeric co-receptor domains to provide necessary co-stimulation (Zhang et al. 2017). Most of the successful clinical trials involving CAR-T cells so far have been conducted against blood cancers. The first two therapies involving Tisagenlecleucel and Axicabtagene were against B-cell lymphoma and targeted CD19, a marker specific for B lymphocytes (Ahmad et al. 2020). However, the therapy has not been proven to be very efficient against solid tumors (Lim and June 2017), and even in case of lymphoma, the emergence of specific marker-negative clones has caused long-term survival rates to drop (Schultz and Mackall 2019). Figure 16.3 shows the difference between normal cells and CAR-T cells.

16.4.4 Antibody-Based Immunotherapy: Additional Forces Recruited

Antibody therapy against cancer is a form of passive therapy where tumor antigen-specific monoclonal antibodies are introduced from outside into the circulation, which act as flagship molecules to drive the immune machinery of the patients toward and against the cancer tissue. While the Fab region of the antibody binds the target epitope on the tumor antigen, the remaining Fc region can act as a platform to support multiple tumoricidal branches of the immune system. Fc receptors on NK cells detect these Fc regions and consequently activate exocytosis of cytotoxic molecules such as perforin and granzyme from the killer cells. These molecules can then kill the tumor cells in the vicinity of the antibody-bound cell, in a process termed antibody-dependent cell cytotoxicity (ADCC) (Kiessling et al. 1975). The Fc region also activates the classical complement pathway, another potent system functioning for direct target cell lysis and induction of immune cell chemotaxis (Thielens et al. 2017). Rituximab (Weiner 2010), an anti-CD20 chimeric monoclonal IgG1 molecule, acts in this way against various B-cell malignancies like diffuse large B-cell lymphoma and B-cell chronic lymphocytic leukemia (Keating 2010).

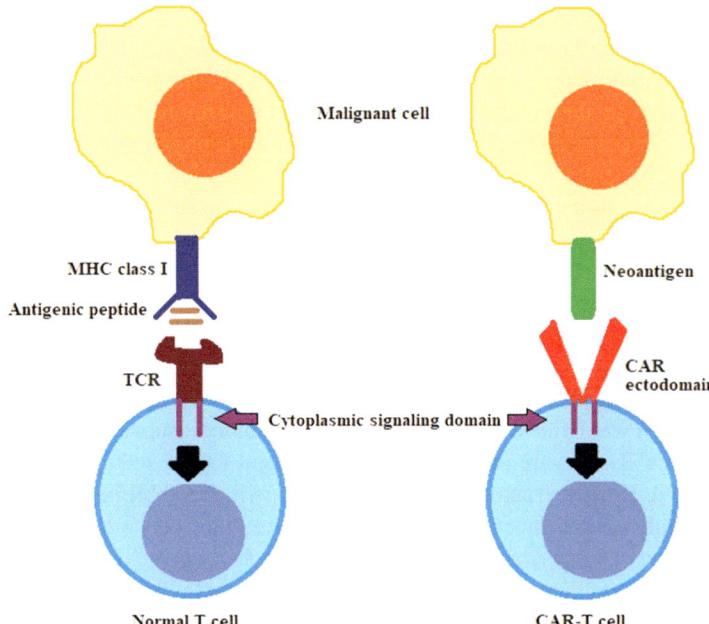

Fig. 16.3 Difference between normal and CAR-T cells. Normal T cells have T-cell receptors (TCRs), which recognize antigenic peptides only when displayed along with MHC class I molecules on the target cell surface, and transduce an activation signal to the nuclei of the T cells. CAR-T cells, on the other hand, have a chimeric receptor with an ectodomain for direct binding to neoantigens, and an endodomain for relaying a subsequent activation signal to the T-cell nuclei

Toxins can be conjugated to these monoclonal antibodies for targeted delivery of the toxin molecule to the tumor milieu and specific tumor killing with reduced off-target effects. Another mechanism of antibody-mediated immunotherapy is targeting and inactivating immune checkpoint mechanisms responsible for T-cell exhaustion. Activated T cells express molecules like CTLA-4 and PD-1, which sequester the binding pockets of co-stimulatory molecules on antigen-displaying cells and put the T cells in an anergic state (Buchbinder and Desai 2016). Ipilimumab and pembrolizumab, two monoclonal IgG antibodies, target these two molecules, respectively, and maintain the T cells in an activated state for prolonged periods and enhance their anticancer functionalities. Ipilimumab has been effective against metastatic melanoma (Lipson and Drake 2011), while pembrolizumab's efficacy has covered non-small cell lung carcinoma (Garon et al. 2015), head-and-neck carcinoma (Chow et al. 2014), gastric carcinoma (Muro et al. 2015), etc.

16.4.5 Other Modes of Immunotherapy in Practice

Another offshoot of immunotherapy deals with the administration of cytokines to patients to activate their immune system in a directed fashion. This kind of cytokine therapy encompasses different cytokines, based on their physiological role and need of individual patient, where they stimulate specific sets of immune cells to proliferate and migrate toward their site of action. One of the most used cytokine against cancer tissues is type I interferon: IFN-α and IFN-β (Dunn et al. 2006). They are extensively applied in hairy cell leukemia, AIDS-related Kaposi's sarcoma, follicular lymphoma, chronic myeloid leukemia, and malignant melanoma. Additionally, interleukin-2 is used in the treatment of malignant melanoma and renal cell carcinoma (Coventry and Ashdown 2012). In fact, the very first approach toward cancer immunotherapy by Dr. William B. Coley made use of bacterial infection into cancer patients, which was believed to evoke a pro-inflammatory response, and the resultant cytokines like IL-12 were at the heart of subsequent cancer reduction (Tsung and Norton 2006). However, actual utility of this therapy is debatable because "available scientific evidence does not currently support claims that Coley's toxins can treat or prevent cancer" (*What is Coley's toxins treatment for cancer?* 2012). Sometimes, when the cytokines themselves suffer from short in vivo half-life, their agonists are administered as an alternative treatment. For example, ALT-803, a super agonist of IL-15, was applied via intravenous and subcutaneous route into patients receiving allogenic hematopoietic cell transplants as a therapy for hematologic malignancies. The subcutaneous, but not intravenous, treatment showed increased number of circulating $CD56^{hi}$ NK cells and $CD8^+$ T cells, without elevated count of T_{reg} cells altogether, compared to IL-15 therapy (Romee et al. 2018).

Oncolytic viruses are an effective tool in treating different types of cancer. These viruses selectively replicate in malignant cells and subsequently kill their host cells. As a result of altering their host cell, they also stimulate the host's immune system to fight against the cancer cells (Fukuhara et al. 2016). T-VEC is the first FDA-approved oncolytic virus, which is selective against melanoma. It is basically an oncolytic herpes virus, which is genetically modified to evade the cellular immune defense and at the same time activate the immune system against the cancer cells and produce GM-CSF to promote immune cell proliferation (Liu et al. 2003).

16.4.6 Critical Challenges to Counter

In today's scenario, immunotherapy is one of the strongest, if not the strongest, weapons in fighting cancer on a personalized level. However, no rose comes without thorns, and immunotherapeutic approaches against cancer also come with certain drawbacks (Ventola 2017). A major one of these is the limited number of cancer types against which these therapies work. Cohort-to-cohort, and even patient-to-patient, variations in efficacy of a single treatment regimen are also a serious concern. Since tumors are highly heterogeneous, efficiency of any treatment in targeting the tumor cells has to take multiple variables into account (Pastan et al.

2006). Pinpointing the clinically relevant biomarkers among the ever-increasing malignancy-associated mutations is another tough task, mainly due to technical and financial issues. Certain biomarkers, despite high correlation with a specific malignancy, are absent from multiple regional populations (Zugazagoitia et al. 2016). Specific improvisations are necessary in clinical trial designs, because the traditional methods do not detect particular therapeutic endpoints. Last but not the least, these therapies come with a significantly high expense, hence are difficult to be afforded by anybody. Tiding over these lacunae is essential to improve survival rate and remission rate of any cancer against which the therapy needs to be applied.

16.4.7 Footsteps for Future

New techniques based on the existing ones are always needed in order to improve therapeutic success. With this intention, several emerging concepts of engineering, physics, chemistry, and biology are being combined with conventional immunotherapy. Amalgamation of immunotherapy with ablation techniques such as photodynamic (Schumacher and Schreiber 2015), cryogenic (Wang et al. 2014), or radiofrequency ablation (Sidana 2014) therapy has resulted in an increase in selectivity and efficacy than any one of the therapy used alone. Application of nanomedicine has also broadened the range of immunotherapy (Fagnoni et al. 2008). Targeting cancer stem cells instead of whole tumor populations has led to a significant drop in relapse rates (Shi and Lammers 2019; Badrinath and Yoo 2019). Standing at this point of time, the ultimate goal of immune therapists should be to personalize therapeutic protocols. Integration of computer-based patient datasets and therapy designing with the help of predefined algorithms should be the master plan to avoid the variables across patients and populations.

16.5 Personalized Therapy

Recently, precision medicine has become a potentially more effective approach to treat cancer. This approach not only conglomerates different targeted cancer therapies but also offers a better way to anticipate the best suitable treatment type for an individual, depending on their relevant *genetic makeup* and the *environmental and lifestyle factors* (Mirnezami et al. 2012; Cesuroglu et al. 2012). Extensive research throughout the world has delineated multiple specific intrinsic pathways and potential targets that ultimately helped to design several biological (Seidel et al. 2018) or chemical inhibitors (Asati et al. 2017) to combat cancer. However, none of them were found suitable weapon to target even for multiple patients having similar cancer types, rather their *sensitivity of promiscuous substrate specificity* made them "*toxic*" than "*savior.*" Investigations of underlining reason for this defeat revealed that *microenvironmental heterogeneity* is the responsible factor (Natrajan et al. 2016; Yuan 2016). Surprisingly, this heterogeneity exists at every level like *cellular, genetic mutational, epigenetic, transcriptomic, and proteomic stage* and varies in patient-to-patient having similar cancer to a large extent. This scenario urges a revolutionary modification in clinical cancer research and ultimately arguments to establish a clinical trial that will review the status of above-mentioned facts of each cancer patients and will advise the best possible therapy to combat cancer: personalized cancer therapy (Collette and Tombal 2015).

The journey of personalized therapy gained the highest priority when former US president Barrack Obama had invested $215 million dollars for this new field of medical treatment for cancer in the year 2015 (Rossi et al. 2014). The reason for this investment was the medical misuse and waste due to ineffective, useless, and redundant treatment that amounted to $75 million in every year, which is close to 30% healthcare expenditure in the USA (The Precision Medicine Initiative n.d.). These misuses of cancer drugs are due to the diversity of individuals in the gene, lifestyle, environment-induced variability of the treatment response, and resistance to medication, which have been recognized as long-term challenges in oncology. The personalized therapy augments the need for the introduction of revolutionary technology to create a new era of medicine. This includes the investigation in the field of cellular, genetic mutational, epigenetic, transcriptomic, and proteomic level to look at abnormalities that may govern as an index of cancer biomarkers and based on that an individual can receive a medical treatment accordingly. A short while after this announcement by US president, China invested about 60 million Chinese currencies in search for investigation on personalized cancer therapy, and preventive measure development (Wong and Deng 2015). Later on, the most of the European countries adopted this approach and have shown immense interest in this new era of medical treatment against cancer. In India, the focus of cancer treatment has just started to switch toward personalized medicine from traditional ways. Some renounced cancer treatment centers have already been started in adopting personalized medicine in Gall Bladder Cancer (Ghosh 2019).

The milestones of these approaches are to (i) find out the genetic, epigenetic, or expressional abnormalities in individuals having cancer, (ii) analyze and compare all the obtained data and relate to the function of available drugs to select suitable anticancer therapy either singly or in a combination with other therapy, (iii) predict new cancer markers that are prevalent in a particular ethnicity, (iv) built an immune and omics landscape to predict perfect therapies for the future cancer treatment based on previous medical records having patient's overall survival and quality of life, (v) establish a perfect platform to associate clinicians, scientists, bioinformaticians, statisticians, and patients in order to bring them under a single umbrella to achieve maximum success in cancer therapy so as to make this world stronger and more compatible to combat cancer. Therefore, a long-term goal of this personalized therapy is to design individualized therapy based on the genomic sequence, epigenetic modifications, RNA/protein expression profiling, and cellular behavior of components of tumor microenvironment and tumor-infiltrating immune cell profiling of the patient, to maximize overall survival and minimize the toxic threat of anticancer drugs and to develop a sophisticated, modern technology-based setup at the international level and to offer maximum benefits to poor cancer patients at the

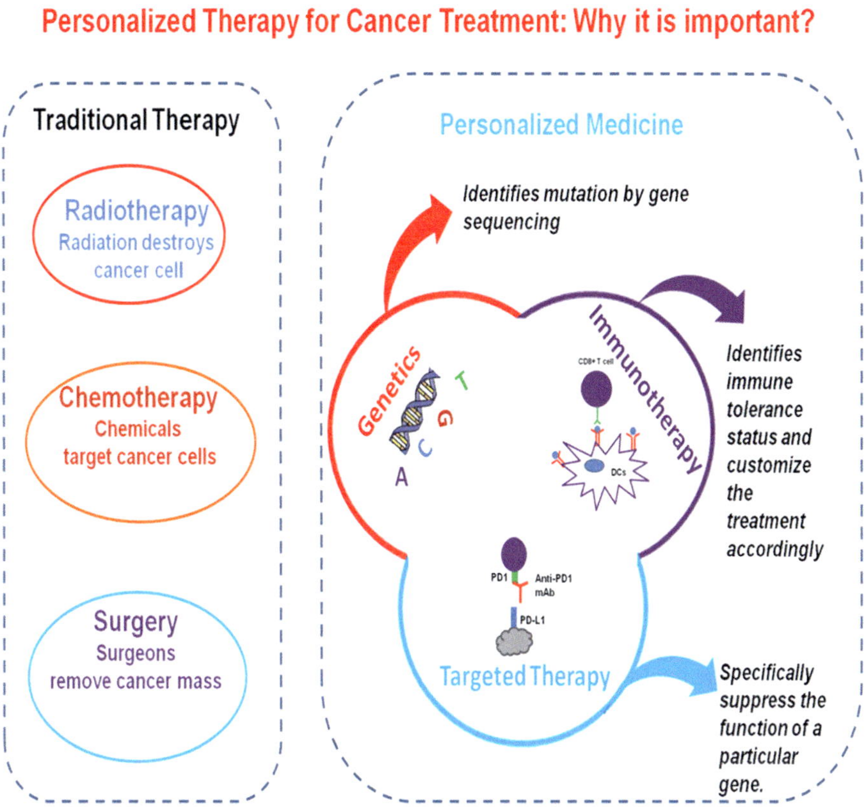

Fig. 16.4 Traditional therapy vs. personal therapy. Traditional therapy of cancer uses radiotherapy, chemotherapy, and surgery either alone or in a combination. In personalized therapy, treatments are customized based on patient genetics, transcriptomics, and proteomic profiling

minimal cost. Figure 16.4 summarizes the difference between the traditional therapy and the personalized therapy for cancer treatment.

In fact, glioblastoma management by the use of personalized therapy is one of the noticeable examples. Glioblastoma, a tumor associated with brain, possesses both intertumoral and intratumoral heterogeneity (Ene and Holland 2015). In fact, intratumoral heterogeneity makes the situation extremely complex. These circumstances clearly indicate that the strategy for glioblastoma treatment for multiple patients cannot be the same and should not rely only on the certain chemotherapeutic drugs as the nature of the disease not only varies from patient-to-patient but also the clonal diversity within each patient's tumor. As per WHO guidelines, gliomas are classified into four types namely pilocytic astrocytoma, diffuse astrocytoma, anaplastic astrocytoma, and glioblastoma multiforme. Among them, the last one has the worse disease outcome and considered as the most common form of primary brain tumors in adults.

At present, there is no FDA-approved drug available for personalized therapy of glioblastoma multiforme (GBM) patients. It has been found that when GBM is treated in a combination with temozolomide (a DNA alkylating agent) and radiation, the patient survival is improved (Hegi et al. 2005) However, GBM with MGMT gene suppression/hypermethylation is more sensitive to temozolomide. Moreover, the molecular classification in glioblastoma indicated the intratumoral heterogeneity that may directly influence the personalized therapy of a patient having distinct sets of genetic alteration. Based on genetic mutations, glioblastoma can be further classified into four groups namely proneural (TP53, IDH1, PIK3R, EGFR, and PDGFRA), neural (EGFR, TP53, PTEN, ERBB2, NF-1, and PIK3R), classic (EGFR, PTEN, and EGFRvIII), and mesenchymal (NF-1, TP53, PTEN, and RB1) (Ene and Holland 2015). The genes indicated in the parentheses are the mutation-prone genes related to the respective subclasses. However, no such randomized clinical trials have been executed using these subtypes. In future, retrospective analysis of prior trials may be explored to have clinical relevance of these subtypes for therapy.

16.6 Concluding Remarks

In summary, the implementation of the rational therapy for cancer does not rely on a specific treatment modality rather depends on multiple therapeutic approaches that can be used in a combination to combat the devastating disease. The complexity of the disease at different level such as cellular, genetic mutational, epigenetic, transcriptomic, and proteomic stage and their heterogeneous existence indicates that the approaches to combat cancer should consider each patient as an individual clinical trial. In fact, this job is not confined to clinicians only, rather seeks the direct involvement of basic researchers, biostatisticians, bioinformaticians, pharmacists, and people in different branches related to health and their cumulative effort may indicate the perfect approach/approaches to treat a particular cancer type of an individual. This in turn argues for the prediction of the best possible drug/treatment of choice. Moreover, the perfect selection of drug of choice is possible when a national facilitated cancer registry with a complete database of use of different drugs/therapy, their outcome, in terms of patient survival and disease relapse, is available.

References

Aboody KS, Brown A, Rainov NG, Bower KA, Liu S, Yang W, Small JE, Herrlinger U, Ourednik V, Black PM, Breakefield XO (2000) Neural stem cells display extensive tropism for pathology in adult brain: evidence from intracranial gliomas. Proc Natl Acad Sci 97 (23):12846–12851

Ahmad A, Uddin S, Steinhoff M (2020) Car-t cell therapies: an overview of clinical studies supporting their approved use against acute lymphoblastic leukemia and large b-cell lymphomas. Int J Mol Sci 21(11):3906

Ahmed AU, Thaci B, Alexiades NG, Han Y, Qian S, Liu F, Balyasnikova IV, Ulasov IY, Aboody KS, Lesniak MS (2011) Neural stem cell-based cell carriers enhance therapeutic efficacy of an oncolytic adenovirus in an orthotopic mouse model of human glioblastoma. Mol Ther 19(9):1714–1726

Al-Hajj M, Wicha MS, Benito-Hernandez A, Morrison SJ, Clarke MF (2003) Prospective identification of tumorigenic breast cancer cells. Proc Natl Acad Sci 100(7):3983–3988

Armstrong AC, Eaton D, Ewing JC (2001) Cellular immunotherapy for cancer. BMJ 323(7324):1289–1293

Asati V, Mahapatra DK, Bharti SK (2017) K-Ras and its inhibitors towards personalized cancer treatment: pharmacological and structural perspectives. Eur J Med Chem 125:299–314

Ashrafi GH, Haghshenas MR, Marchetti B, O'Brien PM, Campo MS (2005) E5 protein of human papillomavirus type 16 selectively downregulates surface HLA class I. Int J Cancer 113(2):276–283

Badrinath N, Yoo SY (2019) Recent advances in cancer stem cell-targeted immunotherapy. Cancers 11(3):310

Bagó JR, Sheets KT, Hingtgen SD (2016) Neural stem cell therapy for cancer. Methods 99:37–43

Barker HE, Paget JT, Khan AA, Harrington KJ (2015) The tumour microenvironment after radiotherapy: mechanisms of resistance and recurrence. Nat Rev Cancer 15(7):409–425

Becker, A.J., McCulloch, E.A. and Till, J.E., 1963. Cytological demonstration of the clonal nature of spleen colonies derived from transplanted mouse marrow cells

Berd D, Maguire HC Jr, McCue P, Mastrangelo MJ (1990) Treatment of metastatic melanoma with an autologous tumor-cell vaccine: clinical and immunologic results in 64 patients. J Clin Oncol 8(11):1858–1867

Bernardo ME, Fibbe WE (2013) Mesenchymal stromal cells: sensors and switchers of inflammation. Cell Stem Cell 13(4):392–402

Bernhard H, Disis ML, Heimfeld S, Hand S, Gralow JR, Cheever MA (1995) Generation of immunostimulatory dendritic cells from human CD34+ hematopoietic progenitor cells of the bone marrow and peripheral blood. Cancer Res 55(5):1099–1104

Binsfeld M, Muller J, Lamour V, De Veirman K, De Raeve H, Bellahcène A, Van Valckenborgh E, Baron F, Beguin Y, Caers J, Heusschen R (2016) Granulocytic myeloid-derived suppressor cells promote angiogenesis in the context of multiple myeloma. Oncotarget 7(25):37931

Biswas SK, Gangi L, Paul S, Schioppa T, Saccani A, Sironi M, Bottazzi B, Doni A, Vincenzo B, Pasqualini F, Vago L (2006) A distinct and unique transcriptional program expressed by tumor-associated macrophages (defective NF-κB and enhanced IRF-3/STAT1 activation). Blood 107(5):2112–2122

Bolpetti, A., Silva, J.S., Villa, L.L. and Lepique, A.P., 2010. Interleukin-10 production by tumor infiltrating macrophages plays a role in human papillomavirus 16 tumor growth. BMC Immunol 11(1), pp.1–13

Bonnet D, Dick JE (1997) Human acute myeloid leukemia is organized as a hierarchy that originates from a primitive hematopoietic cell. Nat Med 3(7):730–737

Bots M, Medema JP (2006) Granzymes at a glance. J Cell Sci 119(24):5011–5014

Buchbinder EI, Desai A (2016) CTLA-4 and PD-1 pathways: similarities, differences, and implications of their inhibition. Am J of Clin Oncol 39(1):98

Carosella ED, HoWangYin KY, Favier B, LeMaoult J (2008) HLA-G–dependent suppressor cells: diverse by nature, function, and significance. Human Immunol 69(11):700–707

Cesuroglu T, Van Ommen B, Malats N, Sudbrak R, Lehrach H, Brand A (2012) Public health perspective: from personalized medicine to personal health. Pers Med 9(2):115–119

Chen Y, Cheng M, Tian Z (2006) Hepatitis B virus down-regulates expressions of MHC class I molecules on hepatoplastoma cell line. Cell Mol Immunol 3(5):373–378

Chow LQ, Burtness B, Weiss J, Berger R, Eder JP, Gonzalez EJ, Pulini J, Johnson J, Dolled-Filhart M, Emancipator K, Lunceford JK (2014) A phase Ib study of pembrolizumab (pembro; MK-3475) in patients (pts) with human papilloma virus (HPV)-positive and negative head and neck cancer (HNC). Ann Oncol 25:v1

Collette L, Tombal B (2015) N-of-1 trials in oncology. Lancet Oncol 16(8):885–886

Coventry BJ, Ashdown ML (2012) The 20th anniversary of interleukin-2 therapy: bimodal role explaining longstanding random induction of complete clinical responses. Cancer Manag Res 4:215

Croft NP, Shannon-Lowe C, Bell AI, Horst D, Kremmer E, Ressing ME, Wiertz EJ, Middeldorp JM, Rowe M, Rickinson AB, Hislop AD (2009) Stage-specific inhibition of MHC class I presentation by the Epstein-Barr virus BNLF2a protein during virus lytic cycle. PLoS Pathog 5(6):e1000490

Dai P, Wang W, Yang N, Serna-Tamayo C, Ricca JM, Zamarin D, Shuman S, Merghoub T, Wolchok JD, Deng L (2017) Intratumoral delivery of inactivated modified vaccinia virus Ankara (iMVA) induces systemic antitumor immunity via STING and Batf3-dependent dendritic cells. Sci Immunol 2(11)

Dastmalchi, F., Karachi, A., Mitchell, D. and Rahman, M. (2018) Dendritic cell therapy. eLS. American Cancer Society, pp. 1–27

De Ruysscher D, Niedermann G, Burnet NG, Siva S, Lee A, Hegi-Johnson F (2019) Radiotherapy toxicity. Nature reviews. Dis Primers 5(1):13

Di Nicola M, Carlo-Stella C, Magni M, Milanesi M, Longoni PD, Matteucci P, Grisanti S, Gianni AM (2002) Human bone marrow stromal cells suppress T-lymphocyte proliferation induced by cellular or nonspecific mitogenic stimuli. Blood 99(10):3838–3843

Dilmore TC, Moore DW, Bjork Z (2013) Developing a competency-based educational structure within clinical and translational science. Clin Transl Sci 6(2):98–102

Dinarello CA (2006) The paradox of pro-inflammatory cytokines in cancer. Cancer Metastasis Rev 25(3):307–313

Dunn GP, Bruce AT, Ikeda H, Old LJ, Schreiber RD (2002) Cancer immunoediting: from immunosurveillance to tumor escape. Nat Immunol 3(11):991–998

Dunn GP, Old LJ, Schreiber RD (2004) The three Es of cancer immunoediting. Annu Rev Immunol 22:329–360

Dunn GP, Koebel CM, Schreiber RD (2006) Interferons, immunity and cancer immunoediting. Nat Rev Immunol 6(11):836–848

Ene CI, Holland EC (2015) Personalized medicine for gliomas. Surg Neurol Int 6(Suppl 1):S89–S95

English K, Ryan JM, Tobin L, Murphy MJ, Barry FP, Mahon BP (2009) Cell contact, prostaglandin E2 and transforming growth factor beta 1 play non-redundant roles in human mesenchymal stem cell induction of CD4+ CD25Highforkhead box P3+ regulatory T cells. Clin Exp Immunol 156(1):149–160

Eseonu OI, De Bari C (2015) Homing of mesenchymal stem cells: mechanistic or stochastic? Implications for targeted delivery in arthritis. Rheumatology 54(2):210–218

Eun K, Ham SW, Kim H (2017) Cancer stem cell heterogeneity: origin and new perspectives on CSC targeting. BMB Rep 50(3):117

Fagnoni FF, Zerbini A, Pelosi G, Missale G (2008) Combination of radiofrequency ablation and immunotherapy. Front Biosci: A J Virtual Library 13:369–381

Fanger NA, Maliszewski CR, Schooley K, Griffith TS (1999) Human dendritic cells mediate cellular apoptosis via tumor necrosis factor–related apoptosis-inducing ligand (TRAIL). J Exp Med 190(8):1155–1164

Fathi E, Farahzadi R, Valipour B, Sanaat Z (2019) Cytokines secreted from bone marrow derived mesenchymal stem cells promote apoptosis and change cell cycle distribution of K562 cell line as clinical agent in cell transplantation. PLoS One 14(4):e0215678

Fukuhara H, Ino Y, Todo T (2016) Oncolytic virus therapy: a new era of cancer treatment at dawn. Cancer Sci 107(10):1373–1379

Garon EB, Rizvi NA, Hui R, Leighl N, Balmanoukian AS, Eder JP, Patnaik A, Aggarwal C, Gubens M, Horn L, Carcereny E, Ahn MJ, Felip E, Lee JS, Hellmann MD, Hamid O, Goldman JW, Soria JC, Dolled-Filhart M, Rutledge RZ (2015) Pembrolizumab for the treatment of non-small-cell lung cancer. N Engl J Med 372(21):2018–2028

Ghosh B. (2019) Towards personalized therapy for Indian gallbladder cancer patients. Health & Medicine, Molecular Biology and Research

Golubeva V, Mikhalevich J, Novikova J, Tupizina O, Trofimova S, Zueva Y (2014) Novel cell population data from a haematology analyzer can predict timing and efficiency of stem cell transplantation. Transfus Apher Sci 50(1):39–45

Guha P, Morgan JW, Mostoslavsky G, Rodrigues NP, Boyd AS (2013) Lack of immune response to differentiated cells derived from syngeneic induced pluripotent stem cells. Cell Stem Cell 12 (4):407–412

Gupta PB, Onder TT, Jiang G, Tao K, Kuperwasser C, Weinberg RA, Lander ES (2009) Identification of selective inhibitors of cancer stem cells by high-throughput screening. Cell 138 (4):645–659

Hanke N, Alizadeh D, Katsanis E, Larmonier N (2013) Dendritic cell tumor killing activity and its potential applications in cancer immunotherapy. Crit Rev™ Immunol 33(1)

Hegi ME, Diserens AC, Gorlia T, Hamou MF, De Tribolet N, Weller M, Kros JM, Hainfellner JA, Mason W, Mariani L, Bromberg JE (2005) MGMT gene silencing and benefit from temozolomide in glioblastoma. N Engl J Med 352(10):997–1003

Hirayama M, Nishimura Y (2016) The present status and future prospects of peptide-based cancer vaccines. Int Immunol 28(7):319–328

Jangamreddy JR, Ghavami S, Grabarek J, Kratz G, Wiechec E, Fredriksson BA, Pariti RKR, Cieślar-Pobuda A, Panigrahi S, Łos MJ (2013) Salinomycin induces activation of autophagy, mitophagy and affects mitochondrial polarity: differences between primary and cancer cells. Biochim Biophys Acta (BBA)-Mol Cell Res 1833(9):2057–2069

Jin L, Lee EM, Ramshaw HS, Busfield SJ, Peoppl AG, Wilkinson L, Guthridge MA, Thomas D, Barry EF, Boyd A, Gearing DP (2009) Monoclonal antibody-mediated targeting of CD123, IL-3 receptor α chain, eliminates human acute myeloid leukemic stem cells. Cell Stem Cell 5 (1):31–42

Jin C, Fotaki G, Ramachandran M, Nilsson B, Essand M, Yu D (2016) Safe engineering of CAR T cells for adoptive cell therapy of cancer using long-term episomal gene transfer. EMBO Mol Med 8(7):702–711

Jordan CT, Upchurch D, Szilvassy SJ, Guzman ML, Howard DS, Pettigrew AL, Meyerrose T, Rossi R, Grimes B, Rizzieri DA, Luger SM (2000) The interleukin-3 receptor alpha chain is a unique marker for human acute myelogenous leukemia stem cells. Leukemia 14(10):1777–1784

Keating GM (2010) Rituximab: a review of its use in chronic lymphocytic leukaemia, low-grade or follicular lymphoma and diffuse large B-cell lymphoma. Drugs 70(11):1445–1476

Khurana S, Margamuljana L, Joseph C, Schouteden S, Buckley SM, Verfaillie CM (2013) Glypican-3–mediated inhibition of CD26 by TFPI: a novel mechanism in hematopoietic stem cell homing and maintenance. Blood, J Am Soc Hematol 121(14):2587–2595

Kiessling R, Klein E, Pross H, Wigzell H (1975) "Natural" killer cells in the mouse. II. Cytotoxic cells with specificity for mouse Moloney leukemia cells. Characteristics of the killer cell. Eur J Immunol 5(2):117–121

Knutson KL, Disis ML, Salazar LG (2007) CD4 regulatory T cells in human cancer pathogenesis. Cancer Immunol Immunother 56(3):271–285

Król M, Pawłowski KM, Majchrzak K, Szyszko K, Motyl T (2010) Why chemotherapy can fail. Pol J Vet Sci 13(2):399–406

Kryczek I, Zou L, Rodriguez P, Zhu G, Wei S, Mottram P, Brumlik M, Cheng P, Curiel T, Myers L, Lackner A (2006) B7-H4 expression identifies a novel suppressive macrophage population in human ovarian carcinoma. J Exp Med 203(4):871–881

Kuang DM, Zhao Q, Peng C, Xu J, Zhang JP, Wu C, Zheng L (2009) Activated monocytes in peritumoral stroma of hepatocellular carcinoma foster immune privilege and disease progression through PD-L1. J Exp Med 206(6):1327–1337

Lapidot T, Sirard C, Vormoor J, Murdoch B, Hoang T, Caceres-Cortes J, Minden M, Paterson B, Caligiuri MA, Dick JE (1994) A cell initiating human acute myeloid leukaemia after transplantation into SCID mice. Nature 367(6464):645–648

Le Blanc K, Rasmusson I, Götherström C, Seidel C, Sundberg B, Sundin M, Rosendahl K, Tammik C, Ringden O (2004) Mesenchymal stem cells inhibit the expression of CD25 (interleukin-2 receptor) and CD38 on phytohaemagglutinin-activated lymphocytes. Scand J Immunol 60(3):307–315

Leventhal A, Chen G, Negro A, Boehm M (2012) The benefits and risks of stem cell technology. Oral Dis 18(3):217

Lewis CE, Leek R, Harris A, McGee JD (1995) Cytokine regulation of angiogenesis in breast cancer: the role of tumor-associated macrophages. J Leukoc Biol 57(5):747–751

Li S, Tokuyama T, Yamamoto J, Koide M, Yokota N, Namba H (2005) Bystander effect-mediated gene therapy of gliomas using genetically engineered neural stem cells. Cancer Gene Ther 12(7):600–607

Li C, Heidt DG, Dalerba P, Burant CF, Zhang L, Adsay V, Wicha M, Clarke MF, Simeone DM (2007) Identification of pancreatic cancer stem cells. Cancer Res 67(3):1030–1037

Lim WA, June CH (2017) The principles of engineering immune cells to treat cancer. Cell 168(4):724–740

Ling X, Marini F, Konopleva M, Schober W, Shi Y, Burks J, Clise-Dwyer K, Wang RY, Zhang W, Yuan X, Lu H (2010) Mesenchymal stem cells overexpressing IFN-β inhibit breast cancer growth and metastases through Stat3 signaling in a syngeneic tumor model. Cancer Microenviron 3(1):83–95

Lipson EJ, Drake CG (2011) Ipilimumab: an anti-CTLA-4 antibody for metastatic melanoma. Clin Cancer Res 17(22):6958–6962

Liu BL, Robinson M, Han ZQ, Branston RH, English C, Reay P, McGrath Y, Thomas SK, Thornton M, Bullock P, Love CA (2003) ICP34. 5 deleted herpes simplex virus with enhanced oncolytic, immune stimulating, and anti-tumour properties. Gene Ther 10(4):292–303

Lu T, Ramakrishnan R, Altiok S, Youn JI, Cheng P, Celis E, Pisarev V, Sherman S, Sporn MB, Gabrilovich D (2011) Tumor-infiltrating myeloid cells induce tumor cell resistance to cytotoxic T cells in mice. J Clin Investig 121(10):4015–4029

Luqmani YA (2005) Mechanisms of drug resistance in cancer chemotherapy. Med Principles Pract 14(Suppl. 1):35–48

Makita S, Yoshimura K, Tobinai K (2017) Clinical development of anti-CD 19 chimeric antigen receptor T-cell therapy for B-cell non-Hodgkin lymphoma. Cancer Sci 108(6):1109–1118

Martin MD, Badovinac VP (2018) Defining memory CD8 T cell. Front Immunol 9:2692

Mirnezami R, Nicholson J, Darzi A (2012) Preparing for precision medicine. N Engl J Med 366(6):489–491

Miska J, Lesniak MS (2015) Neural stem cell carriers for the treatment of glioblastoma multiforme. EBioMedicine 2(8):774–775

Moss PA, Rosenberg WM, Bell JI (1992) The human T cell receptor in health and disease. Annu Rev Immunol 10(1):71–96

Muranski P, Boni A, Wrzesinski C, Citrin DE, Rosenberg SA, Childs R, Restifo NP (2006) Increased intensity lymphodepletion and adoptive immunotherapy—how far can we go? Nat Clin Pract Oncol 3(12):668–681

Muro, K., Bang, Y.J., Shankaran, V., Geva, R., Catenacci, D.V.T., Gupta, S., Eder, J.P., Berger, R., Gonzalez, E.J., Ray, A., Dolled-Filhart, M. (2015) Relationship between PD-L1 expression and clinical outcomes in patients (pts) with advanced gastric cancer treated with the anti-PD-1 monoclonal antibody pembrolizumab (Pembro; MK-3475) in KEYNOTE-012

Nagahiro S (2002) Translational study in cancer research. Intern Med 41(10):770–773

Nagaraj S, Gupta K, Pisarev V, Kinarsky L, Sherman S, Kang L, Herber DL, Schneck J, Gabrilovich DI (2007) Altered recognition of antigen is a mechanism of CD8+ T cell tolerance in cancer. Nat Med 13(7):828–835

Najar M, Raicevic G, Boufker HI, Kazan HF, De Bruyn C, Meuleman N, Bron D, Toungouz M, Lagneaux L (2010) Mesenchymal stromal cells use PGE2 to modulate activation and proliferation of lymphocyte subsets: combined comparison of adipose tissue, Wharton's jelly and bone marrow sources. Cell Immunol 264(2):171–179

Natrajan R, Sailem H, Mardakheh FK, Arias Garcia M, Tape CJ, Dowsett M, Bakal C, Yuan Y (2016) Microenvironmental heterogeneity parallels breast cancer progression: a histology–genomic integration analysis. PLoS Med 13(2):e1001961

Nauta AJ, Kruisselbrink AB, Lurvink E, Willemze R, Fibbe WE (2006) Mesenchymal stem cells inhibit generation and function of both CD34+-derived and monocyte-derived dendritic cells. J Immunol 177(4):2080–2087

Nouri M, Caradec J, Lubik AA, Li N, Hollier BG, Takhar M, Altimirano-Dimas M, Chen M, Roshan-Moniri M, Butler M, Lehman M (2017) Therapy-induced developmental reprogramming of prostate cancer cells and acquired therapy resistance. Oncotarget 8(12):18949

O'Brien CA, Pollett A, Gallinger S, Dick JE (2007) A human colon cancer cell capable of initiating tumour growth in immunodeficient mice. Nature 445(7123):106–110

O'Leary JG, Goodarzi M, Drayton DL, von Andrian UH (2006) T cell–and B cell–independent adaptive immunity mediated by natural killer cells. Nat Immunol 7(5):507–516

Ong HT, Federspiel MJ, Guo CM, Ooi LL, Russell SJ, Peng KW, Hui KM (2013) Systemically delivered measles virus-infected mesenchymal stem cells can evade host immunity to inhibit liver cancer growth. J Hepatol 59(5):999–1006

Palucka K, Banchereau J (2013) Dendritic-cell-based therapeutic cancer vaccines. Immunity 39 (1):38–48

Park SA, Ryu CH, Kim SM, Lim JY, Park SI, Jeong CH, Jun J, Oh JH, Park SH, Oh W, Jeun SS (2011) CXCR4-transfected human umbilical cord blood-derived mesenchymal stem cells exhibit enhanced migratory capacity toward gliomas. Int J Oncol 38(1):97–103

Pastan I, Hassan R, FitzGerald DJ, Kreitman RJ (2006) Immunotoxin therapy of cancer. Nat Rev Cancer 6(7):559–565

Patel GK, Yee CL, Terunuma A, Telford WG, Voong N, Yuspa SH, Vogel JC (2012) Identification and characterization of tumor-initiating cells in human primary cutaneous squamous cell carcinoma. J Investig Dermatol 132(2):401–409

Pettit SJ, Ali S, O'Flaherty E, Griffiths TRL, Neal DE, Kirby JA (1999) Bladder cancer immunogenicity: expression of CD80 and CD86 is insufficient to allow primary CD4+ T cell activation in vitro. Clin Exp Immunol 116(1):48–56

Plosker GL (2011) Sipuleucel-T. Drugs 71(1):101–108

Prasanna PG, Stone HB, Wong RS, Capala J, Bernhard EJ, Vikram B, Coleman CN (2012) Normal tissue protection for improving radiotherapy: where are the gaps? Transl Cancer Res 1(1):35

Repmann R, Wagner S, Richter A (1997) Adjuvant therapy of renal cell carcinoma with active-specific-immunotherapy (ASI) using autologous tumor vaccine. Anticancer Res 17 (4B):2879–2882

Romee R, Cooley S, Berrien-Elliott MM, Westervelt P, Verneris MR, Wagner JE, Weisdorf DJ, Blazar BR, Ustun C, DeFor TE, Vivek S (2018) First-in-human phase 1 clinical study of the IL-15 superagonist complex ALT-803 to treat relapse after transplantation. Blood 131 (23):2515–2527

Rossi A, Torri V, Garassino MC, Porcu L, Galetta D (2014) The impact of personalized medicine on survival: comparisons of results in metastatic breast, colorectal and non-small-cell lung cancers. Cancer Treat Rev 40(4):485–494

Schantz JT, Chim H, Whiteman M (2007) Cell guidance in tissue engineering: SDF-1 mediates site-directed homing of mesenchymal stem cells within three-dimensional polycaprolactone scaffolds. Tissue Eng 13(11):2615–2624

Schatton T, Murphy GF, Frank NY, Yamaura K, Waaga-Gasser AM, Gasser M, Zhan Q, Jordan S, Duncan LM, Weishaupt C, Fuhlbrigge RC (2008) Identification of cells initiating human melanomas. Nature 451(7176):345–349

Schauz D (2014) What is basic research? Insights from historical semantics. Minerva 52 (3):273–328

Schultz L, Mackall C (2019) Driving CAR T cell translation forward. Sci Transl Med 11(481)

Schumacher TN, Schreiber RD (2015) Neoantigens in cancer immunotherapy. Science 348 (6230):69–74

Seidel JA, Otsuka A, Kabashima K (2018) Anti-PD-1 and anti-CTLA-4 therapies in cancer: mechanisms of action, efficacy, and limitations. Front Oncol 8:86

Shi Y, Lammers T (2019) Combining nanomedicine and immunotherapy. Acc Chem Res 52 (6):1543–1554

Shimamura H, Cumberland R, Hiroishi K, Watkins SC, Lotze MT, Baar J (2002) Murine dendritic cell-induced tumor apoptosis is partially mediated by nitric oxide. J Immunother 25(3):226–234

Sidana A (2014) Cancer immunotherapy using tumor cryoablation. Immunotherapy 6(1):85–93

Singh SK, Clarke ID, Terasaki M, Bonn VE, Hawkins C, Squire J, Dirks PB (2003) Identification of a cancer stem cell in human brain tumors. Cancer Res 63(18):5821–5828

Stem cells: What they are and what they do (2019) Mayo Clinic, https://www.mayoclinic.org/tests-procedures/bone-marrow-transplant/in-depth/stem-cells/art-20048117

Storkus, W.J., Finn, O.J., DeLeo, A. and Zarour, H.M. (2003) Categories of tumor antigens. Holland-Frei Cancer Medicine

Takahashi K, Yamanaka S (2006) Induction of pluripotent stem cells from mouse embryonic and adult fibroblast cultures by defined factors. Cell 126(4):663–676

The Precision Medicine Initiative, n.d.. https://obamawhitehouse.archives.gov/precision-medicine

Thielens NM, Tedesco F, Bohlson SS, Gaboriaud C, Tenner AJ (2017) C1q: a fresh look upon an old molecule. Mol Immunol 89:73–83

Tirapu I, Huarte E, Guiducci C, Arina A, Zaratiegui M, Murillo O, Gonzalez A, Berasain C, Berraondo P, Fortes P, Prieto J (2006) Low surface expression of B7-1 (CD80) is an immunoescape mechanism of colon carcinoma. Cancer Res 66(4):2442–2450

Tobias AL, Thaci B, Auffinger B, Rincón E, Balyasnikova IV, Kim CK, Han Y, Zhang L, Aboody KS, Ahmed AU, Lesniak MS (2013) The timing of neural stem cell-based virotherapy is critical for optimal therapeutic efficacy when applied with radiation and chemotherapy for the treatment of glioblastoma. Stem Cells Transl Med 2(9):655–666

Translational science spectrum, 2018, NIH, https://ncats.nih.gov/translation/spectrum

Tsung K, Norton JA (2006) Lessons from Coley's toxin. Surg Oncol 15(1):25–28

Twumasi-Boateng K, Pettigrew JL, Kwok YE, Bell JC, Nelson BH (2018) Oncolytic viruses as engineering platforms for combination immunotherapy. Nat Rev Cancer 18(7):419–432

Van de Laar L, Coffer PJ, Woltman AM (2012) Regulation of dendritic cell development by GM-CSF: molecular control and implications for immune homeostasis and therapy. Blood 119 (15):3383–3393

Ventola CL (2017) Cancer immunotherapy, part 3: challenges and future trends. Pharm Ther 42 (8):514

Vermorken JB, Claessen AM, Van Tinteren H, Gall HE, Ezinga R, Meijer S, Scheper RJ, Meijer CJ, Bloemena E, Ransom JH, Hanna MG Jr (1999) Active specific immunotherapy for stage II and stage III human colon cancer: a randomised trial. Lancet 353(9150):345–350

Vyas D, Laput G, Vyas AK (2014) Chemotherapy-enhanced inflammation may lead to the failure of therapy and metastasis. OncoTargets Ther 7:1015

Wang Y, Yang J, Zheng H, Tomasek GJ, Zhang P, McKeever PE, Eva YHL, Zhu Y (2009) Expression of mutant p53 proteins implicates a lineage relationship between neural stem cells and malignant astrocytic glioma in a murine model. Cancer Cell 15(6):514–526

Wang C, Xu L, Liang C, Xiang J, Peng R, Liu Z (2014) Immunological responses triggered by photothermal therapy with carbon nanotubes in combination with anti-CTLA-4 therapy to inhibit cancer metastasis. Adv Mater 26(48):8154–8162

Weiner, G.J., 2010. Rituximab: mechanism of action. In Seminars in hematology, vol. 47(2), 115–123. WB Saunders

What is Coley's toxins treatment for cancer? 2012, Cancer Research UK

Wong AH, Deng CX (2015) Precision medicine for personalized cancer therapy. Int J Biol Sci 11:1410–1412

Yamashita Y, Shimokata K, Mizuno S, Yamaguchi H, Nishiyama Y (1993) Down-regulation of the surface expression of class I MHC antigens by human cytomegalovirus. Virology 193 (2):727–736

Yang SH, Park MJ, Yoon IH, Kim SY, Hong SH, Shin JY, Nam HY, Kim YH, Kim B, Park CG (2009) Soluble mediators from mesenchymal stem cells suppress T cell proliferation by inducing IL-10. Exp Mol Med 41(5):315–324

Yu Y, Ramena G, Elble RC (2012) The role of cancer stem cells in relapse of solid tumors. Front Biosci (Elite Ed) 4(2):1528–1541

Yuan Y (2016) Spatial heterogeneity in the tumor microenvironment. Cold Spring Harbor Perspect Med 6(8):a026583

Zhang S, Balch C, Chan MW, Lai HC, Matei D, Schilder JM, Yan PS, Huang TH, Nephew KP (2008) Identification and characterization of ovarian cancer-initiating cells from primary human tumors. Cancer Res 68(11):4311–4320

Zhang C, Liu J, Zhong JF, Zhang X (2017) Engineering CAR-T cells. Biomark Res 5:22

Zhao J (2016) Cancer stem cells and chemoresistance: the smartest survives the raid. Pharmacol Ther 160:145–158

Zhao T, Zhang ZN, Rong Z, Xu Y (2011) Immunogenicity of induced pluripotent stem cells. Nature 474:212–215

Zugazagoitia J, Guedes C, Ponce S, Ferrer I, Molina-Pinelo S, Paz-Ares L (2016) Current challenges in cancer treatment. Clin Ther 38(7):1551–1566

Cancer Cell Lines: Its Implication for Therapeutic Use

17

Sen Pathak

> Chromosomes never lie.
> —(Sen Pathak, 1972).

Abstract

The purpose of this mini-review, which is developed for the education of general public, is to make readers acquainted with the usefulness of cancer cell lines in numerous disciplines of biomedical research. Special attention is also given to describe the importance of cell lines in cancer research and treatment, a discipline popularly known as oncology. In addition, by narrating the developmental history of only three human cancer cell lines, HeLa, MCF-7, and MDA-MB-468, the significance of cell lines and the pitfalls the way cell lines are being used in biomedical research are described.

Keywords

Hayflick limit · HeLa cells · MCF-7 cells · MDA-MB-468 cell line

---

S. Pathak (✉)
Department of Genetics, The University of Texas M.D. Anderson Cancer Center, Houston, TX, USA
e-mail: spathak@mdanderson.org

© The Author(s), under exclusive license to Springer Nature Singapore Pte Ltd. 2022
S. K. Basu et al. (eds.), *Cancer Diagnostics and Therapeutics*,
https://doi.org/10.1007/978-981-16-4752-9_17

17.1 Introduction

The foundation of biomedical research is based on three major components: researcher, hypothesis, and reagents. One of the most important reagents is the continuous supply of live cells, whether derived from normal or cancerous tissues, needed for various experimentations. The cell lines derived from human cancer biopsies or from animal tumors are of paramount importance in biomedical research in general and cancer research in particular. It is well known that normal human somatic cells may survive in tissue culture dish for up to 50 passages (known as Hayflick limit), whereas cancer cells can survive several hundred passages in continuous cultures. Cancer cells are generally referred to as being immortal, but in reality, nothing is immortal in this world because everything has an expiration date including cancer cells (Pathak 2019, in press). In continuous culture conditions, cancer cells may survive longer in comparison with normal cells but finally undergo senescence and may die, unless their genome is further reprogrammed and acquires pluripotency (Niu et al. 2016). Long-term cell lines offer a constant supply of the same cellular material for a variety of research projects and needed for the confirmation of earlier results. Such cancer cells as a reagent are particularly useful when tissues in vivo, such as human cancers, cannot be obtained repeatedly because patients either have succumbed to the disease or become uncooperative to give another biopsy because they are doing well. Established human cancer cell lines have played crucial role(s) in the discovery of new antitumor drugs. In addition, specific genetic alterations (mutations) identified in these cancer cell lines have helped in the production of animal models, particularly mouse models, for studying the mechanism involved in development and aggressiveness of specific human tumors and also development of new cancer therapies. Animal models, known as Avatar, have been experimented for the personalized cancer treatment by a particular chemotherapeutic regimen (Couzin-Frankel 2014). This is the era of personalized cancer therapy and an avatar animal model may come under that particular discipline of treatment.

The purpose of this mini-review, which is developed for the education of general public, is to make readers acquainted with the usefulness of cancer cell lines in numerous disciplines of biomedical research. Special attention is also given to describe the importance of cell lines in cancer research and treatment, a discipline popularly known as oncology. In addition, by narrating the developmental history of only three human cancer cell lines, HeLa, MCF-7, and MDA-MB-468, I will describe the significance of cell lines and the pitfalls the way cell lines are being used in biomedical research.

17.2 Dawn of Cell Culture and the Establishment of First Human Cancer Cell Line

Before the formation of the American Tissue Culture Association (ATCA) in 1946, only a few trained tissue culturists started a revolution in the field by lecturing, giving workshops, and training future cell culturists in the country. One of the pioneers in the field was Wilton Earle in Bethesda, Maryland, whose group was involved in developing the chicken embryo extract for culturing rodent cells and succeeded in establishing many long-term cell lines in the early 1940s, including the famous L cells from a C3H mouse (Earle 1943). However, it was very difficult to culture human cells in culture vessels and only a handful of some "great souls ventured in the field but came out with bloody nose." In the early 1950s, George Gey and his group succeeded in establishing the first "immortal" human cancer cell line, designated as HeLa, from the biopsy of a cervical carcinoma taken from a 30-year-old African American female, a farmworker named Henrietta Lacks (Gey 1952). However, earlier it was possible for Warren Lewis in whose laboratory Gey had trained for more than 30 years, to culture rodent cancer cells, but to culture human cancer cells was a far cry (Lewis 1936). Today, tissue culturists have many advantages including defined culture media, better culture techniques, sterile culture hoods, peripheral lymphocytes, fibroblast and epithelial cells stimulating growth factors, and sterilized disposable culture flasks, pipettes, and petri dishes. Therefore, culturing normal and cancer cells of human and animal origin has become rather easy. Still one question remained unanswered until recently. Why has it been impossible or extremely difficult to culture human normal cells as long-term cell lines but possible to culture long-term rodent cell lines? It is now known that rodent cells have telomerase activity in all their somatic cells, but most human somatic cells do not. Telomerase is an enzyme that keeps on adding nucleotides at the end of chromosomes known as telomeres, and therefore, cells in vitro do not undergo senescence. Exogenous addition of telomerase in culture medium can induce prolong growth and transformation of normal human cells in vitro (Hanahan and Weinberg 2011).

17.3 HeLa Cells and their Usefulness in Developing Tissue Culture Techniques

In early days, virologists were in dire need of a cell line as substrate to grow viruses. HeLa cells provided that help to launch a new very active field of virology. After the establishment of HeLa and its free distribution to the interested scientists, important scientific advances were made around the world in a short period of time. Using HeLa cells, tissue culturists developed the techniques of cell freezing and thawing, shipping ampules containing frozen cells to other parts of the world, storing cells between experiments without worrying about continuously feeding them, and collecting cells at various stages of cell cycles. For example, HeLa cells were used to collect enough metaphase spreads by mitotic shake to be used in the production of premature chromosome condensation (PCC) technique for the visualization of interphase chromosomes (Johnson et al. 1970). HeLa cells also helped in studying the effects of various clastogens (chromosome breaking agents) such as chemicals, X-radiations, and microorganisms including viruses on chromosomes of human and other mammalian species. HeLa cells played an important role in developing various defined culture media including Parker 199, Difco, Eagle's, McCoy's 5a, and RPMI 1640. It also helped in testing the toxicity and quality of glasswares and plasticwares used in cell cultures and syringes used in obtaining blood samples. This was also true for testing the suitability of culture media and the fetal calf serum (FCS), including the stoppers used to cap the culture flasks. One of the most important techniques HeLa cells helped develop was the cell cloning methodology. Development of cell cloning technique has been the foundation stone for studying the cancer heterogeneity and instability. Numerous clones derived from the HeLa cell line alone are being used in biomedical research around the world. It has also helped in growing a single cell progeny including the isolation and propagation of stem cells. Another contribution of HeLa cells was in making somatic cell hybrids for the chromosomal assignment and mapping of individual genes responsible for different traits. Currently, scientists are in search of stable cell lines, which are able to express foreign genes, inserted into their chromosomes. Such stable cell lines are extremely useful reagents for expressing specific protein products and are highly useful in molecular biology and biotechnology applications (Lo et al. 2017).

17.4 MDA-MB-468 Cell Line which Questioned the Histopathology of HeLa Cells

In March 1976, a 51-year-old African American female noted a mass in her right breast and came to the University of Texas M.D. Anderson Cancer Centre at Houston for evaluation. A fine needle biopsy of her breast showed diagnosis of adenocarcinoma. Other details of her case history including biochemical isoenzyme and chromosome analyses, and treatment modalities are published earlier (Pathak et al. 1979). She developed bilateral pleural effusion after 3 months postchemotherapy and radiotherapy. On November 4, 1977, approximately 1000 ml. of very bloody pleural effusion was drawn from her and distributed for direct chromosome preparation, isoenzyme analysis, and the establishment of a cell line, later named MDA-MB-468. From both, direct harvest of the effusion and from the cell line, which was established within a month, at least five to six clonal marker chromosomes (rearranged chromosomes) were identified by Giemsa banding (Pathak et al. 1979). These markers were almost identical to the marker chromosomes of HeLa cell line, derived from a cervical carcinoma biopsy (Gey 1952). While I was very surprised to observe the presence of HeLa markers in the MDA-MB-468 breast cancer cell line, I wondered if our patient was injected previously with HeLa cells in the name of vaccine (*in 1980, during a conference in Lake Placid, N.Y., where I was told by the discoverer of polio vaccine, Jonas Salk, that some patients have been accidently injected with HeLa cells in the name of vaccine*).

After extensive inquiries, I was told this breast cancer patient was a housewife who did not receive injection of polio vaccine and never worked in any research laboratory where HeLa cells were in use. I then thought if could it be possible that Henrietta Lacks was suffering with breast cancer which had metastasized to her cervix? I discussed this possibility with T.C. Hsu who was one of the first to do chromosome analysis of HeLa cell before the discovery of banding techniques. He exclaimed, "Sen! histopathology of HeLa cells was in doubt from the very beginning." Even the gynecologist, H. W. Jones, who had examined the original HeLa biopsy, commented that he could never forget that tumor because it was unlike anything he had ever seen. I also checked with my great pathologist friend, Alfred Gropp of Germany, who informed me that breast cancer does metastasize to cervix, but it is a rare phenomenon. While I was gathering such information and speculating about my cytogenetic findings, Mike Siciliano who was working on the isoenzyme pattern of the pleural effusion and established cell line of the same patient came into my office 1 day and asked about the chromosome analysis data. When I told him about the presence of HeLa markers in our patient's effusion and also in the MDA-MB-468 cell line, Mike was equally very surprised. His isoenzyme analysis of 13 polymorphic loci from MDA-MB-468 was identical to that of HeLa. The similarity of the X-linked glucose-6-phosphate-dehydrogenase (G6PDH) expression in HeLa and MDA-MB-468 is also of great interest. Both cell lines were derived from A & B heterozygote individuals and express only G6PDH A (Hsu et al. 1976; Pathak et al. 1979). In other words, the bloody pleural effusion displayed both G6PDH A and G6PDH B, but the cell line MBA-MB-468 showed only A type mobility, which is the characteristic of most African Americans (Gartler 1967, 1968).

The original histopathology report of the biopsy, which was the source of the HeLa cell line considered the tumor an epidermoid carcinoma of the cervix, which was, later on, changed to the adenocarcinoma of the cervix (Jones et al. 1971). If the prevailing concept of a specific chromosome marker(s) associated with a particular neoplasm is true, then other bona fide cervical carcinoma cell lines should share chromosome markers similar to those of HeLa cells. Many studies with this intent, however, have failed to show any HeLa-like markers in these other cervical carcinoma cell lines (Atkin and Baker 1979; Pathak and Hsu 1985; Freedman et al. 1982; Herz et al. 1977; Hsu and Pathak 1989). On the other hand, a number of bona fide human breast cancer cell lines have shown the presence of HeLa markers in their genomes (Seman et al. 1976; Pathak et al. 1979; Krizman et al. 1987). Because of the presence of HeLa markers in some of these and many other cell lines of breast cancer origin, these breast cancer cell lines have been identified to be contaminated with HeLa (Nelson-Rees et al. 1974a, 1975; Nelson-Rees and Flandermeyer 1976). This is not to downgrade the widespread incidence of HeLa cells contaminating many other cell lines originating from human lung, skin, pharynx, prostate, bladder, ovary, liver, and many other organs (Nelson-Rees et al. 1974b, 1980, 1981; Lin and Goldstein 1974; Lavappa et al. 1976). Cell line contamination is a real problem where HeLa has played a major role and still the center of focus, but other long-term cell lines are now equally responsible for contaminating and in replacing original cultures of cell lines.

17.5 Denver Episode and Unethical Behavior of a Senior Scientist

Sometime in early 1978, the 29th Annual meeting of the American Tissue Culture Association (TCA) was held in Denver, Colorado, where I had submitted an abstract entitled "Fresh pleural effusion from a patient with breast cancer shows some characteristic HeLa markers," and had requested for its platform presentation (Pathak et al. 1978). My abstract was being reviewed by then, President elect, Walter A. Nelson-Rees, of TCA. I had met Walter earlier in a different Somatic Cell Genetics Conference and had corrected his identification of G-banded human chromosomes. Also, I had the opportunity of mentoring Walter's associate named, R.R. Flandermeyer, during a workshop on human banded chromosomes held at Alton W. Jones Cell Science Centre in Lake Placid, N.Y. While evaluating my abstract submitted to TCA, Walter called me and requested to send him microphotographs of the maker chromosomes present in the effusion and MDA-MB-468 cell line, which I did in good faith. He finally accepted my abstract for its presentation from the platform, as informed by the TCA Program Committee. In the meantime, I had written my entire manuscript, had it corrected by other co-authors, and submitted to the Editor in Chief of the Journal of National Cancer Institute (JNCI) for its possible publication.

Before I received the acceptance letter for my manuscript from the JNCI editorial office, I traveled to Denver to present my paper during the 29th TCA meeting. Under the chairmanship of John Pattriciani, I presented my observations on the breast cancer cell line MDA-MB-468 and made a strong point questioning the histopathology of the HeLa cells. Because of the presence of five marker chromosomes identical to HeLa and the isoenzyme A type glucose-6-phosphate dehydrogenase in the cell line, I made a tangible point that Henrietta Lacks, from whose cervix the biopsy was obtained to establish HeLa cell line, probably was suffering with breast cancer. As stated earlier, after consulting many experienced pathologists including my expert pathology friend, Alfred Gropp of Germany, I was told that breast cancer does metastasize to cervix but it is rare. After my presentation where I had not criticized any one including Walter, my paper was open for discussion. Sitting at the end of the hallway, the only person who raised his hand to ask a question was Walter Nelson-Rees. He stood up and started reading the typed one full page criticism of my paper. The audience present in the conference hall was surprised by Walter's behavior and did not know what was going on. When Walter finished reading his criticism, I was given time by the session chairman to respond although no time was left. Standing on the stage with a choked and dry throat after hearing the criticism made by Walter, I was trying to find a place to hide my face under the podium. I was also furious but tried to control my anger. I had never previously been criticized so badly and unethically in my entire life. I composed myself after swallowing the anger and responded, "Thank you Dr. Walter Nelson-Rees for your comments on my paper and identifying yourself as a referee of my full-length paper. I hope my next couple of papers on the similar subject will be sent to you for your review" and I left the podium. After the break, I heard some scientists, particularly of Indian origin, who might had enjoyed the criticism and taunting me when they saw me nearby. That afternoon, I did not want to show my face to any one, went to my hotel room, and remained alone. The same evening I climbed on the capital building of Denver and contemplated jumping from the top to kill myself. But, the Almighty saved me and my inner-self told me not to do so. After the meeting, I came back to Houston and decided that I will never work anymore on cancer-related projects. When my senior T.C. Hsu heard about the Denver episode, he came to my office on the same floor at Bates-Freeman Building where his office was located and consoled me. I decided to start working on mammalian meiosis projects and developed the simple silver-staining technique for studying synaptonemal complex by light microscopy (Pathak and Hsu 1979).

During the business meeting of the 29th TCA Conference in Denver, senior scientists and tissue culturists raised the question about the unethical behavior of Walter and voted against promoting him to be the President of TCA. Keith Porter who was the sitting President of the TCA was to be replaced by Walter Nelson-Rees after his retirement, but it did not happen. Walter was removed from the editorial board of scientific journals, and his research grants were either cut or not renewed. Within a year or two, his laboratory was closed, and he was forced to retire in 1981 and started selling antiques. Even today, I have great respect and admiration for Walter's scientific contributions. His original contribution of the facultative heterochromatin in mealybug genome is well known.

Now coming back to the fate of my full-length paper, I was informed by John Blair, the Editor in Chief of the Journal of National Cancer Institute (JNCI) that my paper was accepted for its publication. He enclosed also the reviews of two referees: Avery A. Sandberg, Roswell Park Memorial Institute, Buffalo, N.Y., and Uta Francke, Yale University, New York, one of whom had suggested a minor correction in my manuscript. By that time, I had known one of the original reviewers and that was Walter Nelson-Rees. I called John and requested him to send me the written review of Walter. He then told me about the recommendations of Avery and Uta who had accepted the paper for publication and not to worry about the review of Walter, which I already knew. Ultimately, my paper was strongly recommended for acceptance by two reviewers and was published (Pathak et al. 1979).

17.6 Sexing of Long-Term Cell Lines

After Denver Conference, the 30th Tissue Culture Association (TCA) Conference was held in Seattle, Washington, in 1979. I was invited there by the TCA Organizing Committee to organize a round table discussion and talk about cell line contamination per se. To provide additional input for this topic, I had invited my long-time friend, Michael J. Siciliano, an expert on isoenzymes, who was also co-author in the breast effusion paper (Pathak et al. 1979) to talk on A- and B-type G6PD and its usefulness in deciphering cell line contamination. During my talk in Seattle, I commented about the danger of sexing long-term established cell lines in culture. I mentioned the story of Professor Pera who initiated a cell culture from a male Field Vole, *Microtus agrestis* (Rodentia), known to have the longest X and Y chromosomes. After the establishment of culture, he prepared chromosomes and found out the presence of X and Y chromosomes as expected. After growing cells for some additional passages, he observed that most metaphases have lost the Y chromosome and, cells of some additional passages, showed the presence of two X chromosomes. He wanted to study the replication pattern of these two X chromosomes originally derived from the male due to nondisjunction. Both X chromosomes were found to have identical early replication pattern with 3H-thymidine labeling, which was not surprising because they were male-derived. At additional passages when he labeled the growing cells with BrdU (bromodeoxyuridine), the result was very surprising. Now, one X chromosome was early replicating and the other late replicating, mimicking a female cell behavior. (I had reviewed this paper of Pera and had accepted its publication.) When I presented this story to audience in Seattle, George Yerganian who was sitting in the conference hall jumped and said, "Sen, there are some people sitting in this room who are proving their points of cell contamination by sexing the long-term cell lines. Are they not doing injustice in the light of your presentation?" George was pointing out to Walter Nelson-Rees who was sitting in the hall and had, on many occasions, tried to prove his point of contamination by observing two X chromosomes or the absence of a bona fide human Y chromosome in a male-derived cell line. We concluded that by sexing long-term cell lines one could arrive at an erroneous conclusion.

17.7　My Meeting with the Legendry Jonas Salk in Lake Placid

After the TCA conference in Seattle, I was invited to a Vaccine Production Meeting which was held at Lake Placid, N.Y. in 1980. This meeting was called to discuss the extent of chromosome aberrations present in a human cell line, which could be used for the production of polio vaccine. Walter was also invited in this meeting along with a few other scientists where I was very happy and honored to meet the famous Jonas Salk, whose name I had heard earlier while doing my Ph.D. at Banaras Hindu University, Varanasi, India. I vividly remember the statement of Jonas Salk who said, "In the name of Vaccine, some healthy individuals have been injected with HeLa cells in the past." I found Walter seemed to be lonely there but came to join me during the breakfast. He was feeling sorry for his action in Denver and told me, "Sen, I am very sorry what I did to you in Denver. People say one day I will cut my tongue and die of my own blood poisoning." In other word, he was repenting for what he has done to me. I told him, "Walter! I have never criticized you for what you did to me in Denver, but many others have realized your unethical behavior and have become your enemies. I agree and respect your cell contamination story of various human cell lines with HeLa. However, I do not agree that all human breast cancer cell lines are also contaminated with HeLa. Your research work on chromosomes of Mealy bugs and coining of the term - facultative heterochromatin is well-known to me." After our meeting in Lake Placid, N.Y., I never met Walter Nelson-Rees because he was no more in science and passed away on January 23, 2009, at the age of 82.

17.8 My Interview in Houston by the Senior Documentary Producer of BBC London

The Denver episode became viral throughout the world the same way as was the problem of HeLa cell line contamination. It even caught the attention of a Senior Producer, Adam Curtis, of the Documentaries Department of the British Broadcasting Corporation (BBC) of London. Curtis first contacted me by phone on September 23, 1996, and then wrote a letter to visit Houston to interview me about the episode and HeLa contamination. During our telephone conversation, I had mentioned to Curtis that HeLa may be a breast cancer cell line and not the cervical cancer (Satya-Prakash et al. 1981). In his September 1996, letter he wrote, "I would be very grateful if you would agree to film an interview about this project. I was also fascinated by what you told me on the phone – your theory that in fact HeLa may not be a cervical cell (line). I think it is very important that I include this in the film, and would be keen to ask you more about this in the interview." Curtis, with his camera technician, arrived in Houston on October 21, 1996, and interviewed/filmed me for almost four long hours in a huge old building with long glass panels of the Houston Lighting and Power Company located in Downtown Houston. A documentary on HeLa contamination story was made by the BBC, but I never got to see it.

17.9 My Breakfast Meeting in Houston with Members of Lacks Family

In 2016, I was invited by Jessica Moore, one of my former Research Assistants and now a faculty member at M.D. Anderson Cancer Centre, to have breakfast with the granddaughter and the great-granddaughter of the famous Henrietta Lacks on May 4, 2016. They were invited to the University of Texas M.D. Anderson Cancer Centre by the Division of Cancer Prevention to be recognized as members of the Lacks family during a conference entitled "The Power of One: The Legacy of Henrietta Lacks." I carried a copy of my 1979 paper and handed over it to Veronica Robinson and Jeri Lacks Whye, great-granddaughter and granddaughter, respectively, of Henrietta Lacks and told them and other scientists and physicians who were present during the breakfast, about the possible histopathology of HeLa cells (I wished to reproduce a group photograph here in this article with Lacks family, but a permission was not granted by their agent). Next day after the meeting, I was told by Jessica Moore about the good news, which supported my hypothesis of the origin of HeLa cells. Both granddaughter and the great-granddaughter were happy to know about my assessment of the HeLa cells being originated from the breast cancer and not from the cervical carcinoma. The conclusion published in my paper was considered a connecting link of information which was lacking because in Lacks family some members have been diagnosed with breast cancer but none with cervical cancer. Of course, this was welcoming evidence in support of my contention about the breast cancer origin of the HeLa cell line (Pathak et al. 1979; Satya-Prakash et al. 1981; Pathak 2007).

17.10 Human Breast Cancer Cell Line MCF-7 and its Derivative Sublines

In 1973, Herbert D. Soule at the Michigan Cancer Foundation (MCF) was trying to establish a cell line from the biopsy of a chest wall nodule and from the pleural effusion of a metastatic breast cancer patient named Helen Marion (Sister Catherine Frances) who was 66 years old at the time of diagnosis. Prior to the formation of pleural effusion, she had undergone mastectomy, first of her right breast in 1963 and then of left in 1967. Soule used standard protocols to develop the cell line, but the culture initiated from her nodule was soon taken over by fibroblasts and discarded. However, the culture initiated from her pleural effusion grew as continuous culture in a monolayer (Soule et al. 1973). Since this represented the seventh attempt of Soule to generate a cancer cell line, it was named MCF-7. This MCF-7 cell line, which is the most studied human cancer cell line next only to HeLa, has produced approximately 25,000 published articles as compared to HeLa with 80,000 reports (Lee et al. 2015). It was the first human breast cancer cell line found to be very useful in the study of the estrogen receptor (ER) alpha because its cells express ER and for being hormone-responsive (Levenson and Jordan 1997). This cell line was considered to be an excellent biological reagent to isolate human breast cancer viruses. Bulk of the present knowledge on hormone-responsive breast cancer research has been obtained using the MCF-7 cell line. In addition, this cell line was central to the generation of antibodies against the estrogen receptor (Greene et al. 1980). Although the MCF-7 cells are considered as the "work horse" for studies of estrogen action in breast cancer, it is noteworthy that these cells also express other hormones including androgen, progesterone, thyroid hormone, insulin, prolactin, and glucocorticoid receptors (Horwitz et al. 1975; Shafie and Brooks 1977; Lippman et al. 1977; Burke and McGuire 1978).

Over the past many years, just like HeLa, MCF-7 cell line has been distributed all over the world to many investigators for their biological and biomedical investigations. But, some of their observations could not be reproduced on the same cell line obtained from different laboratories. For example, MCF-7 cells obtained from different laboratories have shown different patterns of cell surface markers. In the original MCF-7 cell line, at passage 2, the chromosome number was found to be highly variable, ranging over 70 to 144. However, at passage 39, the chromosome number reduced to 77 to 99 with a modal number of 88 (Soule et al.

1973). The current modal chromosome number in the MCF-7 cells obtained from the American Type Culture Collection (ATCC) is 82, with a range between 66 and 87 (Whang-Peng et al. 1983; Osborne et al. 1987). Furthermore, controversy has arisen about the estrogen responsiveness of the MCF-7 cells in culture (Lee et al. 2015). Cell lines such as MCF-7 (ATCC), MCF-7, MCF-7 (KO), MCF-7 (S), and many others, derived from the original MCF-7 cells in different laboratories, have also shown different cell surface antigen expression, different copy numbers of N-*ras* oncogene, variable tumorigenic properties in nude mice and different restriction fragment polymorphisms (Graham et al. 1985 and 1986; Pathak 2007). Based on the differences in the biological properties, it has been stated that MCF-7 cell line obtained from ATCC, the original MCF-7 (developed at Michigan Cancer Foundation), and its sublines are not derived from the same individual (Whang-Peng et al. 1983; Graham et al. 1986; Osborne et al. 1987). In fact in an addendum of the paper published by Osborne and associates (1987), R.J. Hay of the American Type Culture Collection (ATCC) has indicated that some of the MCF-7 samples which ATCC has distributed were contaminated with a human colon cancer cell line (HT-29). There is a serious take-home message in this important statement. It is, therefore, very important that researchers must authenticate their MCF-7 cells with the original MCF-7 cell line before launching their experiments. This will save them not only their grant money but expensive labor and embarrassment of publishing scientific data on wrong cell lines (Stacey 2000; Pathak and Hsu 1985; Hsu and Pathak 1989).

17.11 How the Inflammatory Breast Cancer Cell Lines Became Human Embryonic Kidney Cancer Line (293 HEK): A Detective Story?

Cancer cell lines particularly those of human breast cancer origin are maximally used in biomedical research. Cell lines established from different histopathologic breast cancer origin, such as ductal infiltrating, mucus secreting, inflammatory, and other types, are in great demand. Of these, most cell lines are of infiltrating breast cancer types, and cell lines of inflammatory origin are very rare. In the year 2006, we received at early passage a newly established so-called inflammatory breast cancer cell line for its cytogenetic characterization. A year later in 2007, 11 different cell cultures (of inflammatory human breast cancer?) were sent to us for karyotyping from the same laboratory to determine if all of them were different or the same. Karyotyping results demonstrated all 11 cell lines had six identical common marker chromosomes (Fig. 17.1) and, therefore, were the same cell line.

At this point, the principal investigator (P.I.) wanted to know if these 11 cell lines were any way similar to the newly established cell line, which was sent to us earlier for karyotyping. However, our results showed that the karyotype of the new cell line established in the year 2006 was quite different from those 11 lines. So the burning question was what was the origin of these 11 cell lines? Just a couple of months ago, I had karyotyped a long-term established HEK 293 cell line derived from human embryonic kidney and the characteristic marker chromosomes were fresh in my

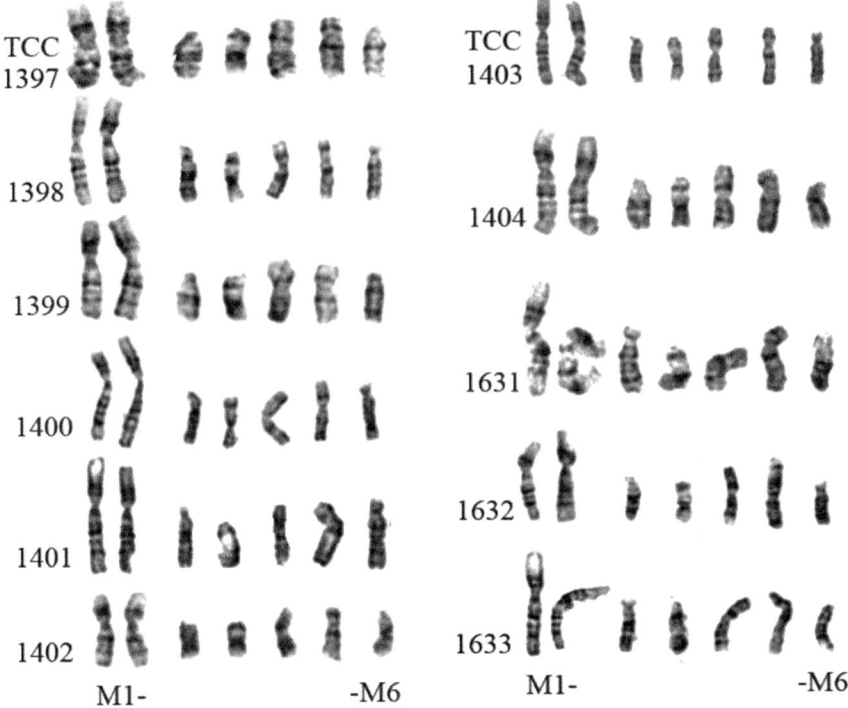

Fig. 17.1 Giemsa-banded partial karyotypes (marker chromosomes only) of 11 so-called human inflammatory breast cancer cell lines (TCC 1397 to 1404; TCC 1631 to 1633) showing six identical common marker chromosomes (M1 ... M6)

mind (Fig. 17.2). All 11 cell lines were found to be the same and actually contaminant of HEK 293. The P. I. was not sure about the newly established cell line(s), so we received another breast cancer cell line (SUM 149), which was in culture in that laboratory.

When we karyotyped SUM 149, we immediately determined that the cell line established in the year 2006 was actually SUM 149 (Fig. 17.3). In conclusion, the newly established cell lines were found not to be a new cell lines but contaminants of two other long-termed established cell lines (HEK 293 and SUM 149). These old cell lines were being cultured in the laboratory at the same time when new cell lines were being established. In other words, a long-termed established cell line can easily contaminate a primary cell culture and may overtake a newly being established cell line in many ways (Pathak 2007). A conclusion from this detective story was that the P.I. was not able to establish inflammatory breast cancer cell line and research dollars went down the drain.

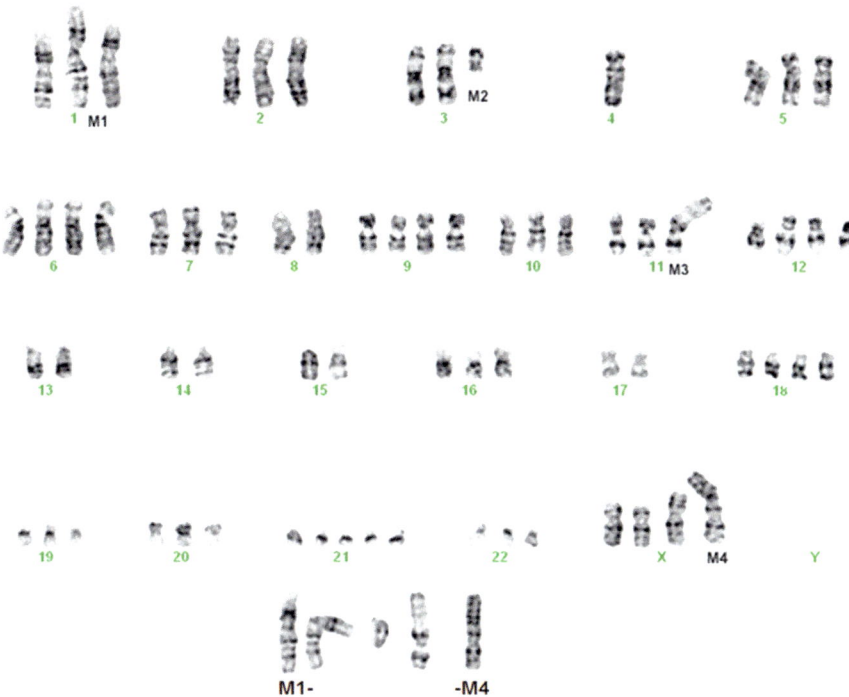

Fig. 17.2 Giemsa-banded complete karyotype of a metaphase spread from human embryonic kidney cell line (293 HEK) showing numerical and structural chromosome abnormalities with four characteristic maker chromosomes (M1 ... M4). Such four marker chromosomes from another metaphase spread of 293 HEK cell line are shown on the bottom row

Fig. 17.3 Giemsa-banded partial karyotypes (marker chromosomes only) of a so-called newly established human inflammatory breast cancer cell line, TCC 1157 (upper row), and marker chromosomes of a long-term established human breast cancer cell line, SUM 149 (TCC 1564) (bottom row). Note the similarity and commonality of all 14 marker chromosomes (M1 ... M14) in these two cell lines

17.12 Source of Cell Line Contamination and Authentication

In one of my earlier publications, the sources and time of cell line contamination have been vividly described (Pathak 2007). Today, one of the main sources of human and mouse cell line contaminations is the inoculation of human cell lines into nude mice. There are well-documented publications to show that human cancer cells can transform murine host cells after inoculation (Pathak et al. 1981, 1997; 1998; Pathak and Hsu 1979; Goldenberg and Pavia 1981, 1982; Beattie et al. 1982; Bowen et al. 1983; Price et al. 1998; Nair et al. 2013). Our experiences with mouse xenografts of human primary and metastatic cancers, whether leukemia, lymphoma, or solid tumors, have shown three outcome possibilities as shown in Fig. 17.4.

When human cancer cells are injected into nude mice, what comes out is (1) human tumor in may be 60% of the cases; (2) human and mouse cells mixture in 20% cases, and (3) only mouse tumor in 20%. This could be applicable also in the production of new anticancer drugs against a human tumor injected in mice. Sometime during the month of October 2000, a conference on cancer was held in Crete, Greece, where a speaker from the USA interested in discovering a new anticancer drug presented his data. He had orthotopically inoculated several female mice with cells of a breast cancer cell line. When tumor grew to a certain size and became palpable, he fed these mice with the new drug. His conclusion from these experiments was that in 60% of the mice, there was complete regression of the tumor, and in 20% partial regression and in other 20%, no regression after the treatment with the new drug. He was not very happy with the results because almost

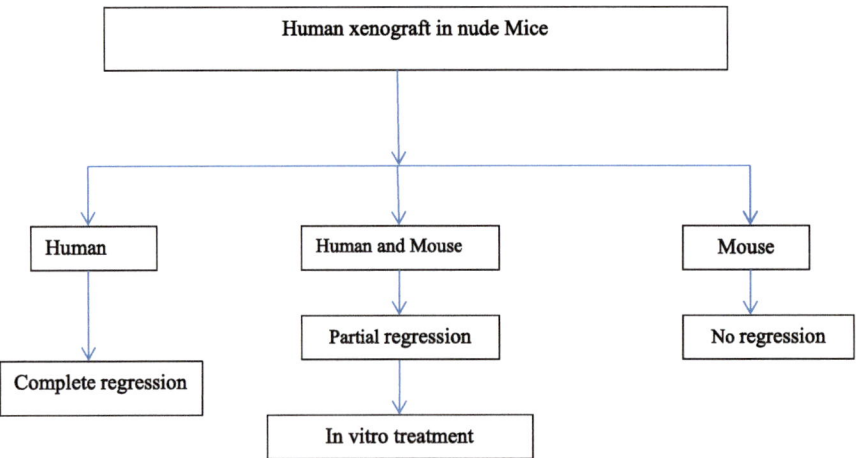

Fig. 17.4 A flowchart showing three possible outcomes when a human cancer cell line is injected in nude mice: (1) only human cells come out; (2) a mixture of human and mouse cells come out; (3) only mouse cell come out. It is also true for testing new anticancer drug experiments: Human tumors can show complete regression in approximately 60% mice, partial regression in 20% mice and no regression in 20% of mice. Histopathology of the so-called human tumor that did not regress in mice (partial or complete) was found to be of mouse origin

40% of treated mice did not show complete regression of the tumor. During the question and answer period, I asked him if he had examined the pathology of the tumor that did not regress. He had not, but thought that my question did not have any merit in his experimental protocol. I told him that the tumor that did not regress may not be of human origin but must be a mouse tumor. He called me after a month and confirmed my prediction. The tumor that did not regress after the treatment was indeed of mouse origin. In conclusion, his new drug was able to kill 100% of the human breast cancer cells in nude mice.

Nowadays, a number of cytological, serological, and DNA-based techniques including the DNA finger printings, cytogenetics, chromosomal bandings, and short tandem repeat (STR) analyses are in use and only some of them are recommended for cell line authentication (Multani and Pathak 1999; Macleod et al. 1992; Masters 2002; Cooper et al. 2007). No one single technique of these is completely qualified for the identification of a variety of intra- and intercell lines contamination because of their inherent limitations. We and many others feel and still practice the gold standard of chromosome analysis for the authentication of intra- and interspecies cell line contamination (Pathak 1976; Price et al. 1998; Nair et al. 2013).

17.13 Recommendations by Eminent Cancer Scientists to the U.S. Secretary of Health

Sometime in 2003, 18 eminent cancer researchers from the USA and UK wrote a letter to the Secretary of Health of U.S. Government regarding the problem in biomedical research using either contaminated or misidentified cell lines. This letter was signed by all those Cancer Researchers including the author of this chapter who was worried about the use of misidentified cell lines and wanted to clean-up the mess (a partial list of some of those signees is presented in Table 17.1 in Appendix). This letter contained two recommendations: (1) Grant money should not be released if the cell lines used in the grant application are not authenticated; and (2) scientific journals will not publish reviewers' friendly articles if cell lines used were not authenticated, and in other words, no cell line authentication, no grant release, and no access to publication. Both recommendations were accepted, and this policy from NIH and NCI was notified to all U.S. Cancer Research Centres including the University of Texas M. D. Anderson Cancer Centre at Houston, Texas, where the present author has been working since May 10, 1972. Following our recommendations, some high-impact journals including NATURE published a commentary stating that papers written on unauthenticated cell lines will not be published (Editorial 2009). Francis S. Collins, Director of the NIH, not only supported the recommendations but also wrote an article—fixing problems with cell lines, which is published in SCIENCE (Lorsch et al. 2014). Now, many U.S. Research Centres have established their own Cell line Authentication Core facilities where cell lines have to be studied for their purity and authentication before being used in research projects or prior to submission of articles for publication.

17.14 Use of Cancer Cell Lines in Indian Scenario for Cancer Treatment and Patient Care

Biologically speaking, each human being is a unique living entity. Because of this, personalized cancer care and treatment has become the talk of today. The University of Texas M. D. Anderson Cancer Centre built a multistory building, which became operational in the year 2015, just for the personalized cancer treatment and care, named after its donor, Sheikh Zayed Bin Sultan Al Nahyan Building (known as Zayed Building). As stated earlier, most cancer therapy drugs that are available in the U.S. market are made either against the cell lines derived from African American cancer patients or from Caucasian patients. It is becoming apparent that chemotherapeutic drugs derived against cell lines from African American cancer patients are not very effective in Caucasian cancer patients and vice versa. Most cancer drugs produced in USA are being commonly prescribed for cancer patients in other countries including India. Indian population is genetically quite different from U.S. and European populations.

In the era of personalized cancer treatment, this form of antitumor drugs and treatment produced against the entirely different genotypic populations may not be very effective for Indian cancer patients. It is, therefore, very crucial and important to produce antitumor drugs against the cell lines produced from Indian cancer patients. In 2015, I visited the National Centre for Cell Science (NCCS) established in the campus of Pune University by the Government of India and delivered talks on this subject. My emphasis was more on the establishment of cell lines from the gall bladder cancer of Indian patients because of its prevalence in northern part of India including the States of Uttar Pradesh and Bihar. This cancer is very rare in the USA, and therefore, no effective chemotherapeutic drug is available in the market. I have also made arrangements with the surgeons of the Institute of Medical Sciences at Banaras Hindu University, Varanasi, Sanjay Gandhi Post Graduate Institute in Lucknow, and with the Delhi State Cancer Institute, Dilshad Garden, Delhi, to ship such biopsies in culture medium to the NCCS in Pune, Maharashtra, for the establishment of permanent cell lines. Many patients known to me in India lost their lives due to the cancer of gall bladder. The only other country in the world, which has a relatively very high incidence of gall bladder cancer, is Chile in South America. Under the Global Oncology Program established at the University of Texas M.D. Anderson Cancer Centre in Houston, researchers of these two countries, India and Chile, are being brought together to research on this particular cancer. The next step would be to discover and manufacture antitumor drugs against these cell lines established from the gall bladder cancer of Indian origin, which would be more effective on such patients. This principle would be applicable to all other histopathologically different cancers that are becoming so prevalent in India including the breast, head and neck, lung, and many other types.

Recently, under the experimental protocol of personalized cancer treatment, whether drug A or drug B will be more effective for a particular cancer patient, some oncologists tried to use "Mouse Avatars." Under animal avatars, pieces of human tumor biopsies are implanted in mice lacking a normal immune system where each animal becomes a minuscule model of a patient's tumor—and a test patient for treatments. Similar cancer "avatars" have recently been created with zebra fish embryos also, which are cheaper, faster to grow, and more efficient compared to the rodents (Leslie 2017). The drawback of such animal avatars is that the patient may not survive for treatment by the time results are obtained either from mouse or from zebra fish experiments.

Most recently, the organoid architect and the production of mini-organs have come into light under the protocol of personalized cancer treatment. Organoids and mini-organs grown in the laboratory from individual patients can help predict their response to new or existing anticancer drugs. In addition, new drugs can be tested on mini-organs to help predict their effects in cancer patients, which may be very useful in their personalized medicine protocols (Sinha 2017).

17.15 Conclusions

Cell line contamination and misidentification, whether inter- or intraspecies, are real problems in biomedical and cancer research. Research scientists working with animal and/or human cell lines must be aware of this problem. Cytogenetic techniques, which are very cheap and easy to perform in addition to the DNA finger-printing techniques, should be used periodically to check the authenticity of their cell lines because "chromosomes never lie." Marker chromosomes can even identify sibling clones isolated from the same parental cell line. Of course, it needs an experienced cytogeneticist to authenticate different clones derived from the same parental line. Early and late passages of every human cell line that has been passed through nude or SCID (severe combined immunodeficient) rodents must be authenticated by species-specific total DNA using FISH technology before launching any big experiments.

Cancer cell lines are of paramount importance in studying the biology and genetics of human cancers. The use of human cancer cell lines in the discovery of new antitumor drugs cannot be overemphasized. This is especially very crucial in the era of personalized cancer therapy. It is recommended that cell lines from native Indian cancer patients should be established and used for the production of medicines to treat their cancers. Precautions described in the present article, if taken into considerations, will avoid errors in cell line contaminations, data collection, and the production of new and effective chemotherapy drugs from authenticated cell lines. Medicines produced against authenticated cancer cell lines will become more effective in saving life of cancer patients. It would also save millions of U.S. dollars, precious research efforts, and minimize the publication of papers on wrong cancer cell lines. The purpose of any research is to find out the truth, and if the substrate (cell line) is not right, then the truth will not come out. Only wrong, unwanted data will be produced and published, and ultimately, the true goal of research will not be achieved.

Acknowledgments It is my pleasure to acknowledge the assistance of my long-term colleague, Asha S. Multani, Ph.D, for preparation of certain figures, and Radha Dixit for her expert editorials. This work was supported in part by the Institutional Molecular Cytogenetics Facility established in the Department of Genetics and partially funded by the University of Texas M. D. Anderson Cancer Centre in Houston, Texas. My apologies are to all those research scientists whose articles could not be cited here due to space limitations.

Appendix

Table 17.1 A partial list of eminent cancer researchers who wrote a white letter to the U.S. Secretary of Education about the danger of using contaminated and/or misidentified human cancer cell lines with two strong recommendations (see the text for these two recommendations)

David Lewis, PhD Head of UK Health Protection Agency Culture Collections Health Protection Agency Porton Down, Salisbury, Wiltshire, UK david.lewis@hpa.org.uk	*signature*
***Glyn Stacey, PhD** United Kingdom Stem Cell Bank Division of Cell Biology and Imaging National Institute for Biological Standards and Control	*signature*
J. Justin McCormick, PhD University Distinguished Professor Department of Microbiology and Department of Biochemistry Michigan State University E. Lansing, MI mccormi1@msu.edu	*signature*
***Stanley M. Gartler, PhD** Professor Emeritus, Medicine and Genome Sciences University of Washington Seattle, Washington gartler@genetics.washington.edu	*signature*
***Sen Pathak, PhD, FNASc** Distinguished Research Professor Department of Cancer Genetics, Unit # 1010 The University of Texas M.D. Anderson Cancer Center Houston, Texas spathak@mdanderson.org	*signature*
***John M. Butler, PhD** Project Leader, Human Identity Project (Forensic DNA Testing) DNA Measurements Group/Biochemical Science Division Chemical Science & Technology Laboratory National Institute of Standards and Technology (NIST) john.butler@nist.gov	*signature*
***Gertrude C. Buehring, PhD** Associate Professor of Virology Division of Infectious Diseases School of Public Health University of California Berkeley, CA buehring@berkeley.edu	*signature*
***Edward J. Massaro, PhD** Editor-in-Chief Cell Biochemistry and Biophysics Durham, NC	*signature*
Anton F. Steuer, PhD Program Management Technical Director BioReliance Corporation	*signature*

References

Atkin NB, Baker MC (1979) Chromosome 1 in 26 carcinomas of the cervix uteri structural and numerical changes. Cancer 44(2):604–613

Beattie GM, Knowles AF, Jensen FC, Baird SM, Kaplan NO (1982) Induction of sarcomas in athymic mice. Proc Natl Acad Sci 79(9):3033–3036

Bowen JM, Cailleau R, Giovanella BC, Pathak S, Siciliano MJ (1983) A retrovirus-producing transformed mouse cell line derived from a human breast adenocarcinoma transplanted in a nude mouse. In Vitro 19(8):635–641

Burke RE, McGuire WL (1978) Nuclear thyroid hormone receptors in a human breast cancer cell line. Cancer Res 38(11 Part 1):3769–3773

Cooper JK, Sykes G, King S, Cottrill K, Ivanova NV, Hanner R, Ikonomi P (2007) Species identification in cell culture: a two-pronged molecular approach. In Vitro Cell Dev Biol Anim 43(10):344–351

Couzin-Frankel J (2014) Hope in a mouse. Science 346:28–29

Earle WR (1943) Propagation of malignancy in vitro. IV. The mouse fibroblast cultures and changes in the living cells. J Natl Cancer Inst:4–165

Editorial (2009) Identity crisis. Nature 457:935–936

Freedman RS, Bowen JM, Leibovitz A, Pathak S, Siciliano MJ, Gallager HS, Giovanella BC (1982) Characterization of a cell line (SW756) derived from a human squamous carcinoma of the uterine cervix. Vitro-Plant 18(8):719–726

Gartler SM (1967) Genetic markers as tracers in cell culture 1, 2. Natl Cancer Inst Monogr 26:167–195

Gartler SM (1968) Apparent HeLa cell contamination of human heteroploid cell lines. Nature 217 (5130):750–751

Gey G (1952) Tissue culture studies of the proliferative capacity of cervical carcinoma and normal epithelium. Cancer Res 12:264–265

Goldenberg DM, Pavia RA (1981) Malignant potential of murine stromal cells after transplantation of human tumors into nude mice. Science 212(4490):65–67

Goldenberg DM, Pavia RA (1982) In vivo horizontal oncogenesis by a human tumor in nude mice. Proc Natl Acad Sci 79(7):2389–2392

Graham KA, Richardson CL, Minden MD, Trent JM, Buick RN (1985) Varying degrees of amplification of the N-ras oncogene in the human breast cancer cell line MCF-7. Cancer Res 45(5):2201–2205

Graham KA, Trent JM, Osborne CK, McGrath CM, Minden MD, Buick RN (1986) The use of restriction fragment polymorphisms to identify the cell line MCF-7. Breast Cancer Res Treat 8 (1):29–34

Greene GL, Nolan C, Engler JP, Jensen EV (1980) Monoclonal antibodies to human estrogen receptor. Proc Natl Acad Sci 77(9):5115–5119

Hanahan D, Weinberg RA (2011) Hallmarks of cancer: the next generation. Cell 144(5):646–674

Herz F, Miller OJ, Miller DA, Auersperg N, Koss LG (1977) Chromosome analysis and alkaline phosphatase of C41, a cell line of human cervical origin distinct from HeLa. Cancer Res 37 (9):3209–3213

Horwitz KB, Costlow ME, McGuire WL (1975) MCF-7: a human breast cancer cell line with estrogen, androgen, progesterone, and glucocorticoid receptors. Steroids 26(6):785–795

Hsu TC, Pathak S (1989) Cell line contamination in biomedical research. Cancer Bull 41:330–333

Hsu SH, Schacter BZ, Delaney NL, Miller TB, McKusick VA, Kennett RH, Bodmer JG, Young D, Bodmer WF (1976) Genetic characteristics of the HeLa cell. Science 191(4225):392–394

Johnson RT, Rao PN, Hughes HD (1970) Mammalian cell fusion III. A HeLa cell inducer of premature chromosome condensation active in cells from a variety of animal species. J Cell Physiol 76(2):151–157

Jones HW, Mc Kusick VA, Harper PS, Wuu KD (1971) George Otto Gey (1899–1970): the HeLa cell and a reappraisal of its origin. Obstet Gynecol 38(6):945–949

Krizman DB, Carpenter NJ, Pathak S, Olivé M, Cailleau R, Hsu TC (1987) HeLa marker chromosomes in human breast tumors: proposal about the origin of the HeLa cell line. J Clin Lab Anal 1(1):93–97

Lavappa KS, Macy ML, Shannon JE (1976) Examination of ATCC stocks for HeLa marker chromosomes in human cell lines. Nature 259(5540):211–213

Lee AV, Oesterreich S, Davidson NE (2015) MCF-7 cells—changing the course of breast cancer research and care for 45 years. J Natl Cancer Inst 107(7)

Leslie M (2017) Zebrafish larvae could help to personalize cancer treatments. Science 357:745

Levenson AS, Jordan VC (1997) MCF-7: the first hormone-responsive breast cancer cell line. Cancer Res 57(15):3071–3078

Lewis WH (1936) Malignant cells. Harvey Lect 31:214–234

Lin CC, Goldstein S (1974) Analysis of Q-banding patterns in human cell lines. J Natl Cancer Inst 53(2):298–304

Lippman ME, Osborne CK, Knazek R, Young N (1977) In vitro model systems for the study of hormone-dependent human breast cancer. N Engl J Med 296(3):154–159

Lo CA, Greben AW, Chen BE (2017) Generating stable cell lines with quantifiable protein production using CRISPR/Cas9-mediated knock-in. BioTechniques 62(4):165–174

Lorsch JR, Collins FS, Lippincott-Schwartz J (2014) Fixing problems with cell lines. Science 346 (6216):1452–1453

MacLeod RA, Haene B, Drexler HG (1992) Cells, lines and DNA fingerprinting. Vitro Cell Dev Biol: J Tissue Culture Assoc 28(9–10):591–594

Masters JR (2002) HeLa cells 50 years on: the good, the bad and the ugly. Nat Rev Cancer 2 (4):315–319

Multani AS, Pathak S (1999) Conventional cytogenetics alone is not sufficient for identifying interspecies cell line contamination. Anticancer Res 19(3A):1753–1754

Nair HB, Bhaskaran S, Pathak S, Ghosh R, Betty D, Moore CM, VandeBerg JL (2013) 1205Lu human metastatic melanoma cells, not human! J Cancer Sci Ther 5(3)

Nelson-Rees WA, Flandermeyer RR (1976) HeLa cultures defined. Science 191(4222):96–98

Nelson-Rees WA, Flandermeyer RR, Hawthorne PK (1974a) Banded marker chromosomes as indicators of intraspecies cellular contamination. Science 184(4141):1093–1096

Nelson-Rees WA, Zhdanov VM, Hawthorne PK, Flandermeyer RR (1974b) HeLa-like marker chromosomes and type-a variant glucose-6-phosphate dehydrogenase isoenzyme in human cell cultures producing Mason-Pfizer monkey virus-like particles. J Natl Cancer Inst 53(3):751–757

Nelson-Rees WA, Flandermeyer RR, Hawthorne PK (1975) Distinctive banded marker chromosomes of human tumor cell lines. Int J Cancer 16(1):74–82

Nelson-Rees WA, Flandermeyer RR, Daniels DW (1980) T-1 cells are HeLa and not of normal human kidney origin. Science 209(4457):719–720

Nelson-Rees WA, Daniels DW, Flandermeyer RR (1981) Cross-contamination of cells in culture. Science 212(4493):446–452

Niu N, Zhang J, Zhang N, Mercado-Uribe I, Tao F, Han Z, Pathak S, Multani AS, Kuang J, Yao J, Bast RC (2016) Linking genomic reorganization to tumor initiation via the giant cell cycle. Oncogenesis 5(12):e281–e281

Osborne CK, Hobbs K, Trent JM (1987) Biological differences among MCF-7 human breast cancer cell lines from different laboratories. Breast Cancer Res Treat 9(2):111–121

Pathak S (1976) Chromosome banding techniques. J Reprod Med 17:25–28

Pathak S (2007) Cell lines: are they really human? Mammology 3:31–38

Pathak S (2019) Healthy ageing and cancer in humans. In models, molecules and mechanisms in biogerontology. Springer, Singapore, pp 395–410

Pathak S, Hsu TC (1979) Silver-stained structures in mammalian meiotic prophase. Chromosoma 70(2):195–203

Pathak S, Hsu TC (1985) Cytogenetic identification of interspecies cell-line contamination: procedures for non-cytogeneticists. Cytobios 43(171):101–114

Pathak, S., Siciliano, M.J. and Cailleau, R., 1978. Fresh pleural effusion from a patient with breast cancer showing some characteristic HeLa markers. In: 29th annual tissue culture Assoc. meeting, Denver, CO (abstract), 14(4), p. 360

Pathak S, Siciliano MJ, Cailleau R, Wiseman CL, Hsu TC (1979) A human breast adenocarcinoma with chromosome and isoenzyme markers similar to those of the HeLa line. J Natl Cancer Inst 62(2):263–271

Pathak S, Hsu TC, Trentin JJ, Butel JS, Panigrahy B (1981) Nonrandom chromosome abnormalities in transformed Syrian hamster cell lines. Genes, Chromosomes, Neoplasia:405–418

Pathak S, Nemeth MA, Multani AS, Thalmann GN, Von Eschenbach AC, Chung LWK (1997) Can cancer cells transform normal host cells into malignant cells? Br J Cancer 76(9):1134–1138

Pathak S, Nemeth MA, Multani AS (1998) Human tumor xenografts in nude mice are not always of human origin: a warning signal. Cancer: Interdiscip Int J Am Cancer Soc 83(9):1891–1893

Price JE, Wolf JK, Pathak S (1998) Distinctive karyotypes and growth patterns in nude mice reveal cross-contamination in an established human cancer cell line. Oncol Rep 5(1):261–267

Satya-Prakash KL, Pathak S, Hsu TC, Olive M, Cailleau R (1981) Cytogenetic analysis on eight human breast tumor cell lines: high frequencies of 1q, 11q and HeLa-like marker chromosomes. Cancer Genet Cytogenet 3(1):61–73

Seman G, Hunter SJ, Miller RC, Dmochowski L (1976) Characterization of an established cell line (SH-3) derived from pleural effusion of patient with breast cancer. Cancer 37(4):1814–1824

Shafie S, Brooks SC (1977) Effect of prolactin on growth and the estrogen receptor level of human breast cancer cells (MCF-7). Cancer Res 37(3):792–799

Sinha G (2017) The organoid architect. Science 357:746–749

Soule HD, Vazquez J, Long A, Albert S, Brennan M (1973) A human cell line from a pleural effusion derived from a breast carcinoma. J Natl Cancer Inst 51(5):1409–1416

Stacey GN (2000) Cell contamination leads to inaccurate data: we must take action now. Nature 403 (6768):356–356

Whang-Peng, J., Lee, E.C., R. Kao-Shan, C.S., Seibert, K. and Lippman, M., 1983. Cytogenetic studies of human breast cancer lines: MCF-7 and derived variant sublines. J Natl Cancer Inst, 71 (4), pp.687–695

Genomics of Cancer

Avnish Kumar Bhatia

Abstract

The problem of cancer has been increasing rapidly for over two decades. This is now recognized that cancer is a disease of genomic changes. It progresses with alterations in DNA in the form of mutations, copy number alterations, structural variants, integration of host–pathogen genomes, and epigenomic modifications. Various initiatives on cancer genomics have been cataloging these genomic changes to decipher genetics of cancer. Methods of genomic analysis such as SNP array, transcriptomics, next-generation sequencing, and whole-genome sequencing are employed to characterize various classes of cancers. Genomic characterization of cancer disease has the potential to diagnose and treat patients with targeted therapies.

Keywords

Cancer genomics · Cancer genome · Cancer therapy

18.1 Introduction

Consideration of genes participation in cancer development was the main motivation for starting human genome project in the year 1986. With the availability of reference human genome sequence in the year 2002, complete range of somatic alterations causing cancer could be deciphered (Wheeler and Wang 2013). After the publication of first draft of genome assembly in 2002, genome sequencing techniques have advanced to the present state to make it possible to sequence whole genome in days at much smaller cost.

A. K. Bhatia (✉)
ICAR—National Bureau of Animal Genetic Resources, Karnal, Haryana, India

Deoxyribonucleic acid (DNA) is the building block of genome for synthesis of various life forms. DNA is formed of alternating units of deoxyribose sugar and phosphate, and the rings are formed of nitrogenous bases—A, T, G, and C (A—adenine, T—thymine, G—guanine, and C—cytosine) termed as nucleotides. A ring involves a base pair where "G" pairs with "C" and "A" pairs with "T." DNA is capable of encoding information in its nucleotide sequence. A group of three bases called "codon" synthesizes an amino acid. A string of amino acids constitutes the basic structure of protein. The redundancy of the genetic code of possible 64 codons decreases to 20 distinct amino acids. This coding in DNA sequence forms messenger ribonucleic acid (mRNA) through transcription process. The mRNA is a single-stranded form of the gene translated into protein molecule. Proteins are fundamental components of genetic events that occur in the DNA. These are the building blocks of enzymes, hormones, bones, muscles, skin, and blood.

Genome comprises the entire sequence of nucleotides (A, T, G, and C) and is partitioned as chromosome(s) of an organism. Entire nucleotide sequence of an organism can be determined using modern sequencing techniques and subsequent application of sequence assembly algorithms on high-performance computers. Genome contains DNA segments called genes distributed across chromosomes. Genes synthesize various proteins essential for functioning of living organisms. Genes include protein-coding regions known as "exons" and noncoding regions known as "introns." Introns are removed from the sequence during protein-coding process.

Human genome size is equal to approximately 3.2 billion nucleotides or bases (3.2 gigabases) split in chromosomes. Human genome comprises of 23 pairs of chromosomes that include one pair of sex chromosome (X and Y) and 22 pairs of autosomes. The recent version of human genome assembly contains around 22,400 protein-coding genes (Micklos et al. 2013).

All the individuals of a species possess almost identical genomes with variations here and there in genomic sequence. Most of the variations in DNA of an organism are present as single nucleotide polymorphisms (SNPs), copy number variations (CNVs), and short insertions and deletions (indels) of bases. SNP consists of two or more different nucleotides at a site in the genome. Indels are one or a few bases in length. CNVs are several kilobases in length.

SNPs appear in the genome due to mutations in DNA where one nucleotide is converted into another nucleotide. SNPs are classified according to their location and the role in coding amino acids. Noncoding SNPs are located in nontranscribed region (NTR), in untranscribed region (UTR), and in introns. Coding SNPs are located in exons and are classified as *nonsynonymous* if the amino acid is changed due to change in codon or *synonymous* if the codon is modified, but the amino acid remains the same. SNPs can also be classified as *transition* or *transversions* on the basis of nucleotides they change. Transitions change (A to G or C to T) and vice versa, that is, a purine to purine or pyrimidine to pyrimidine. Transversions change (A or G to C or T, and C or T to A or G), that is, a purine to pyrimidine and vice versa.

Cancer disease is caused due to changes in the genome, which involves accumulation of somatic mutations, integration of pathogen genome with human genome, epigenomic changes, structural variants (SVs), and copy number alterations (CNAs). Somatic mutations are alterations in genome that appear in non-germline cells. These are not inherited but influence fitness of individuals. Germline mutations appear in germline cells and are inherited by offspring. All these modifications of genome lead to changes in the expression levels or to altered functions of cancer-related genes and their products as proteins. There have been advances in techniques of next-generation sequencing (NGS) aided by modern computational methods that make it possible to analyze cancer genome profiles. Cancer genomic studies reveal abnormal alterations in genes that cause the development of many types of the disease. Genomic studies have been used to develop new methods of diagnosis and treatment of cancer.

18.2 Methods in Cancer Genomics

A number of methods are employed for deciphering cancer genomes that include SNP array, candidate gene sequencing, exome sequencing, transcriptome analysis, and whole-genome sequencing.

High-density SNP chip data have been used for genome-wide studies on structural variation (SV) in cancer genomes. Genomic DNA in tumor and normal cases are compared using SNP chip. SNP chip/arrays are high-density oligo-arrays involving millions of small size probes. It facilitates genotyping of a number of selected SNPs in the chip across the whole genome in a single reaction. Significant amplification or deletion signals are spotted as continuous portions of SNPs that are in deviation to the normal signal strength. The genes in the neighborhoods of modified signals are resequenced to find mutations. These changes are also tested for any alteration in level of gene expression, possibly correlating with copy number alterations (Mardis and Wilson 2009).

Approaches based on next-generation sequencing have been applied to find structural variants at a level of resolution and complexity greater than other methods. Next-generation sequencing platforms generate paired-end reads, which are aligned to obtain the genome assembly of the species. The genome assembly can be studied to identify putative SVs in candidate cancer genome. Read pairs mapping to genome assembly far apart, close together, across chromosomes indicate possible deletion, insertions, inversions, or translocations, respectively. These procedures are suitable for whole-genome sequencing (WGS) followed by the detection of mutation.

Old methods such as polymerase chain reaction (PCR) and Sanger sequencing discovered small number of genes. These methods have been used to find somatic mutations in genomes. These methods have also been applied to characterize a few hundred genes and the entire exome. Exome refers to collection of all the exons in genome.

Understanding effect of somatic alterations on gene expression in tumor cells has become important procedure for cancer study. Next-generation sequencing of RNA

(RNAseq) of cancerous tissues characterizes complete transcriptome of cancer. Transcriptome is the full range of mRNA transcripts created in a certain cell or type of tissue. It also helps in correlating genomic modifications like structural variants, copy number alterations, SNPs, and indels. RNAseq data-set helps in identification of expression of alleles for known mutations.

Whole-genome sequencing (WGS) is a remarkable outcome of next-generation sequencing for studies on genomics of cancer. It makes it possible to re-sequence and align tumor genome with normal genome sequence. Sequences of samples from a number of patients of a cancer type can be generated due to reduced cost of genome sequencing. It requires data generation and careful use of computational tools for analysis. Genome variation is complex in nature. Therefore, entire range of mutations needs to be considered (Mardis and Wilson 2009). WGS for a number of tumors has been studied for structural variants that have revealed remarkable insights into cancer biology.

Each type of tumor type has a characteristic collection of genes, which are mutated more frequently. Higher frequency of mutation is the first indication of the role of these genes on the type of cancer. Exome screening in colorectal and breast cancers revealed mutation in one to three genes in more than 20% of the tumors. Sequencing in a large number of patients is required for adding new low-frequency genes to the list of significantly mutated genes (Wheeler and Wang 2013).

18.3 Cancer Genome Initiatives and Databases

Advanced techniques of DNA sequencing at continuously decreasing cost have helped in understanding the genetics of cancer. Various initiatives have been taken for genotyping, sequencing, and cataloging a large number of cancer patients. Sequencing of more than fifty thousand cancer genomes has already been completed (Nakagawa and Fujita 2018). A few initiatives in this direction include The Cancer Genome Atlas (TCGA), International Cancer Genome Consortium (ICGC), and OncoArray Network for Studies on Germline Variation.

TCGA carried out molecular characterization of more than 20,000 primary cancer patients for 33 types of cancer that have been compared with normal cases. National Cancer Institute and National Human Genome Research Institute started joint effort in the year 2006. A number of scholars from many institutions and various disciplines have formed teams for joint research in cancer genomics. TCGA analyzed samples from around 11,000 cancer patients for a period of 12 years. The Genome Characterization Pipeline involved collection, characterization, and analysis of enormous data. This included clinical information on smoking status, sample portion weight, and corresponding gene expression.

Yang et al. (2015) describe various cancer genomic data depositories. Zou et al. (2015) also list a few genomic databases and tools available online for cancer. A few databases on cancer genomics are listed in Table 18.1 that includes COSMIC, GOBO, GDC, ICGC, and IARC.

18 Genomics of Cancer

Table 18.1 Repositories for cancer genomic data

Sr. no.	Database name	Institution	Contents
1	Genomic Data Commons (GDC) (https://gdc.cancer.gov)	National Cancer Institute (NCI) of United States; TCGA; Cancer Cell Line Encyclopedia (CCLE); Projects on Therapeutically Applicable Research to Generate Effective Treatment (TARGET)	Mutations, copy number variations, expression quantification, post-transcriptional modifications. Data available for 9 programs, 47 projects, 68 primary sites, 33,605 cases
2	Catalogue of Somatic Mutations in Cancer (COSMIC) (https://cancer.sanger.ac.uk/cosmic)	Welcome Sanger Institute, United Kingdom	COSMIC is the biggest source of information on somatic mutation in human cancers. It is manually curated by experts in the field. It contains over 32,000 cases of reviewed large-scale genome screening data
3	Gene expression-based Outcome for Breast cancer Online (GOBO) (http://co.bmc.lu.se/gobo/)	Lund University, Sweden	The tumor data set consists of 1881 samples of breast cancer
4	International Cancer Genome Consortium (ICGC) (http://icgc.org), data portal (https://dcc.icgc.org/)	ICGC The ICGC secretariat was situated at the Ontario Institute of Cancer Research in Toronto. It has now moved to the University of Glasgow in the United Kingdom	Cancer genomic data sets of 86 cancer projects, 22 cancer primary sites, 22,330 donor with molecular data, 24,289 total donors, 81,782,588 simple somatic mutations.
5	The IARC TP53 database http://p53.iarc.fr/	International Agency for Research on Cancer (IARC) of World Health Organization (WHO)	In the current release, R19 data-set stores information on TP53 gene. It includes somatic mutations in case of sporadic cancers, germline mutation in case of familial cancers, gene polymorphisms, functional assessment of p53 mutant proteins, gene status in human cell lines, mouse models with engineered TP53, and mutations induced by experiments
6	MethHC: A database of DNA methylation and gene expression in human cancer (http://methhc.mbc.nctu.edu.tw/php/index.php)	Department of Biological Science and Technology, Institute of Bioinformatics and Systems Biology, National Chiao Tung University, Hsinchu, Taiwan	MethHC-integrated data such as DNA methylation, microRNA methylation, microRNA expression, and gene expression. It also includes correlation of methylation and gene expression from TCGA

Catalogue of Somatic Mutations in Cancer (COSMIC) contains manually curated data that includes gene curation—details of manual curation process, gene fusion curation—details of curation process for gene fusions, genome annotation—information on the annotation of genomes, drug resistance—curation of mutations conferring drug resistance, and mutational signatures—a census of mutation signatures in cancer.

Gene expression-based Outcome for Breast cancer Online (GOBO) is accessible Web-based database that facilitates speedy calculation of gene expression levels, identification of co-expressed genes, and association of breast cancer with result of a set of genes or their signatures using 1881samples. The database helps in the study of levels of gene expression in breast tumors. It also helps in creation of potential patterns of gene expression termed as metagene/model gene.

The International Cancer Genome Consortium (ICGC) is involved in investigating more than 25,000 cancer genomes with commitments of funding from a number of establishments over the globe. It involves 88 project teams in 17 organizations to examine tumors affecting various body parts.

The International Agency for Research on Cancer (IARC) is maintaining P53 database for recording the pattern of P53 mutations in various categories of tumors since 1989 (Brennan and Wild 2015). The database compiles information on human TP53 gene variations from literature and other databanks.

MethHC incorporates data on DNA methylation, microRNA methylation, gene expression, microRNA expression, and the relationship between methylation and gene expression from the TCGA. It comprises 18 human cancers from over 6000 samples and 6548 microarray and 12,567 RNA sequencing data.

Pan-Cancer Analysis of Whole Genomes (PCAWG) Consortium comprises the ICGC and TCGA. The PCAWG Consortium combined whole-genome sequencing data from 2658 donors in 38 types of cancer from various projects over the globe (PCAWG Consortium 2020). It included 2605 cases of primary tumors and 173 cases of metastases.

18.4 Mutational Signatures

Mutational signatures are unique combinations of mutation types that are associated with particular cancer etiology. These emanate as somatic mutations from a number of mutational processes such as inherent infidelity of DNA replication processes, exposures to mutagen, DNA alterations by enzymes, and faulty repair of DNA (Nakagawa and Fujita 2018).

Mutation patterns in TP53 gene differ significantly in lung cancer between smokers and nonsmokers. Smokers have a higher percentage of mutations with a $G > T$ transversion. These mutations could be caused by activity of some chemical compounds present in the smoke. Small-cell lung cancer (SCLC) generally occurs among smokers. Thirty-two percent of 263 p53 mutations recorded from 253 cases are found as $G > T$ transversions. Also, the patterns of p53 mutations in the genome of a single SCLC tumor are nearly similar to those in the TP53 gene in other SCLC tumors (Brennan and Wild 2015).

Whole genome sequencing (WGS) is useful in detecting a number of somatic single nucleotide variants (SNVs) in cancers. COSMIC database stores over thirty mutational signatures of genomes related to cancers. Attempts have been made to relate these signatures with biological and epidemiological processes. Some of the known mutational signatures are linked with particular mutational processes. Signature 1 denotes a clock-like mutational process and is detected in all types of tumors. Signature 24 is linked to aflatoxin. Signature 22 is associated with aristolochic acid found in Chinese herbal products. Signature 4 is associated with exposure to smoking. Signature 3 is associated with the fault of DNA double-strand break repair related to BRCA1/2 mutation. Signature 6 is associated with DNA-mismatch repair defects. Observations on genomic somatic mutational signatures from WGS data help in deducing the etiological factors for development of specific cancer. This is one important step among many internal etiological steps such as aging and inherent DNA repair, and exterior etiological stages such as environmental exposure (Nakagawa and Fujita 2018).

18.5 Mutation Rate

There is variation in the number of mutations in various types of cancer. Basal cell carcinoma (BCC) with around 2200 mutations and melanoma with around 800 mutations in the coding region are the cancers with the highest mutation rate. Some pediatric cancers have less than ten mutations per tumor as observed in rhabdoid cancer. Glioblastoma multiforme (GBM) contains approximately five times mutations as compared to medulloblastoma. BCC, melanoma, and GBM are commonly found cancers in humans. Some less common cancers are rhabdoid and medulloblastoma. So, there exists a relationship between the mutation rate and the risk of cancer. Accumulation of mutations in somatic cells might be the major reason for tumorigenesis (Hao et al. 2016).

Hao et al. (2016) observed a strong correlation observed between the lifespan risk of cancers and the rate of mutation of the corresponding cancers with a Pearson correlation coefficient equal to 0.72. The ratio of risk of cancer to mutation frequency for various categories of cancers was also determined. Four types of cancers show higher ratios. These are cancers related to hormone such as prostate cancer and breast cancer, cancers related to viruses such as liver cancer having hepatitis C infection and head and neck cancer having HPV-16 infection. The ratios of risk of cancer to mutation frequency are 76 times in prostate cancer and 26 times in breast cancer as compared to the median ratio of all cancers. It shows that non-mutagenic factors such as hormones in these cancer types are more powerful in growing the occurrence of the cancers than the number of accumulated mutations.

Shen et al. (2018) considered 137 primary testicular germ cell tumors (TGCTs) and found somatic mutation rate varying by histology. They used high-dimensional assays of genomic, epigenomic, transcriptomic, and proteomic features. Overall median mutation rate equal to 0.5 mutations per Mb of targeted DNA was higher than the stated mutation rate in pediatric tumors. However, the mutation rate was smaller than many of adult tumors. The frequency of nonsynonymous mutations equal to 0.3 mutations per Mb was like those obtained in other studies of exome sequencing.

18.6 Structural Variants

Cancer genomes exhibit frequent alteration in chromosomal structure. This may be because of deletion, translocation, amplification, and/or inversion of chromosomal sections (Mardis and Wilson 2009). These modifications called structural variations (SVs) affect functional genes in many ways critical to initiation of cancer. Three somatic mutational processes create several mutations causing large rearrangement of the genome. These are chromoplexy, kataegis, and chromothripsis. Chromoplexy is the process through which there is disruption in repair of co-occurring double-stranded DNA. Kataegis is a main hypermutation process that causes bunched nucleotide exchanges inclined toward a single DNA strand. Chromothripsis is the process where many DNA breaks occur simultaneously and arbitrarily stitch together (PCAWG Consortium 2020).

Structural variants result in the development of cancer by affecting genes that include changes in gene copy number, disturbing genes responsible for suppressing tumor, fusing genes, or putting the coding part of a gene next to the regulatory part of the other gene. Around 5–10% of cancers in liver are found to have structural variants resulting in activation of the telomerase gene—TERT (Li et al. 2020).

SVs in leukemia and sarcoma result in creation of cancer-specific fusion genes. The chromosome 2p may have a slight inversion that makes the fusion gene—EML4-ALK. This is found in 1–2% of lung adenocarcinomas. A translocation creates SV with ROS1 gene at chromosome 6q22 and RET1 gene at chromosome 10q11.2 that have been identified in some cancers in lungs with distinctive clinical and pathological characteristics. These SVs yield fusion kinases as driver genes and are targets for molecular studies of lung cancer. About 40–70% of prostate cancers have been found to have SV relating ERG gene at chromosome 21q22 and many genes of ETS family. These SVs produce fusions of TMPRSS2-ERG and ETS family genes (Nakagawa and Fujita 2018). It is to be noted here that a chromosome is made up of a short arm (p) and a long arm (q) separated by centromere. When chromosomal segment is lost from the long arm of chromosome 6, it is termed as 6q deletion. The term 10q11.2 indicates chromosome 10 long arm, band 11, and sub-band 2.

18.7 Copy Number Alterations

Copy number variations (CNVs) are typically several kilobases long and encompass the loss or duplication of whole genes or stretches of genes. CNVs are important elements of genomic diversity like single nucleotide polymorphisms (SNPs). Some CNVs do not affect phenotype, while others have been conclusively linked with disease.

Copy number alterations (CNAs) are somatic alterations to the arrangement of chromosome resulting in increase or decrease in copies of fragments of DNA (Bierly 2019). Large changes in genomic segments (10 kb and more) due to CNAs are common signs of cancer genomes. These changes result in activation of cancer genes and inactivation of genes that suppress tumor (Nakagawa and Fujita 2018).

Computational tools for whole-exome sequencing (WES) exhibit low resolution to identify CNA. WES involves sequencing of all the regions in genes that code proteins. Exome is the collection of all the protein-coding regions in the genome. It may be difficult to notice specific CNA as exons are not part of a number of repeated CNA regions. Whole-genome sequencing (WGS) can analyze CNA by counting reads mapped to genomic regions in DNA sampled from diseased and normal cases.

Some examples of focal amplification of CNA-linked cancer genes and tumor-suppressing genes are 8q24.21 for MYC gene, 11q13.3 for CCND1 gene, 7p11.2 for EGFR gene, 17q12 for ERRB2 gene, and 7q31.2 for MET gene. Examples of focal deletion include RB1 gene—13q14.2, CDKN2A gene—9p21.3, and PTEN gene—10q23.31. It requires identification of the cancer gene and tumor-suppressing gene targeted by the driver CNA. CNAs cover a number of genes and thus explain their functional role in genome. Besides protein-coding genes, overexpression of some cancer genes such as KLF5 and MYC is caused by CNAs in the form of copy number gains of noncoding regions near these genes.

Expression of estrogen (ER), progesterone (PR), and human epidermal growth factor receptor 2 (HER2) is absent in triple-negative breast cancer (TNBC). Nedeljković et al. 2019 determined the impact of FGFR1 and c-MYC genes copy number alterations on outcome of TNBC. Analysis of 78 TNBC samples showed that there was increase in FGFR1 copy number in 34(43%) samples and c-MYC copy number in 39 (50%) samples.

18.8 Carcinogenic Pathogens

Viral and bacterial infections are major reasons for the development of some cancer types. Almost 20 percent of cancers are caused by such infections. Carcinogenic pathogens includes seven viruses—hepatitis C virus (HCV), hepatitis B virus (HBV), human T-cell lymphoma virus 1 (HTLV-1), human papillomavirus (HPV), Epstein–Barr virus (EBV), Merkel cell polyomavirus (MCPyV), and human herpesvirus 8 (HHV8); one bacterium—Helicobacter pylori; and three parasites—Schistosoma haematobium, Clonorchis sinensis, and Opisthorchis viverrini. Each of three, viruses—HBV, HPV, and the bacterium—cause 5% of all cancers. HBV causes hepatocellular carcinoma; HPV causes cervical cancer; and H. pylori causes stomach cancer (Vandeven and Nghiem 2014).

Viruses transform protein expression or noncoding RNAs (ncRNAs). They alter cellular biology of key cells (Cantalupo et al. 2018). Viral oncogenes activate and repress signaling pathways that change expression of genes in cells. Viral DNA gets integrated into human DNA leading to the development of cancer.

Cancer genomes might include DNA sequences derived from pathogens leading to inflammation. The technique of whole-genome sequencing (WGS) helps in detecting integration of pathogens to the host genome. Metagenome analysis of gut flora deciphers the interaction of genome and environment in tumor development. It also helps to understand resistance of therapy for tumors in digestive tract. This includes Fusobacterium causing colorectal cancer and Gammaproteobacteria causing pancreatic cancer (Nakagawa and Fujita 2018).

Cantalupo et al. (2018) searched TCGA databases for five virus families. HPV occurred in 98.8 percent of cancer of cervix, some cancers in head and neck, and a few cases of cancer in bladder. HBV or HCV was linked to some cancers of liver. EBV was found in about 20% of stomach cancers.

18.9 Cancer Epigenomics

Epigenetics describes alterations in gene expression, which are not related to the DNA sequences. Epigenetic processes include DNA methylation, histone modification, and regulatory RNAs.

DNA methylation occurs when methylase enzyme adds a methyl group to cytosine nucleotide. Typically, methylation in a promotor presents binding by transcription factor. This does not allow transcription initiation, resulting in silencing of gene. Patterns of methylation can be inherited. The process of inheritance of an epigenetic effect through a parent is termed "imprinting."

General modifications in DNA methylation occurring in cancer cells include demethylation within a number of sections of genome in line with de novo methylation of particular CpG islands. CpG islands are defined as DNA segments of 500–1500 bp size having CG: GC ratio above 0.6. The change occurs on a number of CpG islands that are generally not methylated in each tissue. It appears to be a common process based on a universal mechanism facilitated by polycomb complex. Polycomb is a protein complex, which inhibits expression by causing local heterochromatinization, the process by which euchromatin converts to heterochromatin at the time of inactivation of X chromosomes during embryogenesis. Out of more than 13,000 essentially unmethylated CpG islands in the human genome, almost 2000 are marked with polycomb. This complex may be causing methylases, DNMT3A and DNMT3B in tumors that might cause the abnormal modification (Klutstein et al. 2016).

Su et al. (2018) analyzed 4174 genomic profiles that included whole-genome bisulfite sequencing (WGBS) data from 30 normal tissues and 35 tumors in seven types of cancer. This thorough analysis revealed role of hypermethylation for the activation of homeobox cancer genes.

DNA methylation level is measured as methylation portion at a locus as the unit of beta value in the range (0, 1). For non-seminomatous germ cell tumors (NSGCTs), the total distribution of beta values at recognized CpG sites was bimodal. Peaks in the distribution appeared at unmethylated and methylated CpGs, which is peculiar to most of the primary human tissue samples. But the peak for methylated site was missing in seminomas, which showed midway peaks of DNA methylation along with the unmethylated peak. Samples of seminoma included wholly unmethylated cells and the cells with full methylation at a few loci (Shen et al. 2018).

Histone acetylation modulator proteins (HAMPs) are families of the primary protein that facilitate the alteration and recognition of histone acetylation. The genes for epigenetic regulation are highly modified in cancers, as observed in genomic studies. This indicates the role of certain HAMPs as driver genes during development of cancer. Expression analysis demonstrated universal expression of most HAMPs in all types of cancer (Hu et al. 2019). Various types of cancer could be differentiated by the HAMP expression signatures. It grouped the tumors with related lineage origins together. It indicates importance of cancer type-dependent acetylation status during treatments to target histone acetylation in cancer patients.

Tumors in embryonal carcinoma (EC) were marked by the large expression of a number of microRNAs (miRNAs). Expression of the miRNA–miR-519 genomic group on 19q13.42 appeared 25–50 times higher in EC compared with seminoma and 300–600 times higher compared with other categories of non-seminomatous germ cell tumors (NSGCTs). The miRNAs in this genomic cluster have also been found expressed in embryonic stem cells (ESCs) in other studies. These miRNAs might harmfully regulate the expression of mRNA in EC as their targets show lower expression in EC (Shen et al. 2018).

18.10 Future Directions

Genomic characterization is such an important aspect of cancer diagnosis and treatment that the term "cancer genome" is coined for DNA structure of cancerous cells/tissues. The genomic studies on cancer have led to documentation of biomarkers for diagnosis of cancer.

The occurrence of cell-free DNA (cfDNA) is observed at high levels in cancer cases. Some of cfDNA in cancer cases is circulating tumor DNA (ctDNA), which has been termed as "liquid biopsy" (Brennan and Wild 2015). The fraction of ctDNA compared to the amount of cfDNA may be above 10% in patients with cancer at late stage, whereas the ratio may be 0.1–1% for the disease at early stage.

Panels of microRNAs (miRNAs) from affected tissue may also be possible biomarkers. In lung cancer screening studies, there are reports of panels of a small number of miRNAs found in blood for subsequent risk.

Results from genomic studies on cancer can guide us to devise strategies for specific therapy. There is possibility of reprogramming the epigenome into a hypomethylated state and make immunogenicity by the DNA methyltransferase inhibitors in NSGCT (Shen et al. 2018).

Mutation rate in tumor cells is a trustworthy predictor for cancer occurrence (Hao et al. 2016). Mutations observed in a cancer provide a record of accumulation of mutations due to division of stem cell over the progression of self-renewal and due to environmental or genetic factors.

Mutation signatures indicate toward the cumulative outcome of damage and repair processes for DNA in cancer. It can also show possible responses to a course of therapies without identifying an underlying mutation (Editorial 2019). It helps in exploiting cancer for the development of effective immune therapies. Representative solid tumors hold 30–70 mutations in the affected genes resulting in change in the amino acid sequences of the encoded proteins. These genetic alterations provide opportunity for the development of tumor-specific antigens that could be combined with existing platforms for cancer immune therapy (Vogelstein et al. 2013).

There are nearly 200 types of cancers classified according to organs of origin. Assuming four subclasses for each type of cancer, it makes 800 subtypes. Genomic studies have refined classification of cancer. There were seven subtypes of cancer from microRNA data, five subtypes from copy number alterations, and five subtypes of cancer from methylation data in a study on breast cancer (Song et al. 2015).

The precision medicine requires use of genomics to match patients to targeted therapies. Clinical predictors can be built from knowledge banks of genomic data of thousands of patients with sufficient clinical characterization (PCAWG Consortium 2020). As a result of generation of vast genomic data on cancer, artificial intelligence (AI) techniques are helping to deliver personalized care (Xu et al. 2019). There are several AI tools for targeting mutations, genes, protein–protein interactions, disease treatment, etc.

The best management for cancer will be based on the components of the genome of tumor. Cancer genome studies can be used to improve methods for prevention and early detection of the disease. The preventive care will be needed to decrease cancer morbidity and mortality (Vogelstein et al. 2013).

References

Bierly, A. (2019) Somatic mutations and copy number changes in cancer: finding the right targets. https://www.qiagen.com/us/spotlight-pages/newsletters-and-magazines/articles/reviews-online-copy-number-alteration/ (Accessed June, 2019)
Brennan P, Wild CP (2015) Genomics of cancer and a new era for cancer prevention. PLoS Genet 11(11):e1005522
Cantalupo PG, Katz JP, Pipas JM (2018) Viral sequences in human cancer. Virology 513:208–216
Editorial (2019) Advancing cancer genomics. Nat Genet 51:767
Hao D, Wang L, Di LJ (2016) Distinct mutation accumulation rates among tissues determine the variation in cancer risk. Sci Rep 6(1):1–5
Hu Z, Zhou J, Jiang J, Yuan J, Zhang Y, Wei X, Loo N, Wang Y, Pan Y, Zhang T, Zhong X (2019) Genomic characterization of genes encoding histone acetylation modulator proteins identifies therapeutic targets for cancer treatment. Nat Commun 10:733

Klutstein M, Nejman D, Greenfield R, Cedar H (2016) DNA methylation in cancer and aging. Cancer Res 76(12):3446–3450

Li Y, Roberts ND, Wala JA, Shapira O, Schumacher SE, Kumar K, Khurana E, Waszak S, Korbel JO, Haber JE, Imielinski M (2020) Patterns of somatic structural variation in human cancer genomes. Nature 578(7793):112–121

Mardis ER, Wilson RK (2009) Cancer genome sequencing: a review. Hum Mol Genet 18(R2): R163–R168

Micklos DA, Hilgert U, Nash B (2013) Genome science: a practical and conceptual introduction to molecular genetic analysis in eukaryotes, vol 1. Cold Spring Harbor Laboratory Press, Cold Spring Harbor, NY

Nakagawa H, Fujita M (2018) Whole genome sequencing analysis for cancer genomics and precision medicine. Cancer Sci 109(3):513–522

Nedeljković M, Tanić N, Dramićanin T, Milovanović Z, Šušnjar S, Milinković V, Vujović I, Prvanović M, Tanić N (2019) Importance of copy number alterations of FGFR1 and C-MYC genes in triple negative breast cancer. J Med Biochem 38(1):63

PCAWG Consortium (2020) Pan-cancer analysis of whole genomes. Nature 578(7793):82–93

Shen H, Shih J, Hollern DP, Wang L, Bowlby R, Tickoo SK, Thorsson V, Mungall AJ, Newton Y, Hegde AM, Armenia J, Sánchez-Vega F, Pluta F, Pyle LC, Mehra R, Reuter VE, Godoy G, Jones J, Shelley CS, Feldman DR, Vidal DO, Lessel D, Kulis T, Ca'rcano FM, Leraas KM, Lichtenberg TM, Brooks D, Cherniack AD, Cho J, Heiman DI, Kasaian K, Liu M, Noble MS, Xi L, Zhang H, Zhou W, ZenKlusen JC, Hutter CM, Felau I, Zhang J, Schultz N, Gad Getz G, Meyerson M, Stuart JM, The Cancer Genome Atlas Research Network, Akbani R, Wheeler DA, Laird PW, Nathanson KL, Cortessis VK, Hoadley KA (2018) Integrated molecular characterization of testicular germ cell tumors. Cell Rep 23(11):3392–3406

Song Q, Merajver SD, Li JZ (2015) Cancer classification in the genomic era: five contemporary problems. Hum Genomics 9:27

Su J, Huang YH, Cui X, Wang X, Zhang X, Lei Y, Xu J, Lin X, Chen K, Lv J, Goodell MA (2018) Homeobox oncogene activation by pan-cancer DNA hypermethylation. Genome Biol 19:108

Vandeven N, Nghiem P (2014) Pathogen-driven cancers and emerging immune therapeutic strategies. Cancer Immunol Res 2(1):9–14

Vogelstein B, Papadopoulos N, Velculescu VE, Zhou S, Diaz LA, Kinzler KW (2013) Cancer genome landscapes. Science 339(6127):1546–1558

Wheeler DA, Wang L (2013) From human genome to cancer genome: the first decade. Genome Res 23(7):1054–1062

Xu J, Yang P, Xue S, Sharma B, Sanchez-Martin M, Wang F, Beaty KA, Dehan E, Parikh B (2019) Translating cancer genomics into precision medicine with artificial intelligence: applications, challenges and future perspectives. Hum Genet 138(2):109–124

Yang Y, Dong X, Xie B, Ding N, Chen J, Li Y, Zhang Q, Qu H, Fang X (2015) Databases and web tools for cancer genomics study. Genomics Proteomics Bioinform 13(1):46–50

Zou D, Ma L, Yu J, Zhang Z (2015) Biological databases for human research. Genomics Proteomics Bioinforma 13(1):55–63

Diabetes and Cancer

19

Abhijit Chanda

Abstract

Diabetes mellitus and cancer are two heterogeneous diseases, incidence of both of which is increasing. Epidemiological evidences suggest that cancer coexists with diabetes more often than by chance. Many experimental and observational studies have shown that there may be certain biologic links between these two diseases like obesity, hyperinsulinaemia, oxidative stress and cytokines. Some of the chemotherapeutic agents used to treat cancer can deteriorate glycaemic status. Likewise, some of the anti-diabetics may have some effect on cancer cell proliferation. Proper management of diabetes is an essential component of treatment in cancer patients.

Keywords

Hyperinsulinaemia · Hyperglycaemia · Obesity · Cancer

19.1 Introduction

According to WHO, the number of people with diabetes has risen from 108 million in 1980 to 422 million in 2014. Along with this, the burden of microvascular and macrovascular complications of the disease has also risen over the last few decades. Apart from the known complications of diabetes, it has also been observed that diabetics are more prone to suffer from cancer. Both diabetes and cancer are heterogeneous diseases, and there are multiple possible biologic links between these two diseases like obesity, hyperinsulinaemia, oxidative stress, cytokines and hormonal factors. Cancer mortality also seems to be higher in diabetics than that in

A. Chanda (✉)
Medica Superspecialty Hospital, Kolkata, West Bengal, India

their non-diabetic counterparts. Some of the anticancer drugs can cause hyperglycaemia. On the other hand, there are some evidences that certain antidiabetic drugs can also have some impact on cancer progression.

19.2 Association between Diabetes and Cancer: What Are the Evidences?

Most of the studies linking diabetes and cancer have been done in type 2 diabetes. Commonest cancer in diabetes is hepatocellular carcinoma (El-Serag et al. 2006). Other cancers which are seen more frequently in diabetics are those of pancreas, endometrium, urinary bladder, breast and colorectum. Interestingly, the risk of prostate cancer is reduced in diabetes. However, patients with type 2 diabetes have a lower level of PSA (prostate-specific antigen) (Fukui et al. 2008). In the REDUCE trial (Wu et al. 2011), a multicentre, double-blinded, placebo-controlled trial relationship between diabetes and prostate cancer was assessed. All subjects underwent prostate biopsy irrespective of PSA level. It was shown that prostate cancer is no less common in diabetics in comparison with that in non-diabetics. Possibly, diabetics undergo prostate biopsy less often due to their lower PSA values and as a result prostate cancer is detected less often. Cancer risk in type 1 diabetes is different from that in type 2 diabetes. In a Swedish cohort study, it was shown that type 1 diabetics are at an increased risk of gastric, endometrial and cervical cancer (Zendehdel et al. 2003). Increased incidence of gastric cancer in type 1 diabetes is due to higher prevalence of Helicobacter pylori infection and pernicious anaemia. Increased incidence of uterine malignancy is due to nulliparity, irregular menstrual cycle.

19.3 Increased Cancer Mortality in Diabetes

It has been seen that mortality of at least some of the cancers like breast, colorectal and endometrial is increased in diabetes. Hazard ratio for death in cancer in diabetes is 1.41 (95% CI = 1.28–1.55) in comparison with non-diabetics (Barone et al. 2008). However, diabetes itself is associated with higher risk of mortality. Whether excess mortality seen in diabetic patients with cancer is due to diabetes itself or more aggressive cancer progression in these patients is not known. Moreover, diabetes patients are often less aggressively treated for cancer due to the presence of comorbidities like cardiac and renal diseases. This might also indirectly increase the cancer mortality in diabetics.

19.3.1 Possible Biologic Links

Cancer and diabetes both being heterogeneous disorders, there may be multiple mechanisms linking these two disorders. Hyperinsulinaemia, hyperglycaemia and obesity are possible biologic links.

19.3.2 Insulin/IGF 1 Axis

Many cancer cells express insulin receptor. Insulin receptor has two isoforms, namely isoform A and isoform B. Isoform A is responsible for insulin-induced mitogenesis, whereas isoform B is involved in metabolic process. Cancer cells mainly express the A isoform which is likely to be involved in cancer cell proliferation. The question which arises in this context is if type 2 diabetes is associated with insulin resistance how can it cause cancer cell proliferation. It has been seen that even in presence of insulin resistance the isoform A receptor pathway (mitogenic pathway) continues to function and the resistance is mainly restricted to isoform B pathway (metabolic pathway). Insulin may also act through IGF1 receptor, especially when present at a high level as IGF1 receptor shares 80% homology with the insulin receptor. IGF1 receptor possesses more mitogenic activity than the insulin receptor. Hyperinsulinaemia also reduces the hepatic production of IGF-binding proteins namely IGFBP1 (Powell et al. 1991) and possibly IGFBP2 (Renehan et al. 2006). As a result of the reduction in its binding protein, circulating free IGF1 level rises. IGF1 is a more potent mitogen than insulin and can act through its own receptor, that is IGF1 receptor, insulin receptor and their hybrid receptor. Cancer cells have been shown to express all these receptors abundantly. Thus, IGF1 also seems to play an important role in cancer cell proliferation.

19.3.3 Hyperglycaemia and Cancer

Cancer cells mostly derive energy from glycolysis rather than oxidative phosphorylation (Warburg hypothesis) (Vander Heiden et al. 2009). Glycolysis requires much more glucose than oxidative phosphorylation for ATP generation. So cancer cells have a high requirement for glucose for their survival and proliferation. So, it is likely that hyperglycaemia will promote cancer cell growth. However, it is also true that cancer cells have a very efficient glucose uptake mechanism independent of insulin. So, even in the absence of hyperglycaemia cancer cells get adequate glucose supply. Moreover, in alloxan-induced diabetic rats where beta cells have been destroyed to produce type 1 diabetes (insulin deficient) there is mammary tumour regression (Heuson and Legros 1972). Insulin treatment reversed this effect (Heuson et al. 1972).

19.3.4 Obesity and Cancer

Obese people are more prone to suffer from cancer (Adami and Ttrichopoulos 2003). This might be related to hyperinsulinaemia which is usually seen in obesity. Obesity is also associated with increased oestrogen level and breast cancer risk (Key et al. 2003). Oestrogen level increases in obesity due to increase in aromatase enzyme activity, the enzyme which is responsible for conversion of testosterone to oestradiol. Leptin is a cytokine released from adipocyte, the level of which increases in obesity. High leptin level promotes breast cancer cell proliferation (Hu et al. 2002). In oestrogen-dependent neoplasms like breast and endometrial cancer, leptin can stimulate aromatase enzyme and thus help to stimulate cancer cell proliferation (Boden et al. 1996). Adiponectin is another cytokine derived from adipocytes level of which is inversely associated with body fat content. Addition of adiponectin to breast cancer cell line has been shown to inhibit proliferation of cancer cells (Cleary et al. 2009). Epidemiological studies have also shown reduced cancer incidence with higher adiponectin level (Mantzoros et al. 2004).

19.3.5 Chronic Inflammation

Diabetes is a pro-inflammatory state and associated with increased reactive oxygen species (ROS). This oxidative stress can cause DNA damage or interfere with DNA repair, thus promoting carcinogenesis. Inflammation also increases levels of cytokines like TNF-α and IL-6. They also induce cell proliferation and inhibit apoptosis.

19.4 Effect of Anti-Diabetic Drugs on Cancer

As hyperinsulinaemia is considered to be an important factor for cancer cell proliferation, it was speculated that drugs which increase endogenous insulin or exogenous insulin administration might increase the risk of cancer. In fact, few epidemiological studies showed increased risk of malignancy with the use of insulin glargine (Jonasson et al. 2009; Colhoun 2009). However, the ORIGIN trial did not show any excess risk of cancer with glargine after a median follow-up of 6.2 years (Gerstein et al. 2012). Similarly, although sulphonylureas especially the older ones like glibenclamide were shown to be associated with increased risk of cancer in some observational studies, there were small number of cases of cancer in most of the studies and as a result the power was limited to examine association with cancer of specific sites. There are studies which refute such associations (Yang et al. 2010).

Thiazolidinediones do not increase insulin secretion, but few years back some studies showed increased risk of urinary bladder cancer with the use of pioglitazone (Zhu et al. 2012). However, many of these studies had other confounding factors or detection bias. Although the association between pioglitazone and urinary bladder cancer is not clear, it is better to avoid this drug in patients with active bladder cancer, who give history of bladder cancer or who are at an increased risk of bladder cancer.

Incretin-based therapies (GLP1 analogues and DPP 4 inhibitors) are also not associated with any significant risk of cancer in humans. Metformin is one of the most time tested anti-diabetic drugs. There are experimental and clinical evidences that metformin may reduce cancer incidence or progression. Metformin use was associated with better pathologic response after neoadjuvant chemotherapy in a study on diabetes with early breast cancer (Jiralerspong et al. 2009). In vitro studies have shown that metformin can inhibit breast cancer (Zakikhani et al. 2006), prostate cancer (Sahra et al. 2008) and ovarian cancer (Gotlieb et al. 2008) cell line proliferation. However, it should be noted that this inhibitory effect is observed at a much higher level of concentration of metformin than that is achieved after usual pharmacologic dosing in diabetes (Chong and Chabner 2009). It should also be kept in mind that metformin is the first-line agent used in type 2 diabetes. This is usually given at an early stage of the disease and at a younger age when complications are usually not seen. Cancer incidence is also more at an advancing age. This might be one of the reasons why metformin is associated with lower incidence of cancer. In this context, it must be remembered that it is always difficult to assess the association between one specific anti-diabetic drug and risk of cancer. Type 2 diabetes being a progressive disease, most patients require multiple drugs and the number of drugs keeps on increasing as the disease progresses. Hence, it is almost impossible to assess long-term outcome with a single drug. While choosing an anti-diabetic drug, if possible, it is preferable to choose an agent which will not produce hyperinsulinaemia. However, the risk of cancer should not be a major consideration while choosing an anti-diabetic agent unless there is specific risk like presence of urinary bladder cancer while using pioglitazone or pancreatic cancer while using incretin-based therapies.

19.5 Effect of Anticancer Drugs on Glucose Level

A lot of anticancer drugs can adversely affect diabetic control by either increasing insulin resistance or reducing insulin secretion from the beta cells. Glucocorticoids increase insulin resistance and increase glucose levels, and this effect is more pronounced in diabetics. Androgen deprivation therapy in prostate cancer increases insulin resistance and cardiovascular mortality. Many new anticancer drugs like tyrosine kinase inhibitors or mTor inhibitors like everolimus inhibit insulin receptor/IGF1 receptor or downstream signalling steps in the pathway of insulin action. This increases insulin resistance and causes hyperglycaemia. Cytotoxic T-lymphocyte-associated antigen 4 (CTLA-4) antibodies can rarely cause type 1 diabetes due to immune-mediated beta-cell failure (Gonzalez-Rodriguez and Rodriguez-Abreu 2016). Abdominal irradiation can give rise to diabetes later on. With allogeneic haematopoietic stem cell transplantation, the risk of diabetes can be threefold higher compared to that in sibling donors (Scott et al. 2007). As cancer survivors are at an increased risk of developing diabetes, periodic monitoring of blood glucose should be done.

19.6 Management of Diabetes in Cancer Patients

Certain observational studies reported worse outcome in cancer patients in whom diabetes remains uncontrolled (Lee et al. 2016). Uncontrolled blood glucose level is associated with more cardiovascular events and infection in cancer patients. Although metformin has been shown to arrest the progression of cancer in experimental studies, in most of the cancer patients metformin might not be enough to control blood glucose adequately. In view of frailty, poor intake, chemotherapy-induced increase in blood glucose and limited efficacy and flexibility of non-insulin-based therapies insulin seems to be the safest and best available anti-diabetic in cancer patients with uncontrolled hyperglycaemia. Blood glucose target should be individualized keeping in consideration the life expectancy. In patients with short life expectancy, glycaemic control may be relaxed.

19.7 Conclusion

Diabetes and cancer are both common diseases associated with significant morbidity and mortality. Diabetics are more prone to get certain cancers, and cancer mortality is also increased in diabetic patients. There are evidences of possible biologic links between these two diseases. Hyperinsulinaemia, adipocyte-derived cytokines, oxidative stress play important role linking these two diseases. Better control of blood glucose levels improves efficacy of chemotherapy. Insulin seems to be the best option among all anti-diabetics to achieve better glycaemic control although there are evidences showing metformin to play an important role in preventing cancer cell proliferation. On the other hand, cancer chemotherapy increases blood glucose level which may adversely affect their efficacy in controlling tumour progression. Glycaemic control needs to be individualized keeping in mind the life expectancy of the patient.

References

Adami HO, Trichopoulos D (2003) Obesity and mortality from cancer. N Engl J Med 348(17): 1623–1624
Barone BB, Yeh HC, Snyder CF, Peairs KS, Stein KB, Derr RL, Wolff AC, Brancati FL (2008) Long-term all-cause mortality in cancer patients with preexisting diabetes mellitus: a systematic review and meta-analysis. JAMA 300(23):2754–2764
Boden G, Chen X, Mozzoli M, Ryan I (1996) Effect of fasting on serum leptin in normal human subjects. J Clin Endocrinol Metabol 81(9):3419–3423
Chong CR, Chabner BA (2009) Mysterious metformin. Oncologist 14(12):1178–1181
Cleary MP, Ray A, Rogozina OP, Dogan S, Grossmann ME (2009) Targeting the adiponectin: leptin ratio for postmenopausal breast cancer prevention. Front Biosci (Schol Ed) 1:329–357
Colhoun HM (2009) Use of insulin glargine and cancer incidence in Scotland: a study from the Scottish diabetes research network epidemiology group. Diabetologia 52(9):1755–1765
El-Serag HB, Hampel H, Javadi F (2006) The association between diabetes and hepatocellular carcinoma: a systematic review of epidemiologic evidence. Clin Gastroenterol Hepatol 4(3): 369–380
Fukui M, Tanaka M, Kadono M, Imai S, Hasegawa G, Yoshikawa T, Nakamura N (2008) Serum prostate-specific antigen levels in men with type 2 diabetes. Diabetes Care 31(5):930–931
Gerstein HC, Bosch J, Dagenais GR, (ORIGIN Trial Investigators) et al (2012) Basal insulin and cardiovascular and other outcomes in dysglycemia. N Engl J Med 367(4):319–328

González-Rodríguez E, Rodríguez-Abreu D (2016) Spanish Group for Cancer Immuno-Biotherapy (GETICA). Immune checkpoint inhibitors: review and management of endocrine adverse events. Oncologist 21(7):804–816

Gotlieb WH, Saumet J, Beauchamp MC, Gu J, Lau S, Pollak MN, Bruchim I (2008) In vitro metformin anti-neoplastic activity in epithelial ovarian cancer. Gynecol Oncol 110(2):246–250

Heuson JC, Legros N (1972) Influence of insulin deprivation on growth of the 7, 12-dimethylbenz (a) anthracene-induced mammary carcinoma in rats subjected to alloxan diabetes and food restriction. Cancer Res 32(2):226–232

Heuson JC, Legros N, Heimann R (1972) Influence of insulin administration on growth of the 7, 12-dimethylbenz (a) anthracene-induced mammary carcinoma in intact, oophorectomized, and hypophysectomized rats. Cancer Res 32(2):233–238

Hu X, Juneja SC, Maihle NJ, Cleary MP (2002) Leptin—a growth factor in normal and malignant breast cells and for normal mammary gland development. J Natl Cancer Inst 94(22):1704–1711

Jiralerspong S, Palla SL, Giordano SH, Meric-Bernstam F, Liedtke C, Barnett CM, Hsu L, Hung MC, Hortobagyi GN, Gonzalez-Angulo AM (2009) Metformin and pathologic complete responses to neoadjuvant chemotherapy in diabetic patients with breast cancer. J Clin Oncol 27(20):3297

Jonasson J, Ljung R, Talbäck M, Haglund B, Gudbjörnsdòttir S, Steineck G (2009) Insulin glargine use and short-term incidence of malignancies—a population-based follow-up study in Sweden. Diabetologia 52(9):1745–1754

Key TJ, Appleby PN, Reeves GK, Roddam A, Dorgan JF, Longcope C, Stanczyk FZ, Stephenson HE Jr, Falk RT, Miller R, Schatzkin A, Allen DS, Fentiman IS, Key TJ, Wang DY, Dowsett M, Thomas HV, Hankinson SE, Toniolo P, Akhmedkhanov A, Koenig K, Shore RE, Zeleniuch-Jacquotte A, Berrino F, Muti P, Micheli A, Krogh V, Sieri S, Pala V, Venturelli E, Secreto G, Barrett-Connor E, Laughlin GA, Kabuto M, Akiba S, Stevens RG, Neriishi K, Land CE, Cauley JA, Kuller LH, Cummings SR, Helzlsouer KJ, Alberg AJ, Bush TL, Comstock GW, Gordon GB, Miller SR, Longcope C, Endogenous Hormones Breast Cancer Collaborative Group (2003) Body mass index, serum sex hormones, and breast cancer risk in postmenopausal women. J Natl Cancer Inst 95(16):1218–1226. https://doi.org/10.1093/jnci/djg022. PMID: 12928347

Lee W, Yoon YS, Han HS, Cho JY, Choi Y, Jang JY, Choi H (2016) Prognostic relevance of preoperative diabetes mellitus and the degree of hyperglycemia on the outcomes of resected pancreatic ductal adenocarcinoma. J Surg Oncol 113(2):203–208

Mantzoros C, Petridou E, Dessypris N, Chavelas C, Dalamaga M, Alexe DM, Papadiamantis Y, Markopoulos C, Spanos E, Chrousos G, Trichopoulos D (2004) Adiponectin and breast cancer risk. J Clin Endocrinol Metab 89(3):1102–1107

Powell DR, Suwanichkul A, Cubbage ML, DePaolis LA, Snuggs MB, Lee PD (1991) Insulin inhibits transcription of the human gene for insulin-like growth factor-binding protein-1. J Biol Chem 266(28):18868–18876

Renehan AG, Frystyk J, Flyvbjerg A (2006) Obesity and cancer risk: the role of the insulin–IGF axis. Trends Endocrinol Metab 17(8):328–336

Sahra IB, Laurent K, Loubat A, Giorgetti-Peraldi S, Colosetti P, Auberger P, Tanti JF, Le Marchand-Brustel Y, Bost F (2008) The antidiabetic drug metformin exerts an antitumoral effect in vitro and in vivo through a decrease of cyclin D1 level. Oncogene 27(25):3576–3586

Scott Baker K, Ness KK, Steinberger J, Carter A, Francisco L, Burns LJ, Sklar C, Forman S, Weisdorf D, Gurney JG, Bhatia S (2007) Diabetes, hypertension, and cardiovascular events in survivors of hematopoietic cell transplantation: a report from the bone marrow transplantation survivor study. Blood 109(4):1765–1772

Vander Heiden MG, Cantley LC, Thompson CB (2009) Understanding the Warburg effect: the metabolic requirements of cell proliferation. Science 324(5930):1029–1033

Wu C, Moreira DM, Gerber L, Rittmaster RS, Andriole GL, Freedland SJ (2011) Diabetes and prostate cancer risk in the REDUCE trial. Prostate Cancer Prostatic Dis 14(4):326–331

Yang X, So WY, Ma RCW, Yu LWY, Ko GTC, Kong APS, Ng VWS, Luk AOY, Ozaki R, Tong PCY, Chow C-C, Chan JCN (2010) Use of sulphonylurea and cancer in type 2 diabetes – the Hong Kong diabetes registry. Diabetes Res Clin Pract 90:343–351

Zakikhani M, Dowling R, Fantus IG, Sonenberg N, Pollak M (2006) Metformin is an AMP kinase–dependent growth inhibitor for breast cancer cells. Cancer Res 66(21):10269–10273

Zendehdel K, Nyrén O, Östenson CG, Adami HO, Ekbom A, Ye W (2003) Cancer incidence in patients with type 1 diabetes mellitus: a population-based cohort study in Sweden. J Natl Cancer Inst 95(23):1797–1800

Zhu Z, Shen Z, Lu Y, Zhong S, Xu C (2012) Increased risk of bladder cancer with pioglitazone therapy in patients with diabetes: a meta-analysis. Diabetes Res Clin Pract 98(1):159–163

Oncology Informatics, AI, and Drug Discovery

20

Debarpita Santra

Abstract

Cancer is the deadliest disease across the globe. Statistics says that one in three individuals is affected by this disease. The available treatment options for cancer always do not ensure effectiveness in providing complete protection from the disease. The emergent technologies in the domains of bioinformatics, proteomics, genetics, etc., help in better understanding of different phenotypes of cancer, roles of different cellular genes, and regulatory genetic elements. These technologies intend to offer better treatment efficacies in cancer treatment, which may be greatly enhanced with the help of artificial intelligence and information technology. The techniques of artificial intelligence, in collaboration with the information technology, help in rapid analysis of huge oncology data, extraction of important features from the data, reliable prediction of cancer, drug discovery and design for oncology, and many other issues. These advancements result in increased survival rates and improved quality of life for the patients suffering from cancer.

Keywords

Cancer · Informatics · Artificial intelligence · Drug discovery

---

D. Santra (✉)
Department of Computer Science and Engineering, University of Kalyani, Kalyani, West Bengal, India

© The Author(s), under exclusive license to Springer Nature Singapore Pte Ltd. 2022
S. K. Basu et al. (eds.), *Cancer Diagnostics and Therapeutics*,
https://doi.org/10.1007/978-981-16-4752-9_20

20.1 Introduction

Cancer has significantly high recurrence rate but low median survival rate, with the treatment process being time-consuming and very expensive. Lowering the mortality rate of cancer patients requires appropriate early diagnosis and prognosis. The domain of oncology has seen much advancement in recent years with emergence of stem cell therapy, personalized immunotherapy, targeted therapy, gene analysis, and so on for offering better treatment options for cancer. This rapid progress in oncology continues through following a multidisciplinary approach that involves artificial intelligence (AI) and information technology (IT). Application of AI and IT in oncology for improving the workflow processes and the outcomes in cancer care is termed as oncology informatics. The oncology informatics help in rapid analysis of radiological images with extraction of important features from the images, classification of histopathological data, and analysis of other relevant clinical data and associated assessment. AI helps in reliable prediction of cancer and also in cancer diagnosis and prognosis with greater accuracy. These advancements result in better survival rates and enhanced quality of life for the cancer patients (Basch et al. 2017; Denis et al. 2017).

AI can also be effectively used in discovery of precision drugs for cancer patients. The AI methodologies used for predicting the molecular behavior and the likelihood for obtaining a useful drug help in drastic reduction in the cost and time for cancer treatment. The AI-based process of drug discovery integrates the knowledge about a variety of cancer drugs, drug resistance, sequencing of next-generation drugs, and so on. AI can also be used for prediction of anticancer drug activity or assisting in the development of anticancer drugs. As the same drugs may have varied impacts on different types of cancers, information gathered from the high-throughput screening procedures discloses the relationship between genomic variability of cancer cells and the drug activity. This chapter gives a brief description of oncology informatics and cancer drug discovery using AI. Before this discussion, a brief introduction about AI is given in Sect. 20.2. Section 20.3 gives an overview of oncology informatics, and Sect. 20.4 discusses briefly how AI can be used in drug discovery. Section 20.5 concludes the chapter.

20.2 Introduction to AI

AI was introduced in the year 1956 with the aim to build intelligent computer programs that can analyze complex problems and can deliver solutions to the problems at the level of human experts (Minsky 1961). AI has diverse applications in medicine, spanning from medical diagnosis, medical statistics, human biology to drug discovery, and developing new drug delivery system. There are mainly two branches for AI in medicine: virtual and physical (Hamet and Tremblay 2017). The virtual branch focuses on the medical informatics that deals mainly with constructing, analyzing, and managing electronic health records, as well as developing decision support systems for assisting physicians in proper diagnosis and management of the disease. The physical branch focuses on the development or use of robots for assisting the elderly patients or the surgeons during complicated surgeries. *Da Vinci*, a surgical system approved by the Food and Drug Administration in the USA in the year 2000 for facilitating complex prostatectomies and gynecologic surgical procedures as well as repairing cardiac valve, has already performed more than 5000 operations around the world through minimally invasive techniques and being controlled by a surgeon from a console. The development of drug delivery system also comes under this physical branch of AI in medicine.

The virtual branch of AI in medicine is basically the design of mathematical algorithms that learn from experience over the time, and enhance the knowledge bases. These algorithmic methodologies, called the machine learning or, more specifically, the deep learning techniques, are categorized as supervised learning, unsupervised learning, and reinforcement learning. For supervised learning, a dataset is required that contains input and output labels. From this dataset, a computer program learns in the same way a student learns under the supervision of a teacher. More specifically, during supervised learning, a mathematical function is developed for mapping between the input data pairs and the output labels. On the contrary, unsupervised learning process requires a dataset having no given input and output labels. The unsupervised learning algorithms help to identify the hidden patterns in the form of clusters or specific regions within the dataset. In case of reinforcement learning, there is a software agent who takes actions while interacting with a given environment. The software agent may be thought as a clinician, and the environment may be thought as a clinical treatment. Deep learning is a special branch of machine learning, which uses neural network architecture having more than two hidden layers. The deep learning algorithms have greater excellence in solving pattern recognition and computer vision problems, compared to the classical machine learning approaches.

The physical branch of AI in medicine focuses on the development and use of medical devices and robotic systems for surgical assistance or their solo performance in robotic surgery, as well as drug delivery to targeted organs, tumors, or tissues. But, this drug delivery system sometimes faces difficulty of diffusion of therapeutic agent into the targeted tumor(s), when the tumor is less vascularized, anoxic, but most proliferative in nature. Emergence of nanorobots in this field overcomes the issue. As an intelligent but natural nanoparticle, researchers have identified *magnetococcus marinus*, a special kind of marine coli, that travels through low-oxygenated zones. The nanoparticles are first activated using external magnetic source, and then, the intrinsic properties of the nanoparticles come into action for the entire delivery. These nanorobots are covalently associated with nanoliposomes that carry the therapeutic properties; This leads to a significant increase in the drug delivery into the low-oxygenated zone. The following two sections provide an overview about oncology informatics and the drug delivery.

20.3 Oncology Informatics

Nowadays, transitioning from paper to electronic mode is of significant importance for efficiently archiving and analyzing every patient record covering his/her medical history, examination or pathological test results, diagnosis and treatment decisions, medical administrative information, billing details, insurance information, etc. Manual data entry is very time-consuming and requires huge manpower (Putora et al. 2020). These issues and challenges are overcome with the help of oncology informatics.

Oncology informatics offers an easier means for tracking information of patients. There are numerous stakeholders for oncology informatics such as the caregivers, patients, administrators, and insurance companies, and many others who have diverse data requirements (e.g., data synthesis and data aggregation). The caregivers and administrators need to monitor overall status of every patient including the treatment strategies, patient's response to treatment, adverse impacts of treatment, and information regarding clinical trials. Besides providing support for the collection of relevant information for today's clinical practices, the oncology informatics could be integrated easily with the future advancements in this cancer domain.

Oncology informatics has diverse applications in radiation and surgical oncology. It is effectively used in classification of histopathological data and image processing for radiologic diagnostics and treatment. The use of oncology informatics not only results in the efficacy of oncological procedures, but also increases the survival rates of cancer patients, reduces treatment toxicity, and improves the quality of life for the patients suffering from cancer (Basch et al. 2017; Denis et al. 2017). Also, the shift from offline data (paper) to digitized version helps in extraction of relevant information from electronic clinical records, which could be helpful for predicting clinical outcomes and identifying clinical trial eligibility (Savova et al. 2019). Moreover, the informatics helps in finding out the patterns coming out the digitized information to improve the diagnosis of new cases of cancer and to classify the existing cases based on important features. AI-based technologies in this context, especially the deep learning approaches, have been successfully used in detecting breast cancer, lung cancer, and skin cancers through radiomic analysis of images. These approaches also assist clinicians effectively in decision making (Haibe-Kains et al. 2020). These decision-making methodologies about complex oncological therapies have been further enhanced to incorporate patient preference and experience of physicians, enabling oncology informatics to offer personalized multidisciplinary treatment algorithms.

With the increment in number of biomarkers and the therapeutic options, the human decision-making process for personalized and precision medicine becomes quite difficult and challenging. The AI-based methodologies aid in developing decision support systems in this regard integrating the clinical data, imaging data, biologic, genetic, and cost-related information. The decision support systems conclude with numerous care pathway decisions having optimal efficacy and economy.

Oncology informatics using AI is being actively used in real world. For instance, IBM Watson for Oncology (WFO) has shown great efficacy in making treatment recommendations for patients suffering from cancer (Tseng et al. 2020). WFO has the capability to learn from the extensive knowledge about breast cancer, acquired from a plethora of medical journals, textbooks, and treatment protocols at Memorial Sloan Kettering Cancer Centre (MSKCC). Based on the clinical features of specific cancer patients, the WFO identified the sources of knowledge with the help of a natural language processing mechanism. WFO has also been trained with the data of more than 550 breast cancer cases in MSKCC. The data encompass a number of variables such as patient characteristics, functional status, comorbidities, tumor characteristics, stage, and pathological findings (Somashekhar et al. 2018). The treatment recommendations of WFO get further refined through its enhanced capability for analytical reasoning with accepting the feedback provided by the domain experts. WFO finally comes out with treatment planning such as surgery, immunotherapy, chemotherapy, radiotherapy, and also alternate options within every treatment plan (e.g., drugs or doses) for particular patient. The performance of the system was tested at the Manipal Comprehensive Cancer Centre and has shown great efficacy offering a high concordance (93%) with a multidisciplinary tumor board (Somashekhar et al. 2018).

Imaging diagnosis of a large variety of cancers has been performed with greater efficacy through deep learning algorithms like convolutional neural network, compared to the human expert physicians. An application of convolution neural network was found in distinguishing between the most common and deadliest variations of skin cancers, in which the convolution neural network was trained with a dataset of 12,940 clinical images. The deep learning-based approach outperformed two dermatologists, given the subset of the same validation set (Esteva et al. 2017). In another challenge competition for developing an automated solution to detect lymph node metastasis in breast cancer, the algorithm that was found to be most efficient was developed using deep learning-based approach using convolution neural network. The developed algorithm achieved more accuracy, compared to the diagnostic performance of 11 pathologists in a simulated exercise that mimics a conventional clinical workflow. The AI-based well-trained algorithms not only provide precise diagnosis with significant prediction accuracy, but also offer high efficiency delivering the outputs within a fraction of second and cost-effectiveness making them accessible universally from anywhere, anytime.

AI also acts as an effective tool for molecular targeting through a better understanding of the complicated relationships among genetics and other biological information. CancerSEEK (Cohen et al. 2018), a tool for blood test, can identify eight common types of cancer at the very early stages of the disease and detect the anatomical origins of cancer through assessment of the levels of circulating proteins and the mutations inside the DNA. With early detection of cancers, this kind of test diminishes the rate of deaths due to cancer.

AI is also helpful in analyzing and extracting quantitative features from the radiological images of oncology. This special domain of informatics, called radiomics, uses the digital data extracted from these images to develop diagnostic, predictive, or prognostic models for better understanding the underlying biological phenomena, aiding in personalized medical decisions, and optimizing the individualized treatment planning.

20.4 AI-Assisted Drug Discovery

The process of drug discovery in use till now, in general, is a laborious, time-consuming, and noneffective approach (Vamathevan et al. 2019). Selection of right targets is the main focus of drug discovery. But, according to the literature, one in ten drugs becomes able to enter the phase I clinical trials and reaches to the patients. Advent of AI-based approaches, especially the deep learning-based techniques, has shown great efficiency in the drug discovery for oncology. Overcoming the lack of availability of sufficient data for training deep learning algorithms, the researchers have now been capable to form extensive databases and train the algorithms accordingly. These algorithms can be applied in different stages of drug discovery such as validation of targets, identification of the prognostic biomarkers, and review of the digitized pathological data during clinical trials. As a result, identification of novel targets, finding stronger evidences for target–disease associations, understanding the underlying mechanisms of disease, conceptualizing the disease and nondisease phenotypes, development of novel prognostic biomarkers, etc., become much easier and better.

Target identification and validation is an important stage of drug discovery. Target identification refers to the selection of targets that bind to drugs based on available evidences. Based on the initial choice of the targets, target validation process starts, in which the role of the selected target is validated using ex vivo and in vivo models. The deep learning-based methods have potential applicability in target identification. With the first step in target identification being the establishment of a causal relationship between the target and the disease, the learning algorithms are used here for analyzing huge datasets for making predictions about likely causality, driven by the properties of true targets. In this context, a metaclassifier was designed by Costa et al. (2010) which was based on decision tree, for prediction of the druggable genes related to morbidity. The metaclassifier was trained on network topologies representing metabolic and transcriptional interactions, protein–protein interactions, etc. In another study by Jeon et al. (2014), support vector machine classifier was developed from a variety of genomic datasets for classifying proteins into nondrug and drug targets for three types of cancer such as ovarian, pancreatic, and breast cancers. Classification of drug and nondrug targets was based on key features like essentiality of genes, DNA copy number, expression of mRNA, occurrence of mutation, and the network topology for protein–protein interaction. This classification methodology was found to be helpful for identifying 266, 462, and 355 targets for breast, pancreatic, and ovarian cancers, respectively. Among these targets, two were validated with peptide inhibitors having strong antiproliferative capability in cell culture models. Molecular targets for antiaging therapies specific to tissues were identified by Mamoshina et al. (2018). In this study, comparative analysis of the signatures of gene expression for young and old muscles was done. Effects of cancer-specific drugs can be predicted through machine learning algorithms. In a study by Iorio et al. (2016), 990 cancer cell lines against 265 anticancer drugs were screened. In addition to it, it was investigated how the genome-wide gene expression, methylation of DNA, and gene copy number put impact on drug response. The target identification and validation predict the success of future clinical trials for target-based programs for drug discovery.

During drug discovery, blocking or activating the target protein of interest involves extensive experimental and virtual screening of huge number of compounds. Further refinement and modification of the structures of the target proteins are done for improvement of target selectivity and specificity, and also for optimization of the pharmacokinetic, pharmacodynamic, and the toxicological properties. The deep learning methods are helpful in finding out and designing therapeutically important molecules. Design of molecules can be either structure-based or ligand-based. Structure-based methods require extensive knowledge about the structures of both the ligand and the target, including the knowledge about molecular dynamics, protein–ligand docking, and the mechanisms for determining the free energy of binding. Ligand-based mechanisms require only the information regarding the ligand for predicting the biological response based on the historical data about the ligands that are known to be active or inactive. There exists a lot of work for ligand-based virtual screening using deep learning approaches. Provided a lead compound, the compounds having similar chemical structure can be computationally detected. Though classical statistical methods are useful in this case, deep learning-based approaches have been proved to be more effective. Literature reports combining machine learning algorithms to Markov state models for identification of previously unfamiliar mechanisms for binding of opiate to the µ-opioid receptor, thereby helping reveal an allosteric site involved in the activation of the receptor (Farimani et al. 2018).

Another important application of deep learning, in association with the emergent tree search algorithms in the field of AI, is the planning of efficient routes for chemical synthesis of a target molecule. The plan requires the formal decomposition of the target molecule in a chain of reactions, so that the chain can be executed in the forward direction in laboratory-based experiment for synthesis of the target. During this process, the challenge lies in systematic application of the knowledge about synthetic chemistry on the process. With the rapid and exponential growth of knowledge of chemistry, there are several constraints in manually incorporating the transformation rules. This issue has been addressed in a study by Segler et al. (2018) with execution of a Monte Carlo tree search algorithm in association with the deep learning approach for efficient extraction of the rules from the Reaxys database having almost 11 million reactions and 300,000 rules. The qualitative tests for predicted synthesis routes have been performed double-blindly, in which the organic chemists were to select between the synthesis routes which are predicted and the literature-based routes, without making the chemists informed about the mechanisms behind obtaining the routes. The chemists evaluated the quality of the predicted routes, on average, as good as the literature-based rotes.

The machine learning, especially the deep learning-based approaches, also significantly contributes to identifying the right drug for the right patient (Li et al. 2015; van Gool et al. 2017; Kraus 2018). Moreover, these algorithms are very much helpful in improving the clinical success rates through discovering the biomarkers and developing the drug sensitivity prediction models. With the late-stage clinical trials being very expensive and time-consuming, use of preclinical or early-stage clinical trial data for building, validating, and applying predictive models earlier is very advantageous in this context.

Tissue analysis is also an important part of drug development, in which the deep learning-based algorithms are appropriately used for discovery of novel biomarkers in a reproducible, precise, and high-throughput manner, thereby reducing the time for drug development and for giving patients quicker access to the therapies beneficial to them. With the number of clinical trials being increased day by day, discovery of novel biomarkers becomes very important to find out the patients meant for particular therapies. Before selection of a candidate for clinical trial, it is required to know the impact of drug treatments on selected tissues and cells; therefore, thousands of compounds need to be tested. In a study by Lee et al. (2017), computational analysis has been done on the benign tissue adjacent to the tumor in case of prostate cancer, and the obtained information from the analysis reveals important information relevant for progression-free survival, which is missed or ignored typically by the pathologists. In another study by Mani et al. (2016), numerous markers for lymphocytes helped in better understanding the heterogeneity of the populations suffering from breast cancer. In the experiments similar to the previous one, lot of tissue requires to be experimented with scrutinizing thousands of features and huge number of cell–cell interactions. In this case, machine learning is used for making the experiments easier and time-efficient.

Though AI has a versatile prospect to become a good tool for drug discovery, the main challenge lies in obtaining the datasets on which the learning algorithms would be trained. Some big pharma companies have taken initiative in sharing their own huge preclinical datasets archived since the 1980s or earlier. But these datasets are, in many cases, unstructured in nature, thereby requiring them to be digitized in order to be useful for drug discovery and other purposes (Smalley 2017).

20.5 Conclusions

AI, being associated with IT, is being used in different industrial settings and in everyday life. The utility of machine learning and deep learning algorithms in the fields of oncology informatics and at all the stages of drug discovery are gaining extreme popularity day by day for making better diagnostic and management decisions, as well as for offering enhanced therapeutics through identification of novel biomarkers and reducing the side effects of the therapies. The multidisciplinary approaches would surely enhance the quality of care for the patients suffering from cancer. Further, the insights gained from the approaches would be helpful in addressing the complicated challenges ahead in the domain of oncology. With the progressive development of multidisciplinary approach for providing better healthcare solutions for oncology, the existing technologies are expected to continue to evolve and mature over the time.

References

Basch E, Deal AM, Dueck AC, Scher HI, Kris MG, Hudis C, Schrag D (2017) Overall survival results of a trial assessing patient-reported outcomes for symptom monitoring during routine cancer treatment. JAMA 318(2):197–198

Cohen JD, Li L, Wang Y, Thoburn C, Afsari B, Danilova L, Douville C, Javed AA, Wong F, Mattox A, Hruban RH (2018) Detection and localization of surgically resectable cancers with a multi-analyte blood test. Science 359(6378):926–930

Costa PR, Acencio ML, Lemke N (2010) A machine learning approach for genome-wide prediction of morbid and druggable human genes based on systems-level data. BMC Genomics 11(5):1–15. BioMed Central

Denis F, Lethrosne C, Pourel N, Molinier O, Pointreau Y, Domont J, Bourgeois H, Senellart H, Trémolières P, Lizée T, Bennouna J (2017) Randomized trial comparing a web-mediated follow-up with routine surveillance in lung cancer patients. J Natl Cancer Inst 109(9). https://doi.org/10.1093/jnci/djx029

Esteva A, Kuprel B, Novoa RA, Ko J, Swetter SM, Blau HM, Thrun S (2017) Dermatologist-level classification of skin cancer with deep neural networks. Nature 542(7639):115–118

Farimani AB, Feinberg E, Pande V (2018) Binding pathway of opiates to μ-opioid receptors revealed by machine learning. Biophys J 114(3):62a–63a

Haibe-Kains B, Adam GA, Hosny A, Khodakarami F, Waldron L, Wang B, McIntosh C, Goldenberg A, Kundaje A, Greene CS, Broderick T (2020) Transparency and reproducibility in artificial intelligence. Nature 586(7829):E14–E16

Hamet P, Tremblay J (2017) Artificial intelligence in medicine. Metabolism 69:S36–S40

Iorio F, Knijnenburg TA, Vis DJ, Bignell GR, Menden MP, Schubert M, Aben N, Gonçalves E, Barthorpe S, Lightfoot H, Cokelaer T (2016) A landscape of pharmacogenomic interactions in cancer. Cell 166(3):740–754

Jeon J, Nim S, Teyra J, Datti A, Wrana JL, Sidhu SS, Moffat J, Kim PM (2014) A systematic approach to identify novel cancer drug targets using machine learning, inhibitor design and high-throughput screening. Genome Med 6(7):1–18

Kraus VB (2018) Biomarkers as drug development tools: discovery, validation, qualification and use. Nat Rev Rheumatol 14(6):354–362

Lee G, Veltri RW, Zhu G, Ali S, Epstein JI, Madabhushi A (2017) Nuclear shape and architecture in benign fields predict biochemical recurrence in prostate cancer patients following radical prostatectomy: preliminary findings. Eur Urol Focus 3(4–5):457–466

Li B, Shin H, Gulbekyan G, Pustovalova O, Nikolsky Y, Hope A, Bessarabova M, Schu M, Kolpakova-Hart E, Merberg D, Dorner A (2015) Development of a drug-response modeling framework to identify cell line derived translational biomarkers that can predict treatment outcome to erlotinib or sorafenib. PLoS One 10(6):e0130700

Mamoshina P, Volosnikova M, Ozerov IV, Putin E, Skibina E, Cortese F, Zhavoronkov A (2018) Machine learning on human muscle transcriptomic data for biomarker discovery and tissue-specific drug target identification. Front Genet 9:242

Mani NL, Schalper KA, Hatzis C, Saglam O, Tavassoli F, Butler M, Chagpar AB, Pusztai L, Rimm DL (2016) Quantitative assessment of the spatial heterogeneity of tumor-infiltrating lymphocytes in breast cancer. Breast Cancer Res 18(1):1–10

Minsky M (1961) Steps toward artificial intelligence. Proc IRE 49(1):8–30

Putora PM, Baudis M, Beadle BM, El Naqa I, Giordano FA, Nicolay NH (2020) Oncology informatics: status quo and outlook. Oncology 98(6):329–331

Savova GK, Danciu I, Alamudun F, Miller T, Lin C, Bitterman DS, Tourassi G, Warner JL (2019) Use of natural language processing to extract clinical cancer phenotypes from electronic medical records. Cancer Res 79(21):5463–5470

Segler MH, Preuss M, Waller MP (2018) Planning chemical syntheses with deep neural networks and symbolic AI. Nature 555(7698):604–610

Smalley E (2017) AI-powered drug discovery captures pharma interest. Nat Biotechnol 35:604–605

Somashekhar SP, Sepúlveda MJ, Puglielli S, Norden AD, Shortliffe EH, Kumar CR, Rauthan A, Kumar NA, Patil P, Rhee K, Ramya Y (2018) Watson for oncology and breast cancer treatment recommendations: agreement with an expert multidisciplinary tumor board. Ann Oncol 29(2):418–423

Tseng HH, Wei L, Cui S, Luo Y, Ten Haken RK, El Naqa I (2020) Machine learning and imaging informatics in oncology. Oncology 98(6):344–362

Vamathevan J, Clark D, Czodrowski P, Dunham I, Ferran E, Lee G, Li B, Madabhushi A, Shah P, Spitzer M, Zhao S (2019) Applications of machine learning in drug discovery and development. Nat Rev Drug Discov 18(6):463–477

van Gool AJ, Bietrix F, Caldenhoven E, Zatloukal K, Scherer A, Litton JE, Meijer G, Blomberg N, Smith A, Mons B, Heringa J (2017) Bridging the translational innovation gap through good biomarker practice. Nat Rev Drug Discov 16(9):587–588

Radiomics: Cropping More from the Images

21

Sounak Sadhukhan

Abstract

To reduce the variability in cancer diagnosis from radiological images, quantitative methods are introduced in radiology. In recent years, quantitative imaging biomarkers are augmented with the visual assessment by the radiologists that produces a better outcome. The advancement in machine learning processes gives an extra mileage to the rise of radiomics as it can extract quantitative information from the medical images. In this chapter, we focus on the methodology and the critical issues in the workflow to develop a clinically meaningful radiomics signature.

Keywords

Radiomics · Quantitative image analysis · Machine learning · Deep learning

21.1 Introduction

Radiomics is an emerging translational field of research that finds relationships between qualitative and quantitative features extracted from clinical images and data using characterization algorithms, with or without associated gene expression to support evidence-based clinical decision-making (Gillies et al. 2016).

In the past few decades, medical imaging has been used as a tool for detecting, staging, and evaluating treatment response of cancer. It has a great ability to visualize cancer's appearance, such as macroscopic tumor heterogeneity for primary as well as metastatic tumor in noninvasive way. However, its potential for precision

---

S. Sadhukhan (✉)
Department of Computer Science, Institute of Science, Banaras Hindu University, Varanasi, Uttar Pradesh, India

medicine is not fully utilized as the medical practitioners have been drawn their conclusions by perceiving and recognizing image patterns and the past experiences. Hence, a degree of subjectivity and variability exists in medical image interpretation (Papanikolaou et al. 2020). To reduce the variability, the concept of quantitative evaluation of medical imaging has come into the picture to investigate the images and record more objective measurements. On the other hand, clinicians often profile tumors through invasive biopsy and molecular assay to capture its biological diversity and disease evolution (Bedard et al. 2013). However, repetitive tumor sampling is expensive as well as burdensome for patients. While biopsies often produce erroneous results as they capture the heterogeneity within a small portion of the tumor, radiomics captures the heterogeneity of entire volume of a tumor (Mayerhoefer et al. 2020). Researchers also found that the extracted features through radiomics are strongly correlated with heterogeneity at the cellular level (Moon et al. 2019; Choi et al. 2016). It also provides the expression of the genotype, the tumor microenvironment, and susceptibility of the cancer treatments (Mesci et al. 2019; Sala et al. 2017) and easily identifies the substantial phenotypic differences in tumors (Gillies et al. 2016; Hosny et al. 2018).

The extracted radiomic features may predict patterns of tumor evolution, progression, treatment responses, and survival chances of the patient with the combination of genomic, transcriptomic, or proteomic characteristics (Gillies et al. 2016; Sala et al. 2017; Yip and Aerts 2016). This prediction can be done through mining algorithms from the sufficiently large historical data. The datasets are mostly unstructured from different sources like, computed tomography (CT), positron emission tomography (PET), and magnetic resonance imaging (MRI) (Rizzo et al. 2018), which may be combined with the clinical, laboratory, histological, and genomic data. The data mining algorithms accept the medical images and other unstructured data as input, carefully check the quality, and then find the optimal parameters as reliable and robust output (Mayerhoefer et al. 2020).

However, a lot of challenges exist in the multiphase radiomics processes. Nowadays, scanners are mainly designed for acquiring images rather than quantitative image analysis (Papanikolaou et al. 2020). Most of the acquired images lack desired microscopic resolutions; also, some are affected by ionizing radiation exposure. As a result, the outcome cumulatively varies significantly. Moreover, extraction of quantitative parameters is time-consuming process. Therefore, it is very clear that improvements are required regarding reproducibility and diagnostic accuracy, which are the main driving forces of radiomics. In this chapter, an overview of this multistep process is presented; specifically, the considerations and the best practices for developing radiomics signature are discussed that produce clinical value to cancer patients.

The rest of this chapter elaborates the processes of cancer cell motility and metastasis in detail and is organized as follows: Section 21.2 gives the idea of the workflow of radiomics; Sections 21.3–21.10 illustrate different phases of radiomics; Section 21.11 concludes the chapter.

21.2 Radiomics Process Cascade

Radiomics is a medical image investigation process that consists of multiple phases (Fig. 21.1). At the primary stage, a group of oncologists define a clinical problem that the developed model is taken care of. They have also decided that what types of images are preferred to develop the radiomics model. After that, image acquisition and preprocessing part have been done optimally and it should be ensured that the quality of images is high enough for image analysis. If the size of the data is very large, then deep learning techniques may useful over the conventional machine learning processes. At last, these data are used for training and validation purposes.

From the next subsection, we are going to discuss every phase of the radiomics in detail.

21.3 Defining the Clinical Problem

In the primary phase of radiomics, an appropriate clinical problem is defined by a group of physician. The problem must fulfill some specified criteria. It should have addressed an unfulfilled need in cancer treatment, and the successful radiomic signature could perform a treatment plan for the patient.

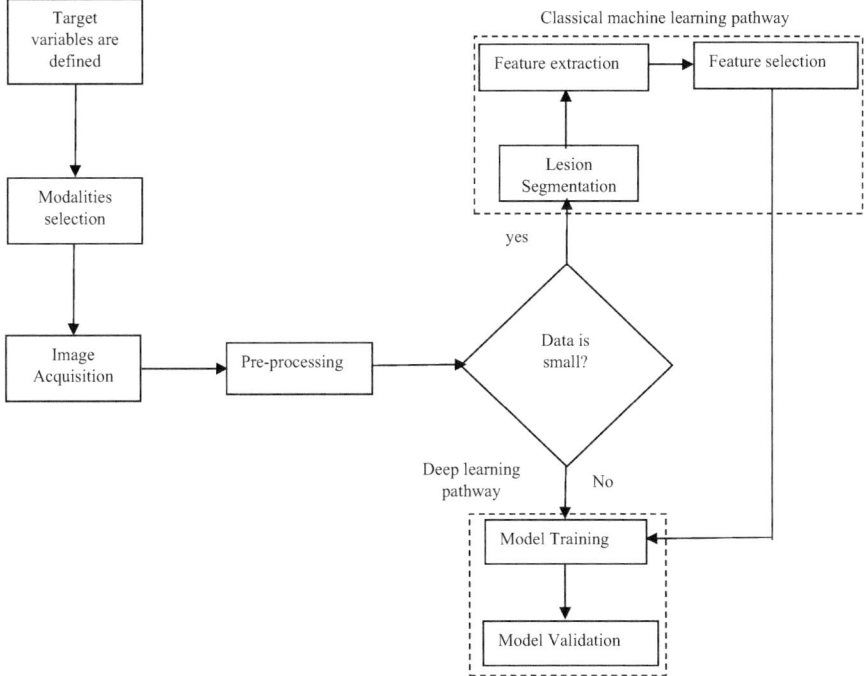

Fig. 21.1 Different phases of radiomics

The size, quality, and diversity of the data are very important for justification of the specified hypothesis. In today's world, although imaging data are quiet easily available, still it is very difficult to collect and curate high-quality comprehensive image data. The complexity of the clinical problem is the major criteria to decide the minimum number of patients that are required to develop a radiomics signature (Kumar et al. 2012). For example, in disease detection problems, image contrast resolution is the sufficient factor to distinguish the abnormal tissue from the image data. As the problem is straightforward, it requires fewer patient's record. However, forecasting of treatment response or prediction of disease-free survival is more complex problems and these require a large number of patient records. According to Chalkidou et al. (2015), in two class classification problem, only 10–15 patient records are required for every features to train the model. If the developed radiomics model is based on small data, then there are high chances of over fitting and instability. The available data should be divided into 60:40 or 70:30 ratio for developing a radiomics model where 60–70% of the data are used for training purpose and rest 40–30% of data are used for validating/testing purpose.

To develop a radiomics signature, the sample size is an important criterion. Sample to develop radiomics signature defines the group of patients or representatives of the target population in terms of occurrence of the disease. The diseases that occur with lower possibility in a society require a large sample size to develop radiomics signature that is useful to prediction tool. In this context, a priori knowledge of the class size of different subtypes of the diseases is needed that will be classified by the analysis of radiomics signature. This would be helpful to observe whether the investigation produces the useful information or not. If the investigation is not useful, then sample or the group of patients should be improved with sufficient cases of each subtype of the disease. However, in practical scenario, several subtypes may often consist of a limited number of cases for radiomics analysis. Therefore, it is an important task to validate the classification model with the available dataset before analyzing the full performance analysis of the system.

21.4 Image Modalities for Radiomics

One of the main issues in image analysis is the absence of any standard guideline for image acquisition. Therefore, high variability exists in quantitative imaging (Raunig et al. 2015; O'Connor et al. 2017). Moreover, all the imaging modalities available in markets are developed only for generating images but not for assessing the images quantitatively. These imaging instruments are designed for radiologist to assess the images through their perceptions. Therefore, the quantitative assessment varies due to human perception. As the devices, made by the different companies, are not able to quantitative analysis of images, this leads to serious problems in treatment. The reason behind this fact is the quantitative measurement of images varies due to the usage of different hardware and software platform in the scanner. Furthermore, hybrid imaging systems are now available in the market, which produce images with varying contrasts. Hence, a careful selection of the imaging modalities is an important issue for getting meaningful outcomes. For an example, CT data are less varied than MRI data, which is the main reason for wide usage of CT images (Papanikolaou et al. 2020).

21.5 Image Acquisition

It is an impossible task to know every specific problem regarding the image acquisition a priori. Therefore, a standard protocol should be required for image acquisition for ensuring accuracy, repeatability, and reproducibility. As the level of standardization is not known initially, hence, a trial and error procedure is required to estimate such requirement. Primarily, it is advised to be more liberal for collecting images with a certain amount of heterogeneity. If it is not able to locate useful features in the collected data, then certain restrictions are imposed like company, sequence parameters, MRI field strength, etc. as the standardization.

Image acquisition protocol changes with time due to the competency of scanners that is getting improved due to upgradation and updation (Park and Kim 2018). The model is trained based on the group of heterogeneous patients' data obtained throughout the year, while testing part is done with the recently acquired exam data. In this way, the model becomes robust although the method needs a higher number of patient data. In the case of multicentric or in the event of multicompany single center studies, mainly two strategies are followed. Either use the model train with the data from one site (or company) and test with data from other sites (or companies) or use mixed data for performing training as well as validation. Effective strategies between these two are decided by basis of the highest performance and generalization.

21.6 Preprocessing of Images

Preprocessing in radiomics is required for removing noise and artifacts in the data. It enhances quality of the data to improve the outcome of the model (Kontos et al. 2017). Several filtering mechanisms have been used to reduce the noise in the images. However, these filters may cause information loss. Therefore, one should use these filters carefully. For example, motion correction methods can handle 4-D data like diffusion weighted images or dynamic contrast-enhanced MRI and CT images of lung by removing patient motion from the images collected from different modalities. But the use of filters and motion correction techniques should be used on the images as a last option in the image acquisition phase. MRI images often suffer from the heterogeneous spatial signals that cannot represent the properties of biological tissue but the patient body habitus, the geometric characteristics of external surface coils, the imperfect profile of rf pulse, and etc. Therefore, to remove such heterogeneity, bias field correction algorithms can be applied (Song and Zhang 2019). Normalization must be a part of preprocessing, while the collected images are from the different modalities in which an underlying heterogeneity exists in the data.

Normalization of signals is bringing the signal intensities into a common range without altering differences in the range of values. It improves the stability and performance of the model and decreases the training duration (Um et al. 2019).

21.7 Lesion Segmentation

The next phase of the radiomics cascade is lesion segmentation, in which image segmentation is executed on the one or all the fragments of the target lesion. The basic methodology includes the tracing of lesion boundary manually that involves high variability that leads to derivation of unstable radiomic features. To build a stable model, more stable features are required. One of the ways to identify stable features is to execute radiological segmentations on the same lesion at least twice. Then, these two lesions are analyzed using correlation analysis to recognize the stable features (Peerlings et al. 2019).

Nowadays, researchers have been trying to automate the lesion segmentation process. The automation of lesion segmentation reduces the data variability and the efforts of radiologists that lead to analyze large dataset easier. However, verification of the final segmented outcomes by the certified radiologists is crucial. Recently, scientists preferred the delineation approach that resects physiologically distinct regions in the tumor based on functional imaging measurement (especially for MRI). From each of these distinct parts (or regions), radiomic features are extracted. In recent studies, it is also found that radiomic features of peritumoral spaces also produce unique information that reflects the treatment effects and disease outcomes.

21.8 Feature Extraction

One of the objectives of the radiomics is to perform complete evaluation of the lesion images using automated data extraction algorithms. These algorithms determine a several quantitative features that capture a large number of variety of phenotypic traits. These features are divided into two groups: semantic feature and agnostic features (Gillies et al. 2016). Semantic features are used for describing lesion by measuring diameters, volume, and morphology. However, agnostic features are the extracted quantitative measures that are not included in radiologists' lexicon. Agnostic features in the image data are first, second, or even higher-order statistical determinants, shape-based feature, and fractal features. These features are recognized by the computational procedures.

The first-order statistics is often use to represent voxel values by ignoring their spatial relations. The mean and median of the voxel values denote the overall tumor intensity or density or variations. There exist some features that represent shape- or location-specific features that identify the shape characteristics in 3-D space. However, the second-order statistics represents the spatial relationships among the voxels of image contrast. These are mainly referred to as texture features (Depeursinge et al. 2014), which are represented as repetitive elements or patterns on the surface with the properties of brightness, color, size, and shape. Wavelet and Laplacian of Gaussian filters are applied on the images to enrich intricate patterns than often cannot quantify through the naked eye.

In deep learning pathway, deep features can be extracted through the use of deep neural networks in the training phase. Deep neural networks are more powerful than the classical machine learning methods in terms of mapping nonlinear representation. It works as a black-box but requires large training data, suffers from low interpretability, and is difficult to conceptualize.

21.9 Feature Selection

The extracted features from the images to develop a radiomic signature are high dimensional in nature. Often, the developed models based on these data suffer from model overfitting (Altman and Krzywinski 2018), which means the model successfully identify the data that belong to the training dataset with high accuracy but fails to classify the unknown data. The reduction of overfitting is a challenging problem in order to develop a robust model that can generalize the objective function and easily identify the unknown samples with high accuracy rate. Several procedures have been used for minimizing the overfitting issue. Usually the models train with large sized of data. There exists a reverse relationship between the sizes of the training datasets and overfitting. If the training dataset is very high then it is more prone to overfitting. In this type of case, dimensionality reduction is the only way to avoid overfitting problem and generalized the objective function by selecting minimum number of features through feature selection methods.

Feature selection methods recognize the redundant, unstable, and irrelevant extracted features and remove them to develop the model. In this way, a robust radiomic signature can be constructed with features that will be highly informative (Aerts et al. 2014; Wu et al. 2016; Aerts 2016; Parmar et al. 2015). The set of features can be evaluated in terms of temporal and spatial stability. Temporal stability is assessed for consistency through test-retest setting, whereas spatial stability is measured in terms robustness to variations in tumor segmentation (Aerts et al. 2014). In the case of split-validation scheme, the distribution of values of every features in the training set and test set is compared. Finally, keep only those features in radiomic signature that show no significance difference between the two and strictly remove the anomalies or system errors. This type of feature selection method is unsupervised in nature as it does not require the class variable during this process. After identifying the stable features, a correlation-based feature removing algorithm is used by developing a correlation heat map to eliminate the redundant features (Wu et al. 2016). Each block in the heat map represents a correlation coefficients. The blocks with correlation coefficient higher than a predefined value are identified and eliminated.

After removing a certain number of features, it is further required to reduce the dimensionality with more sophisticated methods and construct radiomics signature with a least number of features. There exist several methods for this task and are classified into three groups: filter, wrapper, and embedded methods. Filter methods ranked the features based on statistical measures. The selection of features has been done depending upon their ranks and characterized them by their computational efficiency, generalization, and robustness to overfitting. Filter methods can be univariate or multivariate. Univariate method provides scores for each of the features by neglecting the relationship between the features, while, in multivariate methods, consider the dependency between the features. In wrapper, identification of subsets with relevant and nonredundant features is performed and evaluates each of the subsets depending upon the performance of the model with the corresponding subset. These types of methods are computationally expensive. Embedded methods are computationally more efficient than wrapper. Both the selection and classification are performed concurrently in embedded methods and create models with more accuracy.

21.10 Training and Performance Measurement

After building the feature selection set of nonredundant, robust and relevant features, it is used for developing a model that generalizes the objective functions. Depending upon the continuous or discrete decision, variable different methods are used. If the decision variables are continuous, then linear regression, regression trees, or other regression methods are used otherwise, and several classification methods such as logistic regression, Naïve Bayes, support vector machines, decision trees, and random forests are used (Parmar et al. 2015).

Performance of the model is estimated for two nonintersect group of patients; among them, one is large and the other is comparative small dataset. The large one is used for the training purpose so that the relative error is minimized, while the other small one is used for validating the model. The small dataset is responsible for estimating more realistic performance metric. Therefore, the radiomics signature should be developed with many institutional data as this generalizes the objective function across all the institutions. But in reality, very few studies are found that considered data belong to different institutions (Kim et al. 2019).

Most of the studies have been done on the single institutional dataset only. In the case of small datasets, the validation method cannot work in right way and provide biasness in the model that results either over optimistic or pessimistic performance. To overcome this problem, cross-validation method is introduced that divides the dataset into k randomly equal subsets (in k-fold cross-validation) (Aerts 2016). One of the k-subsets is retained for validating the model, and the remaining k-1 are used for training the model. This process is repeated k times for every subset that is used once for validation. The method produces total k results, one for every iteration. Then, the average performance and standard deviation are determined that are the final measure, which indicates reproducibility and robustness.

In the case of machine learning, multiple models are compared and one model is chosen among them on the basis of performance. In general, the chosen model has lower standard deviation, which indicates more robustness and reproducibility.

There are two types of learning methods in radiomics: supervised learning and unsupervised learning. In supervised learning, the model is developed depending on the input features and class variable, while in unsupervised learning, the model tries to reveal the correlations on the input features only. Often, radiomics features can be inspected along with other types of data that include proteomics, metabolomics, and others. In fact, inclusion of clinical features with radiomics variable is the new trend in for predictive modeling.

21.11 Conclusion

The requirement for radiomics signature often arises from unfulfilled clinical needs in detection of diseases, staging, treatment response prediction, and chance of survival. The methods applied in radiomics are generic to data science. They also considers the domain-related conditions in the case of smaller dataset that is very common to cancer imaging. Radiomics in cancer needs interdisciplinary strategy that includes knowledge of several fields like oncologists, radiologists, data science, and image processing.

Acknowledgement I would like to thank Dr. Swapan Kumar Basu, Department of Computer Science, Banaras Hindu University, India, for his help and cooperation in preparation of this chapter.

References

Aerts HJ (2016) The potential of radiomic-based phenotyping in precision medicine: a review. JAMA Oncol 2(12):1636–1642

Aerts HJ, Velazquez ER, Leijenaar RT, Parmar C, Grossmann P, Carvalho S, Bussink J, Monshouwer R, Haibe-Kains B, Rietveld D, Hoebers F (2014) Decoding tumour phenotype by noninvasive imaging using a quantitative radiomics approach. Nat Commun 5:4006

Altman N, Krzywinski M (2018) The curse (s) of dimensionality. Nat Methods 15(6):399–400

Bedard PL, Hansen AR, Ratain MJ, Siu LL (2013) Tumour heterogeneity in the clinic. Nature 501 (7467):355–364

Chalkidou A, O'Doherty MJ, Marsden PK (2015) False discovery rates in PET and CT studies with texture features: a systematic review. PLoS ONE 10(5):e0124165

Choi ER, Lee HY, Jeong JY, Choi YL, Kim J, Bae J, Lee KS, Shim YM (2016) Quantitative image variables reflect the intratumoral pathologic heterogeneity of lung adenocarcinoma. Oncotarget 7(41):67302

Depeursinge A, Foncubierta-Rodriguez A, Van De Ville D, Müller H (2014) Three-dimensional solid texture analysis in biomedical imaging: review and opportunities. Med Image Anal 18 (1):176–196

Gillies RJ, Kinahan PE, Hricak H (2016) Radiomics: images are more than pictures, they are data. Radiology 278(2):563–577

Hosny A, Parmar C, Quackenbush J, Schwartz LH, Aerts HJ (2018) Artificial intelligence in radiology. Nat Rev Cancer 18(8):500–510

Kim DW, Jang HY, Kim KW, Shin Y, Park SH (2019) Design characteristics of studies reporting the performance of artificial intelligence algorithms for diagnostic analysis of medical images: results from recently published papers. Korean J Radiol 20(3):405

Kontos D, Summers RM, Giger M (2017) Special section guest editorial: radiomics and deep learning. J Med Imaging 4(4)

Kumar V, Gu Y, Basu S, Berglund A, Eschrich SA, Schabath MB, Forster K, Aerts HJ, Dekker A, Fenstermacher D, Goldgof DB (2012) Radiomics: the process and the challenges. Magn Reson Imaging 30(9):1234–1248

Mayerhoefer ME, Materka A, Langs G, Häggström I, Szczypiński P, Gibbs P, Cook G (2020) Introduction to radiomics. J Nucl Med 61(4):488–495

Mesci A, Lucien F, Huang X, Wang EH, Shin D, Meringer M, Hoey C, Ray J, Boutros PC, Leong HS, Liu SK (2019) RSPO3 is a prognostic biomarker and mediator of invasiveness in prostate cancer. J Transl Med 17(1):1–11

Moon SH, Kim J, Joung JG, Cha H, Park WY, Ahn JS, Ahn MJ, Park K, Choi JY, Lee KH, Kim BT (2019) Correlations between metabolic texture features, genetic heterogeneity, and mutation burden in patients with lung cancer. Eur J Nucl Med Mol Imaging 46(2):446–454

O'Connor JP, Aboagye EO, Adams JE, Aerts HJ, Barrington SF, Beer AJ, Boellaard R, Bohndiek SE, Brady M, Brown G, Buckley DL (2017) Imaging biomarker roadmap for cancer studies. Nat Rev Clin Oncol 14(3):169

Papanikolaou N, Matos C, Koh DM (2020) How to develop a meaningful radiomic signature for clinical use in oncologic patients. Cancer Imaging 20:1–10

Park JE, Kim HS (2018) Radiomics as a quantitative imaging biomarker: practical considerations and the current standpoint in neuro-oncologic studies. Nucl Med Mol Imaging 52(2):99–108

Parmar C, Grossmann P, Rietveld D, Rietbergen MM, Lambin P, Aerts HJ (2015) Radiomic machine-learning classifiers for prognostic biomarkers of head and neck cancer. Front Oncol 5:272

Peerlings J, Woodruff HC, Winfield JM, Ibrahim A, Van Beers BE, Heerschap A, Jackson A, Wildberger JE, Mottaghy FM, DeSouza NM, Lambin P (2019) Stability of radiomics features in apparent diffusion coefficient maps from a multi-centre test–retest trial. Sci Rep 9(1):1–10

Raunig DL, McShane LM, Pennello G, Gatsonis C, Carson PL, Voyvodic JT, Wahl RL, Kurland BF, Schwarz AJ, Gönen M, Zahlmann G (2015) Quantitative imaging biomarkers: a review of statistical methods for technical performance assessment. Stat Methods Med Res 24(1):27–67

Rizzo S, Botta F, Raimondi S, Origgi D, Fanciullo C, Morganti AG, Bellomi M (2018) Radiomics: the facts and the challenges of image analysis. Eur Radiol Exp 2(1):1–8

Sala E, Mema E, Himoto Y, Veeraraghavan H, Brenton JD, Snyder A, Weigelt B, Vargas HA (2017) Unravelling tumour heterogeneity using next-generation imaging: radiomics, radiogenomics, and habitat imaging. Clin Radiol 72(1):3–10

Song J, Zhang Z (2019) Brain tissue segmentation and Bias field correction of MR image based on spatially coherent FCM with nonlocal constraints. Comput Math Methods Med 2019

Um H, Tixier F, Bermudez D, Deasy JO, Young RJ, Veeraraghavan H (2019) Impact of image preprocessing on the scanner dependence of multi-parametric MRI radiomic features and covariate shift in multi-institutional glioblastoma datasets. Phys Med Biol 64(16):165011

Wu W, Parmar C, Grossmann P, Quackenbush J, Lambin P, Bussink J, Mak R, Aerts HJ (2016) Exploratory study to identify radiomics classifiers for lung cancer histology. Front Oncol 6:71

Yip SS, Aerts HJ (2016) Applications and limitations of radiomics. Phys Med Biol 61(13):R150

Part V

Epidemiology and Statistics of Cancer

Statistics in Cancer: Diagnosis, Disease Progression, Treatment Efficacy, and Patient Survival Studies

22

Satyendra Nath Chakrabartty and Gopesh Chandra Talukdar

Abstract

This article proposes a simple nonparametric measure for diagnosis of cancer and assessing cancer intensity for an individual, without resorting to group data or reduction of dimensionality or scaling or finding weights. The measure also identifies the critical areas/variables requiring attention, can be applied for all non-nominal data, can be used to find mean, variance, and confidence interval for group data, and facilitates statistical tests of hypothesis.

The cancer intensity facilitates ranking/classifying a group of patients along with quantifying progress of treatment at individual and group level. Using suitably designed group data, attempt can be made to find a small interval of values of cancer intensity for each type of cancer, which may be associated with Stage IV cancer or metastatic cancer. The proposed measure of cancer intensity offers an alternative approach for estimation of survival function of cancer patients. This study leads to a number of new areas of statistical analysis in cancer treatment. An empirical study will be of vital interest based on this theoretical study.

Keywords

Cancer diagnosis · Disease progression · Cancer intensity · Cancer statistics

---

S. N. Chakrabartty
Indian Ports Association, Indian Maritime University, Noida, Uttar Pradesh, India
e-mail: chakrabarttysatyendra3139@gmail.com

G. C. Talukdar (✉)
ESI Institute of Pain Management, ESI Hospital Sealdah Premises, Kolkata, West Bengal, India

© The Author(s), under exclusive license to Springer Nature Singapore Pte Ltd. 2022
S. K. Basu et al. (eds.), *Cancer Diagnostics and Therapeutics*,
https://doi.org/10.1007/978-981-16-4752-9_22

22.1 Introduction

Cancer is a dreadful disease, where growth of cells is uncontrolled. It may invade other organs causing fatal outcome over time. Usually, cancer is perceived as growth of tumors. However, all tumors are not cancerous (malignant). Some tumors are noncancerous, i.e., benign. Human body contains trillions of cells, and cancer may start in one or a small group of cells. Cancers could be divided into groups on the basis of types of cells depending on their origin such as *carcinomas* (start in the skin or in tissues that line or cover internal organs), *lymphomas* and *myeloma* (the cells of the immune system and plasma cells in the bone marrow), *leukemias* (the cells of bone marrow and produce abnormal blood cells), brain and spinal cord cancer, and *sarcomas* (connective or supportive tissues like bone, cartilage, fat, muscle, or blood vessels). Each type has several subtypes, and treatments for subtypes are different.

No single test can accurately diagnose cancer. The symptoms for cancer provide a possible indication of occurrence of cancer. Commonly used tests depend on type of cancer and may include blood tests, biopsy, tissue sample, imaging procedures like CT scan, nuclear scan, Ultrasound, PET scan, MRI, X-rays, etc. Diagnostic procedures for cancer are a series of diagnostics tests for deciding whether a patient is having cancer of a particular type including Cytogenetic and immunohistochemistry techniques.

However, low value of platelet count may not always imply Blood cancer. Hence, in addition to complete blood count (CBC), other types of tests like bone marrow biopsy, urine test, blood protein testing, lymph node biopsy, and other tests may be recommended for diagnosis of blood cancer. The results of pathological tests provide clues to the doctor who may decide to conduct other tests necessary to make the diagnosis. However, uses of pathological tests are changing with addition of knowledge.

Cancer informatics has emerged encompassing disciplines like oncology, bioinformatics, genomics, proteomics, metabolomics, pharmacology, statistics and computer science, computational biology, quantitative epidemiology, etc. (Jiang et al. 2014). Thus, it is challenging to unify various disciplines to identify the underlying causes and decide effective diagnosing and treatment plans along with evaluation and monitoring. Statistical approaches may complement usual pathological/clinical inferences and shed better light in comprehending cancer research to allow evidence-based decision makings. Use of statistical techniques in cancer research depends on types of studies, objectives, nature of variables, satisfaction of assumptions involved for each such techniques, etc. Shi and Sargent (2015), gave a review of application of statistical concepts in oncology research. Points to be kept in mind are that there are substantial complexities due to multidimensional variations caused by biological, genetics, environmental, behavioral, and sociological factors. Many of the factors or variables are uncontrollable or may even be unobservable.

Measurable variables primarily emerging from clinical findings have different units and may be independent or dependent with various degrees. The pathological data may involve ratio scale or interval scale data (age, size of tumor, etc.), nominal scale data (gender, blood group, etc.), or categorical/ranking data (opinion, perceptions, etc.). Method of combining them into one unit free similarity measure needs considerable studies.

Distribution of a variable is extremely difficult to find and may not be known. Thus, it makes sense to use nonparametric approaches to find similarity/dissimilarity measures. Oncology researches are primarily observational studies and clinical trials. Application of statistical/quantitative methods depends heavily on types of studies and nature of data. In statistics, the term *error* means variation in patient outcomes due to unknown or uncontrollable factors. Two types of errors are.

Random error: Due to chance, sample fluctuations, measurement error, and other noises. Magnitude and likelihood of the errors resulting from chance can be estimated with the help of probability and statistical theory.
Bias: It is a distortion and not due to chance alone. For example, different initial disease characteristics in two treatment groups can cause a bias. Various statistical methods can be used to correct for bias. However, if bias from unknown sources exceeds the same from known sources, the methods may fail.

The situation where both the risk factors and outcomes are categorical, significance of association between them may be investigated by Chi-square test (statistical notes are given in the appendix). The chi-square test tries to find the probability that the cell frequencies are produced randomly, assuming that both the dimension A and dimension B are independent and find association using expected values. For more than one risk factor, possible techniques could be regression analysis or logistic regression or multiway contingency table using log-linear model or generalization of Pearson's Chi-square test or likelihood-ratio Chi-square test.

Association studies involving moderate or large sample of patients of a particular type of cancer fail to depict prediction of risk of an individual patient. Usual questions in individual level cancer research are diagnosis of cancer, assessing severity or intensity of cancer and grading patients in terms of assessed strength or extent of cancer, measuring potentiality of possible cancer, deciding appropriate dose of therapies and their effects for cancer patients, quantifying success of treatment and relationship between patient characteristics along with clinical outcomes (e.g., response to treatment, toxicity, and survival), etc.

The above motivates us to review the issues and existing methods of cancer research and their shortcomings and to propose a nonparametric measure for diagnosis of cancer and assessing cancer intensity for an individual. The cancer intensity is used to facilitate ranking/classifying a group of patients along with quantifying progress of treatment and estimating survival function.

The remaining parts of this article are organized as follows. Types of studies at group level and at individual level are described in Sect. 22.2. Issues related to diagnosis of cancer, proposed method, and properties along with classification, assessment of progress, and survival curve are deliberated in Sect. 22.3. This article concludes in Sect. 22.4 with usefulness and benefits emerging from the suggested method.

22.2 Types of Studies

Three different types of studies have been discussed in this section as follows.

22.2.1 Population Level Studies

Observational studies are of three types.

Cross-sectional: Data (disease status and relevant factors) are collected at a single time point. The association or similarity measures are calculated. Various measures of similarity need to be compared for their advantages and limitations.

Cohort: Subjects are selected on the basis of risk factors and are followed for disease status until a future point. Popular association measure is Relative Risk (RR) and is expressed as

$$\text{RR} = \frac{\text{disease incidence rate in exposed}}{\text{unexposed subjects}}.$$

Very large sample size and long follow-up time are required for such studies especially when disease incidence is low.

Case-control: It is preferred where subjects are selected on the basis of disease status, with or without matching known risk factors. The measure of association between the risk factor and the disease rate is Odds Ratio (OR), which is expressed as

$$\text{OR} = \frac{\text{odds of exposure among disease cases}}{\text{non} - \text{disease cases}}$$

Note that computation of OR is easy and may be taken as an approximation of RR. Other important measures in cancer research include among others incidence (new cases among previously unaffected individuals) and mortality rates (rate of deaths occurring among those who remain alive). In the absence of national register of cancer patients (for each type) and their survival, death is a major hindrance to calculate such rates. People often resort to different estimation processes.

It may be noted that existence of a causal relationship may result in significant value of association, but the converse is not true. It has been observed that a case-control study may be biased due to inappropriate subject selection, systematic error in data collection, and inordinate amount of random error in the data gathering (Breslow et al. 1982). Such studies with bias may fail to provide a reasonable estimate of the strength of cause-effect relationship. Each type of study has advantages and limitations too.

22.2.2 Prognostic Studies

In oncology, prognostic factors are a set of clinical measures to find risk or chance of a patient to develop outcome like recurrence of disease after primary treatment. Such prognostic factors are important in clinical practice and classification of patients into groups with different risks. Prediction of patient survival depends on a large number of factors like the cancer type and stage, patient's age, overall health, etc. Prognostic studies attempt to identify disease-related biomarkers to assess this marker-prognosis association and to decide a patient's prognosis. It may be noted that *Time-outcomes* curve showing time from the beginning of observation (date of diagnosis/initiation of treatment/surgery) to the occurrence of the relevant outcomes (recurrence or death) is a continuous function. Even if an event is not observed for some subjects, partial information out of *censored data* that the subject was event-free until the last known date still can be valuable. For comparing time-to-event outcomes between different patient groups, the existing approaches use *log-rank test* for determining whether hazard rates are different for between groups. The association between a prognostic factor and the survival outcome can be quantified by taking ratio of the hazard rates of two populations and can be estimated by the Kaplan-Meier curve (1958), which helps to obtain an estimate of the event-free rate at any point during the follow-up period. Commonly used descriptive statistics associated with survival analysis including Kaplan-Meier curves are *median survival time* and *survival rates* for a specific time. However, median survival time cannot always be estimated in a unique manner. If the largest observation is censored and the contribution of uncensored observations is less than 50%, the median survival time cannot be computed. Other important measures commonly used in cancer research include incidence of new cases among previously unaffected individuals, mortality rates of those cancer patients who remain alive, etc. In the absence of national register of cancer patients (for each type) and their survival, death is a major hindrance to calculate such rates. People often resort to different estimation processes.

22.2.3 Individual-Level Studies

Association studies involving moderate or large sample of patients of a particular type of cancer fail to depict absolute risk prediction for any particular patient. Usual questions in individual level cancer research are diagnosis of cancer, assessing strength or intensity of cancer, grading patients in terms of assessed strength or intensity of cancer (This could be different from what doctors call Cancer staging), measuring potentiality of possible cancer, deciding appropriate dose of therapies and their effects for cancer patients, assessing similarities of donor and receiver in the case of bone marrow transfer, quantifying impact of treatment, finding relationship between patient characteristics and patient status, etc.

22.3 Diagnosis

A physician is more concerned about an individual patient's treatment plan. The first step is to accurately diagnose. When a patient first visits a hospital, a number of diagnostic tests are performed to confirm the diagnosis and accordingly decide treatment plan for the patient. Diagnosis plays an important role throughout the cancer treatment.

22.3.1 Method of Diagnosis

Problem of diagnosis of cancer is essentially comparing profile of normal tissues with diseased tissues (Tan and Gilbert 2003). However, high dimensionality is involved in biological and gene expression data. Thus, reduction of dimensionality of feature space was attempted by many researchers (Ramaswamy et al. 2001; Aliferis et al. 2010; Wang et al. 2005). Usual methods like Factor Analysis, Principal Component Analysis, etc. did not yield satisfactory results primarily due to small sample size for specific cancer type, different feature sets associated with each type of cancer, possible nonlinear relationships between expressions of different genes, noncompliance of assumptions, etc.

Morphologic characteristics of biopsy specimens are common in diagnosis of tumor. However, such a diagnostic method fails to provide information on rate of proliferation, capacity for invasion and metastases, etc. Suggestions emerged for use of microarrays in combination with other diagnostic methods for more information about the tumor specimen by considering simultaneously a large number of genes for classifying tumors and also to explore new diagnostic and therapeutic markers along with identification of new subtypes that may correlate with treatment outcome.

22.3.2 The Proposed Method (Statistical Diagnosis)

To answer whether a person undergoing a series of diagnostic tests is having cancer of a particular type, all the above attempted to find answer via techniques used for analysis of group data with different sets of assumptions. The proposed method offers a simple solution without resorting to group data or reduction of dimensionality.

Common point of all clinical tests for diagnosis of cancer of a particular type is selection of a finite number of key variables, measure them, and compare each observed value with a standard. It may be noted that the variables are of different units and may be independent or dependent with various degrees. Few variables are in percentages, which are, in general, not additive. Each variable has a standard. A standard usually has a range, that is, it gives a lower value and a higher value. If the obtained value of one or more variables for a person is lower or higher than the standards, he/she may be diagnosed as a cancer patient or not. All variables may not be equally important in diagnosis of cancer.

22.3.2.1 Data Preprocessing

Step 1: This is to make all variables positively related to cancer (PRC).

Variables like Platelet count, WBC count, etc. whose lower value indicates higher risk to cancer, reciprocal of such variables may be taken. For example, standard for Platelet count may be taken as $\frac{1}{450,000}$ to $\frac{1}{140,000}$ cells/mcL^{-1} or $\frac{1}{450}$ to $\frac{1}{140}$ thousand/mm^3. For variable like Basophils (DLC), where instead of a range, a single value is given in the reference range; the value (2 in the instant case) may be taken as the lower value of the standard.

Step 2: Diagnosis is essentially a comparison of values of the variables obtained from a person with the standards. Such comparison would be easier, if the range of values of a standard is replaced suitably by a single value. As a measure of abundant precaution, lower value of a standard of a variable, after converting it to be positively related to cancer, may be considered and replace the standard by the lower value of that PRC.

22.3.2.2 Statistical Diagnosis and Cancer Intensity

Assuming there are n-key variables, let us denote the observed values of the i-th person as a vector $X_i = (X_{i1}, X_{i2}, \ldots, X_{in})^T$ and the corresponding standards (lower value of PRC) as another vector $X_0 = (X_{01}, X_{02}, \ldots, X_{on})^T$. Statistical diagnosis of cancer can be made through the "Ratio with standards" approach described below:

Consider the ratio vector $Y = \frac{X_i}{X_0}$, where $Y_{ij} = \frac{X_{ij}}{X_{oj}}$ for $j = 1, 2, \ldots, n$. Each ratio is unit free and positive. Value of a ratio may be less than or greater than one. Since all the variables have been converted to be positively related to cancer, value of Y_{ij} exceeding unity for a particular value of j implies that the ith person has exceeded the standard value of the jth variable. Thus, the ith person has the risk of cancer with respect to the jth variable. If value of $Y_{ij} > 1$ for more than one value of j, it would imply the person's risk of cancer with respect to more than one variable. If the value of $Y_{ij} = 1$ or close to 1 from left (say ≥ 0.85) for the jth variable, the ith person may not be diagnosed as cancer patient but may be taken as a potential cancer patient with respect to the jth variable.

However, quantification of overall risk of cancer or intensity of cancer of the ith person could be assessed by finding similarities or deviations between the values of the vector X_i and X_0 by the following method:

22.3.2.3 Geometric Mean

Consider the unit free values of $Y_{i1}, Y_{i2}, Y_{i3}, \ldots, Y_{in}$ for the ith person. Geometric mean of those values will indicate cancer intensity. Thus, cancer intensity (CI$_i$) of the ith person

$$\text{CI}_i = \sqrt[n]{\prod_{j=1}^{n} Y_{ij}}. \qquad (22.1)$$

Alternatively, avoiding the nth root, Cancer intensity of the ith person (CI$_i$) can be taken as

$$\mathrm{CI}_i = \prod_{j=1}^{n} Y_{ij}. \tag{22.2}$$

Equations (22.1) and (22.2) have one-one and onto correspondence. Better measure of statistical diagnosis of cancer for the ith person is CI_i obtained from (22.2). If value of $\mathrm{CI}_i > 1$, the person can be diagnosed as cancer patient. Relative importance of the variables is reflected in the values of $Y'_{ij}s$ for $j = 1, 2, \ldots, n$. Higher value indicates higher relative importance requiring attention and helps to decide or modify the choice of treatment.

The proposed index CI_i reflects cancer intensity of the ith person and thus helps to rank a group of patients in terms of cancer intensity. For general convention, the Geometric Mean may be multiplied by 100 to reflect readily percentage changes. The measure satisfies all the scientific aggregation rules, as observed by Ebert and Welsch (2004). Properties of Geometric Mean as a composite measure were enumerated by Chakrabartty (2018), in a different context and are given below:

(i) Simple, can consider all chosen variables and depicts overall distance the ith person has with respect to the standards (PRC).
(ii) Symmetric over its arguments, that is, independent order of the chosen indicators.
(iii) Represents a continuous function that is unit free.
(iv) Monotonically increasing. Increase of say 1% in the jth variable for the ith person results in 1% increase in the Geometric Mean (GM), if all others remain unchanged. In other words, curve showing gain in a variable and gain in CI is linear.
(v) Differentiable everywhere.
(vi) Independent of change of scale.
(vii) Applicable even for skewed longitudinal data and also for snapshot data.
(viii) Does not require scaling of variables and avoids calculation of weights, for which there are no best methods.
(ix) Avoids computation of variance–covariance matrix or correlation matrix and subsequent data reduction techniques, viz., Principal Component Analysis (PCA) or Factor analysis (FA), involving set of assumptions.
(x) Not affected much by extreme values (outliers) and thus produces practically no bias for measuring disease intensity of a patient.
(xi) Low value of one key variable does not get linearly compensated by high values in another key variable and thus reduces the level of substitutability between key variables.
(xii) Facilitates formation of chain indices, that is $\mathrm{CI}_{20} = \mathrm{CI}_{21} \cdot \mathrm{CI}_{10}$, where 0 denotes the first time (or base period) the ith patient was diagnosed; 1 and 2 denote the subsequent time periods when CI was measured for the ith patient. This may help to obtain graph of the CI over a long period to reflect path of improvement over time for the patient.

(xiii) It is possible to have population estimate of GM, estimate of standard error of the GM, and confidence interval of GM. Thus, testing of null hypotheses $H_0 : GM_1 = GM_2$ can be performed by onventional t-tests on the logarithms of the observations.

(xiv) Assumes positive value for each variable for all periods. If a particular variable attains zero or negative value, the method fails.

Sample mean of GM can be computed by writing Eq. (22.1) as $\log GM = \frac{1}{n} \times \sum \log Y_{ij}$. Population estimate of GM can be taken as the sample GM for large data involving sample size of N. Estimate of Standard error of the GM is $\frac{GM \cdot GSD}{\sqrt{N-1}}$ (Norris 1940), where Geometric standard deviation (GSD), denoted by SGM, is calculated using

$$\log SGM = \left[\frac{1}{n}\sum(\log X_i - \log GM)^2\right]^{\frac{1}{2}}. \tag{22.3}$$

Thus, log (GSD of$(X_1, X_2, \ldots, X_n)$) = usual SD of $\log X_{1,2}, \ldots, \log X_n$. While GSD indicates how spread out are a set of numbers (ratios, in our case) whose preferred average is GM, the ratio $\frac{GSD}{GM}$ indicates consistency of data. Considering S_m as the sample estimate of the standard error of the mean of the logarithms of the sample values and defined as

$$S_m = \left[\frac{n \cdot \sum(\log X)^2 - (\sum \log X)^2}{n^2(n-1)}\right]^{\frac{1}{2}}. \tag{22.4}$$

Alf and Grossberg (1979), have shown that for $(1 - \alpha)$ % confidence limits, upper limit and lower limit of the confidence interval of GM are as follows: Upper limit is e^U, where

$$U = \log GM + S_m \cdot t_{\left(\frac{\alpha}{2}, df\right)}. \tag{22.5}$$

Lower limit is e^L, where

$$L = -S_m \cdot t_{\left(\frac{\alpha}{2}, df\right)}. \tag{22.6}$$

22.3.3 Classification

Principle of statistical classification is to minimize the ratio of within group variance and between group variance and is known as *Wilks' Lambda criterion*, value of which may reflect effectiveness of statistical classification. However, for classification of cancer on the basis of gene expression data, a number of machine learning and data mining algorithms have been proposed. However, identification of relevant genes for different types of cancer from several thousands of genes is extremely difficult. Jiang et al. (2014), observed that there is no single classifier that outperforms others universally and clustering is a more difficult problem than classification due to unknown number of cancer types and lack of learning set of labeled observations.

Popular methods include Bayesian network, k-nearest neighbors, Neural Network, nearest shrunken centroids, logistic regression, random forest, and support vector machine. However, the methods differ in terms of sets of assumptions and also in terms of accuracy of the classifiers, which depend on good number of factors including the nature of datasets. Thus, there is need to have nonparametric method of allocating patterns in the individual data with predefined cancer types. Cluster analysis is commonly used for classification of a number of subjects. Literature survey shows classification of cancer by primary site of origin or by histological or tissue types. Wu et al. (2003), considered Mass Spectrometry (MS) data obtained from serum samples and used several statistical methods to distinguish ovarian cancer patients from normal individuals. The methods considered are linear discriminant analysis, quadratic discriminant analysis, k-nearest neighbor classifier, bagging and boosting classification trees, support vector machine, and random forest (RF). RF and methods similar to RF were found to be preferred over other methods for classification of individuals in the sample.

It could be more important to classify cancer patients in terms of cancer intensity, that is, the extent of the disease, primarily to decide treatment plan. The proposed measure is in terms of CI_i, which combines the relevant n-key variables Y_{i1}, Y_{i2}, Y_{i3}, ..., Y_{in} into a single index for the ith person as a monotonic function and reflecting overall intensity or severity of an individual's cancer. Thus, CI_i's can be well used for undertaking Cluster analysis and classification of cancer patients and decide accordingly the prognosis and treatment plan.

This provides a general method of classifying cancer patients for each cancer type and avoids the usual staging methods like tumor size, node involvement, and distant metastases (TNM) system of classification. Note that membership of a cancer patient to a class may get changed subsequently with time and treatment effect, which results in changes in overall intensity. Such analysis with large volume of data emerging from representative sample under each type of cancer may help to find norms (class boundaries) for the classes.

22.3.4 Assessment of Progress

Need is felt to assess progress made by a cancer patient during treatment period primarily to see effect of treatment on the person's disease characteristics. However, definition of the term "Progress" differs. In addition to patient's physical parameters, psychological and social functioning of the patient have also been considered as the indicative variables of progress. Likert type questionnaires on Quality of life (QOL), viz., QLQ-C30 covering a range of QOL issues relevant to cancer patients was used by the European Organization for Research and Treatment of Cancer (EORTC) (Sprangers et al. 1993). The questionnaire consists of Likert items mostly with four response categories and a few items relating to the patient's perceptions about his/her overall health and quality of life in a seven point scale. Other approaches include trends over time in both morbidity and mortality by age and sex. However, such approaches shift the focus from the effect of treatment and need to further change treatment plan. Moreover, ordinal data generated by a Likert scale have many limitations. For example, statistics like mean and standard deviation of Likert score may not be meaningful since data are not continuous and successive intervals of the response categories are not equidistant (Jamieson 2004). Chakrabartty (2014), came out with a comprehensive view on limitations of Likert Scale.

The proposed measure in terms of geometric mean helps to find answer in this context. Let CI_{i_1} be the cancer intensity of the ith person at time point 1, that is, the first time when the patient was examined in hospital and treatment started. So,

$$CI_{i_1} = \prod_{j=1}^{n} Y_{ij_1} = \prod_{j=1}^{n} \frac{X_{ij_1}}{X_{oj}} \quad (22.7)$$

Subsequently, cancer intensity of the same patient can be assessed and denoted by $CI_{i_2}, CI_{i_3}, \ldots$ and so on. In general, cancer intensity of the ith patient at time point t- may be denoted as CI_{i_t} for $t = 1, 2, 3, \ldots$, and so on. $CI_{i_t} < CI_{i_{(t-1)}}$ will imply that the patient had improved during the tth time from the $(t − 1)$th time. Similarly, if $CI_{i_t} > CI_{i_{(t-1)}}$, the patient has deteriorated in the tth time from the $(t − 1)$th time. Thus, $CI_{i_t} − CI_{i_{(t-1)}}$ will reflect efficacy of treatment of the ith patient during the tth period. The same may also be used to find effect of surgery or a clinical trial. The critical variables where deterioration took place can also be observed by comparing the values of Y_{ij} for the period t and $(t − 1)$, that is those variables for which $Y_{ij_t} > Y_{ij_{(t-1)}}$. Extent of deterioration in the identified variables can be assessed by difference of values of corresponding $X'_{ij}s$.

Alternatively, progress of a patient during tth time point over $(t − 1)$th period can be reflected by the ratio $\frac{CI_{i_t}}{CI_{i_{(t-1)}}}$. Value of the ratio exceeding unity will indicate improvement in the tth period from $(t − 1)$th time point. Note that the ratio.

$$\frac{CI_{i_t}}{CI_{i_{(t-1)}}} = \frac{\prod_{j=1}^{n} Y_{ij_t}}{\prod_{j=1}^{n} Y_{ij_{(t-1)}}} = \frac{\prod_{j=1}^{n} \frac{X_{ij_t}}{X_{oj}}}{\prod_{j=1}^{n} \frac{X_{ij_{(t-1)}}}{X_{oj}}} \quad (22.8)$$

Equation (22.8) may be preferred. It also helps to see progress of a patient from the beginning (i.e., time zero) since the measure CI_{i_t} facilitates formation of chain indices.

Deterioration of overall value of cancer intensity along with identification of critical areas of deterioration will help to decide treatment plan. If the graph of CI_{i_t} and t is nonincreasing, then the ith patient is improving. An increasing graph of CI_{i_t}

and t will indicate steady deterioration of the patient. Attempt can be made to find a small interval of values of CI_{i_t} for each type of cancer, which may be associated with Stage IV cancer or metastatic cancer.

22.3.5 Survival Curve

Concept of life tables, estimation of survival rate, and survival function for cancer patients have been attempted by researchers (Gehan 1969; Berkson and Gage 1950). Collett (2003), has given a review of modeling of Survival Data in Medical Research. The most frequently used method on the development in the field of clinical trials is the Kaplan-Meier method used for estimating the survival function, which is rather similar to the life table technique. Here, survival function is estimated via computation of probabilities of occurrence of an event in a certain interval of time period. This method appears to be an inefficient estimate of the survival probabilities especially when the patients under study refuse to cooperate or when some of the patients in the study continued to survive till the end of the study. Thus, the method may not give a reliable estimator in the case of heavy censoring. Also, at the end points, the Kaplan-Meier survival curve fails to provide reliable estimates. In the case of heavy censoring, weighted Kaplan-Meier Estimator with zero weight to the last censored observation was suggested by Jan et al. (2005). On the other hand (Shafiq et al. 2007) proposed Modified Weighted Kaplan-Meier estimator involving all nonzero weights.

Cancer intensity of a patient at a time point (CI_{i_t}) as per Eq. (22.7) above can be used for estimating survival function of cancer from suitably designed study involving large volume of data. Empirical probabilities can be calculated for improvement or deterioration of cancer intensity from $(t-1)$th period to tth period. Estimated probability that the patient will survive for a time span $t > 0$ or more will help to find monotonically decreasing survival function $S(t) = $ Probability $(T > t)$, where T denotes the time till survival of the patient with $S(0) = 1$. The survival curve will be the estimated graph of $S(t)$ and t.

Thus, the proposed measure of cancer intensity offers an alternative approach to estimate the survival function of cancer patients. Such a function need to be compared with the existing methods like Kaplan-Meier survival curve, Weighted Kaplan-Meier estimator by Jan et al. (2005) giving zero weight to the last censored observation, and Modified Weighted Kaplan-Meier by Shafiq et al. (2007) giving nonzero weight to the last censored observation and a small probability of survival.

22.4 Conclusions

To answer whether a person undergoing a series of diagnostic tests is having cancer of a particular type, the existing statistical methods use techniques for analysis of group data. The proposed method offers a simple solution without resorting to group data or reduction of dimensionality or scaling or finding weights. The method, in terms of function of GM, is nonparametric, satisfies many desired properties, identifies the critical areas/variables requiring attention, can be applied for all non-nominal data. Besides helping statistical diagnosis of cancer for individual patient, the measure can be used to find, mean, variance, and confidence interval for group data and facilitates statistical tests of hypothesis.

The measure can be further used to assess cancer intensity of individual patients, which can be used for classification or staging and quantification of progress during the treatment, and decide need of modification of treatment plan, if any. In case, the graph of cancer intensity and time is nonincreasing, then the patient is improving whereas an increasing graph will indicate deterioration of the patient. Using group data, attempt can be made to find a small interval of values of cancer intensity separately for each type of cancer, which may be associated with Stage IV cancer or metastatic cancer. The proposed measure of cancer intensity offers an alternative approach to estimate the survival function of cancer patients.

Appendix

Statistical Notes

1. Chi-square test is a nonparametric test that makes comparisons (usually of cross tabulated data) between two or more samples on the observed frequency of values with expected frequency of values and also, used as test of goodness of fit of log linear models.
2. Regression analysis establishes relationship of the dependent variable with one or more of the independent variables.
3. Logistic regression analysis deals with dependent variable in binary and one or more independent variables in nominal, ordinal, interval, or ratio level.
4. Factor Analysis (FA)/Principal Component Analysis (PCA) are both multivariate statistical techniques for reduction of data/variables. PCA considers linear combination of weighted observed variables to minimize the variance of the observed variables, while FA explains the covariance between the variables.
5. Wilks' Lambda is the ratio of the within group sum of squares to the total sum of squares. When observed group means are nearly equal, Wilks' Lambda will be high and a small lambda occurs when group means differ.
6. A Bayesian network consists of a set of nodes (random variables) and a set of directed edges (direct dependencies between the variables). Major difficulties are specifying prior probability (a priori *probability*) and computing a posterior probabilities (a posteriori *probability*).
7. The *k*-nearest neighbors (KNN) algorithm is used primarily in classification problems.
8. Neural network starts with a set of variables X_i and associated weights W_i for all $i = 1, 2, \ldots, n$. A function f is determined whose domain is the sums of the weights and range is an output Y. Neural Networks, which have multiple

solutions associated with local minima, may not be robust over different samples.
9. Nearest shrunken centroid classification calculates a standardized centroid for each class in terms of ratio of average gene expression for each gene and the within-class standard deviation for that gene.
10. Random forest or random decision forest is a machine learning algorithm used for classification, regression, and other tasks by constructing multitude of ensemble of decision trees and merging them together for more accurate and stable prediction.
11. Support vector machine (SVM) is a nonparametric method for analysis consisting of both classification of tissue samples and explorations of the data for mislabeled or questionable tissue results. Here, the marginal contribution of each component ratio to the score is variable. Moreover, the choice of the input variables has a decisive influence on the performance results.
12. Cluster analysis groups a set of objects in such a way that objects in the same group are more similar to each other than to those in the other groups.
13. When the dependent variable is categorical and the independent variables are in interval scale or in ratio scale, Discriminant analysis develops Discriminant functions (df) that discriminates between the categories of the dependent variables.
14. Log-rank test compares the survival distributions of two samples when the data are right skewed and censored.

References

Alf EF, Grossberg JM (1979) The geometric mean: confidence limits and significance tests. Percept Psychophys 26(5):419–421

Aliferis CF, Statnikov A, Tsamardinos I, Mani S, Koutsoukos XD (2010) Local causal and Markov blanket induction for causal discovery and feature selection for classification Part I: Algorithms and empirical evaluation. J Mach Learn Res 11(1):171–234

Berkson J, Gage RP (1950) Calculation of survival rates for cancer. In: Proceedings of the staff meetings. Mayo Clinic, 25(11), 270–286

Breslow NE, Day NE, Schlesselman JJ (1982) Statistical methods in cancer research. Volume 1—the analysis of case-control studies. J Occup Environ Med 24(4):255–257

Chakrabartty SN (2014) Scoring and analysis of Likert scale: few approaches. J Knowl Manage Inform Technol 1(2)

Chakrabartty SN (2018) Better composite environmental performance index. Interdiscip Environ Rev 19(2):139–152

Collett D (2003) Modeling of survival data in medical research. Chapman Hall, London, UK

Ebert U, Welsch H (2004) Meaningful environmental indices: a social choice approach. J Environ Econ Manage 47(2):270–283

Gehan EA (1969) Estimating survival functions from the life table. J Chronic Dis 21(9–10): 629–644

Jamieson S (2004) Likert scales: how to (ab) use them. Med Educ 38:1212–1218

Jan B, Shah SWA, Shah S, Qadir MF (2005) Weighted Kaplan Meier estimation of survival function in heavy censoring. Pak J Stat 21(1):55–63

Jiang H, An L, Baladandayuthapani V, Auer PL (2014) Classification, predictive modeling, and statistical analysis of cancer data (a). Cancer Inform 13(2):1–3

Kaplan EL, Meier P (1958) Nonparametric estimation from incomplete observations. J Am Statist Assoc 53(282):457–481

Norris N (1940) The standard errors of the geometric and harmonic means and their application to index numbers. Ann Math Statist 11(4):445–448

Ramaswamy S, Tamayo P, Rifkin R, Mukherjee S, Yeang CH, Angelo M, Ladd C, Reich M, Latulippe E, Mesirov JP, Poggio T (2001) Multiclass cancer diagnosis using tumor gene expression signatures. Proc Natl Acad Sci 98(26):15149–15154

Shafiq M, Shah S, Alamgir M (2007) Modified weighted Kaplan-Meier estimator. Pak J Statist Oper Res 3(1):39–44

Shi Q, Sargent DJ (2015) Key statistical concepts in cancer research. Clin Adv Hematol Oncol: H&O 13(3):180–185

Sprangers MAG, Cull A, Bjordal K, Groenvold M, Aaronson NK (1993) The European Organization for Research and Treatment of cancer approach to quality of life assessment: guidelines for developing questionnaire modules. Qual Life Res 2(4):287–295

Tan AC, Gilbert D (2003) Ensemble machine learning on gene expression data for cancer classification. Appl Bioinform 2

Wang Y, Tetko IV, Hall MA, Frank E, Facius A, Mayer KF, Mewes HW (2005) Gene selection from microarray data for cancer classification—a machine learning approach. Comput Biol Chem 29(1):37–46

Wu B, Abbott T, Fishman D, McMurray W, Mor G, Stone K, Ward D, Williams K, Zhao H (2003) Comparison of statistical methods for classification of ovarian cancer using mass spectrometry data. Bioinformatics 19(13):1636–1643

Epidemiology of Cancer: Asian Perspective Revised

23

Prasanta Ray Karmakar

Abstract

Worldwide cancer is the second major cause of death. Among men, lung, prostate, colorectal, stomach, and liver cancers are the most common types, while breast, colorectal, lung, cervix uteri, and thyroid cancer are the most common among women. Cancer incidence, prevalence, and mortality vary from place to place, and the rates are different in different age and sex groups. The incidence of cancer is similar in both the sexes in the 0–14-year age group, but as age advances it is more in females up to 50 years; afterwards it is more in males. Various carcinogenic agents like tobacco use, alcohol consumption, harmful chemicals, infections, ionizing radiations, pollution in land/air/water, and pharmaceutical products acting for a long period usually give rise to cancer. Cancer incidence is more in developed countries, mostly due to increased longevity, improved methods of diagnosis, and change in lifestyle with socioeconomic development. Survival in cancer patients has increased over the years. Early diagnosis and improved treatment are mainly responsible for this. Cancer control is based on measures undertaken to prevent the occurrence of new cases and early diagnosis and prompt treatment of cancer by different modalities like surgery, radiotherapy, and chemotherapy.

Keywords

Epidemiology · Cancer · Asia · Risk factors

P. R. Karmakar (✉)
Department of Community Medicine, Raiganj Government Medical College, Raiganj, West Bengal, India

© The Author(s), under exclusive license to Springer Nature Singapore Pte Ltd. 2022
S. K. Basu et al. (eds.), *Cancer Diagnostics and Therapeutics*,
https://doi.org/10.1007/978-981-16-4752-9_23

23.1 Introduction

The term "cancer" is not used to represent a single disease. Cancer is a group of diseases characterized by the abnormal growth of cells, not only in their usual boundaries but also in the adjoining parts of the body and/or distant organs (Park 2019). Cancer is also known by other common terms such as malignant tumors and neoplasm. Each and every organ of our body may be affected by cancer and one cancer type may have many anatomic and molecular subtypes. Management strategies are according to the type of cancer.

Description of cancer-like condition was found from Egypt in 1600 BC. The origin of the word cancer is from carcinoma. Hippocrates (460–370 BC), the Greek physician, coined the term carcinoma. He used it to describe tumors, both nonulcer forming and ulcer forming. Symptoms which can be assumed to be of malignant disease were found in Arabic and Chinese medical literature. There is also mention of similar condition in Ayurveda, the ancient Indian medical literature, as "Granthi" (minor neoplasm) and "Arbuda" (major neoplasm) (Manohar 2015).

Globally, cancer is the second leading cause of death which accounted for 8.8 million deaths in the year 2015 (World Health Organization Factsheet 2018) and estimated to cause 9.5 million deaths in 2018. Among men, lung and prostate cancers occupied the first two positions, with incidences of 31.5 per 100,000 and 29.3 per 100,000, respectively. Among women, the two most common cancers are breast and lung cancers with the incidence rate of 46.3 per 100,000 and 14.6 per 100,000, respectively (GLOBOCAN 2018).

Highest incidence rates for all cancers combined (excluding non-melanoma skin cancer) were recorded in rich countries of North America and Western Europe. Japan, the Republic of Korea, Australia, and New Zealand also recorded high incidence. Africa, Asia, and Central and South America recorded more than 60% of the world's cancer cases, and about 70% of the cancer deaths of the world occurred in these regions (WHO 2014). Particular cancers show sharp differences in different geographical areas. Such difference helps to understand the causation, and to institute appropriate preventive measures (WHO 2014).

In 2018, the estimated number of diagnosed new cancer cases was 18 million; estimated deaths were 9.5 million (GLOBOCAN 2018). These estimates gave an *age-standardized* incidence and *mortality* rates of 198 and 101 per 100,000, respectively, globally. In GLOBOCAN (2012) incident cases were slightly more in males (53% of the total) and death was also more in males (57% of the total) (WHO 2018a). Globally, cancer burden and common cancer types are associated with the value of the *Human Development Index* (HDI) for that country. As the HDI value increases, there is overall increase of cancer burden. In rich countries with high or very high Human Development Index (HDI), the lung, breast, prostate, and colorectum cancers are the major incident types. Poor countries with low or medium HDI, colorectum, breast, and lung cancers have become more frequent. In these countries, cancers of the stomach, liver, cervix, and esophagus are also common. These cancers are thought to be related to poverty and infection; in countries which are in transitional phase, there is a decline in the incidence of cervical cancer, but concomitant increases in the incidence of female breast cancer. With the improvement in economic condition, the incidence rates of colorectal cancer increase markedly in a given country (WHO 2014).

23.2 Epidemiology

Epidemiology has been defined by John M Last (2001) as follows: "Epidemiology is the study of the distribution and determinants of health-related states or events in specified populations, and the application of this study to the control of health problems." And cancer epidemiology is the study of distribution and determinants of cancer (Dos Santos Silva 1999).

Cancer is not uniformly distributed in relation to place, person, and time. Cancer incidence, prevalence, and mortality vary from place to place. Incidence means number of new cases in a defined population in a defined period of time (Gordis 2013). Cancer incidence is usually expressed as the number of cases per 100,000 people per year. The rates are different in different age and sex groups. The rates are also variable depending upon different biological and social factors. The study of cancer in relation to place, time, and person characteristics is called descriptive epidemiology of cancer. The study of association of cancer with etiological factors or risk factors is analytical epidemiology of cancer.

Incidence of cancer is more in males than females. The overall age-standardized cancer incidence was higher in men. Cancer incidence in males is 218 and the corresponding rate for females is 182 per 100,000. Across the different regions of the world, cancer incidence in males varies almost fivefold; it was as low as 95 per 100,000 in Western Africa to as high as 571 per 100,000 in Australia/New Zealand. High rate of prostate cancer is responsible for the high rate of cancer in male population. In case of females, variation is less; incidence ranges from 95 per 100,000 in South-Central Asia to 362 per 100,000 in Australia/New Zealand (GLOBOCAN 2018).

Lung, prostate, colorectum, stomach, and liver cancers are the first five common cancers globally observed in males. Cancer of these five sites is also the most common cause of mortality in males. Cancers of breast, colorectum, lung, corpus uteri, and thyroid are the five most common cancers in females (WHO 2018).

Asia is the largest continent in terms of area and population. According to GLOBACAN 2018, Asia with a population of 4.54 billion had 8.7 million new cancer cases. The age-standardized incidence rate was 164.5 per 100,000. There were 5.4 million deaths due to cancer in Asia in that period. The prevalence of cancer was more than 17.03 million. In Asia, when both sexes taken together, the top five most frequent cancers excluding non-melanoma skin cancer were lung, colorectum, breast, stomach, and liver. In males, the top five cancers were lung, colorectum, stomach, liver, and esophagus and top five for females were breast, colorectum, lung, cervix uteri, and thyroid (WHO 2018a, b, c) (Table 23.1).

China, the largest country of the world, also has largest population. Japan, with substantial proportion of elderly population, has high cancer incidence, while Nepal,

Table 23.1 Cancer situation in selected Asian countries, 2018

	China	India	Japan	Nepal
Population	1.42 billion	1.35 billion	127.18 million	29.62 million
No. of new cases	4,285,033	1,157,294	883,395	26,184
No. of deaths	2,865,174	784,821	409,399	19,413
No. of prevalent cases	7,827,961	2,258,208	2,127,559	43,816
Age-standardized incidence rate (world) per 100,000	201.7	89.4	248.0	103.7
Top five cancers (ranked by cases)	Lung	Breast	Colorectum	Lung
	Colorectum	Lip, oral cavity	Lung	Cervix uteri
	Stomach	Cervix uteri	Stomach	Breast
	Liver	Lung	Prostate	Gallbladder
	Breast	Stomach	Breast	Stomach

Data source: https://gco.iarc.fr/today/fact-sheets-populations

a poor country with relatively young population, has low cancer incidence. China and India lie in between. The top five cancers are also different in those countries. In India, cancer of lip and oral cavity is very high due to the widespread use of chewable tobacco. In India and Nepal, cancer of cervix uteri is high, while in China and Japan colorectum cancer is high. This difference may be attributed to different cultural and behavioral practices. Prostatic cancer is high in Japan due to its high proportion of elderly population.

In 2018, the most common cancers in India (GLOBOCAN 2018) were cancers of breast, lip and oral cavity, cervix uteri, lung, and stomach as represented in Fig. 23.1.

The incidence of cancer is same in both the sexes in 0–14-year age group, but as age advances then it is more in females up to 50 years than males; afterwards it is more in males. In females, cancer cervix and breast cancers are responsible for higher rate in females in younger years. Afterwards cancers of lung and prostate are responsible for higher rate in males (WHO 2014).

Cancer usually has a long latent period. Various agents acting for a long period usually give rise to cancer. So, incidence increases with increase of age. Incidence is as low as 10 per 100,000 in 10–14-year age group; then the rate increases reaching 150 per 100,000 in 40–44-year age group, and more than 500 per 100,000 in 60–64-year age group (WHO 2014).

Cancer incidence is not equal throughout the globe. With a few exceptions, it is more in North America and Western Europe. It is also high in Japan, the Republic of Korea, Australia, and New Zealand. The lowest rate is observed in most parts of Africa and West and South Asia. This is mostly due to increased longevity, improved methods of diagnosis, and change in lifestyle with socioeconomic development (WHO 2014).

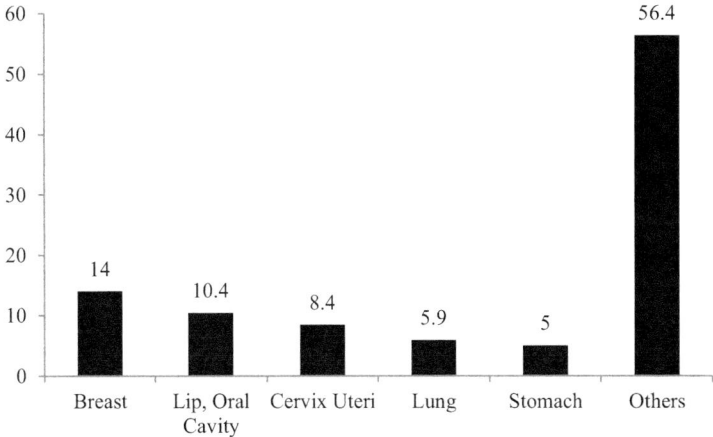

Fig. 23.1 Cancer incidence proportion (both sexes), India, 2018
Data source: http://cancerindia.org.in/globocan-2018-india-factsheet/

In GLOBOCAN 2018, breast cancer incidence rates range from 27 per 100,000 in Central Africa and Eastern Asia to 92 in Northern America. So, there is about fivefold increase across the world regions. Cervix and uterine cancer rates are the lowest in Australia/New Zealand (5.5) and Western Asia (4.4). Cervical cancer is the most common cancer in women in Eastern and Central Africa. Lung cancer has been the most common cancer in the world for several decades. In 2012, out of the total cancers, 12.9% was lung cancer (estimated to be 1.8 million new cases). Less developed regions account for 58% of the total cases. In 2018 also, lung cancer was the commonest cancer in males, accounting for 11.6% of all cancers in males. There is a large variation in the incidence in different regions of the world. It is as high as 49.2 per 100,000 and 47.2 per 100,000 in Central and Eastern Europe and Eastern Asia, and it is as low as 3.4 and 2.4 per 100,000 in Eastern Africa and Western Africa, respectively. This difference may be due to different historical exposures to tobacco in different populations. In all regions, the lung cancer incidence is lower in females. It is the highest in North America (30.7 per 100,000) and lowest in Western Africa (1.2 per 100,000). Lung cancer is responsible for 18.4% of all cancer deaths. Lung cancer has very high fatality. In lung cancer, mortality-to-incidence ratio is 0.87. This is uniform throughout the world. Cancer of prostate also shows a wide variation in different zones of the world. It is high in Australia and New Zealand (86.4 per 100,000), Northern Europe (85.7 per 100,000), Western Europe (75.8 per 100,000), and North America (73.7 per 100,000). The incidence is low in Asia and Africa. It is the lowest in South-Central Asia (5.0 per 100,000).

Cancer is the second major cause of death after heart attack. If we follow the natural history of cancer, most cancers result in mortality within a few years of diagnosis (WHO 2008a, b) (Table 23.2).

Mortality experience is not the same for different types of cancer. It is important to study the common cancer in terms of incidence, mortality, and survival.

Table 23.2 Prognosis of different types of cancer

Cancers that are curable when detected early	Cancers that are disseminated or not amenable to early detection but potentially curable	Cancers that are treatable but not curable
Breast Cervix Colon and rectum Oral cavity Nasopharynx Larynx Stomach Skin melanoma Other skin cancers Urinary bladder Prostate Retinoblastoma Testis	Metastatic seminoma Acute lymphatic leukemia Hodgkin lymphoma Non-Hodgkin lymphoma Osteosarcoma	Advanced breast cancer Advanced cutaneous melanoma Advanced Hodgkin lymphoma Advanced non-Hodgkin lymphoma

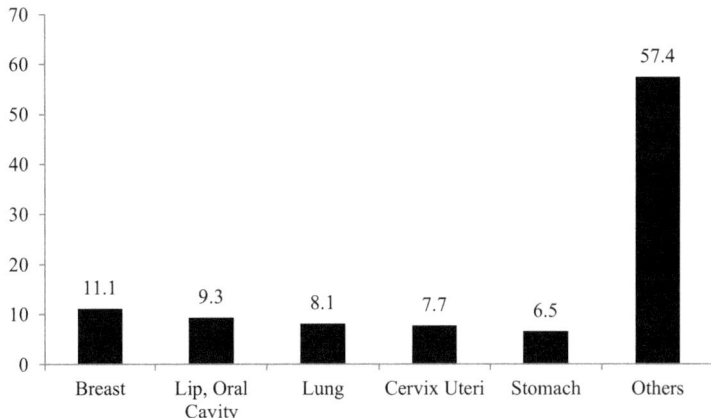

Fig. 23.2 Cancer mortality proportion (both sexes), India, 2018
Data source: http://cancerindia.org.in/globocan-2018-india-factsheet/

Difference in mortality rate is not as marked as in the incidence rate. For all cancers together, mortality is the highest (124.4 per 100,000) in Melanesia and lowest in Central America (65.3).

The highest mortality reported due to cancer was in men in Central and Eastern Europe (173 per 100,000) and mortality was the lowest in Western Africa (69 per 100,000) in GLOBOCAN 2012. The highest mortality due to cancer among women was recorded in Melanesia (119) and Eastern Africa (111), and the lowest mortality was recorded in Central America (72) and South-Central Asia (65).

In India, the most frequent causes of mortality in caners are cancers of breast, lip and oral cavity, lung, cervix uteri, and stomach (Fig. 23.2).

Cancer survival has improved immensely over the years. In early 1930s, less than 20% of cancer patients survived more than 5 years. Now more than 65% can expect to survive more than 5 years (Pastides 2018). Improved survival is due to several factors. Early diagnosis of cancer and successful management are mainly responsible for this. Because of better awareness of the people about the danger signals of cancer and screening programs, cancers are diagnosed early. Better treatment facility of early stages of cancer increases 5-year survival. This has resulted in increased prevalence of cancer.

An important measure of cancer burden in a population is prevalence. Prevalence means the total number of diagnosed cancer cases (both old and new) and are still living at a specific time. In 2018, 5-year prevalent cases were 43,846,302. Among them, 21,014,830 cases were males and 22,826,472 cases were females. Breast cancer was the most prevalent cancer, both sexes combined together. The second and third most prevalent cancers were prostate cancer and cervix uteri cancer. Colorectal cancer came next. Prevalence can be expressed as $P = incidence \times duration\ of\ diseases$ (period of survival after diagnosis). Incidence being the same, the prevalence will be more with cancer with more survival rates and vice versa. In lung cancer, because of its very poor survival, the 5-year prevalence is very close to the annual incidence (Table 23.3, 23.4 and 23.5).

For most of the cancer, the incidence is increasing in developing and underdeveloped countries. Survival rate for most of the cancer has increased in developing countries. It is due to early diagnosis and improved treatment. It was reported about the status of cancer in USA during 1975–2001 that the 5-year survival rate for top 15 cancers was increased from 42.7% to 64.0% among males and 56.6% to 64.3% among females (Jemal et al. 2004). If we omit lung cancer, cancer mortality rate has also decreased with the passage of time in some developed countries.

23.3 Etiological (Risk) Factors of Cancer

Cancer is not a single disease. And one particular cancer is also not caused by a single agent; it has multifactorial causation. Multiple factors acting at different levels ultimately give rise to cancer. Different types of epidemiological studies helped us to understand the etiology of cancer. It is not possible to conduct experiment on human subjects to establish the cause of different types of cancer. Descriptive studies have identified the distribution of cancer in relation to person, place, and time. Analytical studies have identified the association between different factors and cancer. Risk factors are the attribute and exposure that are significantly associated with the development of a disease (WHO 1980). These factors are present more in cancer patients than a comparable group of persons not suffering from that disease. If these risk factors can be modified in the population by different interventions, then the possibility of development of that condition will reduce.

The major etiological (risk) factors for cancer (WHO 2014) are tobacco use, alcohol consumption, infections, reproductive and hormonal factors, diet, obesity and physical activity, and occupation. Other important risk factors are radiation

Table 23.3 Global estimated incidence, mortality, and 5-year prevalence: men, 2018

Cancer	Incidence				Mortality				5-year prevalence		
	Number	%	ASR(W)		Number	%	ASR (W)		Number	%	Prop
Lung	1,368,524	14.5	31.5		1,184,947	21.9	27.1		1,313,092	6.2	34.1
Prostate	1,276,106	13.5	29.3		358,989	6.6	7.6		3,724,658	17.7	96.7
Colorectum	1,026,215	10.9	23.6		484,224	9	10.8		2,595,326	12.3	67.4
Stomach	683,754	7.2	15.7		513,555	9.5	11.7		1,025,232	4.9	26.6
Liver	596,574	6.2	13.9		548,375	10.1	12.7		471,525	2.2	12.2
Bladder	424,082	4.5	9.6		148,270	2.7	3.2		1,296,826	6.2	33.7
Esophagus	399,699	4.2	9.3		357,190	6.6	8.3		378,932	1.8	9.8
Non-Hodgkin lymphoma	284,713	3	6.7		145,969	2.7	3.3		744,561	3.5	19.3
Leukemia	249,454	2.6	6.1		179,518	3.3	4.2		669,292	3.2	17.4
Kidney	254,507	2.7	6		113,822	2.1	2.6		634,383	3	16.5
Lip, oral cavity	246,420	2.6	5.8		119,693	2.2	2.8		628,799	3	16.3
Pancreas	243,033	2.6	5.5		226,910	4.2	5.1		148,493	0.7	3.9
Brain, nervous system	162,534	1.7	3.9		135,843	2.5	3.2		395,680	1.9	10.3
Larynx	154,977	1.6	3.6		81,806	1.5	1.9		425,697	2	11.1
Melanoma of skin	150,698	1.6	3.5		34,831	0.6	0.78		504,509	2.4	13.1
Thyroid	130,889	1.4	3.1		15,557	0.3	0.35		432,092	2	11.2
Gallbladder	97,396	1	2.2		70,168	1.3	1.6		110,454	0.5	2.9
Nasopharynx	93,416	1	2.2		54,280	1	1.3		252,381	1.2	6.6
Multiple myeloma	89,897	0.9	2.1		58,825	1.1	1.3		207,929	1	5.4
Oropharynx	74,472	0.8	1.8		42,116	0.8	0.99		219,128	1	5.7
Testis	71,105	0.8	1.7		9507	0.2	0.23		284,073	1.3	7.4
Hypopharynx	67,496	0.7	1.6		29,415	0.5	0.69		97,029	0.5	2.5
Hodgkin lymphoma	46,559	0.5	1.1		15,770	0.3	0.37		157,379	0.7	4.1
Penis	34,475	0.4	0.8		15,138	0.3	0.35		93,850	0.4	2.4

Salivary glands	29,256	0.3	0.69	13,440	0.2	0.31	67,727	0.3	1.8
Kaposi's sarcoma	28,248	0.3	0.68	13,117	0.2	0.32	60,258	0.3	1.6
Mesothelioma	21,662	0.2	0.48	18,332	0.3	0.4	20,735	0.1	0.54
All cancers	9,456,418	100	218.6	5,385,640	100	122.7	21,014,830	100	545.7

Incidence and mortality data for all ages. 5-year prevalence for adult population only
Age-standardized rates (world) and proportions per 100,000
Data source: http://gco.iarc.fr/today/online-analysis-table

Table 23.4 Global estimated incidence, mortality, and 5-year prevalence: women, 2018

Cancer	Incidence			Mortality			5-year prevalence		
	Number	%	ASR (W)	Number	%	ASR(W)	Number	%	Prop
Breast	2,088,849	24.2	46.3	626,679	14.9	13	6,875,099	32.7	181.8
Colorectum	823,303	9.5	16.3	396,568	9.5	7.2	2,194,309	9.6	58
Lung	725,352	8.4	14.6	576,060	13.8	11.2	816,872	3.6	21.6
Cervix uteri	569,847	6.6	13.1	311,365	7.4	6.9	1,474,265	6.5	39
Thyroid	436,344	5.1	10.2	25,514	0.6	0.49	1,565,754	6.9	41.4
Corpus uteri	382,069	4.4	8.4	89,929	2.1	1.8	1,283,348	5.6	33.9
Stomach	349,947	4.1	7	269,130	6.4	5.2	564,520	2.5	14.9
Ovary	295,414	3.4	6.6	184,799	4.4	3.9	762,663	3.3	20.2
Liver	244,506	2.8	4.9	233,256	5.6	4.6	203,685	0.9	5.4
Non-Hodgkin lymphoma	224,877	2.6	4.7	102,755	2.5	2	608,712	2.7	16.1
Leukemia	187,579	2.2	4.3	129,488	3.1	2.8	505,141	2.2	13.4
Pancreas	215,885	2.5	4	205,332	4.9	3.8	134,081	0.6	3.5
Esophagus	172,335	2	3.5	151,395	3.6	3	168,172	0.7	4.4
Kidney	148,755	1.7	3.1	61,276	1.5	1.1	391,347	1.7	10.3
Brain, nervous system	134,317	1.6	3.1	105,194	2.5	2.3	375,430	1.6	9.9
Melanoma of skin	137,025	1.6	2.9	25,881	0.6	0.5	461,114	2	12.2
Bladder	125,311	1.5	2.4	51,652	1.2	0.87	351,656	1.5	9.3
Gallbladder	122,024	1.4	2.4	94,919	2.3	1.8	123,366	0.5	3.3
Lip, oral cavity	108,444	1.3	2.3	57,691	1.4	1.2	284,715	1.2	7.5
Multiple myeloma	70,088	0.8	1.4	47,280	1.2	0.89	168,076	0.7	4.4
Vulva	44,235	0.5	0.88	15,222	0.4	0.27	132,269	0.6	3.5
Nasopharynx	35,663	0.4	0.82	18,707	0.4	0.41	109,838	0.5	2.9
Hodgkin lymphoma	33,431	0.4	0.8	10,397	0.2	0.22	118,568	0.5	3.1
Salivary glands	23,543	0.3	0.51	8736	0.2	0.18	55,733	0.2	1.5

Larynx	22,445	0.3	0.48	12,965	0.3	0.27	63,203	0.3	1.7
Oropharynx	18,415	0.2	0.4	8889	0.2	0.19	61,380	0.3	1.6
Vagina	17,600	0.2	0.37	8062	0.2	0.16	43,877	0.2	1.2
Kaposi's sarcoma	13,551	0.2	0.33	6785	0.2	0.16	28,121	0.1	0.74
Hypopharynx	13,112	0.2	0.29	5569	0.1	0.12	22,101	0.1	0.58
Mesothelioma	8781	0.1	0.18	7244	0.2	0.14	10,515	0.04	0.28
All cancers	8,622,539	–	182.6	4,169,387	100	83.1	22,826,472	100	603.5

Incidence and mortality data for all ages. 5-year prevalence for adult population only
Age-standardized rates (world) and proportions per 100,000
Data source: http://gco.iarc.fr/today/online-analysis-table

Table 23.5 Global estimated incidence, mortality, and 5-year prevalence (both sexes, 2018)

Cancer	Incidence			Mortality			5-year prevalence		
	Number	%	ASR(W)	Number	%	ASR(W)	Number	%	Prop
Breast	2,088,849	11.6	46.3	626,679	6.6	13	6,875,099	14	181.8
Prostate	1,276,106	7.1	29.3	358,989	3.8	7.6	3,724,658	7.6	96.7
Lung	2,093,876	11.6	22.5	1,761,007	18.4	18.6	2,129,964	4.3	27.9
Colorectum	1,849,518	10.2	19.7	880,792	9.2	8.9	4,789,635	9.8	62.8
Cervix uteri	569,847	3.2	13.1	311,365	3.2	6.9	1,474,265	3	39
Stomach	1,033,701	5.7	11.1	782,685	8.1	8.2	1,589,752	3.2	20.8
Liver	841,080	4.7	9.3	781,631	8.1	8.5	675,210	1.4	8.8
Corpus uteri	382,069	2.1	8.4	89,929	0.9	1.8	1,283,348	2.6	33.9
Thyroid	567,233	3.1	6.7	41,071	0.4	0.42	1,997,846	4.1	26.2
Ovary	295,414	1.7	6.6	184,799	1.9	3.9	762,663	1.6	20.2
Esophagus	572,034	3.2	6.3	508,585	5.3	5.5	547,104	1.1	7.2
Bladder	549,393	3	5.7	199,922	2.1	1.9	1,648,482	3.4	21.6
Non-Hodgkin lymphoma	509,590	2.81	5.7	248,724	2.6	2.6	1,353,273	2.8	17.7
Leukemia	437,033	2.4	5.2	309,006	3.2	3.5	1,174,433	2.4	15.4
Pancreas	458,918	2.5	4.8	432,242	4.5	4.4	282,574	0.6	3.7
Kidney	403,262	2.2	4.5	175,098	1.8	1.8	1,025,730	2.1	13.4
Lip, oral cavity	354,864	1.9	4	177,384	1.8	2	913,514	1.9	12
Brain, nervous system	296,851	1.6	3.5	241,037	2.5	2.8	771,110	1.6	10.1
Melanoma of skin	287,723	1.6	3.1	60,712	0.6	0.63	965,623	2	12.7
Gallbladder	219,420	1.2	2.3	165,087	1.7	1.7	233,820	0.5	3.1
Larynx	177,422	0.9	2	94,771	1.2	1	488,900	1	6.4
Multiple myeloma	159,985	0.8	1.7	106,105	1.1	1.1	376,005	0.8	4.9
Testis	71,105	0.4	1.7	9507	0.09	0.23	284,073	0.6	7.4
Nasopharynx	129,079	0.7	1.5	72,987	0.8	0.84	362,219	0.7	4.7

Oropharynx	92,887	0.5	51,005	0.5	0.57	280,508	0.6	3.7
Hodgkin lymphoma	79,990	0.4	26,167	0.3	0.3	275,947	0.6	3.6
Hypopharynx	80,608	0.4	34,984	0.4	0.39	119,130	0.2	1.6
Vulva	44,235	0.2	15,222	0.2	0.27	132,269	0.3	3.5
Penis	34,475	0.2	15,138	0.16	0.35	93,850	0.2	2.4
Salivary glands	52,799	0.3	22,176	0.2	0.24	123,460	0.3	1.6
Kaposi's sarcoma	41,799	0.2	19,902	0.2	0.24	88,379	0.2	1.2
Vagina	17,600	0.09	8062	0.09	0.16	43,877	0.1	1.2
Mesothelioma	30,443	0.16	25,576	0.3	0.26	31,250	0.06	0.41
All cancers	18,078,957	100	9,555,027	100	101.1	43,841,302	100	574.4

Incidence and mortality data for all ages. 5-year prevalence for adult population only
Age-standardized rates (world) and proportions per 100,000
Data source: http://gco.iarc.fr/today/online-analysis-table

(ionizing, ultraviolet, electromagnetic); pollution of air, water, and soil; pharmaceutical products; naturally occurring chemical carcinogens, etc. Throughout the globe, tobacco is used in various forms, both in smoked and smokeless forms. Out of 7000 chemical compounds present in tobacco smoke, many of those are known carcinogens. Carcinogens present in tobacco smoke lead to carcinogenesis through multiple pathways, including DNA binding and mutations, inflammation, oxidative stress, and epigenetic changes (WHO 2014). Smokeless tobacco also contains more than 3000 chemicals; many of them are carcinogens. In smokeless tobacco use, DNA binding and mutations are among the mechanisms clearly implicated in carcinogenesis (WHO 2014). There is sufficient epidemiological evidence for causal associations between tobacco smoking and various types of cancer. Epidemiological studies have shown that smokeless tobacco causes cancers of the oral cavity and pancreas (WHO 2014). Rise in tobacco use in a country is followed by rising cancer rate. In both the sexes, the increase and decrease in the percentage of people smoking are followed by an increase and a decrease in tobacco-related deaths from diseases including cancers 20 years later. This long latent period is needed to develop the tobacco-related disease among tobacco users (WHO 2014).

Since the beginning of the twentieth century, the relationship between alcohol consumption and cancer risk has been known to us. There is evidence that alcohol consumption causes cancers of the mouth, pharynx, larynx, esophagus, liver, colorectum, and female breast. A dose–response association has been established for cancer types caused by alcohol consumption. Cancers attributable to alcohol were seen more among men, and among them liver cancer accounts for the majority of death (WHO 2014).

Infections have been identified as strong risk factors for specific cancers. Various viruses, bacteria, and macro-parasites have been identified with specific cancers. In 2008, 16% of the total new cancer cases were attributable to infections. This is not uniform throughout the globe. It is the highest in sub-Saharan Africa and lowest in North America, Australia, and New Zealand. *Helicobacter pylori* is responsible for gastric cancer. Hepatitis B and C viruses are important risk factors for liver cancer, and human papillomaviruses are risk factors for cervical cancer. Infection with human immunodeficiency virus (HIV) increases the risk of virus-associated cancers, through suppression of immunity (WHO 2014).

Reproductive and menstrual factors are important for the development of breast, endometrial, and ovarian cancers. Obesity is a risk factor for these cancers in women. The influence of obesity is most likely mediated through hormonal mechanisms. Oral contraceptives use appears to increase the risk of breast and cervical cancers but substantially reduces the risk of endometrial and ovarian cancers. Early marriage, early childbirth, and multiple sex partners increase the risk of cervical cancer (WHO 2014).

Risk of cancers of esophagus, colon, pancreas, endometrium, kidney, and postmenopausal breast increases with excess body fat. Regular physical activity helps in weight control, which thereby reduces risks of many cancers. There is an association between high consumption of red meat and colorectal cancer.

From the late eighteenth century, large numbers of cases of scrotal cancer have been reported among chimney sweepers in the Western world. Large numbers of lung cancer cases were reported among miners, and bladder cancer cases among workers in coal tar production unit. It pointed toward the relationship of occupational exposure and cancer. Many occupational agents have been identified as carcinogenic to humans and many more agents and occupational exposures are probably carcinogenic (WHO 2014).

A striking example of cancer due to radiation is the Japanese cities of Nagasaki and Hiroshima, where cancer rate was much higher among the bomb-exposed persons than the general population. Exposure to both natural and man-made ionizing radiation increases the risk of various types of malignancy. The risk of cancer is higher if the exposure occurs early in life. Cancer incidence rates among patients exposed to radiation for diagnostic and therapeutic reasons, workers in nuclear plants, and others are substantial. Exposure to ultraviolet radiation—both from the sun and from tanning devices—is established to cause all types of skin cancers, including melanoma. Associations between extremely low-frequency magnetic fields and cancer are restricted to increased risk of childhood leukemia, but causal relationship has not been recognized. More research is needed to establish the associations between heavy use of mobile phones and certain brain cancers (WHO 2014).

Environment gets polluted by different known, probable, and possible carcinogenic substances and they can gain entry to human body, and all people carry traces of those pollutants in their bodies. Air pollution from vehicle emissions, power generation, and household combustion of solid fuels and from a range of industries include known carcinogens such as diesel emissions, polycyclic aromatic hydrocarbons, and benzene, together with inorganic carcinogens such as asbestos, arsenic, and chromium compounds. Outdoor air pollution in general is an important cause of lung cancer. In poor countries, indoor air pollution due to combustion of solid fuels like coal and wood causes lung cancer and it is also associated with other diseases and cancers.

In drinking water usually derived from deep tube well, inorganic arsenic is a recognized carcinogen. Other contaminants in water, such as disinfection by-products, organic solvents, nitrates, nitrites, and some pesticides, may also contribute to an increased cancer burden. In newly and rapidly industrializing countries, monitoring and regulation mechanisms are not well established. Pollution levels can be particularly high in those countries. As the occupational exposure is not the same in different regions of the world, pollution of air, water, and soil contributes to the world's cancer burden differently (WHO 2014).

Pharmaceutical agents may also have the potential to induce cancer. Antineoplastic agents used in cancer therapy can induce second cancers in patients who are apparently cured; this is most readily attributable to the genotoxicity of these agents. Immunosuppressants, hormonal agents, and phenacetin are also carcinogenic to humans apart from antineoplastic agents. Some drugs are implicated, but not established, as carcinogenic to humans due to limited epidemiological data (WHO 2014).

Some of the chemical products from plants, fungi, lichens, and bacteria have unique pharmacological effects; these are termed as naturally occurring chemical carcinogens. Human beings are exposed to many of these bioactive substances in their daily life through food and water. Many natural products are used as pharmaceuticals and herbal remedies (WHO 2014).

One of the important functions of epidemiology is to control the disease condition with the available knowledge. Cancer registry is an important measure to understand the distribution of cancer and outcome of different cancer.

23.4 Cancer Registry

The cancer registry is a repository, where data about cancer are collected. These collected data are stored and analyzed using different data mining techniques to have realistic information about various aspects of cancers in the population. Distribution patterns of different cancers in relation to time, place, and person are made available from these cancer registries. It also provides data about the outcome of different types of cancer treatments. Cancer registry is of two main types: hospital-based cancer registry and population-based cancer registries.

Hospital-based cancer registries have information about the patients treated in a particular hospital. It is useful in trend analysis and also for treatment outcome. But incidence cannot be calculated from it because there is no defined catchment population, that is, the populations from which all the cases arise.

In population-based cancer registries, data are collected about all the new cases of cancer occurring in the population of a well-defined geographical or political boundary. Here, we can calculate the different rates (incidence rate, prevalence rate, etc.) of cancer in that population. The first cancer registry came up in the European countries. In 1900, the first attempt was made in Germany to register all cancer patients. The population-based cancer registry was also first set up in Hamburg in Germany in 1926. Nurses visited hospitals and doctors in the city to collect information about cancer patients. Population-based cancer registries started in other places also. At present, there are more than 200 population-based cancer registries in various parts of the world. But these registries cover only a small percentage of the world population. The coverage is much more in the developed countries compared to that in the developing countries. Moreover, in developing countries, registries are mostly hospital-based registries in urban areas; rural areas are less represented in those countries.

In some countries, population-based cancer registry covers the total population of the country. Countries like the UK, Canada, Australia, New Zealand, Israel, and Cuba have nationwide cancer registry. In most countries, population-based cancer registries do not cover the total population; only a proportion of the population (e.g., India) is covered. In some other countries, specialized cancer registry covers special age groups or special cancer types. In the UK, childhood cancer registry operates in Oxford. In Dijon, France, gastrointestinal cancer registry has been established (Dos Santos Silva 1999).

23.5 Prevention

Primary prevention means action taken before the development of cancer. Cancer is not caused by a single agent; it is multifactorial. We have sufficient knowledge to prevent around 40% of cancers. Many of the cancer risk factors are also risk factors for other chronic noncommunicable diseases like cardiovascular disease and diabetes. Cancer prevention should be integrated with prevention of other chronic noncommunicable diseases. Important identified risk factors are tobacco use, alcohol use, dietary factors including low fruit and vegetable intake, physical inactivity, overweight and obesity, infections (sexually transmitted human papillomavirus infection, hepatitis B, hepatitis C), air pollution (outdoor and indoor), reproductive factors, and occupational carcinogens. Activities aimed at reducing the levels of the above risk factors in the population will reduce the incidence of cancer along with the other conditions that share these risks. Measures which would mitigate the possibility of cancer are reduction in tobacco and alcohol consumption, following healthy diet, and increasing physical activity. Avoiding environmental and occupational carcinogens and shielding from ionizing radiation will definitely be helpful in reducing the cancer incidence.

For utilization of knowledge into preventive action, we need behavioral change and communication. It will be more effective if we can use interpersonal communication. Effective health education should start in early childhood at the family level, to be continued in schools and colleges. Legal measures are being used to restrict the use of harmful substances like tobacco, alcohol, and junk food, along with advertisement to disfavor its use, etc. For reducing occupational and environmental exposure, legal measures are very important and must be tried specially in developing and underdeveloped countries. Use of modern technology is also important to minimize occupational and environmental exposure.

Specific protection from some cancers may be achieved through vaccination, like human papillomavirus (HPV) vaccine and hepatitis B vaccine (WHO 2018a, b, c). Two types of HPV vaccines are now available: a bivalent and another quadrivalent vaccine. There are various serotypes of HPV virus and available vaccines are highly effective in preventing infection with virus types 16 and 18. These two virus types are responsible for approximately 70% of cervical cancer cases in the world. They are also highly effective in preventing precancerous cervical lesions caused by these two virus types. The primary target group is young adolescent girls, aged 9–14 years.

Hepatitis B vaccine should be given immediately after birth to the newborn baby to prevent perinatal transmission from mother to baby. This is called birth dose of hepatitis B. This must be given as early as possible, preferably within 24 h of birth. This should be followed by 2 or 3 more doses to complete the primary vaccination against hepatitis B. There is no need for a booster dose of hepatitis B vaccine in these children. This schedule possibly gives lifelong protection against hepatitis B.

Secondary prevention means diagnosing the condition at the earliest, ideally before clinical manifestation if not possible in precancerous stage. Early detection of cancer will reduce suffering and disabilities; early detection and proper treatment may lead to cure. It is likely to prolong life; treatment will also be less costly. Early detection is possible by creating awareness about the danger signal of the cancer and making the diagnostic facilities available and accessible to the people at affordable costs.

23.6 Warning Signals of Cancer

The major warning signals of cancer are as follows:

(i) Lumps in the breast, (ii) nonhealing ulcer, (iii) a mole that changes in shape, size, or color or bleeds in unusual circumstances, (iv) an ongoing cough or hoarseness that lasts longer than three weeks, (v) indigestion or difficulty in swallowing, (vi) a change in bowel or bladder habits for no good reason, (vii) unexplained loss of appetite, (viii) unexplained weight loss or tiredness, and (ix) unexplained bleeding from any natural orifice (Irish Cancer Society 2018).

Cancer screening means detection of cancer or precancerous lesion before clinical manifestations by clinical examination and laboratory investigation. Cancer screening program must be coupled with effective treatment of cancer. Screening program is available and routinely practiced only for a few conditions. Some of the screening methods are (WHO Factsheet 2018) visual inspection with acetic acid (VIA) for cervical cancer, mammography screening for breast cancer, HPV testing for cervical cancer, and PAP cytology test for cervical cancer. Any screening programs should consider the following:

Sensitivity: The ability of a screening procedure to detect cancer in those who have the disease.
Specificity: The ability of the screening test to identify correctly the persons who do not have the disease.
Positive predictive value: This gives the percentage of persons who have the disease among those tested positive in a screening test.
Negative predictive value: This gives the percentage of persons who are free of the disease among those tested negative in a screening test.
Acceptability: Whether the test or procedure is acceptable to intended persons.

23.7 Treatment

For adequate and effective treatment correct diagnosis of cancer is essential. As cancer is not a single disease, treatment is different for different types of cancer. For treatment of cancer, one or more modalities such as surgery, radiotherapy, chemotherapy, and immunotherapy may be employed. For successful cancer control, service should be organized at different levels of health care.

At **primary level** medical officers and community care workers should work for early referral of suspected cases. They are also involved in the follow-up care of patient and retrieval of patients who abandon treatment. They are instrumental in organizing patient support groups, and education and training of patients and community caregivers. Rehabilitation of patients is an important work at the primary level. Community-level health workers can be involved in the management of cancer patients.

Secondary level of care is usually rendered at district hospitals. Here specialist and improved imaging (X-ray, ultrasonography, mammography, endoscopy) and laboratory services (cytology including fine-needle aspiration, hematology, biopsy, routine histopathology) are available. For treatment of patients moderately complex surgery and chemotherapy (mainly outpatient clinics) are made available. At secondary care facility also services such as rehabilitation, psychosocial support, self-help groups, and patient education programs should be made available.

Tertiary-level care can be organized at regional or national level. Here all the services like secondary level should be available. These centers are also involved in training and research in cancer. Most sophisticated diagnostic and treatment modalities should be present in this level. Cancer treatment is highly specialized and requires the involvement of various disciplines (WHO 2008a, b).

Surgery for cancer should be made available from the district level onwards. Surgery for common cancer can be done at the secondary level; complex surgery can be done at the tertiary-level hospitals, like medical colleges or other regional or national level organizations. Surgical treatment of cancer is usually done in conjugation with other modalities like chemotherapy or (and) radiotherapy.

Radiotherapy is usually available at the tertiary-level organization. It requires highly trained personnel and costly equipment. These instruments are usually imported from developed countries and their continuous servicing is a challenge.

Chemotherapy can also be made available from the secondary level onwards. Chemotherapeutic agents are made by a few companies and are usually very expensive. Chemotherapy may have severe side effects and often requires patients to undergo a prolonged period of treatment. Psychosocial support should be an integral part of the treatment services for cancer patients.

When the disease is in advanced stage, cure is not always possible. Cancer usually causes death in months or years. During this period, the patient needs palliative measures including pain management (WHO 2008a, b). Teams having medical, other professional, and community care workers are required for providing palliative care. Psychosocial support for patients and caregivers is also required.

Diagnosis of cancer brings a devastating change in the patients and their family. Cancer palliative cares provide psychosocial and supportive care for patients and their families. This is more so for the patients in advanced stages with little chance of cure. Cancer and its management have profound emotional, spiritual, social, and economic consequences. Only diagnosis and treatment of cancer are not sufficient. Palliative care is very important in cancer. Palliative care is essential for terminally ill cancer patients and their family members. Palliative care is very important for quality of life for the patients and their family members. It helps them to effectively deal with the situation after diagnosis of the cancer. It includes pain relief and it emphasizes on home-based care. Access to oral morphine has made the life less painful for the terminally ill patients. In many countries, legislative measures have been advocated to make it available for the cancer patients for pain relief.

23.8 Conclusion

With increase in life expectancy, change of lifestyle, and environmental pollution, cancer is rising throughout the globe. But with early diagnosis, many cancers are curable and in others life can be prolonged. With identification of cancer risk factors, we can institute different preventive strategies to prevent cancer and other noncommunicable diseases in the population.

References

Dos Santos SI (1999) Cancer epidemiology: principles and methods. International Agency for Research on Cancer, WHO, Lyon, France

GLOBOCAN (2018). Global Cancer Observatory (2019, March 02). Retrieved from: http://gco.iarc.fr/today/data/factsheets/cancers/39-All-cancers-fact-sheet.pdf

Gordis L (2013) Epidemiology, 5th edn. WB Saunders Company, Philadelphia

Irish Cancer Society (2018, 02 February) Retrieved from https://www.cancer.ie/reduce-your-risk/mens-health/early-warning-signs-you-cannot-ignore#sthash.BZ6FuEp6.dpbs

Jemal A, Clegg LX, Ward E, Ries LA, Wu X, Jamison PM, Wingo PA, Howe HL, Anderson RN, Edwards BK (2004) Annual report to the nation on the status of cancer, 1975–2001, with a special feature regarding survival. Cancer 101(1):3–27

Last JM (2001) A dictionary of epidemiology, 4th edn. Oxford University Press, New York. NY

Manohar PR (2015) Description and classification of cancer in the classical Ayurvedic texts. Indian Journal of history of Sciences 50(2):187–195

Park K (2019) Park's textbook of preventive and social medicine, 25th edn. M/S Banarsidas Bhanot, Jabalpur, India

Pastides H (2018) The descriptive epidemiology of cancer. Jones and Bartlett Publishers. Available from: http://www.jblearning.com/samples/076373618X/3618X_CH01_001_028.pdf

WHO (1980). WHO Chron 34(5), 189. Geneva, WHO

WHO (2008a) Cancer control. In: Knowledge into action: WHO Guide for effective programme. Palliative care. WHO, Geneva

WHO (2008b) Cancer control. In: Knowledge into action: WHO Guide for effective programme. Diagnosis and treatment. WHO, Geneva

WHO (2014) In: Stewart BW, Wild CP (eds) World cancer report 2014. International Agency for Research on Cancer, Lyon, France

WHO (2018a, February 02). GLOBOCAN 2012: Estimated cancer incidence, mortality and prevalence worldwide. International Agency for Research on Cancer, Lyon, France. Retrieved from: http://globocan.iarc.fr

WHO (2018b, 02 February) Human Papilloma Virus (HPV). Retrieved from: http://www.who.int/immunization/diseases/hpv/en/

WHO (2018c, February 12) Hepatitis B Virus. Retrieved from: http://www.who.int/immunization/diseases/hepatitis B/en/

WHO Factsheet. (2018, June 28). Retrieved from: www.who.int/en/news-room/fact-sheets/detail/cancer

Cancer Genomics and Diagnostics: Northeast Indian Scenario

24

Sharbadeb Kundu, Raima Das, Shaheen Laskar, Yashmin Choudhury, and Sankar Kumar Ghosh

Abstract

Cancer genomics, which deals with the study of the sum of DNA sequence and differences in patterns of gene expression between cancerous cells and healthy cells, helps to unfold the genetic origin and evolution of cancer genome, ultimately creating new therapeutic interventions. The era of cancer genome will take advantage of P4 medicine to quantify wellness and demystify disease. Future advancements in this field will follow extensive genetic testing, helping scientists to prepare drugs directly targeting those changes causing cancer. The ultimate goal is to provide proper treatment with precise dose—having negligible or no toxicity—for the specific patient to meet the need of the hour.

Keywords

Cancer genomics · Diagnostics · Northeast India · P4 medicine

24.1 Introduction

Cancer can be defined as a severe situation when a cluster of anomalous cells divide and multiply uncontrollably without following the standard rules of cell division. If this proliferation is continued and spread to other parts of the body through blood circulation or lymph vessels (metastasis), where they again start growing, it can be fatal and these phenomena actually are observed in almost 90% of cancer-related

---

S. Kundu
Genome Science, School of Interdisciplinary Studies, University of Kalyani, Nadia, West Bengal, India

R. Das · S. Laskar · Y. Choudhury · S. K. Ghosh (✉)
Department of Biotechnology, Assam University, Silchar, Assam, India

© The Author(s), under exclusive license to Springer Nature Singapore Pte Ltd. 2022
S. K. Basu et al. (eds.), *Cancer Diagnostics and Therapeutics*,
https://doi.org/10.1007/978-981-16-4752-9_24

deaths (Hejmadi 2009). The Greek physician Hippocrates (460–370 B.C.) first coined the term *karkinos* (crab or crayfish) to explain the crab-like feature of the carcinoma tumour or cancer, but he was not the first to discover this disease. Evidence of human bone cancer was found in some ancient Egyptian manuscripts dated about 1600 B.C. and also in ancient mummies. Interestingly, around 1500 B. C., the oldest recorded breast cancer case on this planet came from ancient Egypt (Sudhakar 2009). Besides, John Hill (a Physician) labelled tobacco snuff as the reason of nose cancer in 1761 (Hajdu 2011) and the microscopic view of metastasis was first articulated by Campbell De Morgan, an English surgeon between 1871 and 1874 (Grange et al. 2002).

Globally, after cardiovascular diseases, carcinoma/cancer has become the second major cause of death and it is estimated that every sixth person dies in the world due to cancer (Schutte 2017). In the year 2016, around more than 1.7 million people residing in the United States are anticipated to take a cancer diagnosis in the United States (NCI 2018). India is possibly looking at over 17 lakh new cancer cases and over 8 lakh cancer deaths by the year 2020 according to a report (Katoch 2016). The north-eastern region of India has become the Cancer Capital (*https://blog.onco.com/ north-east-india-cancer-capital/*) with the age-adjusted incidence rates of cancers of the gall bladder, stomach cancer, oesophageal cancers, tongue cancer in females, and nasopharyngeal cancers highest in this part of India (Krishnatreya and Kataki 2016). In India, an average number of cancer incidence cases are being reported as 80–110 per one lakh inhabitants, whereas in the case of Northeast (NE) India, this average number of reported cases is raised to around 150–200 per one lakh people (*https:// blog.onco.com/north-east-india-cancer-capital/*).

24.2 Types of Cancers

There are diverse forms of cancer on the basis of cell category, and hence, the morphology of tumour cells is assumed to be the source of the tumour growth (*https://www.news-medical.net/health/Cancer-Classification.aspx*). Different types of cancers can be divided as follows:

(a) **Carcinoma:** Carcinoma is generally instigated in the epithelial layers that make up the lining of exterior body parts or the interior liners of organs within the body. It is of two types – *adenocarcinoma* and *squamous cell carcinoma (SCC)*.
(b) **Sarcoma:** This form of cancer is derived from the connective and supportive tissues (that is, cartilage, bone, nerve, and fat), which generally develop in the *mesenchymal cells* outside the bone marrow.
(c) **Myeloma:** This category of cancer is generally initiated in the plasma cells, a type of leukocyte formed in the bone marrow.
(d) **Lymphoma and leukaemia:** These both types of blood cancer originate from the haematopoietic (blood-forming) cells leaving the marrow and have a tendency to mature in the lymph nodes and blood, respectively. Unlike leukaemia or "liquid cancer" that affects the blood, lymphomas are referred as "solid

cancers". These may create disturbances in lymph nodes at the particular places like brain, stomach, intestines, etc.
(e) **Germ cell tumour:** Malignancies originating from the pluripotent cells, most frequently present in the testicle or the ovary (seminoma and dysgerminoma, respectively) usually fall under this category.
(f) **Blastoma:** Cancers resulting from the immature *precursor* cells or embryonic tissue.

24.3 Cancer Genomics

Cancer refers to assemblage of related ailments and is characterised by the uncontrolled progress that often invades surrounding tissue and can metastasise to distant sites. This disease can affect almost any part of the body (*www.who.int/topics/cancer/en/; www.cancer.gov/about-cancer/understanding/what-is-cancer*). Also known as the *disease of the genome*, it arises from a series of changes in DNA that modifies cellular behaviour, causing uncontrollable growth, ultimately leading to malignancy (Macconaill and Garraway 2010). All cancer types arise as a result of changes in the DNA of cancer cells such as mutations, polymorphisms, rearrangements, deletions, and amplifications and some epigenetic modifications of genomic DNA or histone proteins such as methylation, acetylation, and other mechanisms known as the key mediators of cancer (Macconaill and Garraway 2010). Some of the commonly affected genes mutated in cancer are given in Table 24.1.

Genetic changes that promote cancer can be of both innate and acquired type. The innate genetic changes that are present in the germ cells can be inherited from parents and also known as the germline variations, whereas when the genetic changes are acquired during an individual's lifetime as a result from exposure to various carcinogens, they are termed as somatic (or acquired) variations that account for almost 90–95% of all the cases of cancer (*www.cancer.gov/about-cancer/causes-prevention/genetics*). Evidence suggested that carcinogenesis, a hereditary disorder, was identified in tumour-specific translocations within lymphoma and leukaemia. This, in turn, indicated the importance of transcriptional regulations of oncogenes and other related genes in cancer causation. Furthermore, over around 30 genes commonly mutated in hereditary cancer cases have already been cloned (Kasahara and Tsukada 2004). Those mutant genes were mostly recognised as the tumour suppressor genes, and a few of them were identified as the DNA repair or oncogenes. Additional somatic mutations also play a crucial role in the expression of genes and penetrance at some stage in carcinogenesis. Apart from genetic factors, cancer is also caused by many external factors like various dietary factors such as consumption of smoked fish and meat, salted fish, and fermented foods and non-dietary factors such as tobacco chewing and smoking, alcohol habits, and occupational sources that ultimately cause DNA damage (*https://www.cancer.gov/about-cancer/causes-prevention/risk/substances*). Some examples of carcinogens resulting from occupational measure are given in Table 24.2.

Table 24.1 Some known cancer susceptibility genes and regions (Collins and Politopoulos 2011)

Sl. No.	Screening/ mapping technique	Mapped location	Candidate gene/genetic region	Allele frequency	Known/possible function
1	Linkage	17q21	*BRCA1*	Rare	Genome stability/DNA repair
2	Linkage	13q13.1	*BRCA2*	Rare	Recombinational repair
3	Linkage	17p13.1	*TP53*	Rare	Apoptosis and li–Fraumeni syndrome
4	Candidate gene resequencing	11q22.3	*ATM*	Rare	DNA repair
5	Candidate gene resequencing	17q23.2	*BRIP1*	Rare	Associated with *BRCA1* and DNA repair
6	Candidate gene resequencing	22q12.1	*CHEK2*	Rare	DNA repair/cell cycle
7	Candidate gene resequencing	16p12.2	*PALB2*	Rare	Associated with *BRCA2*
8	Candidate gene resequencing	17q22	*RAD51C*	Rare	Homologous recombination repair
9	Linkage	10q23.3	*PTEN*	Rare	Cell signalling and Cowden disease
10	Linkage	19p13.3	*STK11 (LKB1)*	Rare	Cell cycle arrest and Peutz–Jeghers syndrome
11	Linkage	16q22.1	*CDH1*	Rare	Intercellular adhesion: Lobular BC
12	GWAS	10q26	*FGFR2*	Common	Fibroblast growth factor receptor
13	GWAS	16q12	*RBL2/TOX3 (TNRC9)*	Common	Cell cycle/chromatin structure
14	GWAS	5q11.2	*MAP3K1*	Common	Cellular response to growth factors
15	GWAS	11p15.5	*LSP1*	Common	Neutrophil motility
16	GWAS	8q24	8q24	Common	Intergenic, enhancer of MYC proto-oncogene[a]
17	GWAS	2q35	2q35	Common	Transcriptional regulation
18	GWAS	2q33	*CASP8*	Common	Apoptosis
19	GWAS	3p24.1	*NEK10/SLC4A7*	Common	Cell cycle control[a]
20	GWAS	17q22	*STXBP4/COX11*	Common	Transport[a]
21	GWAS	5p12	*MRPS30*	Common	Apoptosis[a]
22	GWAS	1p11.2	*FCGR1B/NOTCH2*	Common	Signalling/immune response[a]
23	GWAS	14q24.1	*RAD51L1*	Common	Homologous recombination repair[a]
24	GWAS	9p21	*CDKN2B/CDKN2A*	Common	Cyclin-dependent kinase inhibitors[a]

(continued)

Table 24.1 (continued)

Sl. No.	Screening/ mapping technique	Mapped location	Candidate gene/genetic region	Allele frequency	Known/possible function
25	GWAS	11q13	CCNDL/ MYEOV	Common	Cell cycle control/ fibroblast growth factors[a]
26	GWAS	10q21.2	ZNF365	Common	Zinc finger protein gene
27	GWAS	10p15.1	FBXO18/ ANKRD16	Common	Helicase[a]
28	GWAS	10q22.3	ZMIZ1	Common	Regulates transcription factors[a]

[a]Possible gene or function in context to cancer development

Cancer genomics is the study of total alterations of DNA sequence and outlines of gene expression between cancerous cells and typical cells. It helps to comprehend the genetic background of tumour cell propagation and the evolutionary context of the cancer genome brought about by mutation and selection in the body environment, with the help of the immune system and therapeutic interventions (*https://www.nature.com/subjects/cancer-genomics*). By the late 1990s, the study of cancer genome era commenced with the *oligonucleotide microarrays* and high-throughput DNA sequencing that provided extraordinary insights into the entire genome along with powerful experimental and analytic methodologies, and computational approaches. This enabled a huge acceleration in finding genomic alteration-related discoveries. One such evidence recognised mutation at codon 12 of oncogene like *HRAS* gene-related to bladder cancer. This particular milestone first gave clue about the genomic basis of cancer (Tabin et al. 1982).

The research area of cancer genomics takes advantage of technological advances for studying the molecular basis of cancer growth, metastasis, and drug resistance. The sequencing of the DNA and RNA of cancer cells as well as normal cells helps to identify the genetic differences that may cause cancer (*https://www.cancer.gov/about-nci/organization/ccg/cancer-genomics-overview*). Moreover, advanced technologies such as *next-generation sequencing* have also been used to explain the background of cancer genome and to ascertain new disease-associated genetic changes. These valuable data up-and-coming from cancer genomics studies should be coupled with the clinical history of the patients and medical/therapeutic data for more tailored approaches, resulting in improved way of foreseeing cancer risk and treatment responsiveness (*https://www.cancer.gov/research/areas/genomics*).

24.4 Molecular Diagnostics in Cancer

Molecular diagnostics refers to a collection of techniques that are used to analyse biological markers in the genome and proteome. These techniques assist in diagnosis, disease monitoring, and risk detecting and help to select which therapies will

Table 24.2 Examples of biological, biophysical, and biochemical association with human cancers (Parsa 2012)

Sl. no.	Carcinogenic agents	Cancer sites	Occupations/sources
1	Arsenic	Lungs and skin	Electricians, smelters, and medical imaging procedures
2	Asbestos	Mesothelioma and lungs	Roof and floor tiles and asbestos
3	Benzene	Blood and lymph nodes	Petroleum, painting, detergent, and rubber
4	Beryllium	Lungs	Missile fuel and nuclear reactor
5	Cadmium	Prostate	Battery, painting, and coating
6	Chromium	Lung	Preservatives, pigments, and paints
7	Ethylene oxide	Blood	Ripening agent for fruits and gases
8	Nickel	Nose and lungs	Battery, ceramics, and ferrous alloys
9	Radon	Lung	Uranium decay, mines, and cellars
10	Vinyl chloride	Liver	Refrigerator and glues
11	Smoke	Lungs and colon	Cigar, air pollution, and car smoke
12	Gasoline	Lung and blood	Oil petroleum
13	Formaldehyde	Nose and pharynx	Hospital/laboratory workers
14	Hair dyes	Bladder	Hairdresser and barber
15	Soot	Skin	Chimney cleaners
16	Ionizing radiation	Bone marrow	Radiology technician
17	Hepatic virus B and C	Liver	Hospital workers and drug users
18	HPV/herpes viruses	Cervix, skin, and head/neck	Multiple sexual partners
19	Burkitt's virus	Lymph node	Black people in South Africa, carrier of c-myc oncogene, carcinogenic Epstein-Barr virus variants, and malaria-resistance genes
20	Helicobacter pylori	Stomach	Chronic bacterium infection

work best for individual patients (Poste 2001; Burtis and Bruns 2014). In cancer, one of the significant ways is the detection of DNA sequence variations for the assessment of cancer risk. For example, the BRCA1/2 test by Myriad Genetics evaluates women for lifetime risk of breast cancer (Khodakov et al. 2016). Several major disciplines in cancer medicine are found (Fig. 24.1), which apply molecular-based tests. These include.

- **Hereditary cancer syndromes:** Detection of germ-line mutations in corresponding genes has made the genetic diagnosis of familial cancer possible. Mutations in *BRCA1/BRCA2/PALB2*, in breast and ovarian cancer, and in

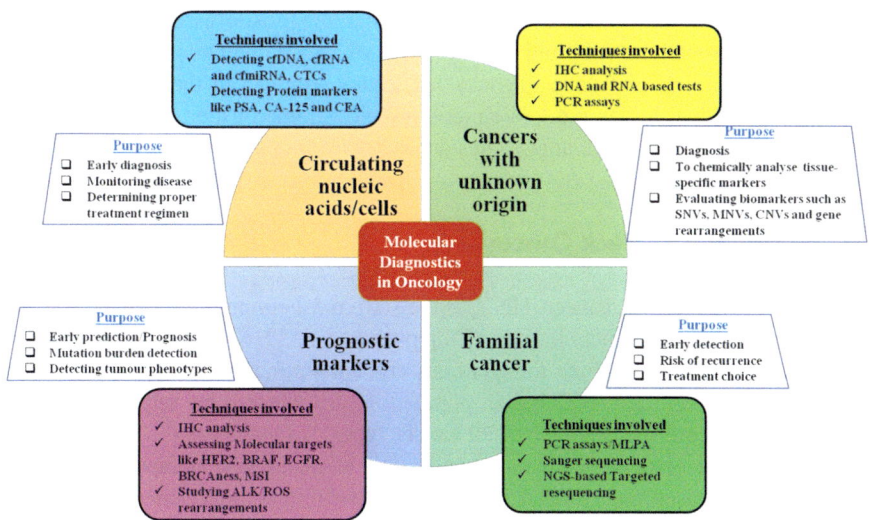

Fig. 24.1 Molecular diagnostics in oncology [adapted and modified from Sokolenko and Imyanitov (Sokolenko and Imyanitov 2018)]

hereditary non-polyposis colorectal cancer (HNPCC) and alterations in *MSH6/MSH2/MLH1/PMS2* and *EPCAM* genes contribute to major disease disposition. These mutations also play an important role in classifying persons at risk and in individualised comprehensive treatment assessment.

- **Molecular markers in cancer therapy:** Screening of oestrogen receptor expression, HER2 amplification and over-expression, and *ALK* and *ROS1* rearrangements of *KRAS/NRAS/BRAF/EGFR/MTOR/TSC1/TSC2* mutations may potentially serve as suitable predictive markers.
- **Liquid biopsy:** Tumours are almost accompanied by shedding of their remains into peritumoural space as clusters of circulating tumour cells (CTCs) (malignant in nature), fragments of nucleic acids as circulating nucleic acids (CNAs), proteins, and other small molecules, which, in turn, act as tumour markers and can be examined from different body fluids, viz. urine, saliva, serum, etc. For example, osimertinib, a new lung cancer drug, has been established with the aim of acting against T790M mutation of *EGFR* gene (Lamb and Scott 2017). This requires regular monitoring to see the effect of the drug upon the mutation, which can be achieved by regular monitoring of the patient through liquid biopsy instead of tissue biopsies.
- **Cancer diagnosis in unknown primary site:** In around 3–5% od recently diagnosed cancer patients, the metastasis of unknown tissue origin or organ is generally observed. One such example is the occurrence of TKI-sensitizing somatic *EGFR* mutation in tumour tissue, which indicates existence of lung cancer. Besides, the presence of *BRCA1/BRCA2* germline alterations in an adenocarcinoma female with unknown primary site indicates breast or ovarian cancers as the most possible tumour category (Sokolenko and Imyanitov 2018).

The overview and elevated use of molecular diagnostic assessments to spot cancer risk as well as care and management of cancer patients proved to be a major breakthrough and pave the way for future progress in the contest against this disease (AdvaMedDx n.d.). We mention below some of the types of cancer in which molecular diagnostic approaches are in use.

24.4.1 Head and Neck Cancer

Globally, head and neck cancer (HNC) is the sixth most common cancer and it is the second most common cancer in India (Tuljapurkar et al. 2016). Nearly 90–95% head and neck cancer cases belong to squamous cell carcinoma (SCC) type (Mao et al. 2004). Within Northeast (NE) Indian population, this cancer type seemed to be the most familiar acquiring 30–40% of all cancer types and stands as the seventh most common reason of death in females and sixth in males (Choudhury and Ghosh 2015b). HNC cases are induced by carcinogens or viral infection together with multiple genetic alterations in various gene groups (Riaz et al. 2014). Interaction of polymorphisms in *XRCC1* (Arg399Gln) and *XRCC2* (Arg188His) genes and tobacco exposure increases an individual's susceptibility to head and neck squamous cell carcinoma (HNSCC) in the NE Indian population. It has been observed that the alternate homozygous AA genotype of *XRCC1* Arg399Gln and heterozygous GA genotype of *XRCC2* Arg188His has an augmented disease risk. It has also been seen that null genotypes of tobacco-metabolising genes like *GSTM1* and *GSTT1* act as markers in determining the genetic vulnerability of HNSCC patients and their first-degree relatives. In addition, *GSTM1* null genotype and tobacco chewing interaction have been reported as the strongest gene-environment model for predicting HNSCC (Choudhury et al. 2015). Polymorphisms in genetic region of cytochrome P450 1A1 (*CYP1A1*) at T3801C in combination with tobacco-betel quid intake smoking and altered activity in glutathione S-transferases (GSTs) genes further modulate HNC risk (Choudhury and Ghosh 2015a). Epigenetic changes along with genetic alterations, tobacco habits like smoking and chewing, and human papilloma virus (HPV) infection are also linked with an enhanced risk of HNSCC. Hypermethylation in the promoter region of the tumour suppressor genes results in the transcriptional insufficiency and reduced gene expression. Promoter methylation profiling of ten tumour-linked genetic regions like *BRCA1, DAPK, GSTP1, ECAD, RASSF1, MLH1, p16, MINT1, MINT2*, and *MINT31* showed connection with tobacco usage, HPV infection, survival status, and genetic variation that may act as a marker to determine subtypes and patient outcome in HNSCC (Choudhury and Ghosh 2015b).

The revolutionary area of cancer research involving liquid biopsy-based development of cell-free DNA (cf-DNA) has been considered as a novel method for detection of biomarkers via next-generation sequencing (NGS) of tumour-originated cf-DNA rather than tissue biopsy for cancer-associated biomarkers (Kumar et al. 2018). The concept of liquid biopsy provides a method for minimal-invasive, real-time supervising, with prospective usability in risk assessments, therapeutic responsiveness, and onset of treatment resistance by using cf-DNA into analysis.

There were investigations that revealed association of smokeless tobacco, smoking, and alcohol habits with the differences in copy number of cell-free mitochondrial DNA (cf-mtDNA), and the levels of cf-DNA were found significantly higher among HNSCC cases than controls pointing towards its greater assurance, thus bearing the key features of diagnostic/prognostic biomarkers, with minimal invasiveness, elevated sensitivity, and specificity (Kumar et al. 2017).

Another study attempted to give the global status of the germline variations within the DNA repair gene family by considering the Human Gene Mutation Database (HGMD) and Exome Aggregation Consortium (ExAC) database for assessing the disease. The ExAC DNA repair dataset comprises 30.4% missense variants, of which 5.6% carried the deleterious SIFT and Polyphen-2 score. However, around 1.2% of those variants found to be cancer-associated as per HGMD (Das and Ghosh 2017). Furthermore, Whole Exome Sequencing (WES) revealed that the total variants in the NE Indian population were distributed among 199 DNA repair genes, and the Fanconi Anaemia (FA) and Double-Strand Break Repair (DSBR) pathway were the pathways, having the highest number of polymorphisms. The stratified association test also identified that the intronic variations in the *RAD52* and *HLTF* genetic region were notably correlated with the disease (OR > 5; $P < 0.05$), whereas the intronic variants in *PARP4, EXO1, PER1,* and *RECQL5* genetic regions and the exonic variant in *TDP2* gene showed protection against HNC (OR < 1; $P < 0.05$). Such variants probably conferring protection upon differential influence on transcriptional regulation, because either the incorporation of new transcription factor binding sites (TFBSs) may act as tumour suppressor or any impairment of TFBSs may aid in making the cancer therapy more efficient by acting as opponent against different cancer types. Additionally, the gene-gene interaction study using multifactor dimensionality reduction (MDR) study proposed that the missense variants in *BRCA2, TP53, MSH6,* and *PALB2* genes and the intronic variant in *RECQL5* genetic region collectively affect DNA repair mechanism for the development of HNC. The exonic variants have been predicted to have an impact on the structural basis and solvent availability of the respective proteins. Apart from this, few 3'-UTR variations were also noticed to cause modifications in the target sites of miRNA like Let-7, the principal family of miRNA, and known to be associated with HNC and thus were potential biomarkers for the risk assessment and constant monitoring of the treatment regimen among the cancer patients (Das et al. 2018).

24.4.2 Oral Cancer

Oral cancer ranks as the 15th most frequent cancer globally, whereas in India, it is the third most occurring cancer type. Northeastern part of India harbours one of the world's uppermost occurrences of oral carcinogenesis (Cheong et al. 2017). In the Indian subcontinent (India, Pakistan, Sri Lanka, and Bangladesh), Oral squamous cell carcinoma (OSCC) predominantly has the uppermost incidence rate, thus comprising up to 25% of the new cases (Warnakulasuriya 2009). Glutathione

S-transferase (GST) gene polymorphisms, HPV infection, and Tobacco consumption were found to be the main causative factors for the expansion of OSCC. Alterations in mtDNA are connected with different forms of cancers, signifying that they could be important causative factors in the *carcinogenesis*. Investigations showed that tobacco habits, betel quid chewing, HPV infection, presence or absence of *GSTM1-GSTT1*, and mitochondrial D-loop mutations as well as mtDNA copy number have been associated with OSCC (Mondal et al. 2013a). The association of GST null genotypes and hotspot mutations in the D-loop region can serve as the promising biomarker to detect OSCC at an early stage and to take precautionary measures against it among those persons, who are the prolonged tobacco consumers (Mondal et al. 2013b). Additionally, in OSCC patients, the mtDNA copy number within the tumour tissues fluctuates with various tumour stages and smokeless tobacco habits. As low mtDNA content recommends invasiveness, it can be utilised as an effective biomarker for the detection of OSCC (Mondal et al. 2013a, b). The complete mitochondrial genome sequencing determined the hot spot mutations in oral cancer. In total, 26 mtDNA mutations of somatic origin were revealed, out of them nine mutations are found in the D-loop region, while 17 mutations occurred in the coding region. Some hotspot mutations in mtDNA among oral cancer cases are at nucleotide positions 16,463, 16,325, and 16,294 in the D-loop and 13,869, 13,542, and 4136 in the mitochondrial coding region (Mondal et al. 2013b; Mondal and Ghosh 2013a, b). The knowledge on the mode of action, patterns, and copy number variations in mtDNA may serve to enable medical implications, and hotspot mutations in mtDNA may be supportive to evaluate risk of carcinogenesis (Mondal et al. 2013b). Another study identified one nonsense-mediated mRNA decay transcript variant associated with Benzo(a)pyrene, in the *DFNA5* gene (rs2237306), conferring protection (OR $= 0.33$; $P = 0.009$) and four harmful intronic variations (OR > 2.5; $P < 0.05$), rs1670661 in *NELL1* and rs169724, rs290974, and rs182361 in *SYK* genetic regions, related to oral cancer susceptibility together with tobacco- and HPV-mediated carcinogenesis within NE Indian population. Among the OSCC cases, 12.6% was found to be HPV infected (of them, HPV16 subtype was observed in 45.5%, HPV18 subtype was observed in 27.3%, and HPV16/18 subtype was observed in 27.3%). MDR analysis revealed that the interactions among *NELL1* variants rs1670661 and HPV with gender and age amplified the threat of both tobacco- and non-tobacco-related OSCC, respectively. These recommend HPV infection as a significant risk factor for OSCC in NE Indian population. Finally, it was emphasised that the *DFNA5*gene variant probably provided defence through nonsense-mediated mRNA decay (NMD) pathway against the tobacco-related OSCC, which was newly reported in this study. Thus, the investigative approach used in that study could be suitable for predicting the significant OSCC-associated variants, specifically for a heterogeneous population (Kundu et al. 2018).

24.4.3 Nasopharyngeal Carcinoma

Nasopharyngeal carcinoma (NPC) is an infrequent type of cancer globally. It also remains uncommon in Indian subcontinent except in the Northeastern part of India. NPC is instigated by the collective effects of genetic predisposition factors, Epstein-Barr virus (EBV) infections, and environmental carcinogens. In NE Indian population, an increased risk of NPC was seen among those frequent tobacco smokers, betel nut chewers, and alcohol drinkers, who were also having the kitchen inside their living room and regularly consume smoked fish and meat, salted as well as fermented fish. Besides, *GST-null* genotypes, decreased mtDNA copy number, and EBV infection were also observed in those people. A significant difference was also found predominantly among the patients having EBV infection and *GST*-null genotypes along with altered copy number of mtDNA. The perception of gene-environmental risk factors and their responsibility in the aetiology of NPC would be useful for taking preventive measures and for early screening of the disease (Ghosh et al. 2014). A significant link was also found between major lifestyle factors and EBV infection towards NPC development (Singh and Ghosh 2014). It has also been seen that interactions of *XRCC1* Arg399Gln and *XRCC2* Arg188His polymorphisms and environmental factors modulate susceptibility to NPC in the NE India. The *XRCC1* Gln/Gln genotype showed increased risk, and individuals carrying both *XRCC1* and *XRCC2* polymorphic variants had elevated NPC risk. An enhanced risk of NPC was also observed among individuals who were frequent consumers of smoked meat and fermented fish and those having habits of tobacco-betel quid chewing with *XRCC1* polymorphic variants. These observations might facilitate in assessing cancer risk (Singh and Ghosh 2016). It has also been speculated that the association of null genotypes in *GST*s and mutations of metabolic neutralising gene, *CYP1A1* T3801C along with the various aetiological practice (tobacco chewers and smokers, fermented fishes, and smoked meat), can assist in early detection and preventive measure of NPC by using these as a possible biomarker (Singh and Ghosh 2019).

24.4.4 Oesophageal Cancer

Oesophageal cancer is the eighth most frequent cause of cancer deaths globally, while in India, it stands fourth in terms of cancer mortality. There is an increased incidence of oesophageal squamous cell carcinoma (ESCC) in India with approximately 47,000 new cases being reported each year and the reported deaths reach up to 42,000 each year (Samarasam 2017). A very high incidence of oesophageal cancers has been reported in the NE India (Kaur et al. 2017). Complex interactions of epigenetic, genetic, and environmental factors result in ESCC development. Frequent epigenetic modifications such as promoter hypermethylation of multiple tumour suppressor genes are thought to result from certain habit-related carcinogens capable of inducing aberrant methylation. Studies have also identified that interactions of lifestyle-related factors with the *GSTM1/GSTT1* gene polymorphism induce promoter hypermethylation of multiple tumour suppressor genes. Study recognised probable interactions between carcinogen metabolising gene variations and tobacco usage towards controlling the methylation patterns in the promoter regions of tumour suppressor genes in ESCC (Talukdar et al. 2013). Infection of HPV has been strongly correlated with ESCC or squamous cell carcinoma of upper aerodigestive tract (SCC of UADT). However, probable functions of HPV infection on abnormal methylation pattern in these tumours have not been fully understood. Association of HPV with aberrant methylation pattern of many tumour-associated genes comprising the conventional CpG Island Methylator Phenotype (CIMP) panel markers (*p16, MINT1, MINT2, MINT31,* and *hMLH1*) as well as other recurrently methylated cancer-associated genes (*BRCA1, GSTP1, DAPK1, RASSF1,* and *ECAD*) and also the continued subsistence of UADT cancer patients was found by undertaking many investigations. Although HPV occurrence has no effect over the survival of total UADT cancer patients, it has been identified to be connected with a proper prognosis for the detection of HNSCC cases. Besides, three separate clusters with diverse methylation profile and incidence of HPV infection revealed that the subgroup of CIMP-high displayed the maximum HPV-infected individuals. Additionally, noteworthy synergistic interactions of tobacco and HPV towards altering the promoter hypermethylation patterns among UADT cancerous individuals were also anticipated. This study by Talukdar et al. (2015) proposed a crucial effect of HPV in driving unusual methylation pattern in some precise tumour-linked loci, which, in turn, might cause the onset and development of SCC of UADT.

24.4.5 Breast Cancer

Worldwide, around 1.35 million people are diagnosed with cancer of breast each year. One in every 17 women in India develops the disease. The majority of ovarian/breast cancer is caused by mutations in the Breast Cancer 1 (*BRCA1*) gene. A study had been performed among the NE Indian population to provide incidence of *BRCA1* germline mutations among people having relation in terms of individual as well as family history of breast cancer, where *BRCA1* mutations were identified in 6.25% and 12.5% cases, respectively. Three mutations, namely, 3889DelAG, 1014DelGT, and 185DelAG, were observed in exons 2 and 11 among the patients, resulting in generation of the truncated BRCA1 protein by introducing premature stop codons discretely at the 1265, 303, and 39 amino acid positions in the NE Indian population. These observations are recommended for a general mutation screening procedure for the high-risk breast cancer cases in this population, which, in turn, would offer an improved, decisive surgical and clinically protective option (Hansa et al. 2012). Additionally, polymorphism in *GSTM1* and *GSTT1* genes inducing mutations in various types of cancer was known, but its role in bringing alteration in the *BRCA1* gene was uncertain. In breast carcinoma patients, a noteworthy association was seen amongst the polymorphism of GSTM1/T1 and mutation in the *BRCA1* gene. The presence of *GSTM1* and *GSTT1*-null genotypes was significantly correlated with the higher rate of mutation in breast cancer patients (OR > 8;

$P < 0.05$). In summary, this work proposed that the *GSTM1/GSTT1* gene loss probably serves as an interesting predisposition indicator for the *BRCA1* mutation and also the need of genetic screening for *BRCA1* after identification of recognised polymorphism in xenobiotic metabolising genes in breast carcinoma patients (Hansa et al. 2015).

24.4.6 Colorectal Cancer

Colorectal cancer (CRC) is the third most frequently diagnosed malignancy and the fourth leading cause of cancer-related deaths in the world (Arnold et al. 2017). In India, it is the fourth major cause of cancer in males and third most common cause of cancer in females (Sharma and Singh 2017). This heterogeneous disease develops through genetic and epigenetic alterations in multiple pathways. India carries relatively higher percentage of rectal cancers as well as early-onset CRC. In a hospital-based study from the Southern Assam, high proportion of young age rectal cancer patients was reported, where distinct clinicopathological differences with the older patients were also observed (Laskar et al. 2014). Furthermore, a group of researchers studied genetic alterations in genes like *BRAF* V600E, *KRAS*, and *TP53* by direct sequencing and epigenetic alterations like methylation in CpG islands in 10 tumour-associated genetic regions as well as Microsatellite instability (MSI) using mononucleotide markers like BAT 25 and BAT 26, along with the related pathological characteristics and survival tendency of the rectal cancer victims in India. Additionally, Methyl-specific polymerase chain reaction (MS-PCR) revealed the methylation patterns of the promoter region within the typical CIMP panel markers and various tumour-linked genes (*RASSF1, DAPK, GSTP1,* and *BRCA1*). Although *BRAF* mutations and MSI were infrequent, occurrence of high incidence of overall mutations in *KRAS* (67.5%) including the G15S mutation and mutations in codon 12 along with the methylation pattern in the *RASSF1* gene in the early onset cases was worth mentioning. In addition, three separate categories of cancer victims, having distinct age at onset, clinicopathological, molecular, and survival characteristics, were observed such as (a) a subgroup of CIMP-high alongside *KRAS* variation, (b) a subgroup of CIMP-low with Microsatellite stability (MSS), *TP53* mutation plus differential *KRAS* mutations, and (c) a subgroup of CIMP-negative with mutated *TP53*. Such genetic and epigenetic outlining of rectal cancer patients may help in identifying distinct subtypes in Indian population (Laskar et al. 2015a). Moreover, among the HPV-associated and HPV-non-associated CRC patients in India carrying the various genetic variations like epigenetic dysregulation like CpG island methylation, oncogenic mutations and MSI were found. HPV DNA was noticed in tumour tissues, with HPV18 type being the predominant high-risk type. Occurrence of HPV infection was found not to be linked with age, tumour grade/stage, MSI, or mutations in tumour suppressor genes mainly *KRAS, TP53*, or *BRAF* genes. In HPV-infected tumours, reasonably elevated methylation patterns of all genetic regions were seen except *RASSF1*, whereas considerably more CIMP-high features were also noticed than the negative CRC cases. Infection with HPV in correlation with epigenetic as well as genetic factors might be a potential threat and also a determining factor for CRC in the Indian population (Laskar et al. 2015b).

24.5 The "P4" Medicine

P4 Medicine is *Predictive, Preventive, Personalised,* and *Participatory*. The two most important objectives of this are to measure wellness and demystify disease (Fig. 24.2). The vision of medicine 'P4' has long been advocated by Leroy Hood and other pioneers of systems medicine (Flores et al. 2013).

Fig. 24.2 Overview of P4 medicine (adapted and modified from https://systemsbiology.org/research/p4-medicine/)

24.5.1 Predictive

This part of P4 medicine utilises specific laboratory and genetic tests to determine the possibility if an individual will develop a disease. The use of biomarkers is a common measure in the field of oncology to predict recurrence of cancer, but the aim is also to increase its use to predict the more common clinical disorders in everyday life (Jen and Varacallo 2020). One such example is *UGT1A1*, which has a crucial application in forecasting as well as in the therapeutic interventions against cancer. More importantly, at present, the marker *UGT1A1*28* is an comprehensively studied genetic loci in various tumours like leukaemia and colon cancer (Cheng and Zhan 2017).

24.5.2 Preventive

The preventive actions include deviations in lifestyle, pressure for quitting hazardous habits like cigarette smoking, extreme alcohol consumption, and following useful drives that are meant at applying specific active treatments, as well as designed with the aim of avoiding the instigation, growth, spread, and resistance status strongly associated with many common cancers. Gaining knowledge about the factors causing genetic instability and the numerous epigenetic alterations promoting the development of chemo-resistant lineages, that is, cell population yield and heterogeneity in tumours, can be employed to improve approaches to avert cancer progression (Hochberg et al. 2013). The followings are a few instances of chemotherapeutic medicines used for cancer control. Researchers have studied a way to lower risk of oestrogen receptor-positive breast cancer and also in those who are *BRCA1/2* carriers by means of administering drugs like Tamoxifen (Soltamox) and raloxifene (Evista). This type of breast cancer expansion is dependent on the hormone oestrogen. Tamoxifen acts by blocking the action of oestrogen on tumour progress. It has also been shown that recurrence rate of breast cancer is also lowered by this approach, while Raloxifene was shown to lower the hazard of breast cancer in post-menopause women. Medications such as use of Aspirin and other non-steroidal anti-inflammatory drugs (NSAIDs) combat the occurrence of many types of cancer in individuals having average chance of cancer development (*https://www.cancer.net/navigating-cancer-care/prevention-and-healthy living/chemoprevention*).

24.5.3 Personalised

The goal of personalised medicine is to provide proper treatment with precise dose – having negligible or no toxicity – for the specific patient to meet the need of the hour (Verma 2014). Apart from genetic factors, it involves socio-economic, behavioural, and environmental factors, also contributing to the expansion of any disease and treatment non-responsiveness. For example, all-trans retinoic acid (ATRA) is known to be very much useful in the treatment of acute promyelocytic leukaemia (APL) due

Table 24.3 Some selective drugs for treating cancer in the presence of specific biomarkers (Verma 2014)

Sl. No.	Name of some Selective Drugs	Biomarker(s)	Cancer Type(s)
1	Irinotecan	*UGT1A1* gene	CRC
2	Cetuximab	*KRAS* and *EGFR* genes	CRC, HNC
3	Getifinib	*EGFR*-TK mutations	NSCLC
4	Busulfan	*Ph+*	CML
5	Denileukin diftitox	*CD24+*	CTL
6	Imatinib	*Ph* + and C-kit	CML, GIST
7	Trastuzumab	*EbR2* over-expression	BC, GIC
8	Mercaptopurine	*TPMT*	ALL, CML
9	Decatinib	*Ph+*	ALL, CML
10	Thioguanine	*TPMT*	ALL, CLL
11	Erlotinib	*EGFR+*	NSCLC, PC
12	Nilotinib	*Ph+*	CML
13	Arsenic trioxide	*PMAL* and *RAR*-α	AML
14	Lapatinib	*HER2+*	BC
15	Panitumumab	*KRAS* and *EGFR* genes	CRC, BC

Note: **AML** acute myeloid leukaemia, **ALL** acute lymphoblastic leukaemia, **BC** breast cancer, **CLL** chronic lymphocytic leukaemia, **CML** chronic myeloid leukaemia, **CRC** colorectal cancer, **CTL** cutaneous T-cell lymphoma, **GIC** gastrointestinal cancers, **GIST** Gastrointestinal stroma tumours, **HNC** head and neck cancer, **NSCLC** non-small cell lung cancer, **PC** pancreatic cancer

to the presence of the *PML–ARRA* fusion gene. In lung cancer, the expression of *EGFR* is affected by the mutations in the tyrosine kinase gene, and translocations of *EML–ALK* fusion gene, likewise in melanoma, and alterations in the *BRAF* gene have been recognised to have certain consequences in personalised medicine. In tumours of some oestrogen receptor-positive breast cancer patients with a lower expression of cytochrome P450 gene (*CYP2D2* variant), response to tamoxifen was reported to be lower. Besides, a few other selective drugs can be administered for treating cancer due to the presence of some precise biomarkers as mentioned in Table 24.3 (Verma 2014).

24.5.4 Participatory

Participatory medicine can be defined as "a movement in which networked patients shift from being mere passengers to responsible drivers of their health, and in which providers encourage and value them as full partners" (Frydman 2010). The proposal of participatory medicine, based on increasing engagement of the patients, also embraces assurance to improve observance to therapy as well as outcomes, increase patient satisfaction, decrease medical errors, and reduce the cost of care (Paperna et al. 2016). The initiatives undertaken with the purpose of making medicine more participatory are by increasing the space provided to the patients and their friends and families, so that they can take some meaningful decisions for themselves by improving certain parameters like medical literacy, by exposing patients with data and information, whereby they can get engaged and understand the points that are relevant to them, as well as to attach with others who might have appropriate expertise or experience, or by enabling mutual support. Transparency is also necessary in terms of commercial stakes and benefits, as well as about how the information would be gained from patients, kept, processed, and applied (Prainsack 2014).

24.6 Present and Future Scenario of P4 Medicine in Cancer Treatment

P4 medicine plays a pivotal role in cancer ailment. To date, although the treatment of patients with matching cancer type, grade, and stage is done with the similar regimens, it has created baffling among doctors in patients' different response to the same treatment regime. At present, any cancer patient has to go through vigorous monitoring to recognise the degree, grade, and stage of cancer, and then, a team of oncologists decides the sequence of treatment based on the testing results. Although it is well-known that some precise genomic changes in cells are responsible for the onset of cancer, but still, this is not observed in every patient. Moreover, there is a lack of efficient methodologies that can suggest the clinicians to choose the most appropriate treatment regime for a particular patient at his/her personalised level. Consequently, they used to proceed through a uniform approach for cancer treatment. However, things will change in the near future with the expectation that treatment will become more personalised on the basis of actual genetic changes within an individual person's cancer. In molecular biology, recent advancements are lending a hand in wide-ranging genetic testing to assist researchers/scientists for preparing drugs directly targeting those genetic changes triggered within the tumour macro-environment or in the cellular pathways. By following these steps, unnecessary trauma, side effects, and risks of cancer surgery can also be evaded. The two promising technologies that support precision medicine are CRISPR/Cas Technology and Cryo-electron Microscopy (Cryo-EM). The former one is the gene editing technology that usually creates model organisms that imitate the genetic variations present in a cancerous individual. These organisms can be examined further systematically to ascertain the plausible treatment regime or personalised drugs. On the other hand, the latter technology is a unique form of transmission electron Microscopy (TEM), where the biological mechanisms can be studied precisely at atomic level in a cryogenic environment, which, in turn, can be helpful in understanding the possible outcome of any genetic variation towards the precise drug response and/or drug resistance. Finally, regarding the efficacy of precision medicine a study had been carried out with two groups of cancer patients. In one group, biomarker-targeted treatment approach or precision medicine was implemented, whereas in the other group, conventional treatment regime was opted. In the first group, noteworthily, better results were noticed, as the response rates of the patients in that group were soaring to 30% in comparison to the other group, where the response rates were only 4.9%. Nowadays, we can get a lot of such instances of accomplishments due to the execution of *precision medicine*. As a consequence, various measures for putting precision medicine into the regular convention of the medical/health practitioners/professionals are expected to be taken for the sake of common populace of India at the earliest (ReferralMD 2018).

Acknowledgements The authors are grateful to Silchar Medical College and Hospital (SMCH) and Cachar Cancer Hospital and Research Centre (CCHRC), Assam, for issuing tissue and blood samples of the patients. They are also appreciative to the Department of Biotechnology (DBT), Government of India for supporting infrastructure facilities and Department of Science and Technology (DST), Government of India as well as University Grant Commission (UGC), Government of India for providing fellowships under the DST-INSPIRE and UGC-MANF scheme, respectively.

Appendix

Adenocarcinoma: It is cancer form that arises in the mucus-secreting glands of the body. It develops in different places, with most prevalent in the lung, resulting in lung cancer. Adenocarcinoma is the most common type of Non-small Cell Lung Cancer (NSCLC) which accounts for 80% of lung cancers.

Intronic variant: The introns are non-coding regions of an RNA transcript, which are removed by splicing before translation. Any type of variations in introns might affect alternative splicing of the mRNA as they are integral to gene expression regulation.

Mesenchymal cells: These are adult stem cells multipotent stromal in nature, which can differentiate into a various cell types such as chondrocytes (cartilage cells), osteoblasts (bone cells), adipocytes (fat cells), and myocytes (muscle cells).

Next-generation sequencing (NGS): It is a powerful platform enabling the sequencing of thousands to millions of DNA molecules simultaneously at high speed and at low cost. Various applications include whole genome sequencing, whole exome sequencing, study of genomic diversity, metagenomics, epigenetics, finding of non-coding RNAs and protein-binding sites, and RNA sequencing for gene-expression profiling.

Oligonucleotide microarrays: It is also normally known as DNA chip or biochip consisting of an assembly of microscopic DNA points, which remains attached to its solid surface. It is used to quantify the expression levels of huge numbers of genes at the same time and also to genotype numerous regions of a genome. It is generated either by in situ synthesis or deposition of pre-synthesized oligonucleotides ranging in size from 25- to 60-mers.

Pluripotent cells: Pluripotent stem cells, also known as human embryonic stem cells, have the potential to go through self-renewal and to create all cells of the tissues of the body. It can be applied to counter a wide range of diseases, from diabetes to spinal cord injury, to childhood leukaemia, to heart disease.

Squamous Cell Carcinoma (SCC): This type of cancer is caused by an uncontrolled growth of abnormal squamous cells. It comprises a number of different types of cancer and was found that about 90% of cases of HNC (cancer of the mouth, throat, nasopharynx, nasal cavity, and associated structures) are caused by SCC.

References

AdvaMedDx (n.d.) The value of molecular diagnostics: advancing cancer treatment and care [online]. Available: https://www.advamed.org/sites/default/files/valuemoleculardiagnostics1.pdf (Accessed January 10, 2019)

Arnold M, Sierra MS, Laversanne M, Soerjomataram I, Jemal A, Bray F (2017) Global patterns and trends in colorectal cancer incidence and mortality. Gut 66(4):683–691

Burtis CA, Bruns DE (2014) Tietz fundamentals of clinical chemistry and molecular diagnostics-e-book, 7th edn. Elsevier Health Sciences

Cheng T, Zhan X (2017) Pattern recognition for predictive, preventive, and personalised medicine in cancer. EPMA J 8(1):51–60

Cheong SC, Vatanasapt P, Yi-Hsin Y, Zain RB, Kerr AR, Johnson NW (2017) Oral cancer in South East Asia: current status and future directions. Transl Res Oral Oncol 2:2057178X17702921

Choudhury JH, Ghosh SK (2015a) Promoter hypermethylation profiling identifies subtypes of head and neck cancer with distinct viral, environmental, genetic and survival characteristics. PLoS One 10(6):e0129808

Choudhury JH, Ghosh SK (2015b) Gene–environment interaction and susceptibility in head and neck cancer patients and in their first-degree relatives: a study of N ortheast I ndian population. J Oral Pathol Med 44(7):495–501

Choudhury JH, Singh SA, Kundu S, Choudhury B, Talukdar FR, Srivasta S, Laskar RS, Dhar B, Das R, Laskar S, Kumar M (2015) Tobacco carcinogen-metabolising genes CYP1A1, GSTM1, and GSTT1 polymorphisms and their interaction with tobacco exposure influence the risk of head and neck cancer in northeast Indian population. Tumour Biol 36(8):5773–5783

Collins A, Politopoulos I (2011) The genetics of breast cancer: risk factors for disease. Appl Clin Genet 4:11

Das R, Ghosh SK (2017) Genetic variants of the DNA repair genes from exome aggregation consortium (EXAC) database: significance in cancer. DNA Repair 52:92–102

Das R, Kundu S, Laskar S, Choudhury Y, Ghosh SK (2018) Assessment of DNA repair susceptibility genes identified by whole exome sequencing in head and neck cancer. DNA Repair 66:50–63

Flores M, Glusman G, Brogaard K, Price ND, Hood L (2013) P4 medicine: how systems medicine will transform the healthcare sector and society. Pers Med 10(6):565–576

Frydman G (2010) A patient-centric definition of participatory medicine. Available: https://participatorymedicine.org/epatients/2010/04/a-patient-centric-definition-of-participatory-medicine.html (accessed June 23, 2019)

Ghosh SK, Singh AS, Mondal R, Kapfo W, Khamo V, Singh YI (2014) Dysfunction of mitochondria due to environmental carcinogens in nasopharyngeal carcinoma in the ethnic group of northeast Indian population. Tumour Biol 35(7):6715–6724

Grange JM, Stanford JL, Stanford CA (2002) Campbell De Morgan's 'observations on cancer', and their relevance today. J R Soc Med 95(6):296–299

Hajdu SI (2011) A note from history: landmarks in history of cancer, part 2. Cancer 117(12):2811–2820

Hansa J, Kannan R, Ghosh SK (2012) Screening of 185DelAG, 1014DelGT and 3889DelAG BRCA1 mutations in breast cancer patients from north-East India. Asian Pac J Cancer Prev 13(11):5871–5874

Hansa J, Ghosh SK, Agrawala SK, Kashyap MP (2015) Risk and frequency of mutations in the BRCA1 in relation to GSTM1 and GSTT1 genotypes in breast cancer. Adv Biores 6:86–89

Hejmadi, M. (2009) Introduction to cancer biology, Bookboon [online], available: https://bookboon.com/en/introduction-to-cancer-biology-ebook (accessed February 16, 2020)

Hochberg ME, Thomas F, Assenat E, Hibner U (2013) Preventive evolutionary medicine of cancers. Evol Appl 6(1):134–143

Jen MY, Varacallo M (2020) Predictive medicine. In: StatPearls, Treasure Island (FL)

Kasahara Y, Tsukada Y (2004) New insights and future advances in cancer diagnostics. In: Cancer diagnostics. Humana Press, Totowa, NJ, pp 15–26

Katoch V (2016) Three-year report of population based cancer registries 2012–2014. National Cancer Registry Program, Indian Council of Medical Research, Bengaluru, India

Kaur, T., Babu, G., Sirohi, B., 2017. Consensus document for management of oesophageal cancer. Available: https://www.icmr.nic.in/sites/default/files/reports/Oesophageal%20Cancer.pdf (accessed February 16, 2020)

Khodakov D, Wang C, Zhang DY (2016) Diagnostics based on nucleic acid sequence variant profiling: PCR, hybridisation, and NGS approaches. Adv Drug Deliv Rev 105:3–19

Krishnatreya M, Kataki A (2016) A way forward to address the cancer burden in north-East India. Int J Health Allied Sci 5(1):61–61

Kumar M, Srivastava S, Singh SA, Das AK, Das GC, Dhar B, Ghosh SK, Mondal R (2017) Cell-free mitochondrial DNA copy number variation in head and neck squamous cell carcinoma: a study of non-invasive biomarker from Northeast India. Tumour Biol 39(10):1010428317736643

Kumar M, Choudhury Y, Ghosh SK, Mondal R (2018) Application and optimisation of minimally invasive cell-free DNA techniques in oncogenomics. Tumour Biol 40(2):1010428318760342

Kundu S, Ramshankar V, Verma AK, Thangaraj SV, Krishnamurthy A, Kumar R, Kannan R, Ghosh SK (2018) Association of DFNA5, SYK, and NELL1 variants along with HPV infection in oral cancer among the prolonged tobacco-chewers. Tumour Biol 40(8):1010428318793023

Lamb YN, Scott LJ (2017) Osimertinib: a review in T790M-positive advanced non-small cell lung cancer. Target Oncol 12(4):555–562

Laskar RS, Talukdar FR, Mondal R, Kannan R, Ghosh SK (2014) High frequency of young age rectal cancer in a tertiary care Centre of southern Assam, north East India. Indian J Med Res 139(2):314–318

Laskar RS, Ghosh SK, Talukdar FR (2015a) Rectal cancer profiling identifies distinct subtypes in India based on age at onset, genetic, epigenetic and clinicopathological characteristics. Mol Carcinog 54(12):1786–1795

Laskar RS, Talukdar FR, Choudhury JH, Singh SA, Kundu S, Dhar B, Mondal R, Ghosh SK (2015b) Association of HPV with genetic and epigenetic alterations in colorectal adenocarcinoma from Indian population. Tumour Biol 36(6):4661–4670

MacConaill LE, Garraway LA (2010) Clinical implications of the cancer genome. J Clin Oncol 28(35):5219

Mao L, Hong WK, Papadimitrakopoulou VA (2004) Focus on head and neck cancer. Cancer Cell 5(4):311–316

Mondal R, Ghosh SK (2013a) Accumulation of mutations over the complete mitochondrial genome in tobacco-related oral cancer from Northeast India. Mitochondrial DNA 24(4):432–439

Mondal R, Ghosh SK (2013b) HPV infection, GSTM 1-GSTT 1 genotypes , mitochondrial mutations and tobacco association with oral cancer from Northeast India. Head Neck Oncol 5(4):46

Mondal R, Ghosh SK, Choudhury JH, Seram A, Sinha K, Hussain M, Laskar RS, Rabha B, Dey P, Ganguli S, NathChoudhury M (2013a) Mitochondrial DNA copy number and risk of oral cancer: a report from Northeast India. PLoS One 8(3):e57771

Mondal R, Ghosh SK, Talukdar FR, Laskar RS (2013b) Association of mitochondrial D-loop mutations with GSTM1 and GSTT1 polymorphisms in oral carcinoma: a case control study from Northeast India. Oral Oncol 49(4):345–353

NCI (2018) 'Cancer Statistics', [online], Available: http://www.cancer.gov/about-cancer/what-is-cancer/statistics (Accessed January 10, 2019)

Paperna T, Staun-Ram E, Avidan N, Lejbkowicz I, Miller A (2016) Chapter 26: Personalised medicine and Theranostics: applications to multiple sclerosis. In: Arnon R, Miller A (eds) Translational Neuroimmunology in multiple sclerosis. Academic Press, San Diego, pp 387–414

Parsa N (2012) Environmental factors inducing human cancers. Iran J Public Health 41(11):1–9

Poste G (2001) Molecular diagnostics: a powerful new component of the healthcare value chain. Expert Rev Mol Diagn 1(1):1–5

Prainsack B (2014) The powers of participatory medicine. PLoS Biol 12(4):e1001837

ReferralMD 2018. The future role of precision medicine in cancer treatment. Available: https://getreferralmd.com/2018/05/the-future-role-of-precision-medicine-in-cancer-treatment/ [accessed June 23, 2019]

Riaz N, Morris LG, Lee W, Chan TA (2014) Unravelling the molecular genetics of head and neck cancer through genome-wide approaches. Genes Dis 1(1):75–86

Samarasam I (2017) Oesophageal cancer in India: current status and future perspectives. Int J Adv Med Health Res 4(1):5–10

Schutte AE (2017) Global, regional, and national age-sex specific mortality for 264 causes of death, 1980–2016: a systematic analysis for the global burden of disease study 2016. Lancet 390 (10100):1151–1210

Sharma D, Singh G (2017) Clinico-pathological profile of colorectal cancer in first two decades of life: a retrospective analysis from tertiary health Centre. Indian J Cancer 54(2):397–400

Singh SA, Ghosh SK (2014) Association of Epstein Barr virus and lifestyle on nasopharyngeal cancer risk among the ethnic population of Northeast India. Sci Technol J 2(2):95–102

Singh SA, Ghosh SK (2016) Polymorphisms of XRCC1 and XRCC2 DNA repair genes and interaction with environmental factors influence the risk of nasopharyngeal carcinoma in Northeast India. Asian Pac J Cancer Prev 17(6):2811–2819

Singh SA, Ghosh SK (2019) Metabolic phase I (CYPs) and phase II (GSTs) gene polymorphisms and their interaction with environmental factors in nasopharyngeal cancer from the ethnic population of Northeast India. Pathol Oncol Res 25(1):33–44

Sokolenko AP, Imyanitov EN (2018) Molecular diagnostics in clinical oncology. Front Mol Biosci 5:76

Sudhakar A (2009) History of cancer, ancient and modern treatment methods. J Cancer Sci Ther 1 (2):1–4

Tabin CJ, Bradley SM, Bargmann CI, Weinberg RA, Papageorge AG, Scolnick EM, Dhar R, Lowy DR, Chang EH (1982) Mechanism of activation of a human oncogene. Nature 300 (5888):143–149

Talukdar FR, Ghosh SK, Laskar RS, Mondal R (2013) Epigenetic, genetic and environmental interactions in oesophageal squamous cell carcinoma from Northeast India. PLoS ONE 8(4): e60996

Talukdar FR, Ghosh SK, Laskar RS, Kannan R, Choudhury B, Bhowmik A (2015) Epigenetic pathogenesis of human papillomavirus in upper aerodigestive tract cancers. Mol Carcinog 54 (11):1387–1396

Tuljapurkar V, Dhar H, Mishra A, Chakraborti S, Chaturvedi P, Pai PS (2016) The Indian scenario of head and neck oncology–challenging the dogmas. South Asian J Cancer 5(3):105

Verma M (2014) Molecular profiling and companion diagnostics: where is personalised medicine in cancer heading? Pers Med 11(8):761–771

Warnakulasuriya S (2009) Global epidemiology of oral and oropharyngeal cancer. Oral Oncol 45 (4–5):309–316